THE LIBRARY
ST. MARY'S COLLEGE OF MARYLAND
ST. MARY'S CITY, MARYLAND 20686

THE INTERNATIONAL SERIES OF MONOGRAPHS ON PHYSICS

GENERAL EDITORS

R. J. ELLIOTT   J. A. KRUMHANSL
W. MARSHALL   D. H. WILKINSON

MOLECULAR BEAMS

NORMAN F. RAMSEY

OXFORD
AT THE CLARENDON PRESS

Oxford University Press, Walton Street, Oxford OX2 6DP
Oxford New York Toronto
Delhi Bombay Calcutta Madras Karachi
Kuala Lumpur Singapore Hong Kong Tokyo
Nairobi Dar es Salaam Cape Town
Melbourne Auckland
and associated companies in
Beirut Berlin Ibadan Nicosia

Oxford is a trademark of Oxford University Press

Published in the United States
by Oxford University Press, New York

© Oxford University Press, 1956

Paperback edition 1985
Reprinted 1990

All rights reserved. No part of this publication may be reproduced, stored in a retrieval system, or transmitted, in any form or by any means, electronic, mechanical, photocopying, recording, or otherwise, without the prior permission of Oxford University Press

This book is sold subject to the condition that it shall not, by way of trade or otherwise, be lent, re-sold, hired out or otherwise circulated without the publisher's prior consent in any form of binding or cover other than that in which it is published and without a similar condition including this condition being imposed on the subsequent purchaser

British Library Cataloguing in Publication Data
Ramsey, Norman F.
Molecular beams. − (The International series
of monographs on physics)
1. Molecular beams.
I. Title II. Series
539'.12   QC173.4.M65
ISBN 0 19 852021 2

Printed in Great Britain by
Antony Rowe Ltd,
Chippenham.

PREFACE

MOLECULAR-BEAM experiments have for many years been among the most fruitful sources of fundamental information about molecules, atoms, and nuclei. Some of the early experiments provided direct experimental evidence for spatial quantization and for the spin of the electron. More recent experiments have led to such important discoveries as the anomalous magnetic moments of protons and neutrons, the deuteron quadrupole moment with its implication of a nucleon tensor interaction, the Lamb shift in the fine structure of atomic hydrogen with its quantum electrodynamic implications, the anomalous magnetic moment of the electron, the existence of nuclear octupole moments, and many others. In addition molecular-beam experiments have produced a wealth of molecular, atomic, and nuclear data including an array of nuclear spin, magnetic moment, and quadrupole moment determinations which, among other things, have provided much of the initial evidence for nuclear shell models.

Despite the scientific importance of the subject, there have been remarkably few books on molecular beams. As discussed in Chapter I, a few review articles and brief monographs have been written but only one detailed book has been published. That book, by R. Fraser (FRA 31), was written twenty-five years ago, prior to almost all of the important discoveries listed in the preceding paragraph.

It is my hope that this book will satisfy the need for a detailed, consistent, and up-to-date discussion of the subject of molecular beams. Although the entire subject is discussed in the present book, experiments prior to 1930 are dealt with more briefly since they are described in greater detail in Fraser's book (FRA 31).

I wish to express my appreciation to the many workers with molecular beams who have supplied me with information and help in the preparation of this book. Special thanks are due to Professors I. I. Rabi, P. Kusch, and J. R. Zacharias and to Drs. H. Kolsky and H. Lew. I should also like to thank John Wiley & Sons for their permission to use some material, especially in Chapter III, from my book on *Nuclear Moments* published by them, and the John Simon Guggenheim Memorial Foundation for the fellowship I held at Oxford University during the completion of this book. Finally, I wish to express special appreciation to my wife, Elinor Ramsey, for her help in making available the time during which the book was written, and to my secretary, Phyllis Brown, for her extensive and excellent help in the preparation of the manuscript.

N. F. R.

Harvard University
Cambridge, Massachusetts
August 1955

CONTENTS

I. MOLECULAR AND ATOMIC BEAMS
1. Introduction 1
2. Typical Molecular-beam Experiment 2
3. Kinds of Molecular-beam Measurements 8

II. GAS KINETICS
1. Molecular Effusion from Sources 11
   1.1 Effusion from thin-walled apertures 11
   1.2. Effusion from long channels 13
2. Molecular-beam Intensities and Shapes 16
   2.1. Beam intensities 16
   2.2. Beam shapes 16
3. Velocities in Molecular Beams 19
   3.1. Velocity distribution in a volume of gas 19
   3.2. Velocity distribution in molecular beams 20
   3.3. Characteristic velocities 20
   3.4. Experimental measurements of molecular velocities 21
4. Molecular Scattering in Gases 25
   4.1. Methods of measurement 25
   4.2. Calculation of scattering cross-sections from attenuation measurements 28
   4.3. Cross-section theory 30
   4.4. Results of atomic-beam scattering experiments 33
5. Interaction of Molecular Beams with Solid Surfaces 35
   5.1. Introduction 35
   5.2. Reflection 35
   5.3. Diffraction 38
   5.4. Inelastic collisions 44
6. Molecular-beam Studies of Chemical Equilibria, Ionization Potentials, and Optical and Microwave Spectroscopy 47
   6.1. Chemical equilibria 47
   6.2. Ionization and excitation 48
   6.3. Optical and microwave absorption spectroscopy 50

III. INTERACTION OF A NUCLEUS WITH ATOMIC AND MOLECULAR FIELDS
1. Introduction 51
2. Electrostatic Interaction 52
   2.1. Multipole expansion 52
   2.2. Theoretical restrictions on electric multipole orders 58
   2.3. Nuclear electric quadrupole interactions 59
   2.4. Multipole moments of arbitrary order 66

3. Magnetic Interaction ... 68
   3.1. Magnetic multipoles ... 68
   3.2. Magnetic dipoles ... 72
   3.3. Magnetic dipole interactions with internal and external magnetic fields ... 76
   3.4. Magnetic interactions in molecules ... 80
4. Intermediate Coupling ... 83
   4.1. Introduction ... 83
   4.2. Solution of secular equation ... 84
   4.3. Perturbation theory ... 87

IV. NON-RESONANCE MEASUREMENTS OF ATOMIC AND NUCLEAR MAGNETIC MOMENTS

1. Deflexion and Intensity Relations ... 89
   1.1. Effective magnetic moment ... 89
   1.2. Deflexion of molecule by magnetic fields ... 90
2. Deflexion Measurements of Atomic Magnetic Moments ... 100
3. Deflexion Measurements of Nuclear and Rotational Magnetic Moments in Non-paramagnetic Molecules ... 102
4. Atomic-beam Deflexion Measurements of Nuclear Moments ... 104
5. Atomic-beam Zero-moment Method for Nuclear Moments ... 107
6. Atomic-beam Refocusing Method for Nuclear Moments ... 111
7. Atomic-beam Measurements of Signs of Nuclear Moments ... 113
8. Atomic-beam Space Focusing Measurements of Nuclear Spins ... 114

V. MOLECULAR-BEAM MAGNETIC RESONANCE METHODS

1. Introduction ... 115
2. Magnetic Resonance Method ... 116
3. Transition Probabilities ... 118
   3.1. Transition probability for individual molecules ... 118
   3.2. Transition probability averaged over velocity distribution ... 123
4. Magnetic Resonance Method with Separated Oscillating Fields ... 124
   4.1. Introduction ... 124
   4.2. Transition probability for individual molecule with separated oscillating fields ... 127
   4.3. Separated oscillating fields transition probability averaged over velocity distribution ... 128
   4.4. Phase shifts in the molecular-beam method of separated oscillating fields ... 131
5. Magnetic Resonance with Focused Beams, with Very Slow Molecules, and with Wide Beams ... 134
   5.1. Space focused beams ... 134
   5.2. Beams of very slow molecules. Wide beams ... 138
6. Distortions of Molecular-beam Resonances ... 139

VI. NUCLEAR AND ROTATIONAL MAGNETIC MOMENTS

1. Resonance Measurements of Nuclear Magnetic Moments 145
2. Interpretation with Rotating Coordinate System 146
   - 2.1. Introduction 146
   - 2.2. Classical theory 147
   - 2.3. Quantum theory 151
3. Spins Greater than One-half 153
4. Signs of Nuclear Moments 155
5. Absolute Values of Nuclear Moments 158
6. Nuclear Magnetic Moment Measurements 159
7. Magnetic Shielding 162
8. Molecular Rotational Magnetic Moments 166
   - 8.1. Experimental measurements of rotational magnetic moments 166
   - 8.2. Theory of rotational magnetic moments 169
9. Results of Nuclear Moment Measurements 170
   - 9.1. Introduction 170
   - 9.2. Nuclear moment tables 171
10. Significance of Nuclear Spin and Nuclear Magnetic Moment Results 171
    - 10.1 Introduction 171
    - 10.2. Relations between nuclear statistics, spins, and mass numbers 178
    - 10.3. Nucleon magnetic moments 179
    - 10.4. Deuteron magnetic moment 179
    - 10.5. H^3 and He^3 magnetic moments 180
    - 10.6. Systematics of nuclear spins and magnetic moments 181
    - 10.7. Nuclear models 184

VII. NEUTRON-BEAM MAGNETIC RESONANCE

1. Introduction 189
2. Polarized Neutron Beams 189
   - 2.1. Neutron beams 189
   - 2.2. Polarization of neutron beams 190
3. Neutron Magnetic Moment 195

VIII. NUCLEAR AND MOLECULAR INTERACTIONS IN FREE MOLECULES

1. Introduction 203
2. Nuclear and Rotational Magnetic Moments 203
3. Nuclear Spin-Spin Magnetic Interactions 204
   - 3.1. Direct spin-spin magnetic interaction 204
   - 3.2. Electron-coupled interactions between nuclear spins in molecules 206
4. Spin-rotational Magnetic Interactions 208
5. Nuclear Electrical Quadrupole Interaction 213
6. Diamagnetic Interactions 228
7. Effects of Molecular Vibration 230
8. Combined Hamiltonian 233

9. Matrix Elements and Intermediate Couplings 234
10. Interactions in Molecular Hydrogens 238
11. Molecular Polymerization 239

IX. ATOMIC MOMENTS AND HYPERFINE STRUCTURES

1. Introduction 241
2. Absolute Scale for Nuclear Moments. The Fundamental Constants 245
3. Nuclear Spins from Atomic Hyperfine Structure 249
4. Atomic Hyperfine Structure Separations for $J = \frac{1}{2}$ 251
5. Atomic Magnetic Moments; The Anomalous Electron Moment 256
   5.1. Atomic magnetic moments 256
   5.2. The anomalous electron magnetic moment 262
6. Hyperfine Structure of Atomic Hydrogen 263
   6.1. Atomic hydrogen experiments 263
   6.2. Theoretical interpretation of atomic-hydrogen experiments 267
7. Direct Nuclear Moment Measurements with Atoms 270
8. Quadrupole Interactions 271
9. Magnetic Octupole Interactions 277
10. Anomalous Hyperfine Structure and Magnetic-moment Ratios for Isotopes 279
11. Frequency Standards and Atomic Clocks 283
12. Atomic-beam Resonance Method for Excited States 285

X. ELECTRIC DEFLEXION AND RESONANCE EXPERIMENTS

1. Introduction 287
2. Molecular Interactions in an Electric Field 289
   2.1. Molecular Hamiltonian 289
   2.2. Energy of vibrating rotator 290
   2.3. Stark effect 292
   2.4. Hyperfine structure interactions 293
3. Electric Deflexion Experiments 296
4. Electric Resonance Experiments 298
   4.1. Introduction 298
   4.2. Electric resonance experiments for molecules with negligible hyperfine structure interactions 301
   4.3. Electric resonance experiments with hyperfine structure 303
   4.4. Electric resonance experiments with change of J 307
5. Molecular Amplifier 309

XI. NUCLEAR ELECTRIC QUADRUPOLE MOMENTS

1. Introduction 310
2. Nuclear Electric Quadrupole Interactions 310
   2.1. Nuclear quadrupole interactions in molecules 310
   2.2. Nuclear quadrupole interactions in atoms 310

3. Electric-field Gradients 313
   3.1. Electric-field gradients in atoms 313
   3.2. Electric-field gradients in molecules 315
   3.3. Effect of atomic core on nuclear quadrupole interaction 316
4. Nuclear Electric Quadrupole Moments 319
5. Theoretical Interpretation of Nuclear Quadrupole Moments 319
   5.1. Theory of deuteron quadrupole moment 319
   5.2. Theories of quadrupole moments of complex nuclei 323

XII. ATOMIC FINE STRUCTURE
1. Introduction 327
2. Hydrogen Atomic-beam Method 329
3. Hydrogen and Deuterium Results 331
4. Fine Structure of Singly-ionized Helium 336
5. Theoretical Interpretation 340

XIII. MOLECULAR-BEAM DESIGN PRINCIPLES
1. Introduction 346
2. Formulae Useful in Design of Molecular-beam Experiments 347
3. Design Considerations 351
   3.1. Optimum collimator position 351
   3.2. Beam widths, heights, and lengths 352
4. Design Illustration 354

XIV. MOLECULAR-BEAM TECHNIQUES
1. Introduction 361
2. Sources 361
   2.1. Sources for non-condensable gases 361
   2.2. Heated ovens 364
   2.3. Sources for dissociated atoms 372
3. Detectors 374
   3.1. Miscellaneous detectors 374
   3.2. Surface ionization detectors 379
   3.3. Electrometers, mass spectrometers, and electron multiplier tubes 381
   3.4. Electron bombardment ionizer 387
   3.5. Stern–Pirani detector 389
   3.6. Radioactivity detection 393
4. Deflecting and Uniform Fields 394
   4.1. Stern–Gerlach and Hamburg deflecting fields 394
   4.2. The Rabi deflecting field 395
   4.3. The two-wire deflecting field 397
   4.4. Two-wire deflecting fields from iron magnets 399
   4.5. Uniform magnetic fields 401
   4.6. Electrostatic fields 402
   4.7. Focusing fields 404

5. Oscillatory Fields 407
   5.1. Oscillators 407
   5.2. Oscillatory field loops and electrodes 407
6. Vacuum System and Mechanical Design 410
7. Miscellaneous Components 415
8. Alignment 415
   8.1. Optical alignment 415
   8.2. Alignment with beam 416

APPENDIXES
   A. Fundamental Constants 418
   B. Vector and Tensor Relations 419
   C. Quadrupole Interaction 421
   D. Table of Velocity-averaged Functions 425
   E. Majorana Formula 427
   F. Theory of Two-wire Field 431
   G. Notes added in Proof 433

REFERENCES 435

AUTHOR INDEX 455

SUBJECT INDEX 461

I
MOLECULAR AND ATOMIC BEAMS

I. 1. Introduction

THE subject of molecular beams is the study of directed beams of neutral molecules at such low pressures that the effects of molecular collisions are for most purposes negligible. The earliest molecular-beam experiments were those of Dunoyer (DUN 11). He used an apparatus illustrated in Fig. I. 1. A glass tube 20 cm long was divided into three separately evacuated compartments. Some sodium introduced into the first chamber was heated until it vaporized, whereupon a deposit of sodium appeared in the third chamber. The deposit was of the form to be expected on the assumption that the Na atoms travelled in straight lines in the evacuated tube.

Subsequent to Dunoyer's pioneering work, many different experiments based on the molecular-beam methods have been performed. Particularly significant from the point of view of the development of molecular-beam techniques have been the experiments of Stern and his collaborators at Hamburg in the nineteen-twenties and early thirties. The papers of Stern and Knauer in 1926 (STE 26, KNA 26, KNA 29) indeed laid down many of the principles of molecular-beam techniques which have been followed even to the present time.

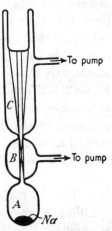

FIG. I. 1. Schematic diagram of Dunoyer's original atomic-beam apparatus. A is the source chamber, B the collimator chamber, and C the observation chamber (DUN 11).

Precision measurements of nuclear, molecular, and atomic properties began in 1938 with the introduction of the molecular-beam resonance method by Rabi and his associates (RAB 38, KEL 39). At first, the resonance method was applied to the measurement of nuclear magnetic moments only, but it was soon extended as a more general technique of radiofrequency spectroscopy initially by Kellogg, Rabi, Ramsey, and Zacharias (KEL 39a) in applications to molecules and later by Kusch and Millman (KUS 40) in the study of atomic states. Since 1938 the history of molecular beams has dominantly been the history of the resonance method. As discussed in subsequent chapters, important improvements and extensions of the molecular-beam resonance method have been made by Lamb (LAM 50a), Ramsey (RAM 50c), Kusch (KUS 51), Rabi (RAB 52), Zacharias (ZAC 53), Lew (WES 53), and others. However, since 1938, most molecular-beam researches have been applications of the resonance technique to physical measurements. The molecular-beam laboratories at Columbia University, Massachusetts Institute of Technology, and Harvard University have been particularly effective in measuring with the resonance method many varied properties of nuclei, atoms, and molecules.

The developments in molecular beams prior to 1937 have been described in detail by Fraser in two different books (FRA 31, FRA 37). Since that time no book devoted exclusively to molecular beams has been published. However, considerable discussion of molecular beams is given in the nuclear moment books of Kopfermann (KOP 40) and Ramsey (RAM 53b, RAM 53e) and in various review articles (HAM 41, BES 42, KEL 46, EST 46, KUS 50, RAM 50, RAM 52, ZZB 1–6). In accordance with existing conventions as to the scope of the subject molecular beams, the present book will be limited to beams of electrically neutral molecules. Ion beam experiments have been discussed extensively by Massey (MAS 50, MAS 52), Burhop (MAS 52), and others.

I. 2. Typical molecular-beam experiment

Although molecular-beam experiments and apparatus vary greatly according to their applications, they also have many features in common. This is especially true of the experiments in recent years, which have almost exclusively been by the molecular-beam resonance method. For this reason a general description of a typical molecular-beam resonance apparatus is given in this section as an introduction to the more detailed considerations which follow in subsequent chapters.

A photograph of a typical apparatus is shown in Fig. I. 2. A schematic diagram of the apparatus is shown in Fig. I. 3. The molecules emerge from a source S into an evacuated source chamber. The source may be either a heated oven or a small chamber cooled with liquid nitrogen, depending on the vapour-pressure properties of the substance being investigated. The source shown in Fig. I. 3 is for the study of gaseous hydrogen and is cooled with liquid air. The pressure inside the source is about 6 mm Hg. A typical source slit width is 0·015 mm and height about 0·8 cm. Sometimes the slits are a few millimetres long to assure that a larger fraction of the emerging molecules goes in the direction of the detector. The molecules then pass through a separately

FIG. I. 2. Photograph of a typical molecular beam apparatus (KOL 50).

Fig. I. 4. View of source end of molecular beam apparatus with separating chamber bulkheads removed. The A deflecting magnet is shown (KOL 50).

pumped separating chamber and into the main chamber. The purpose of the separating chamber is to aid in the attainment of a better vacuum in the main chamber. Typical pressures are 2×10^{-5} mm in the separating chamber, and 5×10^{-7} mm in the main chamber. These pressures are sufficiently low for most of the molecules to travel the entire length of the apparatus without being subjected to collisions.

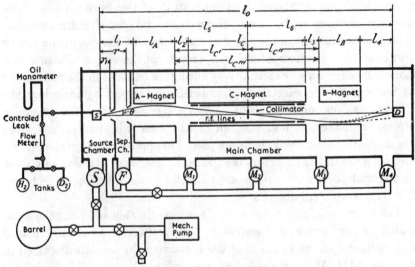

Fig. I. 3. Schematic diagram of a molecular-beam apparatus. The transverse beam displacements are much exaggerated (KOL 50). The S and D on the beam axis indicate source and detector, respectively. Typical dimensions are given in Table XIII.1 on p. 356.

The molecules are detected in the main chamber by any of several different kinds of detector described in Chapter XIV. Two of the most frequently used detectors are the surface ionization detector and the Pirani detector. The former consists of a heated wolfram (tungsten) wire of about 0·015 mm diameter. If an atom, such as potassium, with a low ionization potential (4·3 eV for K) strikes the heated tungsten wire, whose work function is 4·5 eV, the valence electron goes into the tungsten and the atom evaporates from the wire as a positive ion. The wire is surrounded by a negatively charged cylinder and the positive current to the cylinder is determined, thus providing a measurement of the beam intensity. The Pirani detector, used with non-condensable gases like H_2, consists of a small volume chamber (7·5 cm × 0·5 cm × 0·06 cm in a typical case) along whose length are stretched one or more thin platinum strips (typically 7·5 cm × 0·05 cm × 0·0001 cm). The beam is admitted to the Pirani cavity through a long narrow channel (2·5 cm long, 0·8 cm high, and 0·015 cm wide in a typical case).

The purpose of the long channel is to make it much easier for the directed beam molecules from the source to enter the cavity than for the randomly directed molecules inside the cavity to emerge; in this way a given number of incident molecules produces a much greater pressure inside the Pirani cavity than would be the case in the absence of the channel. A current is passed through the platinum strips to heat them about 100° C above the temperature of the cavity walls. The pressure changes in the cavity from the variations of beam intensity then lead to changes in the thermal conduction of the gas in the cavity and hence to variations of the temperature of the strips. These variations can be measured externally by noting the changes in the resistances of the wires. The resistances are usually measured by including the Pt strips as one or more arms of a Wheatstone bridge. To minimize effects from thermal and vacuum fluctuations two identical detectors are ordinarily used, their strips are placed in opposite arms of the bridge, and only one of the detectors is exposed to the beam. In this way most of the undesired fluctuations are partially cancelled. The detector in Fig. I. 3 is a Pirani detector.

The beam is defined by means of a collimating slit of about the same width as the source and detector. The collimator is usually placed approximately in the middle of the apparatus for reasons discussed in Chapter XIII. When the collimator is in place, the beam is fully defined and the full beam intensity will be measured by the detector only when the source, collimator, and detector are in line.

In Fig. I. 3 the molecules pass through two regions of inhomogeneous magnetic field produced by two magnets usually designated by the letters A and B and often called the deflecting and refocusing magnets. Fig. I. 4 is a photograph of a typical one of these magnets for producing an inhomogeneous magnetic field. Electrically neutral molecules or atoms which possess magnetic moments will be deflected by such inhomogeneous magnetic fields (in a uniform magnetic field there would be no net translational force since the forces on the north and south poles would be equal and opposite but in an inhomogeneous field one of these forces exceeds the other). Consequently, the molecules which in the absence of the field would have gone straight to the detector (as shown in the straight line of Fig. I. 3) will be deflected by the field so that they do not even pass through the collimator. On the other hand some molecules which in the absence of the field would have missed the detector will be deflected so as to pass through the collimator as shown by the curved line in Fig. I. 3; the deflexion shown in Fig. I. 3 is, of course, highly exaggerated since in practice it is ordinarily only about 0·006 cm. In the absence of a B field these molecules would miss the detector. On the other hand, if the B field has the direction of its gradient opposite to that of the A field and has its magnitude suitably selected, it can refocus the deflected beam on to the detector. This refocusing is velocity independent since a slow molecule is deflected more than a fast one by both the deflecting and the refocusing fields. In practice, the beam intensity with both the deflecting and refocusing fields is about 95 per cent. of the value in the absence of all inhomogeneous fields.

The refocusing condition given above depends upon the magnitude and orientation of the net molecular magnetic moment (including the nuclear contribution) being the same in the refocusing field as in the deflecting field. As a result if the molecular state changes between the A and B region the beam intensity will in general be reduced. The apparatus then serves as a means for determining when a change of state occurs; it is this use that makes molecular-beam methods of value in magnetic-resonance experiments.

Ordinarily a uniform magnetic field, often called the C field, is placed between the two deflecting fields. The presence of the static C field does not lead to any change in the molecular state or to any consequent reduction in beam intensity. However, if an oscillatory magnetic field is also applied in the C field region, molecular transitions will occur when the oscillator frequency equals one of the Bohr frequencies, ν_{pq}, of the system in the field, where
$$\nu_{pq} = (W_p - W_q)/h, \tag{I. 1}$$
provided that the transition is an allowed one for the kind of oscillatory field used. Since the resultant component of the net molecular magnetic moment is in general different in the two states, the refocusing will not occur after the transition has taken place and there will be a reduction in beam intensity with the minimum occurring at the frequency given by Eq. (I. 1).

If then the oscillator frequency is varied while the beam intensity is observed a minimum will occur each time that the frequency is at a Bohr frequency of Eq. (I. 1) for which a transition is allowed. In this fashion a radiofrequency spectrum for the molecular system is obtained. The width of the resonance line is often 300 c/s or 10^{-8} cm^{-1} as discussed in greater detail in Chapter V. Fig. I. 5 is a typical radiofrequency spectrum, that for molecular H_2 in its ground state. The interpretation of this spectrum will be discussed in Chapter VIII.

A particularly simple application of the magnetic-resonance method occurs when the dominant spin-dependent interaction within the molecule is that of a nuclear magnetic moment with an external magnetic field. In this case

$$W = -\mathbf{\mu}_I \cdot \mathbf{H}_0 = -(\mu_I/I)H_0 m_I, \qquad (I.\,2)$$

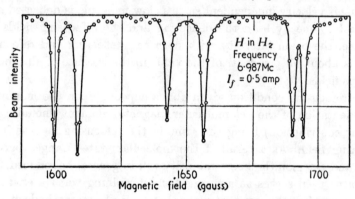

FIG. I. 5. Radiofrequency spectrum of ortho-H_2 molecules arising from transitions of the resultant nuclear spin (KEL 39a).

where $\mathbf{\mu}_I$ is the nuclear magnetic moment, μ_I is the maximum possible z component of $\mathbf{\mu}_I$, I is the nuclear spin quantum number, and m_I is the magnetic quantum number for the spin orientation. For the selection rule $\Delta m_I = \pm 1$ then, a resonance will occur at the frequency

$$\nu_0 = \mu_I H_0/hI. \qquad (I.\,3)$$

Therefore, from a measurement of ν_0 and H_0 and a knowledge of I and h the magnetic moment may be directly calculated.

The dimensions of the apparatus shown in Figs. I. 2, I. 3, and I. 4 are appropriate to the study of molecules with no net electronic magnetic moment, such as $^1\Sigma$ diatomic molecules. In this case the observed interactions are due to such properties as the nuclear magnetic moment, the molecular rotational magnetic moment, and the nuclear quadrupole moment as discussed in Chapter VIII. However, such experiments are by no means the most general ones to which the molecular-beam resonance method may be applied. Paramagnetic atoms may be studied, in which case shorter and weaker deflecting fields can be used. A schematic illustration for such an apparatus is shown in Fig. I. 6. Also in the foregoing experiment the deflecting fields were adjusted to produce maximum intensity in the absence of an oscillatory field so that a resonance provided a reduction in beam intensity (often called a 'flop-out' experiment). On the other hand, the directions and magnitudes of the gradients of the two deflecting fields can be made such that refocusing only occurs for molecules which have undergone a specific transition, in which case the resonance provides an increase rather than a decrease

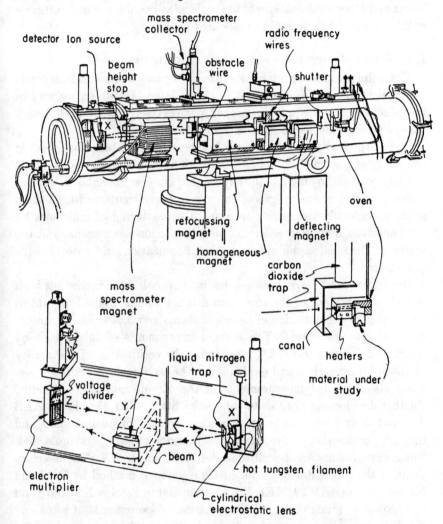

FIG. I. 6. Perspective diagram of atomic-beam apparatus with mass spectrometer. A schematic diagram of the same apparatus is shown in Fig. XIV. 12 (DAV 48b).

in beam intensity (called a 'flop-in' experiment). Likewise a mass spectrometer can be used in conjunction with a molecular-beam apparatus both to identify the isotopes and to increase the detection sensitivity. If, for example, a mass spectrometer is used with a surface-ionization detector, the wolfram wire of the detector serves as a source for the mass spectrometer as shown in Fig. I. 6. A still different modification provides for the use of electric deflecting and oscillatory fields. These and other modifications of the molecular-beam resonance principle will be discussed in detail in subsequent chapters.

I. 3. Kinds of molecular-beam measurements

The different kinds of quantities that can be measured with molecular-beam methods will be described in detail later. However, by way of introduction and to provide proper perspective, a brief preliminary survey of these measurements is desirable.

One of the earliest applications of molecular-beam techniques was in the study of gas kinetics. These applications will be discussed in detail in Chapter II. They include the experimental measurement of molecular velocity distributions in gases, the study of molecular effusion from sources, molecular scattering in gases, the scattering of molecules by solid surfaces, including reflection and diffraction phenomena, and the study of chemical equilibria and related ionization and spectroscopic phenomena.

One of the most important applications of molecular beams has been in the measurement of the spins and magnetic moments of both atoms and nuclei by the deflexion of beams of atoms and molecules in inhomogeneous magnetic fields. The earliest experiments of this nature by Stern and Gerlach (GER 24) provided early verification of the reality of space quantization and evidence for the existence of electron spin. They also provided measurements of the atomic magnetic moments. Further development of this technique by Stern (FRI 33c), Rabi (RAB 33), and their associates led to the measurement of nuclear spins and magnetic moments by the study of the deflexion of atoms and molecules. These experiments are described in detail in Chapter IV. The development of the molecular-beam magnetic resonance method by Rabi and his associates (RAB 39, KEL 39), as discussed in Section I. 2 above and in Chapter V, made possible high precision measurements of spins and magnetic moments of nuclei and atoms. The results of these measurements are discussed in Chapter VI, while Chapter VII discusses the closely related magnetic moment measurements of the neutron, of which the first precision experiments were those of Alvarez and Bloch (ALV 40). Ramsey (RAM 40, HAR 52) and his associates have applied the magnetic-resonance method to the measurement of rotational magnetic moments of molecules.

The magnetic-resonance method may also be used for studying the radiofrequency spectrum of molecules and atoms which arises from the various spin-dependent interactions within the molecules themselves as well as the interactions with an external field. Extensive measurements on H_2, D_2, and HD by Rabi (KEL 39a, KEL 40), Ramsey (KEL 39a, KOL 52, HAR 52), and their associates have provided much information on the spin-spin magnetic interaction between the two nuclei, on the spin-rotational magnetic interaction in the molecule, on the nuclear electrical quadrupole moment, on magnetic shielding of nuclei by the electrons, on diamagnetic susceptibility and its variation with molecular orientation, on internuclear spacings in the molecules, on the effects of zero-point vibration and centrifugal stretching, and on similar molecular properties. Measurements of some of the preceding properties in heavier molecules have been made by Nierenberg and Ramsey (NIE 47) and others (ZEI 52, ZXX 54). Detailed discussions of these measurements in both light and heavy molecules are given in Chapters VIII and XI.

The radiofrequency spectra of atoms were first measured by Kusch, Millman, and Rabi (KUS 40). Such measurements have provided information on atomic hyperfine structure, on the precision values of the atomic magnetic moments, and on nuclear quadrupole moment interactions. One of the most important results of these measurements has been the determination (KOE 51) that the spin magnetic moment of the electron is $(1 \cdot 001145 \pm 0 \cdot 000013)$ Bohr magnetons rather than exactly one Bohr magneton as was supposed previously. It has also been found that the ratios of nuclear magnetic moments of isotopes are not exactly equal to the ratios of the magnetic hyperfine structure for the same isotopes. These and related experiments are described in Chapter IX.

When measurements are made with electric deflecting fields and oscillatory electric fields, as in the experiments of Hughes (HUG 47), Trischka (TRI 48), and others (CAR 52), quantities are obtained such as molecular dipole moments, molecular moments of inertia, rotational-vibrational interactions, nuclear electric quadrupole interactions, spin-rotational magnetic interactions, and effects of molecular vibration. A description of such experiments is given in Chapter X. A general discussion of measurements of nuclear electrical quadrupole moments is given in Chapter XI.

By a quite different kind of atomic-beam resonance experiment Lamb and his associates (LAM 50a) have measured the fine structure of atomic hydrogen. A particularly significant result of these measurements has been the discovery that the $2\,^2S_{\frac{1}{2}}$ level of atomic hydrogen is not exactly degenerate with the $2\,^2P_{\frac{1}{2}}$ level, as expected by previous theories, but instead is higher by $1057{\cdot}77\pm0{\cdot}10$ Mc. This fundamental discovery was the stimulus of most of the advances in quantum electrodynamics and field theory that have been made in recent years. Chapter XII is concerned with these measurements.

Molecular-beam design principles and techniques are described in Chapters XIII and XIV.

II

GAS KINETICS

II. 1. Molecular effusion from sources

II. 1.1. *Effusion from thin-walled apertures.* The sources of molecules in practically all molecular-beam experiments consist of small chambers which contain the molecules in a gas or vapour at a few mm Hg pressure and which have small apertures or narrow slits about 0·02 mm wide but up to 1 cm or so high. The width of the slit and the pressure are adjusted so that there is molecular effusion (KEN 38) as contrasted to hydrodynamic flow. Under such conditions, the number, dQ, of molecules which will emerge per second from the source slit travelling in solid angle $d\omega$ at angle θ relative to a normal to the plane containing the slit jaws is, by the most elementary kinetic theory arguments (KEN 38, ROB 33),

$$dQ = (d\omega/4\pi)n\bar{v}\cos\theta\, A_s, \qquad (\text{II. 1})$$

where n is the number of molecules per unit volume, $\bar{v}$ is the mean molecular velocity inside the source, and A_s is the area of the source slit. For an ideal gas of pressure p and absolute temperature T,

$$p = nkT. \qquad (\text{II. 1 a})$$

The total number Q of molecules that should emerge from the source in all directions can be found by integrating Eq. (II. 1) over the 2π solid angle of the forward direction. In this case and with $d\omega$ taken as $2\pi \sin\theta\, d\theta$ so that the integration goes from θ equals 0 to $\tfrac{1}{2}\pi$, one immediately obtains from the integration that

$$Q = \tfrac{1}{4}n\bar{v}A_s. \qquad (\text{II. 2})$$

Two assumptions are inherent in Eqs. (II. 1) and (II. 2). One of these is that every molecule which strikes the aperture passes through it and does not have its direction changed. This assumption is valid only if the thickness of the slit jaws is negligible. The effect of the appreciable thickness of the slit jaws is discussed later in this section. The other assumption is that the spatial and velocity distributions of the molecules inside the source are not affected by the effusion of the molecules. The strict requirement for this condition is that

$$w \ll \lambda_{Ms}, \qquad (\text{II. 3})$$

where w is the slit width and λ_{Ms} the mean free collision path inside the source. By the usual kinetic theory demonstration,

$$\lambda_{Ms} = \frac{1}{n\sigma\sqrt{2}}, \qquad \text{(II. 4)}$$

where σ is the molecular collision cross-section. This relation can be re-expressed in terms of the source pressure p' in mm of Hg, the absolute temperature T in °K, and σ in cm² with the following result:

$$\lambda_{Ms} = 7\cdot 321 \times 10^{-20} \frac{T}{p'\sigma} \text{ cm}. \qquad \text{(II. 5)}$$

For air at room temperature $\lambda_{Ms} = 0\cdot 3$ mm at 1 mm of Hg pressure and $\lambda_{Ms} = 300$ metres at 10^{-6} mm of Hg. As discussed in Sections II. 2 and II. 4, the mean free paths for the small angle collisions that are significant with well-defined molecular beams are considerably smaller than the above.

In actual practice it is found that for most purposes sources are effective when

$$w \sim \lambda_{Ms}, \qquad \text{(II. 6)}$$

and w is sometimes taken to be even slightly greater than λ_{Ms}. If Eq. (II. 3) or (II. 6) is not satisfied, a partial hydrodynamic flow results with the creation of a turbulent gas jet instead of free molecular flow. It is significant that the above restriction depends only on the width and not on the height of the slit. With circular apertures, on the other hand, it depends on the radius of the aperture. Consequently approximately the same source pressure can be used with a slit whose width equals the radius of a circular aperture. On the other hand, if the slit is a high one it will have a much greater area and produce a correspondingly greater beam intensity. It is for this reason that slits rather than circular apertures are used in almost all molecular-beam experiments (FRA 31).

The effects of source pressure in molecular-beam experiments were first investigated experimentally by Knauer and Stern (KNA 26). They found that with a well-collimated molecular beam at low source pressures the beam intensity was indeed proportional to the source pressure. However, as the pressure reached a value such that Eq. (II. 3) and (II. 6) were violated, the detected beam intensity increased only slightly with source pressure. This, they explained (KNA 26, MAY 29, KNA 30, KRA 35), was due to an increasingly large fraction of the effusing molecules colliding with each other either in the source slit or immediately after they emerge. In this way a cloud of molecules is formed in front of the source slit, and the boundary of the cloud instead of the source slit serves as the effective source of the molecules. As the source pressure is further increased, the cloud increases more in size than in intensity. Consequently the beam is broadened but the ratio of beam intensity to source pressure is diminished.

II. 1.2. *Effusion from long channels.* The cosine law of molecular effusion that is implied by Eq. (II. 1) was established by the pioneer

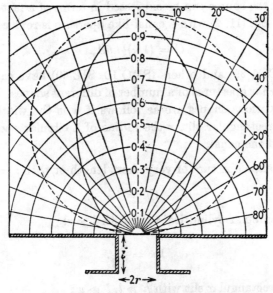

FIG. II. 1. The angular distribution of the molecules effusing through a short circular canal. The full curve is for the length and diameter being equal and the dashed curve is for negligible length (CLA 29).

work of Knudsen (KNU 09) and has been verified independently by others (MAY 29). However, if the thin-walled aperture assumed above is replaced by a canal-like aperture of appreciable length, the molecules which start to emerge at a considerable angle will strike the wall and have a smaller chance of escaping, and even if they do so it will probably be at a different angle. In this way the angular distribution of the emergent beam is changed considerably. Thus for a circular aperture whose length equals twice the radius, Claussing (CLA 29) has calculated the angular distribution indicated in Fig. II. 1. The full curve shows the actual distribution while the dashed curve shows that to be expected from the cosine law.

One consequence of the changed angular distribution with canal-shaped apertures is that the total amount of gas which emerges from the source is diminished while that which emerges in the direction of the canal is undiminished, provided the pressure is sufficiently low for collisions inside the canal to be negligible, that is, provided $\lambda_{Ms} \gtrsim l$, where l is the canal length. This improvement of beam intensity per amount of source material consumed is of great value in many molecular-beam experiments such as those with radioactive isotopes. The effectiveness of the canal in reducing the total number of molecules that emerge from the source can be expressed in terms of a factor $1/\kappa$ which is such that Eq. (II. 2) with a canal-like aperture is replaced by

$$Q = (1/\kappa)\tfrac{1}{4} n \bar{v} A_s. \tag{II. 7}$$

Claussing (CLA 29) and others (SMO 10, SIL 50) have calculated the values of $1/\kappa$ for apertures of a number of different shapes. Their results for slits of various shapes are as follows if w is the width and h the height of a rectangular slit, r the radius of a circular aperture, and l the length of the canal.

(a) Any shape aperture of very short length or with $l = 0$:

$$1/\kappa = 1. \tag{II. 8}$$

(b) Long circular cylindrical tube with $l \gg r$:

$$1/\kappa = \frac{8}{3}\frac{r}{l}. \tag{II. 9}$$

(c) Long rectangular slit with $h \gg l, l \gg w$:

$$1/\kappa = \frac{w}{l}\ln\frac{l}{w}. \tag{II. 10}$$

(d) Long rectangular slit with $l \gg h, l \gg w$:

$$1/\kappa = \frac{1}{lwh}\left\{w^2 h \ln\left[\frac{h}{w} + \sqrt{\left(1+\left(\frac{h}{w}\right)^2\right)}\right] + wh^2 \ln\left[\frac{w}{h} + \sqrt{\left(1+\left(\frac{w}{h}\right)^2\right)}\right] - \frac{(h^2+w^2)^{\frac{3}{2}}}{3} + \frac{(h^3+w^3)}{3}\right\}. \tag{II. 11}$$

(e) Long rectangular slit with $l \gg h, l \gg w, h \gg w$ (a special limiting case of the preceding formula when $h \gg w$):

$$1/\kappa = \frac{w}{2l}\left\{1 + 2\ln\frac{2h}{w}\right\}. \tag{II. 12}$$

Silsbee (SIL 50), with different assumptions, has derived a formula that agrees with the last one except that the 2 in the denominator is replaced by 8/3.

If the aperture is of a shape different from any of the above, a poorer approximation to $1/\kappa$ can often be obtained for large l from the expression of Knudsen (KNU 09)

$$1/\kappa = (16/3)\bigg/\left(A_s \int_0^l \frac{O}{A^2}\, dl\right), \qquad (\text{II. 13})$$

where O is the periphery and A the area of a normal cross-section at position l along the length of the canal. A_s is present just to cancel that in Eq. (II. 7). If the cross-section of the canal remains unaltered along its length, Eq. (II. 13) reduces to

$$1/\kappa = (16/3)A/lO. \qquad (\text{II. 14})$$

It should be noted that Eq. (II. 9) is a special case of Eq. (II. 14).

All of the above expressions except Eq. (II. 8) assume that l is much greater than the narrowest slit dimension. For short canals of infinite height, Claussing (CLA 29) has calculated and tabulated values of $1/\kappa$ for various values of l/w. His results are given in Table II. I.

Table II. I

Effusion of Gas through Canals of Infinite Height and Intermediate Length

Values of the factor $(1/\kappa)$ for use in Eq. (II. 7) with canals of intermediate lengths but infinite height are given in the following table. These values were calculated by Clausing (CLA 29) and additional tabulated values may be found in his original paper (CLA 29, FRA 31). The length of canal is designated by l and the width by w.

l/w	$1/\kappa$	l/w	$1/\kappa$
0	1	1·0	0·6848
0·1	0·9525	1·5	0·6024
0·2	0·9096	2·0	0·5417
0·3	0·8710	3·0	0·4570
0·4	0·8362	4·0	0·3999
0·5	0·8048	5·0	0·3582
0·7	0·7503	8·0	0·2789
		∞	$\frac{w}{l}\ln\frac{l}{w}$

II. 2. Molecular-beam intensities and shapes

II. 2.1. *Beam intensities.* If the pressure in the molecular-beam apparatus is sufficiently low that no appreciable amount of the beam is scattered out and if no collimating slit or similar obstruction intercepts the beam on its way to the detector, the theoretical beam intensity can readily be calculated from the expressions of the preceding section.

Let A_d be the area of the detector, l_0 be the length of the apparatus from source to detector, and I be the beam intensity or number of molecules which strike the detector per second. Then from Eq. (II. 1)

$$I = \frac{1}{4\pi} \frac{A_d}{l_0^2} n \bar{v} A_s. \qquad (II.\ 15)$$

If p' is the source pressure in mm Hg, M the molecular weight (not the weight of the molecule), T the absolute temperature in °K, A_s and A_d in cm², and l_0 in cm, the preceding equation can be re-expressed as

$$I = 1 \cdot 118 \times 10^{22} \frac{p' A_s A_d}{l_0^2 \sqrt{(MT)}} \text{ molecules sec}^{-1}. \qquad (II.\ 16)$$

The above expression for the beam intensity assumes that there is no attenuation of the beam by scattering. If there is such attenuation in any chamber, such as the separating chamber, through which the beam passes, the resultant beam intensity can be obtained by multiplying Eq. (II. 16) by the attenuation factors appropriate to each such chamber. If λ_{kv} is the mean free path for a beam molecule of velocity v in passing through the kth chamber whose length is l_k, the attenuation factor for that chamber and that molecular velocity is $\exp(-l_k/\lambda_{kv})$. With the very narrow beams used in molecular-beam experiments, collisions which deflect a molecule only slightly are sufficient to eliminate the molecule from the beam. As a result the appropriate value for λ_{kv} in molecular-beam experiments is often less than the mean free path appropriate to other less sensitive experiments. A more detailed discussion of beam attenuation by gas scattering is given in Section II. 4.

II. 2.2. *Beam shapes.* If a collimator slit is in position, the apparent beam intensity measured at a specific location is no longer given by Eq. (II. 16), but it depends on how much of the beam is intercepted by the collimator. In general the measured beam intensity depends on the widths of the source, collimator, and detector slits and on their relative positions. However, the beam shape can most easily be calculated by first considering the dependence of the beam intensity on detector position in the limit when the detector is infinitely narrow, i.e. the beam intensity that reaches the detector but is not integrated over

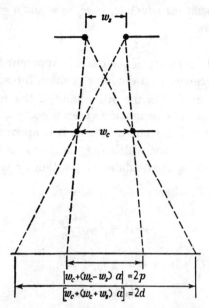

FIG. II. 2. Relation of source and collimator widths to beam shape. In the penumbra region the intensity varies linearly with displacement since the amount of exposed source varies in this way. Consequently the beam is of trapezoidal shape.

a detector slit of appreciable width. This beam shape can easily be determined with the aid of Fig. II. 2 to be that given in Fig. II. 3, where

$$p = \tfrac{1}{2}|w_c+(w_c-w_s)a|, \tag{II. 17}$$

$$d = \tfrac{1}{2}[w_c+(w_c+w_s)a], \tag{II. 18}$$

$$a \equiv l_{cd}/l_{sc} = l_{sd}/l_{sc}-1 = r-1, \tag{II. 19}$$

$$r = l_{sd}/l_{sc}, \tag{II. 20}$$

and where l_{cd} is the distance from collimator to detector, l_{sc} the distance from source to collimator, l_{sd} the distance from source to detector, w_s the width of the source slit, w_c the width of the collimator slit, p the half-width of the top of the trapezoidal beam shape, and d the half-width of the bottom of the trapezoidal beam shape. The absolute value signs are included to make the result valid in the case when the collimator slit is so narrow that there is no position at the detector location for which the source is completely unobscured. A special case, that is often relevant, is that for which $w_s = w_c = w$ and $a = 1$; the preceding equations then give

$$p = \tfrac{1}{2}w; \quad d = \tfrac{3}{2}w. \tag{II. 20 a}$$

The trapezoidal character of the beam is apparent from the fact that full intensity is received at a detector position for which the source is completely unobscured. On the other hand, if the source is partially obscured the amount of obscuration varies linearly with position at the detector. From this linear characteristic it is apparent that the beam intensity per unit detector width $I_0(s_0)$ is given as a function of the observation position s_0 at the detector position by the following:

(a) In region AB, where $-d \leqslant s_0 \leqslant -p$,

$$I_0(s_0) = I_{00}\frac{d+s_0}{d-p}. \tag{II. 21}$$

(b) In region BC, where $-p \leqslant s_0 \leqslant p$,

$$I_0(s_0) = I_{00}. \tag{II. 22}$$

(c) In region CD, where $p \leqslant s_0 \leqslant d$,

$$I_0(s_0) = I_{00}\frac{d-s_0}{d-p}. \tag{II. 23}$$

If the collimator and source widths are such that in the region BC the source is completely unobscured, I_{00} is the beam intensity per unit detector width that would be observed in the absence of a collimator as in Eq. (II. 15). If on the other hand the source is partially obscured in this region, I_{00} is reduced in proportion to the amount of obscuration.

If the beam is measured with a detector of appreciable width, for any specific location of the detector the observed total beam intensity is the integral of all of Fig. II. 3 that is included within the detector slit width. Consequently, the beam intensity pattern that is observed as such a detector is moved across the beam is similar to Fig. II. 3 but the sharp edges are all rounded off by the effect of detector slit width. Fig. II. 4 shows typical experimental beam shapes obtained with sources, detectors, and collimators whose slits are of various widths.

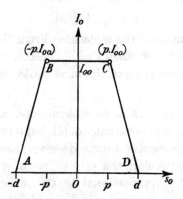

Fig. II. 3. Beam shape with detector of negligible width. The beam intensity I_0 is plotted as a function of the transverse displacement s_0 from the centre of the beam.

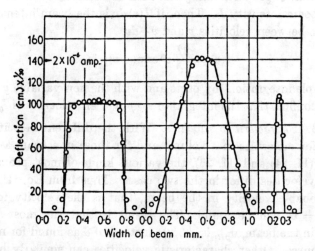

Fig. II. 4. Experimental beam shapes with different combinations of source and collimator slit widths (TAY 30). The solid lines are the geometrical beams constructed as in Fig. II. 3 from the slit dimensions.

II. 3. Velocities in molecular beams

II. 3.1. *Velocity distribution in a volume of gas.* In a volume of gas, such as inside the source, the velocity distribution of the molecules is in accordance with the Maxwell distribution law (KEN 38, ROB 33). If dN is the number of molecules in the velocity interval v to $v+dv$ and N is the total number of molecules, this law gives

$$\frac{dN}{dv} = Nf(v) = \frac{4N}{\sqrt{\pi}}\frac{1}{\alpha^3}v^2\exp(-v^2/\alpha^2), \qquad \text{(II. 24)}$$

where $\qquad \alpha = \sqrt{(2kT/m)},\qquad$ (II. 25)

and where m is the mass of the molecule. From (II. 24) it follows immediately that $d^2N/dv^2 = 0$ when $v = \alpha$. Therefore α may be given the simple physical interpretation of the most probable molecular velocity inside the source.

II. 3.2. *Velocity distribution in molecular beams*. On a superficial consideration one might expect that Eq. (II. 24) would also be applicable to molecules in the molecular beam. However, such is not the case as is easily seen. In particular, the probability of a molecule emerging from the source slit is proportional to the molecular velocity, as is apparent either from a simple physical consideration or from the form of explicit velocity dependence in Eq. (II. 1). The velocity dependence in the beam will then be proportional to Eq. (II. 24) multiplied by v. The constant of proportionality can then be obtained by normalizing to the full beam intensity I_0. Then, if $I(v)\,dv$ is the beam intensity from molecules between velocities v and $v+dv$,

$$I(v) = \frac{2I_0}{\alpha^4} v^3 \exp(-v^2/\alpha^2). \qquad \text{(II. 26)}$$

By a simple integration (by parts and with the new variable $y = v^2/\alpha^2$) one can confirm that the integral of $I(v)$ over all velocities is I_0.

II. 3.3. *Characteristic velocities*. With the different velocity distributions for molecules in a volume and for those in a beam as implied by Eqs. (II. 24) and (II. 26), the various kinds of most probable and average velocities differ in the two cases. Thus from Eq. (II. 26) the most probable velocity in the beam, that is the velocity for which $dI(v)/dv = 0$, can be found. The result is that the most probable velocity in the beam, v_{pB}, is $\sqrt{(3/2)}\alpha$ instead of α as found for molecules in the source. Other characteristic velocities can similarly be found. Let v_p indicate a most probable velocity, $\bar{v}$ an average velocity, V a root mean square velocity, v_{md} a median velocity, the additional subscript B a beam velocity, and the additional subscript V a velocity inside a volume such as the source. Then the following relations apply:

(a) Inside a volume of gas:

Most probable velocity

$$v_{pV} = \alpha = \sqrt{(2kT/m)}. \qquad \text{(II. 27)}$$

Average or mean velocity

$$\bar{v}_V = \int_0^\infty vf(v)\,dv = (2/\sqrt{\pi})\alpha = 1\cdot 13\alpha. \qquad \text{(II. 28)}$$

Root mean square velocity

$$V_V = \left[\int_0^\infty v^2 f(v)\, dv\right]^{\frac{1}{2}} = \sqrt{(3/2)}\alpha = 1\cdot 22\alpha. \tag{II. 29}$$

Median velocity $\left(\text{such that } \int_0^{V_{mdV}} [I(v)/I_0]\, dv = \tfrac{1}{2}\right)$

$$v_{mdV} = 1\cdot 09\alpha.$$

(b) In a molecular beam:
Most probable velocity

$$v_{pB} = \sqrt{(3/2)}\alpha = 1\cdot 22\alpha. \tag{II. 30}$$

Average velocity

$$\bar{v}_B = (3/4)\sqrt{\pi}\,\alpha = 1\cdot 33\alpha. \tag{II. 31}$$

Root mean square velocity

$$V_B = \sqrt{2}\,\alpha = 1\cdot 42\alpha. \tag{II. 32}$$

Median velocity

$$v_{mdB} = 1\cdot 30\alpha. \tag{II. 33}$$

The above characteristic velocities can be related to the temperature of the molecules with the aid of Eq. (II. 25).

II. 3.4. *Experimental measurements of molecular velocities.* Some of the earliest molecular-beam research was for the purpose of checking experimentally the velocity distribution of the emerging molecules. In 1920 Stern (STE 20, STE 20*g*) made a first rough determination with an apparatus whose principle is schematically illustrated in Fig. II. 5. When the apparatus is stationary the beam atoms deposit at position P. If, on the other hand, the apparatus is rotated in the direction indicated by the arrow, the atoms will follow a path which will appear curved to an observer rotating with the apparatus. The atoms will then deposit at a different location P'. The separation between P and P' then measures the velocity of the atoms, and the spread of the deposit indicates the distribution of velocities. A diagram of the actual apparatus used by Stern is shown in Fig. II. 6. The cylinder R and its contents were

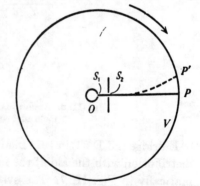

Fig. II. 5. Schematic diagram illustrating principle of Stern's molecular-velocity experiment (STE 20). The distance $S_1 P$ was approximately 10 cm.

rotated. The source of the beam was silver atoms evaporated from the silver-sheathed platinum wire L.

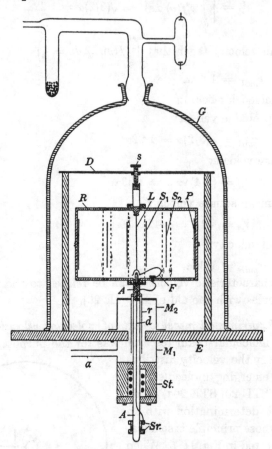

Fig. II. 6. Apparatus used by Stern in measuring molecular velocities (STE 20).

Eldridge (ELD 27) in 1927 made a better determination of the velocity distribution with the aid of the molecular-beam apparatus shown schematically in Fig. II. 7. The system of disks was mounted in a large glass tube, the lowest disk served as the rotor of an induction motor and rotated the whole system of disks at speeds of up to 7,200 revolutions per minute. The other disks each had a hundred radial slots and when rotated served as a velocity filter. Cadmium atoms were used for the beam and they emerged from the slit, which was connected to the cadmium reservoir which could be heated in a furnace. The spread of the cadmium deposit then measures the velocity distribution of the

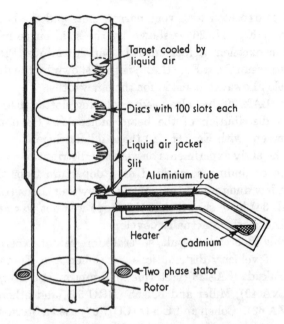

FIG. II. 7. Eldridge's method for molecular-velocity measurement (ELD 27).

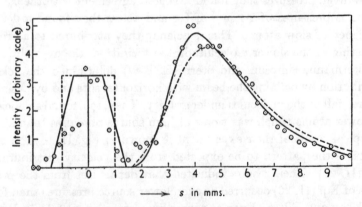

FIG. II. 8. Results of Eldridge's measurements of beam intensity as a function of displacement. To the left is the undeflected trace with the rotor only in very slow rotation; to the right is the spectrum at a high rotor velocity. The full curve at the right is the theoretical distribution on the assumption of negligible parent beam width. The agreement is in part fortuitous as can be seen from the dashed theoretical curve in which allowance is made for the known width of the parent beam (FRA 31).

atoms. The experimental results are shown by the small circles in Fig. II. 8. The left portion is obtained when the disks are rotated very slowly and the right portion is for rapid rotation. The theoretical result that would be expected with very narrow slots and the beam velocity distribution of Eq. (II. 26) is shown by the solid curve in the figure. However, this excellent agreement is in part fortuitous (FRA 31) since the predicted result from Eq. (II. 26) is that shown by the dashed curve when suitable allowance is made for the slit widths.

Lammert (LAM 29) has also used a rotating slotted disk to measure the velocity distribution of the beam molecules. He obtained even better agreement with Eq. (II. 26) than did Eldridge.

Some of the early experiments on velocity distributions were inspired by the hope of finding some kind of a departure from the classical distribution law due to quantum effects. However, Lenz (LEN 29) and others (HAL 34) have shown that this hope was not warranted by the quantum theory which actually developed.

Despite this theoretical result, several more recent experiments have been made of velocity distributions by Cohen and Ellett (COH 37); Saski and Fukuda (SAS 38); Estermann, Simpson, and Stern (EST 47); Knauer (KNA 49); Miller and Kusch (BRO 54), and others (MOO 53, BEN 54, ZZA 63). Cohen and Ellett (COH 37) used a Stern–Gerlach experiment (see Chapter IV) to provide the velocity selection in the beam. At low oven pressures they found excellent agreement with theory, but at source pressures of 1·4 mm of Hg and above they found a definite deficiency of slow atoms. This deficiency they attributed to a greater scattering of the slower molecules in the vicinity of the oven slit.

Estermann, Simpson, and Stern (EST 47) determined the velocity distribution by defining the beam with horizontal slits and by observing the free fall of the molecules under gravity. They, too, found a deficiency of slower atoms which was worse at high source pressures than at low. A typical result of their experiment is shown in Fig. II. 9. The curve indicates the pattern to be expected from the velocity distribution of Eq. (II. 26). Their results indicated that departures from the predictions of Eq. (II. 26) occurred at even lower source pressures than found by Cohen and Ellett. Even at source pressures of 2×10^{-2} mm Hg and with an oven slit 3 mm $\times$ 0·06 mm they found more than a 10 per cent. deficiency of the slower molecules. Similar results have been obtained by Knauer (KNA 49a). Knauer determined the velocity distribution by electronically measuring the time of flight of an interrupted beam. He agreed with the other observers in attributing the deficiency of slow molecules to scattering by the cloud in the neighbourhood of the source slit.

A quite different experiment on molecular velocities has been carried out by Paul (PAU 48). He has measured the mean velocity of the beam with a sensitive balance to determine the impulsive force in combination with a determination of the mass of the deposited silver.

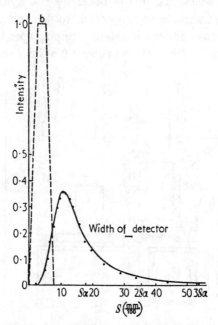

Fig. II. 9. Gravity deflexion of a Cs beam (EST 47).

II. 4. Molecular scattering in gases

II. 4.1. Methods of measurement.
The first rough measurements of molecular scattering by atomic-beam methods were those of Born (BOR 20) and Bielz (BIE 25). They measured the rate of deposition of silver by an atomic beam that passed through air at various pressures and determined the mean free path at pressure p from the relation

$$I_p = I_0 \exp(-l/\lambda_p). \tag{II. 34}$$

As is discussed in the next section, the above relation is only approximate since it does not allow for the variation of λ_p with the different velocities of the beam molecule. However, it is an adequate approximation for the description of the earliest crude experiments. Born and Bielz found values for the mean free path which as to order of magnitude agreed with theoretical expectations, and were not inconsistent with viscosity and diffusion measurements which also determine a mean free path.

Knauer and Stern (KNA 26, KNA 29) measured the mean free path for the non-condensable gas hydrogen by measuring the beam intensity as a function of source pressure, and by attributing the saturation that was observed to scattering of the beam by the hydrogen gas in the apparatus, which increases in proportion to the source pressure for a non-condensable gas and for a constant pumping speed. The value they obtained for the mean free path was about 0·4 times that from standard

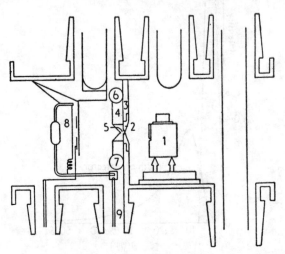

FIG. II. 10. Rosin and Rabi apparatus for effective collision cross-sections (ROS 35).

viscosity and diffusion measurements. They attributed this difference to the fact that a very gentle collision between two molecules which deflects them only slightly is fully effective in attenuating the well-collimated molecular beams, whereas such a collision is relatively ineffective in viscosity or diffusion problems.

Broadway and Fraser (FRA 33, BRO 33, BRO 35, KNA 35) obtained a higher angular resolution by confining the scattering gas to a limited region by using a crossed molecular beam, i.e. the normal molecular beam passed close above the circular opening of an oven from which mercury was streaming molecularly. Fraser particularly investigated the scattering at small angles and confirmed, in agreement with quantum theory (see Section II. 4.4), that the cross-section rose to a high value for small-angle scattering but that this high value was not infinite.

Some of the most extensive measurements of mean free paths have been those of Rosin and Rabi (ROS 35). Their apparatus is shown in Fig. II. 10. In different experiments, beams of neutral atoms of Li,

Na, K, Rb, and Cs from the oven (1) passed through the scattering chamber (4) and were detected with a surface ionization detector. H_2, D_2, He, Ne, and A were used as scattering gases. Both the change in beam shape and the attenuation of the beam by the scattering gas were observed. The beam shapes with and without the scattering gas are shown in Fig. II. 11. Although the beam intensity is reduced by the scattering gas there is no appreciable change in beam shape. The

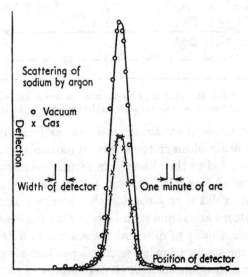

Fig. II. 11. Apparent beam shape with and without argon scattering of sodium (ROS 35).

absence of such a broadening indicates that only a small fraction of the atoms are scattered by only a few minutes of arc. This result is in agreement with the quantum-mechanical interpretation of Section II. 4.4 and in agreement with Broadway's result discussed in the preceding paragraph. The inference of scattering cross-sections from the attenuation measurements of Rosin and Rabi (ROS 35) and of Rosenburg (ROS 39) by the same method will be discussed in the next section.

Estermann, Foner, and Stern (EST 47a) measured gas scattering with the same apparatus they used for measuring velocity distributions by the free fall of the atoms in a beam as described in Section II. 3. Their apparatus is schematically represented in Fig. II. 12. The molecules of different velocities follow different curved trajectories in such an apparatus as they fall gravitationally in the apparatus. Consequently when the detector is in a specific position the detected beam is due to molecules of only a limited velocity range. With such velocity selection, Estermann, Foner, and Stern introduced a scattering gas into the scattering chamber C and observed the attenuation of the beam. Their results will be discussed in Section II. 4.3. Jawtusch (JAW 52) has measured the scattering of K atoms by crossed beams of Hg and of organic molecules.

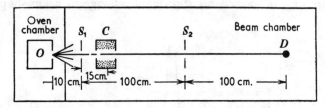

FIG. II. 12. Schematic diagram of molecular-scattering apparatus. The scattering vapour was dominantly in the chamber C (EST 47a).

II. 4.2. *Calculation of scattering cross-sections from attenuation measurements.* The order of magnitude of total molecular scattering cross-section can be related to the observed molecular-beam attenuations by the combination of Eqs. (II. 34) and (II. 4). However, this simple procedure is not valid if any considerable accuracy is required. The added complications arise from such facts as that the beam and scattering molecules are usually of different masses and at different temperatures, and that the attenuation is velocity dependent even for a velocity independent cross-section since a slow molecule is in the scattering region longer and is more likely to be bumped into by the scattering gas.

Let M_B and T_B be the mass and temperature of a beam molecule while M_G and T_G are the corresponding quantities for the scattering gas. The chance that a beam molecule of velocity v will pass a distance l through the scattering gas without collision is given by

$$P(v) = \exp(-l/\lambda_v). \qquad (II.\ 35)$$

Although the free path λ_v in the above equation is for a specific velocity of the beam molecule, it is averaged over all velocities of the gas molecules. λ_v can be related to the mean effective cross-section $\bar{\sigma}_{BG}$ by a suitable average over the velocity distribution of the scattering gas. This relation is derived in most books on kinetic theory of gases (JEA 25, LOE 34, FOW 36); the result is that

$$\lambda_v = \pi^{\frac{1}{2}}(v/\alpha_G)^2/n_G\,\bar{\sigma}_{BG}\,\psi(v/\alpha_G), \qquad (II.\ 35a)$$

where
$$\psi(x) \equiv x\exp(-x^2)+(2x^2+1)\int_0^x \exp(-y^2)\,dy, \qquad (II.\ 36)$$

and where n_G is the number of gas molecules per cm³ and α_G is defined by Eq. (II. 25) for the scattering gas molecules. The ratio I/I_0 of the beam intensities without and with the scattering gas is then equal to the probability $P(v)$ averaged over all beam velocities with the aid of Eq. (II. 26). Therefore

$$I/I_0 = \langle P(v) \rangle$$
$$= 2(\alpha_G/\alpha_B)^4 \int_0^\infty \exp[-\{ln_G\,\bar{\sigma}_{BG}\,\psi(x)/\pi^{\frac{1}{2}}x^2\} - (\alpha_G/\alpha_B)^2 x^2]x^3\,dx, \quad \text{(II. 37)}$$

where x has been written for v/α_G.

The evaluation of $\bar{\sigma}_{BG}$ from Eq. (II. 37) is clearly rather tedious. An approximation that is often used is obtained by taking

$$I/I_0 = \langle P(v) \rangle = \langle \exp(-l/\lambda_v) \rangle \approx \exp(-l/\langle \lambda_v \rangle). \quad \text{(II. 38)}$$

$\langle \lambda_v \rangle$ is then determined from the experimental observations and Eq. (II. 38). However, from Eqs. (II. 35a) and (II. 26),

$$\langle \lambda_v \rangle = (2\pi^{\frac{1}{2}}/\bar{\sigma}_{BG}\,n_G)\mathscr{I}(\alpha_G^2/\alpha_B^2), \quad \text{(II. 39)}$$

where by definition

$$\mathscr{I}(z) = z^2 \int_0^\infty [x^5/\psi(x)]\exp(-zx^2)\,dx. \quad \text{(II. 40)}$$

Numerical values for $\mathscr{I}(z)$ have been calculated by Rosin and Rabi (ROS 35) and are given in Table II. II. From this table and Eq. (II. 39) it is apparent that in the special case when $\alpha_G/\alpha_B = 1$,

$$\langle \lambda_v \rangle = 0.75/\bar{\sigma}_{BG}\,n_G. \quad \text{(II. 41)}$$

TABLE II. II

Values of $\mathscr{I}(z)$ for a Number of Values of z

$\mathscr{I}(z)$ is defined in Eq. (II. 40) (ROS 35)

z	$\mathscr{I}(z)$	z	$\mathscr{I}(z)$	z	$\mathscr{I}(z)$	z	$\mathscr{I}(z)$
0	$1/(2\sqrt{\pi})$	0.545	0.236	3.0	0.155	10	0.094
0.05	0.275	0.8	0.221	4.0	0.140	15	0.0813
0.1	0.265	0.9	0.216	5.28	0.127	21	0.0714
0.2	0.256	1.0	0.211	7.0	0.113	25	0.0643
0.3	0.248	2.0	0.173	8.0	0.107	30	0.0589
						35	0.0548

Although the calculation of $\bar{\sigma}_{BG}$ from Eqs. (II. 38) and (II. 41) is less accurate than by Eq. (II. 37), Rosin and Rabi have shown that calculations by the two methods are usually in approximate agreement.

II. 4.3. *Cross-section theory.* The theoretical values for the total scattering cross-sections that should apply to the atomic-beam experiments described in the two preceding sections have been discussed extensively in the literature (MOT 33, MAS 33, MAS 36, FRA 35, ZXX 54). In addition the theories have recently been summarized in an excellent book by Massey and Burhop (MAS 52). For this reason, no details of the calculations will be given here, but only the major theoretical results will be outlined.

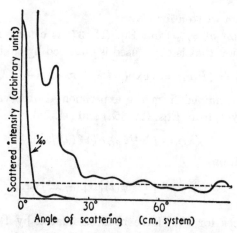

FIG. II. 13. Angular distribution of scattered atoms for a rigid sphere model with $ka = 20$. Dashed curve for classical distribution and full curve for quantum-mechanical distribution (MAS 33).

Massey and Mohr (MAS 33) have calculated the nature of the scattering to be expected when the atoms have been approximated by rigid spheres, i.e. when the potential between the molecules has been taken as zero if the centres of the two molecules are more than a distance a apart and as ∞ when the molecules are less than a distance a apart. The angular distribution of the scattered atoms in the centre of mass system for this model is shown in Fig. II. 13 for the particular case of a relative velocity v such that the de Broglie wave number k of relative motion is $20/a$, where

$$k \equiv 1/\lambda = \mu_m v/\hbar \qquad (\text{II. 42})$$

and μ_m is the reduced mass. As can be seen in the figure, there is a large but finite maximum for very low-angle scattering. On the other hand, the differential cross-section for purely classical scattering with such a model would be a uniform $\tfrac{1}{4}a^2$ per steradian as indicated by the dashed line in the figure. The positions of the subsidiary maxima are in general velocity dependent so that they would be smoothed out by the velocity distribution of the beam (MIZ 31). For this reason and because of the very weak intensity of the scattered beam they have never been observed. As mentioned in the preceding section, almost all atomic-beam scattering studies have been measurements of total cross-sections by the attenuation of the beam in passing through a gas-filled scattering chamber.

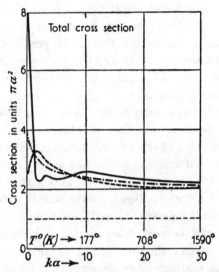

FIG. II. 14. Total collision cross-section as a function of ka. ------ classical theory, — · — quantum theory for dissimilar atoms, ———— quantum theory for identical atoms obeying Bose–Einstein statistics, and ------ quantum theory for identical atoms obeying Fermi–Dirac statistics (MAS 33).

One use of the angular distribution curves such as Fig. II. 14 is that they indicate the angular resolution required in an atomic-beam experiment if the total cross-section as determined from an attenuation experiment is to have real meaning and be independent of the resolution of the apparatus. If $\sigma(\theta)$ is the differential cross-section and θ_0 is the minimum angle of deviation which is counted as a scattering, then the ratio β, where

$$\beta = \int_{\theta_0}^{\pi} \sigma(\theta)\sin\theta\, d\theta \bigg/ \int_{0}^{\pi} \sigma(\theta)\sin\theta\, d\theta, \qquad (II.43)$$

should be as near unity as possible. As shown by Massey and Burhop (MAS 52), β will be 0·9 if

$$\theta_0 \approx 277/a(MT)^{\frac{1}{2}}, \qquad (II.44)$$

where M is the molecular weight of the incident molecule and where a is in Å.

The total cross-sections to be expected with the above model at different velocities are plotted in Fig. II. 14. The temperature scale is that for which the root mean square velocity of He is the indicated velocity if $\frac{1}{2}a = 1\cdot05$ Å. As indicated in the figure caption the different curves are for dissimilar atoms, for identical atoms obeying Bose–Einstein statistics, and for identical atoms obeying Fermi–Dirac statistics, respectively. It should be noted that the total cross-section at short wavelengths is double the classically predicted value, as first pointed out by Massey and Mohr (MAS 33). However, as can be seen by a comparison with Fig. II. 13, the extra scattering above the classical prediction all corresponds to small-angle scattering. Consequently, good angular resolution is required in the total cross-section measurement if that is to be observed. This doubling of the total cross-section above the classical value at short wavelengths is the familiar problem of shadow scattering that arises in diffraction theory and nuclear scattering as well (BLA 52). It corresponds to the fact that not only is there scattering of the incident de Broglie wave which strikes the scatterer but also that a destructive wave of equal amplitude and opposite phase to the incident one must be postulated immediately behind the scatterer to provide the necessary shadow. However, this wave at a sufficient distance from the scatterer spreads out by diffraction, as does any limited wave, and appears to be a small angle additional scattered wave.

For actual atoms and molecules the above rigid sphere model is a poor approximation. As an alternative model, Massey and Mohr (MAS 33) have calculated the total cross-section for an interaction potential which is cut off at small radii but which has the asymptotic form

$$V \sim -Cr^{-s}. \tag{II. 45}$$

In this case they showed that the total cross-section σ was given approximately by (ZZA 64)

$$\sigma = \pi \frac{2s-3}{2s-2} f^{2/(s-1)}(4\pi C/hv)^{2/(s-1)}, \tag{II. 46}$$

where v is the velocity of the atoms and

$$f(s) = \frac{s-3}{s-2} \cdot \frac{s-5}{s-4} \cdots \frac{1}{2} \cdot \frac{1}{2}\pi \quad (s \text{ even and } > 2),$$

$$f(s) = \frac{s-3}{s-2} \cdot \frac{s-5}{s-4} \cdots \frac{2}{3} \quad (s \text{ odd and } > 3), \tag{II. 47}$$

$$f(s) = 1 \quad (s = 3), \qquad f(s) = \tfrac{1}{2}\pi \quad (s = 2).$$

They showed that the above should be applicable to gas kinetic collisions at ordinary temperatures between all atoms other than He. The special case of $s = 6$, or asymptotically of

$$V \sim -C/r^6, \qquad (\text{II. 48})$$

is of particular interest in Eq. (II. 46) since it is what is usually known as a van der Waals attraction and arises from the dynamic polarization of one atom by the other. The inverse sixth power law of attraction arises from the fact that the two electric dipoles interact as r^{-3} but the polarizability process is one in which the dipoles are not free but must be induced. As a consequence of this a second-order perturbation calculation is required so that the dependence is as $(r^{-3})^2$. Some of the results to be presented in the next section will be interpreted with the aid of Eq. (II. 46) in the special case of $s = 6$.

II. 4.4. *Results of atomic-beam scattering experiments.* The earliest important result of atomic-beam scattering experiments was the demonstration by Born (BOR 20) and Bieler (BIE 25) that cross-sections as determined from molecular-beam scattering experiments were of the same order of magnitude as those from other kinetic theory experiments such as viscosity and diffusion measurements. Knauer and Stern (KNA 29) with the experiment described in Section II. 4.1 showed that the mean free path in hydrogen-beam scattering experiments was about 0·4 times that to be expected from the measurements of viscosity and diffusion if the latter were interpreted in terms of a classical rigid-sphere model. However, the experiments could be reconciled with an angular distribution, such as that in Fig. II. 13, which provides a large amount of small-angle scattering since such scattering would be relatively ineffective in viscosity and diffusion phenomena but would be fully effective in scattering molecules from the beam. Fig. II. 13 shows that for a rigid-sphere model the quantum-mechanical distribution is suitable to account for the molecular-beam results whereas the classical predictions with the same model are in definite disagreement with experiment. On the other hand, with an interaction potential as in Eq. (II. 45) a classical theory predicts a forward scattering peak as required by the experiment of Knauer and Stern (KNA 29).

In fact, with a potential of the form of Eq. (II. 48), a classical theory overdoes the prediction of a small-angle scattering peak and predicts one whose differential cross-section becomes infinite as the angle tends to zero. On the other hand the experiments of Broadway and Fraser

(FRA 33, BRO 33, BRO 35, KNA 35) and of Rosin and Rabi (ROS 35) show that the forward-scattering cross-section remains finite as discussed in Section II. 4.1 above. The lack of change in the beam shape of Fig. II. 11 is a particularly effective demonstration of the finiteness of the small-angle scattering. The quantum-mechanical prediction is that for an interaction of the form of Eq. (II. 45) with $s > 3$ the scattering should be finite though large at small angles in agreement with experiment.

A number of effective total cross-sections measured by the atomic-beam method are given in the literature, particularly in the papers of Mais (MAI 34), Rosin and Rabi (ROS 35), Rosenberg (ROS 39), Estermann, Foner, and Stern (EST 47a), and others (DOR 38, ZZA 1). Knauer (KNA 33, KNA 34) and Zabel (ZAB 33, ZAB 34) have observed the influence of temperature on the scattering of molecular beams. In Table II. III are given observed collision cross-sections of alkali metal atoms with rare-gas scattering atoms. The large disagreement between the measurements of Rosin and Rabi (ROS 35) and Estermann, Foner,

TABLE II. III

Observed Collision Cross-sections for Alkali Rare-gas Atom Collisions and derived van der Waals Constants

The van der Waals constant C is defined by Eq. (II. 48). All measurements are by Rosin and Rabi (ROS 35) except for data marked with a dagger (†) which are by Rosenberg (ROS 39), with an asterisk (*) by Mais (MAI 34), and with a double asterisk (**) by Estermann, Foner, and Stern (EST 47a). The calculated values of C have been given by Massey and Burhop (MAS 52). C_σ indicates the value of C inferred from the listed collision cross-section σ while C_p indicates the value of C inferred from atomic polarizability data (MAS 52).

Alkali atom	Rare gas atom	σ (10^{-16} cm^2)	C_σ (10^{-60} cm^6)	C_p (10^{-60} cm^6)
Li	He	106	14·4	17
Li	Ne	120	18·7	32
Li	A	303	188	125
Na	He	130	17·8	(26)
Na	Ne	213	40·4	(51)
Na	A	401	192	(200)
K	He	165, 170·8†, 150*	31·7	35
K	Ne	259, 239*	50·5	68
K	A	580, 587·1†, 400*	356	260
Rb	He	152	25·4	40
Rb	Ne	268	49·0	(77)
Rb	A	572	249	(290)
Cs	He	162, (446**)	29·4 (370**)	44
Cs	Ne	287	56·7	87
Cs	A	572	235	325

and Stern (EST 47a) in the case where they overlap is both puzzling and disturbing. If the interaction between the atoms is assumed to be of the form of Eq. (II. 48), the value of the van der Waals constant C can be derived from Eq. (II. 46). These constants derived from the molecular-beam cross-sections are given in Table II. III. However, the polarizabilities of the alkali atoms have been measured in other molecular-beam experiments (SCH 34a) and the polarizabilities of the rare-gas atoms can be deduced from refractivity measurements. Since the van der Waals interaction is a second-order polarizability effect, the constant C can be calculated from the polarizability data (MAS 52). The values of C so determined are listed in Table II. III for comparison with the values inferred from molecular-beam data. On the whole the agreement is quite good except for the anomalously large C inferred from the Cs–He experiment of Estermann et al. (EST 47a).

II. 5. Interaction of molecular beams with solid surfaces

II. 5.1. *Introduction.* When a molecule strikes a surface it may suffer an elastic collision and rebound immediately with no loss of energy or it may suffer an inelastic collision in which case it may either gain or lose energy. A special case of an inelastic process is for the molecule to lose so much energy that it remains adsorbed on the surface. In the case of adsorption the molecule can either remain permanently on the surface or it may remain there for only short time and then be restituted to the gas. The molecules which rebound immediately and elastically can do so in directions that have a direct probability relation to the direction of incidence. Such molecules are studied in the sections below entitled *Reflection* and *Diffraction*. The remaining cases are discussed in the section on *Inelastic collisions*.

II. 5.2. *Reflection.* The simplest type of reflection of molecules at a surface is specular reflection, that is the incident beam, the normal to the surface at the point of impact, and the reflected beam are all in the same plane and the angle of incidence equals the angle of reflection just as in the corresponding optical case.

Two conditions must be satisfied to achieve specular reflection:

(a) The height of the irregularities of the surface when projected on the direction of the beam must be less than a wavelength. This is just the familiar requirement for reflection in optical experiments, which is illustrated by the fact that a piece of smoked glass is a poor reflector at normal incidence but a good one at glancing incidence. If h is the average height of irregularities and if ϕ_0 is the glancing angle of the incident beam (the complementary angle to the angle of incidence θ_0), this requirement is

$$h \sin \phi_0 < \lambda. \tag{II. 49}$$

(b) The average time the molecule spends on the surface must be small. If this is not so the conditions under which re-evaporation of the molecule will occur will bear no relation to those which prevailed at condensation. Instead they will be determined by the last stages of the history of the molecule on the surface.

The irregularities of the best mechanically polished surfaces are of the order of 10^{-5} cm, while the de Broglie wavelength for hydrogen at room temperature is of the order of 10^{-8} cm. When these values are substituted in Eq. (II. 49) the condition for reflection becomes $\phi_0 < 10^{-3}$ or the glancing angle must be less than a few minutes of arc.

With polished speculum metal Knauer and Stern (KNA 29a) indeed observed a 5 per cent. reflecting power for a hydrogen beam at an angle of 10^{-3} and at room temperature. The reflecting power is taken as the maximum intensity of the reflected beam divided by the maximum intensity of the parent beam. They also found that the reflecting power was increased if the temperature of the beam was lowered, as would be expected from Eq. (II. 49).

The smoothest surfaces that are obtainable are the cleaved surfaces of a crystal. Even these surfaces are not completely smooth because the ions in the surface are subject to temperature vibrations about their positions of equilibrium. The amplitudes of these oscillations are of the order of 10^{-8} cm. Consequently a room-temperature helium beam should show specular reflection for glancing angles less than 20° or 30°.

The result anticipated above has been found by Estermann and Stern (EST 30) in the reflection of a helium beam from a cleaved LiF crystal. Their experimental result is shown in the curve of Fig. II. 15 where the beam intensity is plotted as a function of the glancing angle. It will be noted in a comparison of the curves at two different temperatures that the reflecting power at the lower beam temperature begins to fall off only at larger glancing angles in agreement with Eq. (II. 49). Johnson (JOH 28) has found similar effects with atomic hydrogen and rock salt. The dependence of the reflecting power on the crystal temperature has been studied (KNA 29a, EST 30, JOH 28) in a number of experiments. In all cases the reflecting power is increased by a lowering of the crystal temperature. This is in agreement with previous assumption that in cleaved crystals the h of Eq. (II. 49) was largely due to thermal vibrations of the ions in the crystal lattice.

The importance of the above condition (b) for the production of specular reflection is clearly indicated in an experiment of Taylor (TAY 30, JOS 33). He searched for a specularly reflected Cs beam from the

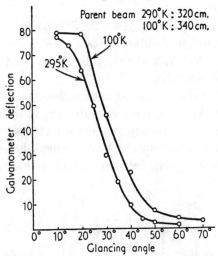

FIG. II. 15. Reflecting power of LiF for a helium beam at two different temperatures of the LiF (EST 30).

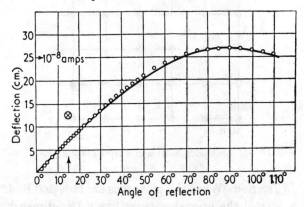

FIG. II. 16. Restitution of Li from cleaved planes of NaCl. The solid curve represents angular distribution of scattering according to Knudsen's law and the circles represent the observed intensities. The incident beam was at a glancing angle of 15° (TAY 30).

cleaved surface of NaCl and LiF. Although the condition of Eq. (II. 49) was fully satisfied he found no specularly reflected beam for glancing angles between 2° and 60°. An example of his results for an incident glancing angle of 15° are shown in Fig. II. 16. No signs of a specularly reflected beam at 15° can be seen. The angular distribution of the restituted beam is in complete accord with the Knudsen cosine law as in Eq. (II. 1) which is plotted as the solid curve in the figure. This is what would be expected for atoms that were adsorbed on the surface and then later re-evaporated.

Knauer and Stern (KNA 29a, HAN 32, JOS 33, ZZA 65) have studied reflection with a number of different gases forming the molecular beam; out of H, H_2, He, Ne, A, and CO_2 only H, H_2, and He gave strong specular reflection with NaCl.

If condition (b) above is satisfied but condition (a) is not, one obtains diffuse reflection. With a macroscopically rough surface the molecules emerge from hills and valleys in random directions even though each one has been reflected on impact. Hence unless the surface is smooth it is difficult to distinguish between reflection and adsorption with re-evaporation.

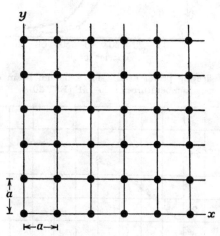

FIG. II. 17. Scattering centres in a cross-grating (MAS 52).

II. 5.3. *Diffraction.* When the molecules are reflected by the cleaved surface of a crystal the crystal appears like a two-dimensional cross-grating as shown in Fig. II. 17. For a one-dimensional grating as in Fig. II. 18, it is apparent that the difference in path for the two rays is

$$a \cos \alpha - a \cos \alpha_0.$$

For this to provide constructive interference it should equal an integral multiple of the de Broglie wavelength λ. A similar analysis applies to the two-dimensional cross-grating of Fig. II. 17. For the incident beam making angles $(\alpha_0, \beta_0, \gamma_0)$ with the coordinate axes the maxima of the diffracted beam should be at angles (α, β, γ) with these axes, where

$$\cos\alpha - \cos\alpha_0 = n_1 \lambda/a,$$
$$\cos\beta - \cos\beta_0 = n_2 \lambda/a, \qquad (II.\ 50)$$

where n_1 and n_2 are any integers, positive or negative. The diffracted rays are often designated by (n_1, n_2).

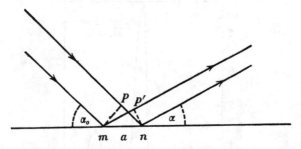

Fig. II. 18. Diffraction by a one-dimensional grating (FRA 31).

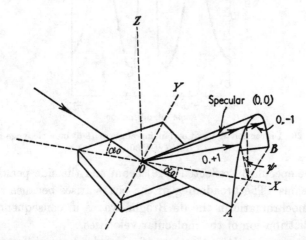

Fig. II. 19. Diffraction of a molecular beam by a cross-grating. The positions of the (0, 0) specularly reflected and the (0, ±1) diffracted rays are shown (MAS 52).

The case of $n_1 = n_2 = 0$ is just the case of specular reflection that was discussed in the preceding section.

In most molecular diffraction experiments the crystal is arranged so that β_0 is zero. Fig. II. 19 illustrates the (0, 0) reflected and (0, ±1) diffracted rays.

The first experimental evidence that molecular rays were diffracted by the cleavage plane of a crystal as from a cross-grating was obtained by Stern (STE 29). This work was extended by Estermann and Stern (EST 30); Estermann, Frisch, and Stern (EST 31, EST 31a); Johnson (JOH 30, JOH 31); and others (ELL 29, ELL 37, ZAH 38, ART 42, COH 37, KEL 32, MIS 51, VES 38, ZAB 32). A typical result is that of Estermann and Stern (EST 30) as shown in Fig. II. 20. The central

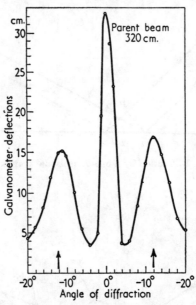

Fig. II. 20. Diffraction of He at a glancing angle of 18·5° on a cleavage face of LiF (EST 30).

peak is the specularly reflected $(0,0)$ beam and the side peaks are the $(0, \pm 1)$ beams. The broadening of the peaks arises because the beam is not monochromatic in the de Broglie waves in consequence of the Maxwell distribution of the molecular velocities.

The most complete investigations of molecular-beam diffraction have been those of Frisch and Stern (FRI 33a). Their apparatus is shown schematically in Fig. II. 21. It provided a convenient means for rotating the crystal about its principal axis OX, i.e. for varying the angle ψ of Fig. II. 19. Their results for helium on LiF are shown in Fig. II. 22. The dashed curve gives the theoretically predicted distribution of intensity for the $(0,1)$ diffracted beam, derived from the Maxwell distribution of velocity (FRI 33a, LEN 34). The agreement with position and shape of the experimental distribution is generally quite good but the two experimental irregularities for ψ's of 20° and 26° that also occur for corresponding negative values of ψ are reproducible irregularities that are smaller than the experimental error. Similar anomalies were found in specular reflection when the angle η between the plane

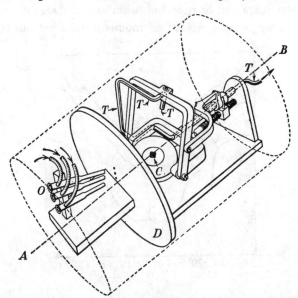

Fig. II. 21. Frisch and Stern's apparatus for studying diffraction of molecular beams (FRI 33a).

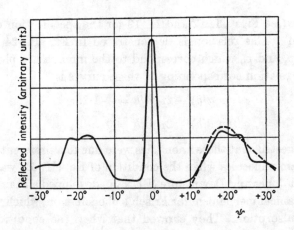

Fig. II. 22. Observed variation with ψ of reflected intensity for a beam of helium atoms incident on the cleavage plane of a LiF crystal. Dashed curve is the theoretical intensity for (0, 1) diffraction (FRI 33a).

of incidence and the principal axis OX of the crystal was varied. A typical result is shown in Fig. II. 23. Similar anomalies were also found with He on NaF and H_2 on LiF.

From an analysis of the conditions under which the anomalous minima appeared Frisch (FRI 33b) was able to show that they arose when, for the diffracted or reflected beam under observation, the components p_z and p_y of the molecules' momenta bore certain relations to

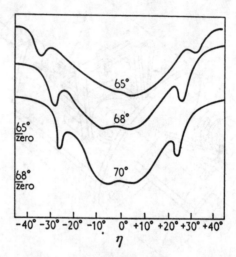

Fig. II. 23. Variation of reflected intensity with η for a beam of helium atoms incident on the cleavage plane of a LiF crystal at angles of 65°, 68°, and 70° (FRI 33a).

each other (see Figs. II. 17 and II. 19 for the specification of the axis orientation). This relation is demonstrated in Fig. II. 24 where the values of p_z and p_y which correspond to the minima are plotted. The analytical relation corresponding to these curves is

$$(p_z a/h)^2 - 2p_y a/h = -1\cdot 25,$$
$$(p_z a/h)^2 - 2p_y a/h = 0. \qquad (II. 51)$$

Frisch suggested that these anomalies were due to a preferential adsorption of beam molecules when the condition of Eq. (II. 51) was satisfied.

Lennard-Jones and Devonshire (LEN 36) developed a beautiful quantum-mechanical mechanism for Frisch's phenomenon, which they called selective adsorption. They showed that when the conditions of Eq. (II. 51) were satisfied, a transfer of energy from motion perpendicular to the surface to motion parallel to the surface may easily take place because of the periodic nature of the lattice. The molecules then slide freely over the crystal surface while vibrating with a finite amplitude in the normal direction. Hence, if the momentum conditions are correct, this variety of adsorption can take place with a corresponding reduction in the reflected or diffracted intensity. Alternatively, if after collision the reflected or diffracted beam satisfies the conditions it may be adsorbed in this way before leaving the crystal.

The details of the analysis of Lennard-Jones and Devonshire may be found in their original paper (LEN 36) or in the book of Massey and Burhop (MAS 52). However, the justification of the relations of Eq. (II. 51) may be seen from a simple qualitative argument. Let unprimed

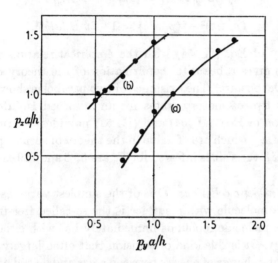

FIG. II. 24. Relation between momentum components p_z, p_y of the diffracted and specularly reflected molecular beams when anomalous minima occur (FRI 33b).

k_x's designate 2π times de Broglie wave-number components before the molecules strike the surface and primed k'_x's the same things afterwards, so $k_x = p_x/\hbar$. Let E_n be the binding energy of the molecule when occupying an allowed energy level as far as motion normal to the surface is concerned. Then, from the conservation of energy,

$$k_x^2 + k_y^2 + k_z^2 = k'^2_x + k'^2_y - 2mE_n/\hbar^2. \qquad (II. 52)$$

From now on let the consideration be limited to transitions for which $k_x = k'_x$. A transition between two different states in most transition probability calculations occurs when there is a resonance between the difference in wave number of the two states and the wave number of the periodic phenomenon inducing the transition, so it is not unreasonable that the selection rule for the transition is

$$k_y - k'_y = \pm 2\pi/a. \qquad (II. 53)$$

A detailed proof of (II. 53) is given in the basic references (LEN 36, MAS 52), but the essential fact is that in the absence of such a coherency the integrals involved in the calculation of the transition matrix integrate to zero owing to the oscillatory character of the integrands. From Eqs. (II. 52) and (II. 53) with the negative choice of sign and from the assumed equality of k_x and k'_x, it follows immediately that

$$k_z^2 - 2k_y(2\pi/a) = (2\pi/a)^2 - 2mE_n/\hbar^2,$$

or
$$(p_z a/h)^2 - 2(p_y a/h) = 1 - 2mE_n(a/h)^2. \tag{II. 54}$$

The identity of Eq. (II. 54) with the empirical relation of Frisch in Eq. (II. 51) gives a beautiful confirmation of the theory of Lennard-Jones and Devonshire. The existence of two parabolas shows that there are at least two oscillatory states for an adsorbed He atom on LiF. A comparison of Eqs. (II. 54) and (II. 51) provides determinations of the values of E_n, which are of value in the theory of surface phenomena. For He on LiF the values inferred for E_n are 57·5 and 129 cals. per mole.

II. 5.4. *Inelastic collisions.* One of the simplest varieties of inelastic collision of a molecule with a surface is the so-called free-free collision in which the molecule rebounds immediately but with either a gain or loss of energy. It is this kind of collision that often largely determines the rate of interchange of energy between a gas and a solid at a different temperature. The rate of interchange depends on Knudsen's thermal accommodation coefficient α', which is defined as follows. Let E_i denote the average energy brought up to the wall by a molecule, E_r the average energy carried away by the restituted molecule, and E_w the average energy of the molecule if it were in thermal equilibrium with the wall. Then the accommodation coefficient α' is defined by

$$\alpha' = (E_i - E_r)/(E_i - E_w). \tag{II. 55}$$

Extensive summaries of the experimental studies on accommodation coefficients and of the theories of free-free inelastic transitions have been given in the book of Massey and Burhop (MAS 52). Since few of these measurements have involved molecular-beam techniques such summaries will not be repeated here. However, it should be noted that one of the most important molecular-beam detectors—the Stern–Pirani detector described in Chapters I and XIV—depends on this process.

Inelastic collisions which lead to adsorption can either be followed by a subsequent restitution of the molecule to the gas or by its permanently remaining on the surface. Both of these processes have been studied with molecular-beam techniques. A theory of adsorption has been developed by Lennard-Jones and Devonshire (LEN 35, LEN 36) and summarized by Massey and Burhop (MAS 52).

In the case of adsorption and subsequent restitution, the earliest studies were devoted to a determination of the angular distribution of the restituted molecules. As discussed in Section II. 5.2, if the molecule has been adsorbed for an appreciable time the conditions under which re-evaporation of the molecule will occur bear no relation to those which prevailed at condensation. Instead they will be determined by the last stages of the history of the molecule on the surface. Consequently one would expect the angular distribution to be that for the simple evaporation of molecules. Knudsen (KNU 15) has pointed out that this angular distribution should be in accordance with a simple cosine law. If N molecules, whose directions lie within a solid angle $d\omega'$, strike a surface for which there is no specular reflection or diffraction, then the number dN of those restituted from the surface within an element of solid angle $d\omega$ making an angle θ with the normal to the surface is, according to Knudsen,

$$dN = (1/\pi)N \cos\theta \, d\omega. \qquad \text{(II. 56)}$$

For a gas and a surface in complete equilibrium, it can be shown that the cosine law for all the molecules that come off the surface is a necessary consequence of the second law of thermodynamics (GAE 13, EPS 24). However, Knudsen's law is more general in that it is assumed to apply to evaporated molecules even in the absence of equilibrium.

The earliest molecular-beam studies of the angular distribution of the restituted molecules were those of Wood (WOO 15). He allowed a mercury beam to fall on plane glass target at the centre of a bulb and then examined the thickness of the deposit on the bulb. He later did the same thing with a cadmium beam. The Knudsen law (Eq. II. 56) was obeyed to within the experimental error.

Knudsen (KNU 15) performed a similar experiment in which the restituting surface itself was part of the receiving sphere as in Fig. II. 25. In that case the mercury deposit on the sphere should be uniform if Knudsen's law were followed and this was found experimentally to be true.

One objection to the above experiments was that a cosine law might result from the roughness of the surface. However, Taylor (TAY 30) showed that this was not the case by observing the restitution of Li, K, and Cs atoms from the cleaved planes of NaCl and LiF. As described above in Section II. 5.2 and illustrated in Fig. II. 16, Taylor obtained a complete confirmation of the Knudsen law in these cases (TAY 30, HUR 54). On the other hand, if He, H_2, or H is used with the same cleaved crystals reflection and diffraction phenomena are observed, as pointed out in Section II. 5.2 (HAN 32).

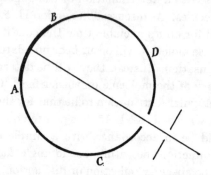

Fig. II. 25. Apparatus for confirmation of Knudsen's law (KNU 15).

Numerous molecular-beam experiments have been made to study adsorption phenomena. Since these experiments and the corresponding theories (LEN 36, ZEN 32) have been summarized in books by Fraser (FRA 31) and Massey and Burhop (MAS 52) they will not be described in detail here. Some of these experiments have been for the purpose of measuring the time of adsorption for the restituted molecules, i.e. the mean time for which they linger on the surface before being restituted to the gas (CLA 30, ZXX 54). In other experiments the conditions for permanent adsorption and the migrations of atoms over the surface while they are adsorbed have been studied (EST 23, COC 28, FRA 31, DEV 52, JOH 33, STA 34, STE 20a, MEL 52, ZXX 54). Measurements of condensation or sticking coefficients have been made and tabulated by a number of observers including Fraunfelder (FRA 50), whose paper contains an extensive list of other references, and Wexler (BRO 54). An important application of these measurements has been in the selection of suitable collector materials for experiments in which moments of radioactive nuclei are measured by molecular-beam methods in which the detection is by the deposition of radioactive materials as in Section XIV. 3.6. Condensation coefficients may vary from less than 0·0001 to almost one, depending on the materials, temperature, surface condition, etc. For example, with Cs on clean W the condensation coefficient is 0·7.

II. 6. Molecular-beam studies of chemical equilibria, ionization potentials, and optical and microwave spectroscopy

II. 6.1. *Chemical equilibria.* The effusion of molecular beams from sources provides a means of measuring vapour pressures as functions of temperature since the beam intensity will depend upon vapour pressure of the source as shown in Eq. (II. 16); this was first pointed

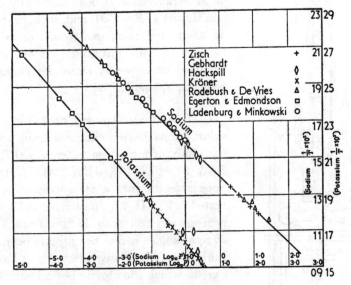

FIG. II. 26. Vapour pressures of sodium and potassium. T is in °K and P in mm of Hg (EDM 27).

out by Knudsen (KNU 09). However, frequently a mixture of both atoms and molecules effuses from the source. For example, a sodium beam is a mixture of Na and Na_2. Therefore the relative contribution of the different components must be distinguished before a significant vapour pressure determination can be deduced (FRA 31). A detailed list of references to vapour pressure measurements determined in this way has been given by Fraser (FRA 31). Typical results are shown in Fig. II. 26 for sodium and potassium (EDM 27).

Molecular beams can also be used as a means for determining the degree of molecular dissociation that occurs, i.e. the fraction of the total possible number of combined molecules in the vapour that are dissociated. There are many means for distinguishing between say the Na_2 molecules and the Na atoms which effuse from the source (FRA 31, WEI 26, ROD 27). One of the simplest is by deflexion in an inhomogeneous magnetic field (see Chapter IV) since the paramagnetic Na atoms are easily deflected and the diamagnetic Na_2 molecules are only slightly deflected (LEU 28). As discussed in Section VIII. 11, molecular beams have been used to study the degree of polymerization of vapours (e.g. NaCl to $(NaCl)_n$).

Fraser (FRA 34, FRA 37a) has pointed out that free radicals such as CH_3 and C_2H_5 can be studied by molecular-beam techniques. Beams of these radicals have been formed by Fraser and Jewitt (FRA 37a), and such properties as their ionization potentials have been measured. The initiation of elementary chemical processes by molecular beams has been studied (MAR 52).

II. 6.2. *Ionization and excitation.* Molecular beams have sometimes been used in the study of ionization potentials (FRA 31, FRA 37, FUN 30, JEW 34). A typical apparatus for such a measurement is illustrated in Fig. II. 27. In this method a homogeneous electron beam is fired from K to C at right angles across the atomic beam, and the ions so formed are collected by a Faraday cage in which the atomic beam itself is condensed. With such an apparatus one can measure both the ionization potentials and the ionization function (the probability of ionization as a function of the electron energy). Photo-ionization has also been studied by molecular-beam methods of a similar nature except that the crossed electron beam is replaced by electromagnetic radiation (FRA 31, DOR 38). Electron bombardment ionization has been used to detect beams in other experiments (WES 53).

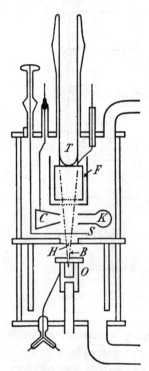

FIG. II. 27. Funk's apparatus for measuring the ionization cross-sections of Na and K by the atomic-beam method. K is the electron source and C the electron collector (FUN 30).

Smyth (SMY 22) and others (DIT 29) have ionized the atoms in an atomic beam by cross bombardment with electrons and have then analysed the ion products with a mass spectrometer. In this way the degree of ionization was determined.

One of the earliest applications of molecular beams was in the study of excitation by Dunoyer (DUN 13). Dunoyer excited a sodium beam to emit resonance radiation and then observed the sharpness with which the region of emitted radiation fell off. From this he could show that the duration of the excitation was less than 10^{-7} sec. Since the atomic beam travels only about 0·01 cm in such a short time, the atomic beam method is in general ill adapted to the study of excited electronic states whose transitions to lower states are allowed. However, as discussed

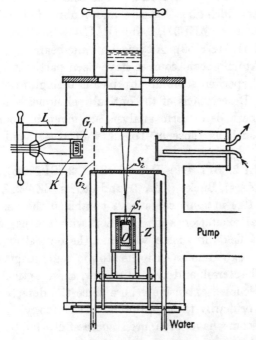

Fig. II, 28. Typical atomic-beam source for optical hyperfine-structure measurements. The beam emerges from oven O and is cross-bombarded by the electron beam from cathode K (PAU 41).

in Chapter IX, even such states as these can be studied to some extent with a suitable molecular-beam resonance method (RAB 52). Since metastable excited states ordinarily have lifetimes that are long compared with the length of time that it takes a molecule to traverse the apparatus, they are well adapted to almost all molecular-beam methods and have frequently been studied (HUG 53, ZXY 54). Frisch (FRI 33d) has detected the recoil of Na atoms when an atomic beam of such atoms is irradiated with resonance radiation; this experiment showed that the momentum of the absorbed photon was indeed acquired by the atom.

II. 6.3. *Optical and microwave absorption spectroscopy.* An atomic beam forms one of the most effective sources for high-resolution optical spectroscopy. Collision broadening of the spectral lines is eliminated by the low density of the beam and Doppler broadening is greatly diminished if the emitted light is viewed in a direction at right angles to the beam so that the component of velocity in the direction of observation is small. Such atomic beam sources have been used by many observers, including Jackson and Kuhn (JAC 34), Minkowski (MIN 35), Meissner (MEI 37), Paul (PAU 41), and others (DUN 13, DOB 28, SUR 51, HEY 38). A typical atomic-beam source is shown in Fig. II. 28. Atomic-beam sources have been particularly effective in the study of hyperfine structure by means of high-resolution optical spectroscopy. Descriptions of the optical techniques employed and of the detailed methods of term analysis are given in books on optical spectroscopy such as those of Tolansky (TOL 48) and Kopfermann (KOP 40).

Kastler (ZZA 2, BIT 49b), Brossel (ZZA 2, BIT 49b), Bitter (BIT 49b), Dicke (ZZA 4), Bucka (ZZA 3), and others (ZZA 2, ZZA 4, ZZA 9, ZZA 10, BIT 49b) in some cases have combined the use of polarized light in optical excitation and detection with the use of oscillatory radiofrequency fields in such a way that the optical experiments are quite similar to experiments with the atomic beam magnetic resonance methods of Chapters I and IX. However, these optical experiments will not be discussed extensively here since the detection techniques employed are primarily those of optical spectroscopy.

Molecular beams have been used successfully by Strandberg and Dreicer (STR 54) in microwave absorption experiments as a means for diminishing Doppler broadening. These experiments are to be distinguished from many of the microwave molecular-beam experiments discussed below, in that the transitions in Strandberg and Dreicer's experiments were detected by their effects on the radiation field rather than by the changed state of the molecule.

III

INTERACTION OF A NUCLEUS WITH ATOMIC AND MOLECULAR FIELDS

III. 1. Introduction

MANY of the most important molecular-beam measurements are determinations of the interactions between a nucleus of a molecule and the molecular fields at the nucleus from the remainder of the molecule. For this reason, detailed considerations of molecular-beam measurements are best preceded by a theoretical discussion of the interaction of a nucleus with atomic and molecular fields.

The interactions between a nucleus and its atom or molecule may be studied in various ways, including the methods of optical spectroscopy, molecular beams, nuclear paramagnetic resonance, microwave spectroscopy, and paramagnetic resonance as described in the author's book on *Nuclear Moments* (RAM 53e). All of these measurements are consistent with the following assumptions concerning atomic nuclei:

(a) A nucleus has a charge Ze confined to a small region of the order of 10^{-12} cm diameter. It is the interaction of this charge with the electrons of an atom that gives rise to the gross features of optical atomic spectra.

(b) A nucleus whose mass number A is odd obeys Fermi–Dirac statistics (the sign of the wave function is reversed if two such identical nuclei are interchanged), and a nucleus whose mass number A is even obeys Bose–Einstein statistics (wave function unaltered on interchange).

(c) A nucleus possesses a spin angular momentum capable of being represented by a quantum-mechanical angular momentum vector $\mathbf{a}$ with all the properties† usually associated with such vectors. In nuclear moment work it is usually most convenient to use a dimensionless quantity $\mathbf{I}$ to measure the angular momentum in units of $\hbar$, where $\mathbf{I}$ is therefore defined by

$$\mathbf{a} = \hbar \mathbf{I}. \qquad (III.\ 1)$$

The spin I of a nucleus is defined as the maximum possible component of $\mathbf{I}$ in any given direction.

(d) The nuclear spin I is half integral if the mass number A is odd and integral if the mass number is even.

† Good discussions of the properties of angular momentum vectors are given in books by Condon and Shortley (CON 35), Wigner (WIG 31), Feenberg and Pake (FEE 53), Racah (RAC 42), Pauli (PAU 32), and Schwinger (SCH 46, SCH 52).

(e) A nucleus has a magnetic moment (PAU 24) μ_I which can be represented as
$$\mu_I = \gamma_I \hbar \mathbf{I} = g_I \mu_N \mathbf{I}, \qquad (III.\ 2)$$
where γ_I and g_I are defined by the above equation and are called the nuclear gyromagnetic ratio and the nuclear g factor, respectively. μ_N is the nuclear magneton defined as $e\hbar/2Mc$, where M is the proton mass.† μ_N has the numerical value of $(5 \cdot 05038 \pm 0 \cdot 00036) \times 10^{-24}$ erg gauss^{-1}. The scalar quantity which is a measure of the magnitude of μ_I and which is called the magnetic moment μ_I is
$$\mu_I = \gamma_I \hbar I. \qquad (III.\ 3)$$
With this Eq. (III. 2) can be written
$$\mathbf{\mu}_I = (\mu_I/I)\mathbf{I}. \qquad (III.\ 4)$$

(f) Many nuclei with spin equal to or greater than 1 possess an electrical quadrupole moment, i.e. have electrical charge distributions which depart from spherical symmetry in a manner appropriate to an electrical quadrupole moment. (A precise definition of an electrical quadrupole moment is given in the next section.)

(g) A nucleus with a spin 3/2 or greater may possess a magnetic octupole moment (JAC 54). (The definition of a magnetic octupole interaction is given in the next section and in Section IX. 9.)

(h) The nucleus is neither infinitely heavy nor of negligible size, but instead with certain experiments allowance must be made for the finite mass and appreciable size of the nucleus (BRE 32, ZXW 54).

(i) The nucleus has a finite polarizability and can be polarized by a strong electric field (KOP 49, BRE 50, GUN 51, RAM 53f).

III. 2. Electrostatic interaction

III. 2.1. *Multipole expansion.* The general electrostatic interaction $\mathscr{H}_E$ between a charged nucleus and the charged electrons and nuclei of the remainder of the molecule is as follows, if the finite extension of the nucleus is taken into account:
$$\mathscr{H}_E = \int_{\tau_e} \int_{\tau_n} \frac{\rho_e \rho_n d\tau_e d\tau_n}{r}. \qquad (III.\ 5)$$

---

† In the literature on the subject of nuclear moments there is considerable confusion as to notation. In different papers μ_0, μ_n, μ_{NM} represent the nuclear magneton, whereas μ_0, μ_1, and β designate the Bohr magneton. Here μ_N will be used for the nuclear magneton and μ_0 for the Bohr magneton. Also in many papers g or g_I is written for $\mu_I/I\mu_N$, but in others g represents $\mu_I/I\mu_0$. In the present book g_I will ordinarily designate $\mu_I/I\mu_N$, but in Chapter IX and on p. 86 it will designate $-\mu_I/I\mu_0$ in order to be consistent with the literature of molecular-beam hyperfine-structure experiments; it is clear from the context which meaning applies.

ρ_e is the charge density of the electrons and of the other nuclei in the volume element $d\tau_e$ at position $\mathbf{r}_e$ relative to the centroid of the nucleus concerned, so that ρ_e is a function of $\mathbf{r}_e$. ρ_n is the nuclear charge density of the nucleus concerned, in the volume element $d\tau_n$ at position $\mathbf{r}_n$ relative to the centroid of the nucleus, whence ρ_n is a function of $\mathbf{r}_n$. In addition r is the magnitude of the radius vector $\mathbf{r}$ drawn from $d\tau_n$ to $d\tau_e$ as shown in Fig. III. 1. The definition of ρ_e is such that it is negative for electrons and positive for positive charges.

Let θ_{en} be the angle between $\mathbf{r}_e$ and $\mathbf{r}_n$ and let r_e be greater than r_n, as may be ensured by limiting consideration to electronic charges more distant than the radius R_N of the nucleus (exceptions to this restriction are discussed in Sections II. 3.2 and IX. 10). This limitation will be indicated explicitly by the use of a superscript e with ρ_e^e to indicate charges external to a small sphere of radius R_N

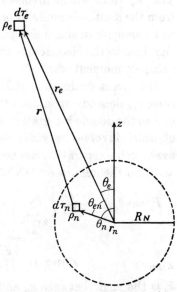

Fig. III. 1. Electrostatic interactions of a nucleus in atoms and molecules.

surrounding the nucleus and centred on the nuclear centroid. Then from the cosine law of trigonometry $1/r$ may be expressed as follows, using the well-known power series expansion in r_n/r_e [cf. MAR 43, p. 96, MOR 53]:

$$1/r = (r_e^2 + r_n^2 - 2r_e r_n \cos\theta_{en})^{-\frac{1}{2}} = 1/r_e + \frac{r_n}{r_e^2}P_1 + \frac{r_n^2}{r_e^3}P_2 + \frac{r_n^3}{r_e^4}P_3 + \dots, \quad (\text{III. 6})$$

where P_k is the kth Legendre polynomial of $\cos\theta_{en}$, so that

$$\begin{aligned} P_0 &= 1, \\ P_1 &= \cos\theta_{en}, \\ P_2 &= \tfrac{1}{2}(3\cos^2\theta_{en} - 1), \\ P_3 &= \tfrac{1}{2}(5\cos^3\theta_{en} - 3\cos\theta_{en}), \\ P_k &= \frac{1}{2^k k!}\frac{d^k}{(d\cos\theta_{en})^k}(\cos^2\theta_{en} - 1)^k. \end{aligned} \quad (\text{III. 7})$$

From Eqs. (III. 5) and (III. 6), one can write

$$\mathscr{H}_E = \sum_k \mathscr{H}_{Ek}, \quad (\text{III. 8})$$

where
$$\mathcal{H}_{Ek} = \int_{\tau_e} \int_{\tau_n} \frac{\rho_e^e \rho_n}{r_e} \left(\frac{r_n}{r_e}\right)^k P_k(\cos\theta_{en}) \, d\tau_e \, d\tau_n. \quad \text{(III. 9)}$$

The $\mathcal{H}_k$ term which involves P_k is said to be the interaction energy from the multipole moment of order 2^k. Thus the first term corresponds to a monopole or single charge, the second to the electric dipole moment, the third to the electric quadrupole moment, the fourth to the electric octupole moment, etc.

The direct evaluation of Eq. (III. 9) would necessarily be difficult since θ_{en} depends upon both the nuclear and the electronic coordinates and so the double integration is not separated into two integrations each of which involves only the electron or the nuclear coordinates. However, such a separation can be achieved by using the spherical harmonic addition theorem [see CON 35, p. 53 or RAC 42] according to which

$$P_k(\cos\theta_{en}) = \frac{4\pi}{2k+1} \sum_{q=-k}^{k} (-1)^q Y_{-q}^{(k)}(\cos\theta_n, \phi_n) Y_{+q}^{(k)}(\cos\theta_e, \phi_e)$$

$$= \sum_{q=-k}^{k} (-1)^q C_{-q}^{(k)}(\theta_n, \phi_n) C_{+q}^{(k)}(\theta_e, \phi_e), \quad \text{(III. 10)}$$

where $C_q^{(k)}(\theta, \phi) = [4\pi/(2k+1)]^{\frac{1}{2}} Y_q^{(k)}(\cos\theta, \phi)$,

θ_n is the angle between r_n and an arbitrary fixed z-axis, ϕ_n the azimuthal angle of $d\tau_n$ from the xz-plane, while θ_e and ϕ_e are the corresponding quantities for $d\tau_e$ as shown in Fig. III. 1. $Y_q^{(k)}$ is the normalized tesseral harmonic

$$Y_q^{(k)}(\cos\theta, \phi) = \Theta_q^{(k)}(\cos\theta) \frac{1}{\sqrt{(2\pi)}} e^{iq\phi}, \quad \text{(III. 11)}$$

where $\Theta_q^{(k)}$ except for normalization is the associated Legendre polynomial defined for $q > 0$ as

$$\Theta_q^{(k)} = (-1)^q \sqrt{\left(\frac{(2k+1)(k-q)!}{2(k+q)!}\right)} \sin^q\theta \frac{d^q}{(d\cos\theta)^q} P_k(\cos\theta) \quad \text{(III.12)}$$

and for $q < 0$ as $\quad \Theta_q^{(k)} = (-1)^q \Theta_{-q}^{(k)}. \quad \text{(III. 13)}$

With these expressions, the 2^k interaction can be written as

$$\mathcal{H}_{Ek} = \mathbf{Q}^{(k)} \cdot \mathbf{F}^{(k)}, \quad \text{(III. 14)}$$

where the use of the heavy dot ($\cdot$) in $\mathbf{Q}^{(k)} \cdot \mathbf{F}^{(k)}$ for the irreducible (RAC 42) tensors $\mathbf{Q}^{(k)}$ and $\mathbf{F}^{(k)}$ of degree k indicates the tensor scalar product defined by

$$\mathcal{H}_{Ek} = \mathbf{Q}^{(k)} \cdot \mathbf{F}^{(k)} = \sum_{q=-k}^{k} (-1)^q Q_q^{(k)} F_{-q}^{(k)}, \quad \text{(III. 15)}$$

where

$$Q_q^{(k)} = \sqrt{\left(\frac{4\pi}{2k+1}\right)} \int_{\tau_n} \rho_n r_n^k Y_q^{(k)}(\cos\theta_n, \phi_n) \, d\tau_n = \int_{\tau_n} \rho_n r_n^k C_q^{(k)}(\theta_n, \phi_n) \, d\tau_n$$

$$\text{(III. 16)}$$

and
$$F_q^{(k)} = \sqrt{\left(\frac{4\pi}{2k+1}\right)} \int_{\tau_e} \rho_e^e r_e^{-(k+1)} Y_q^{(k)}(\cos\theta_e, \phi_e) \, d\tau_e$$
$$= \int_{\tau_e} \rho_e^e r_e^{-(k+1)} C_q^{(k)}(\theta_e, \phi_e) \, d\tau_e. \qquad \text{(III. 17)}$$

The integrals in Eqs. (III. 16) and (III. 17) have the great advantage that they separately concern either the electron or the nuclear coordinates, but not both simultaneously (RAC 42, POU 50).

Although Eqs. (III. 15), (III. 16), and (III. 17) give the general 2^k-pole interaction energy, they are inconvenient to work with in their most general form. However, the specific expressions applicable for the three lowest multipoles may easily be derived as in the following paragraphs.

The monopole interaction is
$$\mathscr{H}_{E0} = Ze\phi^e, \qquad \text{(III. 18)}$$
where
$$Ze = Q_0^{(0)} = \int_{\tau_n} \rho_n \, d\tau_n \qquad \text{(III. 19)}$$
and
$$\phi^e = F_0^{(0)} = \int_{\tau_e} \frac{\rho_e^e}{r_e} \, d\tau_e. \qquad \text{(III. 20)}$$

ϕ^e is the electrostatic potential at the centroid of the nucleus from the electrical charges *external* to a small sphere of radius R_N centred on the nuclear centroid and surrounding all the nuclear charge.

In a similar fashion the electric dipole interaction is
$$\mathscr{H}_{E1} = -p_z E_z^e - \tfrac{1}{2} p_+ E_-^e - \tfrac{1}{2} p_- E_+^e, \qquad \text{(III.21)}$$
where p_z is the z component of nuclear electrical dipole moment in a coordinate system whose origin is at the nuclear centroid so that
$$p_z = Q_0^{(1)} = \int_{\tau_n} \rho_n r_n \cos\theta_n \, d\tau_n = \int_{\tau_n} \rho_n z_n \, d\tau_n \qquad \text{(III. 22)}$$
and $p_\pm$ is defined by
$$p_\pm \equiv p_x \pm i p_y = \pm\sqrt{2} Q_{\pm 1}^{(1)}$$
$$= \int_{\tau_n} \rho_n r_n \sin\theta_n \cos\phi_n \, d\tau_n \pm i \int \rho_n r_n \sin\theta_n \sin\phi_n \, d\tau_n$$
$$= \int_{\tau_n} \rho_n x_n \, d\tau_n \pm i \int \rho_n y_n \, d\tau_n. \qquad \text{(III. 23)}$$

It is then apparent that p_x is the x-component of the nuclear electrical dipole moment. Similarly, E_z^e is the electric field at the nuclear centroid from the electrical charges external to the small sphere of radius R surrounding the nucleus so that

$$E_z^e = -F_0^{(1)} = -\int_{\tau_e} \frac{\rho_e^e}{r_e^2} \cos\theta_e \, d\tau_e = -\int_{\tau_e} \frac{\rho_e^e}{r_e^3} z^e \, d\tau_e = -\left[\frac{\partial \phi^e}{\partial z}\right] \quad \text{(III. 24)}$$

and

$$E_\pm^e \equiv E_x^e \pm i E_y^e = \pm F_{\pm 1}^{(1)} = -\int_{\tau_e} \frac{\rho_e^e}{r_e^2} \sin\theta_e \cos\phi_e \, d\tau_e \mp i \int_{\tau_e} \frac{\rho_e^e}{r_e^2} \sin\theta_e \sin\phi_e \, d\tau_e$$

$$= -\int_{\tau_e} \frac{\rho_e^e}{r_e^3} x^e \, d\tau_e \mp i \int_{\tau_e} \frac{\rho_e^e}{r_e^3} y_e \, d\tau_e = -\left[\frac{\partial \phi^e}{\partial x}\right] \mp i \left[\frac{\partial \phi^e}{\partial y}\right].$$

(III. 25)

It should be noted that in the derivation of the preceding equations such differentiations as $\partial/\partial z$ with respect to the field point can be replaced by $-\partial/\partial z_e$ under the integral sign.

Likewise the electric quadrupole interaction can be written as

$$\mathscr{H}_{E2} = Q_0(\nabla E^e)_0 - Q_{+1}(\nabla E^e)_{-1} + Q_{+2}(\nabla E^e)_{-2} -$$
$$- Q_{-1}(\nabla E^e)_{+1} + Q_{-2}(\nabla E^e)_{+2}. \quad \text{(III. 26)}$$

In the above Q_0 is the zz-component of the nuclear electrical quadrupole moment tensor. The Q_q's are in general defined by

$$Q_0 = Q_0^{(2)} = \tfrac{1}{2} \int_{\tau_n} \rho_n r_n^2 (3\cos^2\theta_n - 1) \, d\tau_n = \tfrac{1}{2} \int_{\tau_n} \rho_n (3z_n^2 - r_n^2) \, d\tau_n,$$

$$Q_{\pm 1} = Q_{\pm 1}^{(2)} = \mp\sqrt{\tfrac{3}{2}} \int_{\tau_n} \rho_n z_n x_{n\pm} \, d\tau_n,$$

$$Q_{\pm 2} = Q_{\pm 2}^{(2)} = \sqrt{\tfrac{3}{8}} \int_{\tau_n} \rho_n x_{n\pm}^2 \, d\tau_n, \quad \text{(III. 27)}$$

where $x_{n\pm}$ is an abbreviation for $x_n \pm i y_n$. The (∇E^e)'s are components of the gradient of electric field at the nuclear centroid arising from the electrical charges external to the sphere of radius R_N. They are given explicitly by

$$(\nabla E^e)_0 = F_0^{(2)} = \tfrac{1}{2} \int_{\tau_e} \frac{\rho_e^e}{r_e^3} (3\cos^2\theta_e - 1) \, d\tau_e = -\frac{1}{6}\left[3\frac{\partial E_z^e}{\partial z} - \nabla \cdot \mathbf{E}\right]$$

$$= -\frac{1}{2}\left[\frac{\partial E_z^e}{\partial z}\right],$$

$$(\nabla E^e)_{\pm 1} = F_{\pm 1}^{(2)} = \mp\sqrt{\tfrac{3}{2}} \int_{\tau_e} \frac{\rho_e^e}{r_e^3} \cos\theta_e \sin\theta_e (\cos\phi_e \pm i\sin\phi_e) \, d\tau_e$$

$$= \pm\frac{\sqrt{6}}{6}\frac{\partial E_\pm^e}{\partial z} = \pm\frac{\sqrt{6}}{6} \partial_\pm E_z^e, \quad \text{(III. 28)}$$

$$(\nabla E^e)_{\pm 2} = F^{(2)}_{\pm 2} = \sqrt{\tfrac{3}{8}} \int_{\tau_e} \frac{\rho^e_e}{r^3_e}\sin^2\theta_e(\cos\phi_e \pm i\sin\phi_e)^2\, d\tau_e = -\frac{\sqrt{6}}{12}\partial_\pm E^e_\pm,$$

where
$$\partial_\pm = \frac{\partial}{\partial x} \pm i\frac{\partial}{\partial y}. \tag{III.29}$$

In the expansions of $\mathscr{H}_{Ek}$ in Eqs. (III. 21) and (III. 26), each of the indicated terms is in general itself a sum of separate terms. As a result, there is no uniqueness concerning the expansion in the indicated terms: the individual terms can be recombined into other quite different indicated terms to yield alternative expansions which superficially appear to be quite different but which on full expansion are completely equivalent. One such alternative can be obtained by combining the expressions $+q$ and $-q$ in Eq. (III. 11) so as to make the expansion in terms of $\cos q\phi$ and $\sin q\phi$ instead of $e^{\pm iq\phi}$. Another alternative is to make the expansion in terms of the Cartesian components of the tensor. In this case

$$\mathscr{H}_{E1} = -\mathbf{p}\cdot\mathbf{E}^e = -p_x E^e_x - p_y E^e_y - p_z E^e_z. \tag{III.30}$$

Also
$$\mathscr{H}_{E2} = -\tfrac{1}{6}\sum_{i=x}^{z}\sum_{j=x}^{z} Q_{ij}(\nabla E^e)_{ij}, \tag{III.31}$$

where
$$Q_{ij} = \int_{\tau_n} \rho_n(3x_{ni}x_{nj} - \delta_{ij}r^2_n)\, d\tau_n,$$
$$(\nabla E^e)_{ij} = -\int_{\tau_e} \frac{\rho^e_e}{r^5_e}(3x_{ei}x_{ej} - \delta_{ij}r^2_e)\, d\tau_e, \tag{III.32}$$

and where x_{nz} denotes z_n, etc. The equivalence of Eqs. (III. 30) and (III. 31) to Eqs. (III. 21) and (III. 26) may be shown by a straightforward expansion of each in Cartesian coordinates. Alternatively, in the author's book on *Nuclear Moments* (RAM 53e) the direct calculation of Eqs. (III. 30) and (III. 31) from Eq. (III. 9) is given.

In quantum-mechanical calculations of the expectation values for the above expressions, ρ_n should be replaced by its expectation value (SCH 46)

$$\rho_n = \int \psi^*(\mathbf{r}_1,...,\mathbf{r}_A)\rho_{op}\psi(\mathbf{r}_1,...,\mathbf{r}_A)\, d\tau_1...d\tau_A, \tag{III.32a}$$

where
$$\rho_{op} = \sum_i e_i \delta(\mathbf{r}_n - \mathbf{r}_i). \tag{III.32b}$$

With this, for example, from Eq. (III. 16),

$$\langle Q^{(k)}_q\rangle = \iint_{\tau_n} \psi^*(\mathbf{r}_1,...,\mathbf{r}_A)r^k_n C^{(k)}_q(\theta_n,\phi_n)\times$$
$$\times \sum_i e_i\delta(\mathbf{r}_n-\mathbf{r}_i)\psi(\mathbf{r}_1,...,\mathbf{r}_A)\,d\tau_1...d\tau_A\,d\tau_n$$
$$= \sum_{i=1}^{A} e_i \int \psi^*(\mathbf{r}_1,...,\mathbf{r}_A)r^k_i C^{(k)}_q(\theta_i,\phi_i)\psi(\mathbf{r}_1,...,\mathbf{r}_A)\,d\tau_1...d\tau_A. \tag{III.32c}$$

III. 2.2. *Theoretical restrictions on electric multipole orders.* If a nuclear multipole moment is associated with a nucleus of definite spin I whose orientation properties are fully determined by the orientation of the spin angular momentum I, the possible varieties of nuclear electric multipole moments are restricted by two general theoretical theorems.

The first theorem arises from a parity consideration and is as follows: If all nuclear electrical effects arise from electrical charges, if there is no degeneracy of nuclear states with different parity, and if the nuclear Hamiltonian is unaltered by an inversion of coordinates (replacement of $\mathbf{r}$ by $-\mathbf{r}$), no odd (k odd) electrical multipole can exist. In particular, a nucleus satisfying the above conditions can have neither an electric dipole moment nor an electric octupole moment.

The proof of this theorem is given in detail in the author's book on *Nuclear Moments* (RAM 53e, p. 23). The essentials of the argument are that, if the above assumptions apply, then the wave function must be of either odd or even parity, as discussed in Section XI. 5.1, i.e.

$$\psi(x_1, y_1, ..., y_A, z_A) = \pm \psi(-x_1, -y_1, ..., -y_A, -z_A). \quad \text{(III. 33)}$$

However, with either odd or even parity,

$$|\psi(x_1, ..., z_A)|^2 = |\psi(-x_1, ..., -z_A)|^2. \quad \text{(III. 34)}$$

On the other hand, for even k, the $Y_q^{(k)}$ of Eq. (III. 32c) is unaltered for the same coordinate inversion, whereas for odd k the sign is reversed upon inversion. Therefore with the latter case in Eq. (III. 16) the integrand is of opposite sign at $(-x_1, ..., -z_A)$ to that at $(x_1, ..., z_A)$. Consequently, when the completed integration is carried out over both these positions, the net result is zero, which proves the theorem.

In atomic problems, exceptions to the above theorem sometimes arise owing to degeneracy of states with different parity. On the other hand, the energy separations of different nuclear energy levels are so large compared to normal interaction energies with nuclear electric moments that such degeneracies are very unlikely. Purcell and Ramsey (PUR 50) have pointed out that the above proof depends on the assumption that the electrical effects of a nucleus arise only from electric charges, or on a related parity assumption, and that these assumptions are not necessarily obvious in the case of little-understood particles like nucleons and nuclei (DIR 48). Therefore, in an experiment with J. Smith (SMI 51a, SMI 51b), they searched with high precision for a possible electric dipole moment of the neutron. They find, however, that if such a dipole moment exists its magnitude must be less than the charge of the electron multiplied by a distance D of 5×10^{-20} cm.

The second theorem is the following: For a nuclear spin I it is impossible to observe a nuclear electrical multipole moment of order 2^k for k greater than $2I$.

This theorem is also derived in detail in the author's *Nuclear Moments* (RAM 53e, p. 24). The essentials of the argument are that if ρ_n in Eq. (III. 16) is written as $\sim \psi_n^* \psi_n$, the equation can be expressed as

$$Q_q^{(k)} \sim \int_{\tau_n} \psi_n^* r_n^k Y_q^{(k)} \psi_n \, d\tau_n. \tag{III. 35}$$

However, $Y_q^{(k)}$ is a possible orbital wave function for an orbital angular momentum of k (CON 35, p. 52). As ψ_n is a wave function of angular momentum I and by the vector addition of angular momenta (CON 35, p. 56), the product $Y_q^{(k)} \psi_n$ can be considered a wave function of a system whose angular momentum is between $k+I$ and $|k-I|$. However, since ψ_n^* and hence $\psi_n^* r_n^k$ correspond to angular momentum I, they will be orthogonal to $Y_q^{(k)} \psi_n$ unless they can correspond to the same angular momentum eigenvalue. Therefore, to prevent Eq. (III. 35) from vanishing, I must lie between $k+I$ and $|k-I|$ or $k \leqslant 2I$, which is the desired theorem. This restriction is often expressed by saying that I, I, and k must satisfy the triangle rule, i.e. it must be possible to form a triangle from three sides proportional to I, I, and k respectively.

Theorems analogous to the above are applicable to the field quantities $F_q^{(k)}$ of Eq. (III. 17) when they arise from an atom or molecule whose angular momentum is J. Thus, $F_q^{(k)}$ vanishes unless $k \leqslant 2J$. Even a nucleus with a large I and a nuclear quadrupole moment can have no electric quadrupole interaction energy with an atom whose $J = \frac{1}{2}$.

III. 2.3. *Nuclear electric quadrupole interactions.* The most important orientation-dependent electrical interaction of a nucleus is that arising from its electric quadrupole moment. It is therefore desirable to study in further detail the effect of a nuclear quadrupole interaction upon the energy levels of an atom or molecule. In general, the calculation of the energy levels is performed either by a perturbation-theory calculation or by the solution of the secular equation which results from the matrix of the Hamiltonian when the quadrupole contributions of Eq. (III. 26) are included. In either case, the matrix elements of the quadrupole terms must be calculated for the representation that is used.

The matrix elements may be calculated in a number of different ways. One procedure is the direct use of a general formula given in the next section for any order of multipole. Another procedure, the one to be followed in this section, depends on the observation that one usually uses a representation based on the angular momentum operators of the atom, molecule, or nucleus. Matrix elements of functions of the angular-momentum operators can usually be calculated easily in such representations from the standard forms (CON 35) for $(Im_I|\mathbf{I}|I'm_I')$, whose only non-vanishing matrix elements are

$$(Im_I|I_z|Im_I) = m_I, \qquad \text{(III. 35a)}$$
$$(Im_I\pm 1|I_\pm|Im_I) = \sqrt{\{(I\mp m_I)(I\pm m_I+1)\}},$$

where $\quad \mathbf{I} = I_z\mathbf{k} + \tfrac{1}{2}I_+(\mathbf{i}-i\mathbf{j}) + \tfrac{1}{2}I_-(\mathbf{i}+i\mathbf{j}),\qquad$ (III. 35b)

where $\mathbf{i}, \mathbf{j}, \mathbf{k}$ are the Cartesian unit vectors. If, therefore, the quadrupole-moment tensor can be expressed as simply proportional to a tensor in the angular-momentum variables the calculation of the matrix elements can be simplified.

The quadrupole moment tensor can be expressed in terms of the angular momentum variables with the aid of a theorem that is most simply proved by group theory or by the tensor-commutation rules (WIG 31, RAC 42, WEY 39, CAS 35, CAS 36, NIE 48). However, it can also be proved directly from simple matrix multiplications and from the matrix elements for vectors. Such a proof is straightforward, though numerically tedious, and is given in the author's book on *Nuclear Moments* (RAM 53f, p. 21). By whatever the means of proof, the theorem applies to all tensors which transform under space rotations in the same way as does Eq. (III. 11). This will be true of tensors that are similar in form (including symmetry with respect to commutation) to Eqs. (III. 11), (III. 16), (III. 17) but which may be based on some vector **A** which is other than the space vector **r** upon which the above equations depend, provided that **A** satisfies the same commutation rules with respect to the total angular momentum operator **I** as does **r** (the dependence of the above equation on the components of **r** is most clearly illustrated when they are expressed in forms similar to Eq. (III. 27)). The significance of the phrase in the preceding sentence 'including symmetry with respect to commutation' is, for example, that one component of the tensor in Eq. (III. 27) is proportional to $z_n x_{n\pm}$, which is also equal to $x_{n\pm} z_n$ since the components of r commute among themselves. On the other hand, a tensor based on the components of the vector **I** would not have this characteristic since these components do not commute among themselves. Consequently, the appropriate form for the corresponding component of a tensor based on **I** is

$$\tfrac{1}{2}(I_\pm I_z + I_z I_\pm)$$

since this expression is symmetric with respect to commutation. The theorem which applies to tensors satisfying the above conditions is that the quantum-mechanical matrix elements diagonal in I of *all* such tensors have the same orientation dependence, e.g. the same dependence on the magnetic quantum number m_1. As discussed and proved in the author's *Nuclear Moments* (RAM 53e), a similar theorem applies to tensors, such as those of Eq. (III. 32), which are symmetric with respect to the interchanges of any two indices and for which all spurs vanish (FAL 50). It also applies to tensors based on more than one vector, as can be shown by the mathematical process of *polarization* (FAL 50, WEY 39).

As a result of the above theorem, the calculation of matrix elements diagonal in I can be simplified by writing the tensor components as proportional to the corresponding expressions in components of $\mathbf{I}$, which by the above theorem have the same dependence on m_I. Thus, from Eq. (III. 27), one can write

$$Q_0 = \tfrac{1}{2}\int \rho_n(3z_n^2 - r_n^2)\, d\tau_n = C \cdot \tfrac{1}{2}(3I_z^2 - \mathbf{I}^2),$$
$$Q_{\pm 1} = \mp\sqrt{\tfrac{3}{2}}C \cdot \tfrac{1}{2}(I_z I_\pm + I_\pm I_z), \qquad\qquad \text{(III. 36)}$$
$$Q_{\pm 2} = \sqrt{\tfrac{3}{8}} C I_\pm^2.$$

The value of C in the above equation can be related to the scalar quantity Q which is conventionally taken as the measure of a nuclear quadrupole moment. By convention the quantity Q, often called simply the quadrupole moment, is defined by

$$Q \equiv \frac{1}{e}\int_{\tau_n} \rho_{nII}(3z_n^2 - r_n^2)\, d\tau_n, \qquad\qquad \text{(III. 37)}$$

where ρ_{nII} is the charge density when the nucleus is in the orientation state with $m_I = I$. If $(Im_I|Q_0|Im_I')$ designates a matrix element of Q_0, the value of C can be related to Q with the aid of Eqs. (III. 36) and (III. 37) as follows

$$Q = \frac{2}{e}(II|Q_0|II) = \frac{C}{e}[3I^2 - I(I+1)] = \frac{C}{e}I(2I-1).$$
$$\text{(III. 38)}$$

Therefore, Eq. (III. 36) can be written as

$$Q_0 = \frac{eQ}{2I(2I-1)}[3I_z^2 - I(I+1)],$$
$$Q_{\pm 1} = \mp\frac{\sqrt{6}}{2}\frac{eQ}{2I(2I-1)}[I_z I_\pm + I_\pm I_z], \qquad\qquad \text{(III. 39)}$$
$$Q_{\pm 2} = \frac{\sqrt{6}}{2}\frac{eQ}{2I(2I-1)} I_\pm^2.$$

Since the orientation of the atom or molecule is specified by the orientation of its angular momentum J, a similar procedure to the above is applicable in the calculation for matrix elements diagonal in J of the tensor components of the gradients of the molecular electric field. Thus, for such calculations, Eq. (III. 28) can be written as

$$(\nabla E^e)_0 = \frac{eq_J}{2J(2J-1)}[3J_z^2 - J(J+1)],$$

$$(\nabla E^e)_{\pm 1} = \mp \frac{\sqrt{6}}{2} \frac{eq_J}{2J(2J-1)}[J_z J_\pm + J_\pm J_z], \quad \text{(III. 40)}$$

$$(\nabla E^e)_{\pm 2} = \frac{\sqrt{6}}{2} \frac{eq_J}{2J(2J-1)} J_\pm^2,$$

where q_J is defined by

$$q_J \equiv \frac{1}{e}\int_{\tau_e} \frac{3\cos^2\theta_e - 1}{r_e^3} \rho_{eJJ}^e \, d\tau_e = \frac{1}{e}\int_{\tau_e} \frac{3z_e^2 - r_e^2}{r_e^5} \rho_{eJJ}^e \, d\tau_e = \frac{1}{e}\left\langle\frac{\partial^2 V^e}{\partial z^2}\right\rangle_{JJ}.$$

(III. 41)

If q_J is due to a single electron, it becomes

$$q_J = -\left\langle\frac{3\cos^2\theta_e - 1}{r_e^3}\right\rangle_{JJ}, \quad \text{(III. 41a)}$$

where the minus sign is due to the negative charge of the electron. The symbol $\langle \ \rangle_{JJ}$ indicates the average potential when the molecule is in the state with $m_J = J$ and ρ_{eJJ}^e has a similar meaning for electronic charge density.

In a similar fashion, when the tensors are written in the form of Eq. (III. 32), they may be re-expressed for calculations of matrix elements diagonal in I and J as

$$Q_{ij} = \frac{eQ}{I(2I-1)}\left[3\frac{I_i I_j + I_j I_i}{2} - \delta_{ij} I(I+1)\right],$$

$$(\nabla E^e)_{ij} = -\frac{eq_J}{J(2J-1)}\left[3\frac{J_i J_j + J_j J_i}{2} - \delta_{ij} J(J+1)\right].$$

(III. 42)

The exact forms for the coefficients are obtained analogously to the above by considering $(II|Q_{zz}|II)$. Details of this simple calculation are given in the author's *Nuclear Moments* (RAM 53e, p. 17).

A third and often useful form for the quadrupole interaction is

$$\mathscr{H}_{E2} = \frac{e^2 q_J Q}{2I(2I-1)J(2J-1)}[3(\mathbf{I}\cdot\mathbf{J})^2 + \tfrac{3}{2}\mathbf{I}\cdot\mathbf{J} - I(I+1)J(J+1)].$$

(III. 43)

A derivation of this equation from Eqs. (III. 31) and (III. 42) is given in Appendix C. The relation can also be proved by expanding Eq. (III. 43) in terms of the I_z and $I_\pm$ components and comparing it with the similar expansion of Eqs. (III. 26), (III. 39), and (III. 40). In either of these two derivations the components of I and J should be commuted among themselves only in accordance with the usual commutation rules (CON 35)

$$[I_z, I_\pm] = I_z I_\pm - I_\pm I_z = \pm I_\pm,$$
$$[I_+, I_-] = 2I_z, \qquad (\text{III. 44})$$
$$[I_i, I_j] = iI_{i \times j}.$$

The subscript $i \times j$ indicates the Cartesian component perpendicular to the i and j components as in a vector product.

In diatomic molecules with either identical or non-identical nuclei the quadrupole interaction can be obtained by applying either Eq. (III. 42) or (III. 43) to each nucleus separately. However, in the case of identical nuclei of spin $\tfrac{1}{2}$ (or spin 1 if J is odd) it is usually more convenient to express the interaction in terms of the resultant angular momentum $\mathbf{I}_R = \mathbf{I}_1 + \mathbf{I}_2$, since with these restrictions I_R is a good quantum number in a rotational state of definite J because of the symmetry requirements on the wave function with identical nuclei. This re-expression is calculated in Appendix C, and two alternative forms for the calculation of matrix elements diagonal in I_R are given. One of these is that the interaction is as given in Eqs. (III. 31) and (III. 42), except that Q_{ij} is replaced in the homonuclear diatomic case by

$$Q_{ij} = \frac{eQ}{I_1(2I_1-1)}\left[1 - \frac{I_R(I_R+1)+4I_1(I_1+1)}{(2I_R-1)(2I_R+3)}\right] \times$$
$$\times \left[3\frac{I_{Ri}I_{Rj}+I_{Rj}I_{Ri}}{2} - \delta_{ij}I_R(I_R+1)\right]. \quad (\text{III. 44a})$$

The other form derived in Appendix C is that the interaction in the diatomic homonuclear case for matrix elements diagonal in I is

$$\mathscr{H}_{E2} = \frac{e^2 q_J Q}{2I_1(2I_1-1)J(2J-1)}\left[1 - \frac{I_R(I_R+1)+4I_1(I_1+1)}{(2I_R-1)(2I_R+3)}\right] \times$$
$$\times [3(\mathbf{I}_R \cdot \mathbf{J})^2 + \tfrac{3}{2}\mathbf{I}_R \cdot \mathbf{J} - \mathbf{I}_R^2 \mathbf{J}^2]. \quad (\text{III. 44b})$$

Since the values of $I(I+1)$ and $J(J+1)$ in Eqs. (III. 43) and (III. 44b) are independent of the molecular and nuclear orientations, and since most molecular-beam experiments measure differences in energy as an orientation is changed, these terms make no net contribution to the final result and are often omitted (or included in the reference energy from which the other energies are expressed). With such an omission the above-mentioned equations can alternatively be written as

$$\mathcal{H}'_{E2} = hb_1 \, 2\mathbf{I}.\mathbf{J}(2\mathbf{I}.\mathbf{J}+1). \tag{III. 44c}$$

The above relations are applicable to the calculation of tensor matrix elements diagonal in J for both atoms and molecules. For atoms, the above forms are also those most frequently used. However, in the case of molecules it is usually convenient to express q_J in a form that reduces the implicit dependence of q_J upon J. Most of this dependence arises from the definition of q_J in Eq. (III. 41) being based on the state with $m_J = J$. For this reason, q_J would vary with J even if the shape of the molecule were unaltered by centrifugal stretching since an angular-momentum vector J in the state $m_J = J$ can lie much closer to the z-axis if J is large than if J is small. This difficulty can be avoided if q_J is expressed in terms of a coordinate system fixed within the molecule.

Consider a symmetric top molecule and let z_0 be a coordinate along the axis of symmetry, while x_0 is perpendicular to the axis of symmetry but in the plane of z and z_0, and y_0 is mutually perpendicular to x_0 and z_0. Then, if θ'' is the angle between z and z_0,

$$q_J = \frac{1}{e}\left\langle\frac{\partial^2 V^e}{\partial z^2}\right\rangle_{JJ} = \frac{1}{e}\left\langle\frac{\partial^2 V^e}{\partial z_0^2}\cos^2\theta'' + \frac{\partial^2 V^e}{\partial x_0^2}\sin^2\theta''\right\rangle_{JJ}, \tag{III. 45}$$

since $\partial^2 V^e/\partial z_0 \, \partial x_0$ vanishes by symmetry and since $\partial x_0/\partial z = \sin\theta''$. However, from Laplace's equation and from the symmetry between x_0 and y_0,

$$\frac{\partial^2 V^e}{\partial x_0^2} = \frac{\partial^2 V^e}{\partial y_0^2} = -\frac{1}{2}\frac{\partial^2 V^e}{\partial z_0^2}, \tag{III. 46}$$

so

$$q_J = \frac{1}{e}\frac{\partial^2 V^e}{\partial z_0^2}\left\langle\frac{3\cos^2\theta''-1}{2}\right\rangle_{JJ}. \tag{III. 47}$$

For linear molecules the indicated average can be calculated from the wave function for a rotating linear molecule in the rotational state with $m_J = J$, which wave function is (CON 35, p. 54)

$$\psi_{JJ}(\theta'', \phi'') = \frac{(-1)^J}{\sqrt{(2\pi)}}\sqrt{\left(\frac{(2J+1)!}{2}\right)}\frac{1}{2^J J!}\sin^J\theta'' e^{iJ\phi''}. \tag{III. 48}$$

The appropriate integral can be evaluated [PEI 99, Eq. 483] with the result that

$$q_J = -\frac{J}{2J+3}\frac{1}{e}\frac{\partial^2 V^e}{\partial z_0^2} = -\frac{J}{2J+3}\frac{1}{e}q, \tag{III. 49}$$

where

$$q \equiv (\partial^2 V^e/\partial z_0^2). \tag{III. 49a}$$

On the other hand, if the wave function for a symmetric top is used (TOW 54, KNI 49a), Eq. (III. 47) becomes

$$q_J = -\left(1 - \frac{3K^2}{J(J+1)}\right)\frac{J}{2J+3}\frac{1}{e}q, \qquad \text{(III. 50)}$$

where K is the quantum number for the projection of J along the axis of symmetry. For $K = 0$ Eqs. (III. 49) and (III. 50) agree as they should since K is necessarily zero for a linear molecule. The forms for asymmetric tops have been given by Bragg (BRA 48) and others (KNI 49a).

With the above, Eq. (III. 43), for example, can be written as

$$\mathscr{H}_{E2} = -\frac{eqQ}{2I(2I-1)(2J+3)(2J-1)}\left[1 - \frac{3K^2}{J(J+1)}\right] \times$$

$$\times [3(\mathbf{I}.\mathbf{J})^2 + \tfrac{3}{2}\mathbf{I}.\mathbf{J} - I(I+1)J(J+1)]. \qquad \text{(III. 51)}$$

Unlike q_J, the quantity q is independent of J except for such small effects as centrifugal stretching.†

It is important to realize that V^e in Eq. (III. 49a) is not the same as the potential V of all electrical charges from the rest of the molecule, including the electron density of the molecule inside the nuclear radius. The difference between these two can readily be evaluated. Let ρ_1 be the electronic charge density at the nucleus, and assume that in the immediate vicinity of the nucleus it is spherically symmetric. Then, as $V - V^e$ arises solely from the spherically symmetric charge distribution ρ_1, from Poisson's equation

$$\frac{\partial^2(V-V^e)}{\partial x_0^2} + \frac{\partial^2(V-V^e)}{\partial y_0^2} + \frac{\partial^2(V-V^e)}{\partial z_0^2} = -4\pi\rho_1 = 3\frac{\partial^2(V-V^e)}{\partial z_0^2},$$

(III. 52)

so

$$q = \frac{\partial^2 V^e}{\partial z_0^2} = \frac{\partial^2 V}{\partial z_0^2} + \frac{4\pi}{3}\rho_1. \qquad \text{(III. 53)}$$

Owing to its spherical symmetry, the ρ_1 term does not affect the nuclear quadrupole interaction.

When the interaction between the vectors **I** and **J** is large compared

† There is much notational confusion in the literature on q_J and related quantities. Kellogg, Rabi, Ramsey, and Zacharias (KEL 40) use the symbol q for what is here called q_J, and Casimir (CAS 36) uses C_s for the same quantity. Feld (FEL 47) and Gordy (GOR 53) use $\partial^2 V/\partial z^2$ for the quantity here designated as q, and Ramsey (RAM 53e) uses $\partial^2 V^e/\partial z_0^2$ for the same quantity. Townes and Bardeen (TOW 54) use q in the same sense as here but the quantity they call q_J is eq_J in the present notation. Nordsieck (NOR 40) and Nierenberg (NIE 47) have sometimes used a different quantity q' which is related to the present notation by $q' = (1/2e)q$. Although q will be the notation usually used in the present book, the more explicit $\partial^2 V^e/\partial z_0^2$ will occasionally be used.

with the interaction of either with any other field, Eq. (III. 51) can be expressed in an alternative form. Let **F** be the vector sum of **I** and **J**, so $\mathbf{F} = \mathbf{I}+\mathbf{J}$ with quantum number F. Then $\mathbf{I}\cdot\mathbf{J}$ can be evaluated from

$$\mathbf{F}^2 = (\mathbf{I}+\mathbf{J})^2 = \mathbf{I}^2+\mathbf{J}^2+2\mathbf{I}\cdot\mathbf{J}, \qquad (\text{III. 54})$$

so

$$\mathbf{I}\cdot\mathbf{J} = \tfrac{1}{2}(\mathbf{F}^2-\mathbf{I}^2-\mathbf{J}^2) = \tfrac{1}{2}[F(F+1)-I(I+1)-J(J+1)] \equiv \tfrac{1}{2}C. \qquad (\text{III. 55})$$

This may be used in Eq. (III. 51), whence the quadrupole energy W_{E2} (when $K = 0$) becomes

$$W_{E2} = -\frac{eqQ}{I(2I-1)(2J+3)(2J-1)}[\tfrac{3}{2}C(C+1)-\tfrac{1}{2}I(I+1)J(J+1)]. \qquad (\text{III. 56})$$

Magnetic interactions with an external field are often large compared with the interactions coupling **I** to **J**. In such cases a representation based on the magnetic quantum numbers m_I and m_J is often most suitable. In such a representation $\mathscr{H}_{E2}$ is not diagonal. However, the appropriate diagonal and non-diagonal matrix elements can be calculated from Eqs. (III. 39), etc., or from Eq. (III. 51). These matrix elements are calculated in Eq. (C. 28) of Appendix C by the procedure used by Kellogg, Rabi, Ramsey, and Zacharias (KEL 40). Their paper (KEL 40) also gives a detailed example of the application of the above quadrupole interaction expressions to a practical problem, the evaluation of the quadrupole moment of the deuteron.

In order that the electrical quadrupole moment Q may be determined experimentally from the observed interaction energy, an evaluation of q_J or of q is required. Various methods for estimating these quantities are available. The methods will be discussed in Chapter XI.

III. 2.4. *Multipole moments of arbitrary order.* Matrix elements, diagonal in I, of general multipole moments can be calculated by rewriting them as proportional to corresponding expressions based on the angular momentum **I** with the use of the general theorem given in the previous section.

On the other hand, it is often more convenient to use direct general expressions for the matrix elements that have been derived by Racah (RAC 42) and Wigner (WIG 31). These expressions also give matrix elements that are non-diagonal in I; they are as follows for the components of, say, $Q_q^{(k)}$ in an $\alpha I m_I$ representation:

$$(\alpha I m_I | Q_q^{(k)} | \alpha' I' m_I') = (-1)^{I+m_I}(\alpha I \| Q^{(k)} \| \alpha' I') V(II'k; -m_I m_I' q), \qquad (\text{III. 57})$$

where

$$V(abc;\alpha\beta\gamma) = [(a+b-c)!(a+c-b)!(b+c-a)!/(a+b+c+1)!]^{\frac{1}{2}} \times$$
$$\times v(abc;\alpha\beta\gamma) \quad \text{(III. 58)}$$

and

$$v(abc;\alpha\beta\gamma) = \delta(\alpha+\beta+\gamma,0) \sum_{z} (-1)^{c-\gamma+z} \times$$
$$\times \frac{[(a+\alpha)!(a-\alpha)!(b+\beta)!(b-\beta)!(c+\gamma)!(c-\gamma)!]^{\frac{1}{2}}}{z!(a+b-c-z)!(a-\alpha-z)!(b+\beta-z)!(c-b+\alpha+z)!(c-a-\beta+z)!}.$$
(III. 59)

The summation parameter z takes on all integral values consistent with the factorial notation, the factorial of negative numbers being meaningless. $(\alpha I \|Q_k\| \alpha' I')$ indicates the proportionality factor which is independent of m_I. Similar expressions are available for matrix elements of $F_q^{(k)}$ except that J and m_J replace I and m_I in the above. The above formulae can be derived directly from the commutation rules for vectors (RAC 42).

Alternatively, Racah (RAC 42) has given formulae for the matrix elements of the tensor scalar products as Eq. (III. 15) in a $\gamma JIFm$ representation, with the result that

$$(\gamma JIFm|\mathbf{F}^{(k)} \cdot \mathbf{Q}^{(k)}|\gamma' J'I'F'm') = \delta(F,F')\delta(m,m')(-1)^{J+I'-F} \times$$
$$\times \sum_{\gamma''} (\gamma J\|F^{(k)}\|\gamma''J')(\gamma''I\|Q^{(k)}\|\gamma'I')W(JIJ'I';Fk), \quad \text{(III. 60)}$$

where $W(abcd;ef)$ is a so-called Racah coefficient defined by

$$W(abcd;ef)$$
$$= \begin{bmatrix} (a+b-e)!(a+e-b)!(b+e-a)!(c+d-e)!(c+e-d)!(d+e-c)! \times \\ \times (a+c-f)!(a+f-c)!(c+f-a)!(b+d-f)!(b+f-d)!(d+f-b)! \\ \hline (a+b+e+1)!(c+d+e+1)!(a+c+f+1)!(b+d+f+1)! \end{bmatrix}^{\frac{1}{2}} \times$$
$$\times w(abcd;ef), \quad \text{(III. 61)}$$

with

$$w(abcd;ef)$$
$$= \sum_{z} (-1)^{z} \frac{(a+b+c+d+1-z)!}{(a+b-e-z)!(c+d-e-z)!(a+c-f-z)!(b+d-f-z)! \times}.$$
$$\times z!(e+f-a-d+z)!(e+f-b-c+z)!$$
(III. 62)

Numerical values of the Racah coefficients for different values of the parameters have been given by several authors (BIE 52, OBI 54, SIM 54, ZZA 5).

III. 3. Magnetic interaction

III. 3.1. *Magnetic multipoles.* All electronic and nuclear magnetic effects presumably arise from circulating currents including the various anomalous moments associated with spins (SCH 37, SCH 48). Therefore, the mutual potential energy of magnetic interaction between the nucleus with nuclear current density $\mathbf{j}_n$ and the electrons with current density $\mathbf{j}_e$ producing a vector potential $\mathbf{A}_e$ at the nucleus may be taken (CAS 36, CAS 42, ROS 48) as

$$\mathcal{H}_M = -\frac{1}{c}\int_{\tau_n} \mathbf{j}_n \cdot \mathbf{A}_e \, d\tau_n = -\frac{1}{c}\int_{\tau_e} \mathbf{j}_e \cdot \mathbf{A}_n \, d\tau_e. \qquad \text{(III. 63)}$$

However, from the equation of continuity for stationary current distributions,
$$\nabla \cdot \mathbf{j} = -\dot{\rho} = 0, \qquad \text{(III. 64)}$$
whence $\mathbf{j}_n$ can be derived from a vector potential $\mathbf{m}_n$ so that

$$\mathbf{j}_n = c\nabla \times \mathbf{m}_n \qquad \text{(III. 65)}$$

and $\mathbf{m}_n$ may be taken as zero outside the nucleus. As a result of this, the following equation applies since the first term equals zero if the surface S_n surrounds but lies outside the nucleus:

$$\mathcal{H}_M = -\int_{S_n} \mathbf{A}_e \times \mathbf{m}_n \cdot d\mathbf{S}_n - \int_{\tau_n} (\nabla_n \times \mathbf{m}_n) \cdot \mathbf{A}_e \, d\tau_n$$

$$= -\int_{\tau_n} \nabla_n \cdot (\mathbf{A}_e \times \mathbf{m}_n) \, d\tau_n - \int_{\tau_n} \mathbf{A}_e \cdot (\nabla_n \times \mathbf{m}_n) \, d\tau_n$$

$$= -\int_{\tau_n} \mathbf{m}_n \cdot \nabla_n \times \mathbf{A}_e \, d\tau_n = -\int_{\tau_n} \mathbf{m}_n \cdot \mathbf{H}_e \, d\tau_n. \qquad \text{(III. 66)}$$

The standard vector and tensor relations summarized in Appendix B are used repeatedly in the preceding and following equations.

For that portion of $\mathbf{H}_e$ which arises exclusively from electron currents in a region τ_e^e external to the nucleus, as indicated by the superscript e in τ_e^e and $\mathbf{H}_e^e$, one can write, if r equals $|\mathbf{r}_e - \mathbf{r}_n|$,

$$\mathbf{H}_e^e = \nabla_n \times \mathbf{A}_e^e = \nabla_n \times \frac{1}{c}\int_{\tau_e^e} \frac{\mathbf{j}_e}{r} \, d\tau_e$$

$$= \nabla_n \times \int_{\tau_e^e} \frac{\nabla_e \times \mathbf{m}_e}{r} \, d\tau_e$$

$$= \nabla_n \times \left[\int_{\tau_e^e} \nabla_e \times \frac{\mathbf{m}_e}{r} \, d\tau_e - \int_{\tau_e^e} \left(\nabla_e \frac{1}{r}\right) \times \mathbf{m}_e \, d\tau_e \right]$$

$$= \nabla_n \times \left[-\int_{S_e' + S_e^*} \frac{\mathbf{m}_e}{r} \times d\mathbf{S}_e + \int_{\tau_e^e} \left(\nabla_n \frac{1}{r}\right) \times \mathbf{m}_e \, d\tau_e \right]. \qquad \text{(III. 67)}$$

S'_e in the above indicates an inner surface defining the electron-current distribution, which for H^e_e is assumed wholly external to the nucleus, and S''_e an outer surface to the electron distribution. If $\mathbf{m}_e$ vanishes over both S'_e and S''_e, the first integral on the right in the last of the above expressions vanishes. Then

$$\begin{aligned}
H^e_e &= \int_{\tau_e} \nabla_n \times \nabla_n \times \left(\frac{\mathbf{m}_e}{r}\right) d\tau_e = \int_{\tau_e} (\nabla_n \nabla_n \cdot - \nabla_n^2) \frac{\mathbf{m}_e}{r} d\tau_e \\
&= \nabla_n \int_{\tau_e} \nabla_n \cdot \frac{\mathbf{m}_e}{r} d\tau_e = \nabla_n \int_{\tau_e} \mathbf{m}_e \cdot \nabla_n \frac{1}{r} d\tau_e \\
&= -\nabla_n \int_{\tau_e} \mathbf{m}_e \cdot \nabla_e \frac{1}{r} d\tau_e \\
&= -\nabla_n \int_{\tau_e} \nabla_e \cdot \frac{\mathbf{m}_e}{r} d\tau_e + \nabla_n \int_{\tau_e} \frac{\nabla_e \cdot \mathbf{m}_e}{r} d\tau_e \\
&= -\nabla_n \int_{S'_e+S''_e} \frac{\mathbf{m}_e}{r} \cdot d\mathbf{S}_e + \nabla_n \int_{\tau_e} \frac{\nabla_e \cdot \mathbf{m}_e}{r} d\tau_e \\
&= \nabla_n \int_{\tau_e} \frac{\nabla_e \cdot \mathbf{m}_e}{r} d\tau_e.
\end{aligned} \qquad \text{(III. 68)}$$

Therefore,

$$\begin{aligned}
\mathscr{H}^e_M &= -\int_{\tau_n} \mathbf{m}_n \cdot \left[\nabla_n \int_{\tau_e} \frac{\nabla_e \cdot \mathbf{m}_e}{r} d\tau_e\right] d\tau_n \\
&= -\int_{\tau_n} \nabla_n \cdot \left[\mathbf{m}_n \int_{\tau_e} \frac{\nabla_e \cdot \mathbf{m}_e}{r} d\tau_e\right] d\tau_n + \int_{\tau_n}\int_{\tau_e} (\nabla_n \cdot \mathbf{m}_n) \frac{\nabla_e \cdot \mathbf{m}_e}{r} d\tau_e d\tau_n.
\end{aligned} \qquad \text{(III. 69)}$$

If the divergence theorem is applied to the first term on the right, it vanishes since $\mathbf{m}_n$ has already been taken equal to zero over a surface surrounding the nucleus. Therefore,

$$\mathscr{H}^e_M = \int_{\tau_e}\int_{\tau_n} \frac{(-\nabla_n \cdot \mathbf{m}_n)(-\nabla_e \cdot \mathbf{m}_e)}{r} d\tau_e d\tau_n. \qquad \text{(III. 70)}$$

In the special case of a magnetic 2^k-pole interaction for an electron state in which $J = L+\tfrac{1}{2}-k/2$, special care must be taken to avoid indeterminacy difficulties, like those discussed following Eq. (III. 83) below for a magnetic dipole interaction in an $S_{\frac{1}{2}}$ state (SCH 54, FER 30, RAM 53e, p. 26).

It should be noted that Eq. (III. 70) is exactly the same as the electric interaction of Eq. (III. 5) except for the correspondence

$$\rho_e \to -\nabla_e \cdot \mathbf{m}_e,$$

$$\rho_n \to -\nabla_n \cdot \mathbf{m}_n. \qquad (III.\ 71)$$

As a result, all the theorems on multipole expansion proved in Section III. 2 are equally applicable here except for the substitutions of Eq. (III. 71). The various multipole moments are then defined by a combination of Eq. (III. 71) with Eqs. (III. 16), (III. 22), and (III. 30). It should be noted, however, that Eq. (III. 70) is correct only when $\mathbf{m}_e$ vanishes throughout the nuclear volume as indicated by the superscript e, whereas Eqs. (III. 5) and (III. 63) apply generally. For the corrections which enter when the contributions from $\mathbf{m}_e$ inside the nuclear volume are included, the reader is referred to Section III. 3.2, to Chapter IX, and to the original papers of Bohr and Weisskopf (BOH 50, BOH 51, BOH 53) and Schwartz (SCH 54).

From the derived correspondence above, the restrictions on multipole order obtained in Section III. 2.2 can be applied to magnetic multipoles. One important modification, however, arises from the fact that the electrical charge density ρ_n is normally assumed to be a scalar, in which case $-\nabla_n \cdot \mathbf{m}_n$ is a pseudo-scalar (since from Eq. [III. 65] $\mathbf{m}_n$ is a pseudo-vector or axial vector). As a result $-\nabla_n \cdot \mathbf{m}_n$ reverses sign upon inversion of coordinates while ρ_n does not. Therefore the parity argument of Section III. 2.2 is exactly reversed and it is even (even k) magnetic multipoles which cannot exist. The restrictions on magnetic multipoles therefore are: (a) For a nuclear spin I it is impossible to observe a nuclear magnetic multipole moment of order 2^k greater than that corresponding to $k = 2I$. (b) No even (k even) magnetic multipole moment can exist if all nuclear magnetic effects arise from the circulation of electrical charges, if there is no degeneracy of nuclear states with different parity, and if the nuclear Hamiltonian is unaltered by an inversion of coordinates. Corresponding theorems apply to the atomic field gradients with which the nuclear multipole moments interact.

Magnetic multipole moments for a definite nuclear spin I can be expressed in terms of the nuclear angular momentum operators with the aid of the general theorem given in Section III. 2.3. Alternatively, matrix elements for magnetic multipole moments or for magnetic multipole interactions can be evaluated with the general expressions given in Section III. 2.4.

From the above theorems, the lowest nuclear magnetic multipole moment is the magnetic dipole moment and the next lowest is the magnetic octupole moment. Pauli (PAU 24) first suggested that atomic hyperfine structure might be due to nuclear magnetic dipole moments, and until recently only magnetic dipole moments have been observed. For this reason the most detailed discussion of individual magnetic multipole interactions will be given for magnetic dipoles, as in the next section. However, Jaccarino, Zacharias, King, Statten, and Stroke (JAC 54, TOL 13, CAS 42, SCH 54) and others (KUS 54, DAL 54) have recently found evidence for magnetic octupole interactions in iodine, indium, and gallium, as discussed in detail in Section IX. 9. The theory of magnetic octupole interactions has been given in detail by Casimir and Karreman (CAS 42) and by Schwartz (SCH 54). A summary of the results of the magnetic octupole theory is given in Section IX. 9.

In the use of an equation such as Eq. (III. 63), an expression for the electron current density j_e is required. The electron current density consists of two parts j_{e1} and j_{e2}, of which the first is the orbital current density which non-relativistically is given (PAU 32) as usual by

$$j_{e1} = -\frac{e\hbar}{2mi}(\Psi^*\nabla\Psi - \Psi\nabla\Psi^*). \qquad (III.\ 71a)$$

The other, j_{e2}, is the spin current density and, from Eq. (III. 65), it is given in a non-relativistic theory (PAU 32) by

$$j_{e2} = c\nabla \times \mathbf{m}_{e\sigma} = c\nabla \times \left[\Psi^*\left(\frac{-\hbar e}{2mc}\boldsymbol{\sigma}\right)\Psi\right]$$

$$= -\frac{e\hbar}{2m}\nabla \times [\Psi^*\boldsymbol{\sigma}\Psi]. \qquad (III.\ 71b)$$

Similar expressions apply to the neutron and proton current densities, but the value of e in Eq. (III. 71a) must be taken as zero for neutrons and $-e$ for protons while Eq. (III. 71b) must be multiplied by appropriate factors to allow for the anomalous nucleon magnetic moments. Relativistically, the current density is given (PAU 32) by

$$j^\mu = -ec\{\Psi, \alpha^\mu \Psi\} = -ec\bar{\psi}\gamma^\mu\psi$$

with the usual notation (PAU 32, WEI 54). But Gordon (WEI 54) has shown that this also may be interpreted as consisting of a convection-current and spin-current part (PAU 32, SCH 49, WEI 54), with each part being similar to the non-relativistic expressions except for an additional factor β.

III. 3.2. *Magnetic dipoles.*
From Eqs. (III. 22), (III. 30), and (III. 71) an expression for the nuclear magnetic dipole moment is

$$\mathbf{\mu}_I = \int_{\tau_n} (-\nabla_n \cdot \mathbf{m}_n)\mathbf{r}_n \, d\tau_n. \tag{III. 72}$$

With the aid of the vector transformation in Appendix B, this can also be written as

$$\mathbf{\mu}_I = -\int_{\tau_n} \nabla_n \cdot (\mathbf{m}_n \mathbf{r}_n) \, d\tau_n + \int_{\tau_n} (\mathbf{m}_n \cdot \nabla_n)\mathbf{r}_n \, d\tau_n$$

$$= -\int_{S_n} d\mathbf{S}_n \cdot (\mathbf{m}_n \mathbf{r}_n) + \int_{\tau_n} \mathbf{m}_n \, d\tau_n$$

$$= \int_{\tau_n} \mathbf{m}_n \, d\tau_n. \tag{III. 73}$$

A different expression can be obtained from the relations in Appendix B since

$$\frac{1}{c}\mathbf{r}_n \times \mathbf{j}_n = \mathbf{r}_n \times (\nabla_n \times \mathbf{m}_n)$$

$$= \nabla_n(\mathbf{r}_n \cdot \mathbf{m}_n) - (\mathbf{r}_n \cdot \nabla_n)\mathbf{m}_n - (\mathbf{m}_n \cdot \nabla_n)\mathbf{r}_n - \mathbf{m}_n \times (\nabla \times \mathbf{r}_n)$$

$$= \nabla_n(\mathbf{r}_n \cdot \mathbf{m}_n) - (\mathbf{r}_n \cdot \nabla_n)\mathbf{m}_n - \mathbf{m}_n$$

$$= \nabla_n(\mathbf{r}_n \cdot \mathbf{m}_n) - \nabla_n \cdot (\mathbf{r}_n \mathbf{m}_n) + 2\mathbf{m}_n. \tag{III. 74}$$

Therefore,

$$\mathbf{\mu}_I = \int_{\tau_n} \mathbf{m}_n \, d\tau_n = \frac{1}{2c}\int_{\tau_n} \mathbf{r}_n \times \mathbf{j}_n \, d\tau_n - \tfrac{1}{2}\int_{\tau_n} \nabla_n(\mathbf{r}_n \cdot \mathbf{m}_n) \, d\tau_n +$$

$$+ \tfrac{1}{2}\int_{\tau_n} \nabla_n \cdot (\mathbf{r}_n \mathbf{m}_n) \, d\tau_n$$

$$= \frac{1}{2c}\int_{\tau_n} \mathbf{r}_n \times \mathbf{j}_n \, d\tau_n \tag{III. 75}$$

since two of the terms in the penultimate step vanish when transformed to surface integrals.

From Eqs. (III. 30) and (III. 72), the magnetic dipole interaction can be taken as

$$\mathcal{H}_{M1} = -\mathbf{\mu}_I \cdot (\mathbf{B}_J + \mathbf{H}_0), \tag{III. 76}$$

where $\mathbf{H}_0$ is the externally applied magnetic field and $\mathbf{B}_J$ the magnetic induction at the nucleus arising from the rest of the atom or molecule which has angular momentum $\mathbf{J}$ in units of $\hbar$. As discussed in the preceding section, the derivation given here of the magnetic interaction implies the assumption that the electron distribution inside the nucleus does not contribute appreciably to the interaction energy. However, by the writing of $\mathbf{B}_J$ instead of $\mathbf{H}_J$ above, the result also applies to a point-sized nucleus inside an atom whose electronic magnetization does not vanish at the nucleus, e.g. in a $^2S_{\frac{1}{2}}$ state; for a current loop of negligible dimensions at an induction of $\mathbf{B}_J$, the potential energy is $-\boldsymbol{\mu}_I \cdot \mathbf{B}_J$ even if the magnetization does not vanish at the loop. Corrections for the variation of $\mathbf{B}_J$ and of the nuclear magnetization over the finite nuclear volume still remain. However, it can be shown, as discussed in Chapter IX (CAS 35, BOH 50, BOH 51), that this contribution is only a few per cent. even in the worst case—heavy nuclei and s electrons. However, this contribution is responsible for the important hyperfine-structure anomaly discussed in Chapter IX.

By the general theorem of Section III. 2.3, the magnetic moment for a nucleus of definite spin I can be written as proportional to $\mathbf{I}$ as already anticipated in Eqs. (III. 2) and (III. 4), so

$$\boldsymbol{\mu}_I = (\mu_I/I)\mathbf{I} \qquad \text{(III. 77)}$$

with the notation of Section III. 1. From this the magnetic dipole energy can be written as

$$\mathscr{H}_{M1} = -(\mu_I/I)\mathbf{I}\cdot(\mathbf{B}_J+\mathbf{H}_0). \qquad \text{(III. 78)}$$

If $\mathbf{B}_J$ is zero as in ^1S atomic states or $^1\Sigma$ states of non-rotating molecules, μ_I may readily be evaluated from a measurement of the energy-level separations, from a knowledge of the applied field $\mathbf{H}_0$, and from Eq. (III. 78). However, in other cases a knowledge of $\mathbf{B}_J$ is required. Just as $\boldsymbol{\mu}_I$ in Eq. (III. 77) could be taken as proportional to $\mathbf{I}$, so here $\mathbf{B}_J$ can be taken as proportional to $\mathbf{J}$ for matrix elements diagonal in J. Therefore if the external field $\mathbf{H}_0$ is zero, Eq. (III. 78) becomes

$$\mathscr{H}_{M1} = ha\mathbf{I}\cdot\mathbf{J}, \qquad \text{(III. 79)}$$

where a contains all the proportionality constants and is defined by

$$ha = -\left(\frac{\mu_I}{I}\right)\frac{\mathbf{B}_J}{\mathbf{J}} = -\left(\frac{\mu_I}{I}\right)\frac{\mathbf{B}_J\cdot\mathbf{J}}{\mathbf{J}\cdot\mathbf{J}}. \qquad \text{(III. 80)}$$

From Eq. (III. 55), the energy for the above may be written in an Fm representation as

$$W_{M1} = \frac{ha}{2}[F(F+1)-I(I+1)-J(J+1)] = \frac{ha}{2}C. \qquad \text{(III. 81)}$$

Instead of a, the symbol $A_1 = IJa$ has sometimes been used (SCH 54).

In order that μ_I may be determined experimentally from the energy levels it is necessary according to Eqs. (III. 80) and (III. 81) to have an estimate for the effective magnetic field at the nucleus due to the rest of the atom. For hydrogen-like atoms this effective magnetic field has been calculated quantum-mechanically by Fermi (FER 30), Goudsmit (GOU 29), and others (KOP 40). The principal results can also be obtained by simple semi-classical calculations as in the next paragraph. The procedure is somewhat different for an s electron and an electron in any other state since the wave function for an s electron does not vanish at the nucleus.

For a single electron in a $^2S_{\frac{1}{2}}$ state the value of a may be calculated as follows. Imagine a sphere drawn about the nucleus at a radius much larger than the nucleus yet sufficiently small for the electron wave function to be approximately constant throughout the sphere. Let $\mathbf{B}_{J1}$ be the magnetic induction at the nucleus from the electron density inside the sphere while $\mathbf{B}_{J2}$ is due to the electron density outside the sphere. Then, as a result of the spherical symmetry of the S state, $\mathbf{B}_{J2} = 0$. Therefore, by Eq. (III. 80),

$$ha = -\frac{\mu_I}{I}\frac{\mathbf{B}_{J1}.\mathbf{J}}{\mathbf{J}.\mathbf{J}}.$$

But the electron-spin magnetic-moment density, by assumption, is uniform inside the sphere with a magnetization of

$$\mathbf{m}_e = -|\psi_{n0}(0)|^2 2\mu_0 \mathbf{S} = -|\psi_{n0}(0)|^2 2\mu_0 \mathbf{J},$$

where $\psi_{n0}(0)$ is the wave function of the s-electron at the position of zero radius. However, for a uniformly magnetized sphere

$$\mathbf{B}_{J1} = \frac{8\pi}{3}\mathbf{m}_e = -\frac{16\pi}{3}\mu_0|\psi_{n0}(0)|^2\mathbf{J}.$$

Therefore, for an s-electron

$$ha = \frac{16\pi}{3}\mu_0\frac{\mu_I}{I}|\psi_{n0}(0)|^2 = \frac{8}{3}\frac{hcR_y\alpha^2Z^3g_I}{n^3(M/m)}, \qquad \text{(III. 82)}$$

where R_y is the Rydberg constant $(me^4/4\pi\hbar^3c \text{ cm}^{-1})$, α is the fine structure constant $e^2/\hbar c$, and M/m is the ratio of the proton to the electron mass.

A semi-classical evaluation of a for non s-states is given in the author's *Nuclear Moments* (RAM 53e, p. 25). The result of this calculation for an electron of orbital angular momentum $L \neq 0$ is

$$ha = \mu_0(\mu_I/I)\langle r_e^{-3}\rangle 2L(L+1)/J(J+1)$$

$$= \frac{hcR_y\alpha^2Z^3g_I}{n^3(L+\frac{1}{2})J(J+1)(M/m)}. \qquad \text{(III. 83)}$$

Although Eq. (III. 82) is numerically consistent with Eq. (III. 83), the former cannot be derived rigorously from the latter since the latter is indeterminate for $L = 0$ as its first form shows.

Small corrections to the above have been introduced by Breit (BRE 28, BRE 30, BRE 33, BRE 48, BRE 49) and Margenau (MAR 40) to allow for the reduced-mass effect of the nucleus and for the relativistic effects of the atomic electron. The reduced-mass effect introduces an extra factor of $(1+m/M)^{-3}$ into Eq. (III. 82). Low and Salpeter (LOW 51) have introduced a small additional reduced-mass correction corresponding to recoil of the nucleus from the virtual photons that are interchanged. The correction to Eq. (III. 82) from the anomalous magnetic moment of the electron is discussed in Chapter IX.

An approximate formula for the calculation of the interaction constant in alkali metals has been derived by Goudsmit (GOU 33) and is identical with Eq. (III. 83) except that Z^3 is replaced by $Z_0^2 Z_i$ and n is replaced by n_0 where Z_i is the effective charge when the valence electron is inside the core of closed electron shells, Z_0 is the effective charge when it is outside the core, and n_0 is the effective principal quantum number outside the core. Therefore, for alkali atoms

$$ha = \frac{2\pi\hbar c R_y \alpha^2 Z_0^2 Z_i g_I}{n_0^3 (L+\tfrac{1}{2}) J(J+1)(M/m)}. \qquad (\text{III. 84})$$

If the fine-structure separation δ of the same term is known, it can be used empirically to eliminate the Z_0 and n_0 dependence, since by atomic theory

$$\delta = \frac{2\pi\hbar c R_y \alpha^2 Z_0^2 Z_i^2}{n_0^3 L(L+1)}; \qquad (\text{III. 85})$$

so

$$ha = \frac{L(L+1)\delta g_I}{(L+\tfrac{1}{2}) J(J+1) Z_i (M/m)}. \qquad (\text{III. 86})$$

Various corrections to Eq. (III. 84), including those for mixing of higher configurations, have been calculated by Fermi and Segré (FER 33), Breit and Wills (BRE 30, BRE 33), Casimir (CAS 36), Racah (RAC 31), Goudsmit and Bacher (GOU 29) and Koster (KOS 52); the results have been summarized by Kopfermann (KOP 40, SCH 54). In addition to Eq. (III. 84) applying for alkali atoms with one valence electron, it can also be used for halogens where just one electron is missing from the outer shell. It can further be applied approximately to atoms of the aluminium group, since the first two valence electrons form a more or less closed shell. More complicated relationships hold for two valence electrons such as in the alkaline earths. These have been summarized by Kopfermann (KOP 40, KOS 52, SCH 54).

All the states of a single hyperfine-structure group have the same values of the quantum numbers I and J in Eq. (III. 81) while F takes on values from $I+J$ to $|I-J|$. The energy difference between states with F equal to F and $F-1$ is, according to Eq. (III. 81),

$$W_{M1}(F) - W_{M1}(F-1) = haF. \qquad (III. 87)$$

This regularity is called the interval rule and for this reason a is often called the interval factor. However, even with magnetic dipole interactions, exceptions to this rule can occur if two atomic states of different J are separated by an amount which is not large compared with the hyperfine structure, in which case J in Eq. (III. 81) is no longer a good quantum number and second-order perturbations must be taken into account. Calculations of the effect of perturbations have been made by Fermi and Segré (FER 33), Casimir (CAS 32), and Goudsmit and Bacher (GOU 33a), Foley (FOL 47a), Ramsey (RAM 53), and others (KOS 52, SCH 54).

If $J = \frac{1}{2}$ and $F = I+\frac{1}{2}$, the separation of the hyperfine-structure levels is, from Eq. (III. 87),

$$\Delta W = W(F) - W(F-1) = haF = ha(I+\tfrac{1}{2}) = ha\frac{2I+1}{2},$$

$$\Delta \nu = |\Delta W|/h. \qquad (III. 88)$$

The quantity $\Delta\nu$ defined in this way is frequently referred to as the hyperfine-structure separation, or the hyperfine-structure $\Delta\nu$.

III. 3.3. *Magnetic dipole interactions with internal and external magnetic fields.* So far in the detailed discussion of Eq. (III. 78) it has been assumed either that there was no intramolecular interaction or that the external field was zero. However, in many of the practical cases which arise in molecular-beam measurements, the magnetic interactions of the electrons and the nuclei with each other and of one or both with the external field are of comparable importance.

The effect of an external magnetic field on the hyperfine structure of an atomic energy level illustrates such a combined interaction. In this case, if the interaction energy with the external field of the electrons as well as of the nucleus is included and if μ_J is the resultant electronic magnetic moment defined analogously to the definitions in Section III. 1, Eqs. (III. 78) and (III. 80) become

$$\mathscr{H}_{M1} = ha\mathbf{I}.\mathbf{J} - \frac{\mu_J}{J}\mathbf{J}.\mathbf{H}_0 - \frac{\mu_I}{I}\mathbf{I}.\mathbf{H}_0. \qquad (III. 88a)$$

(a) *Weak external field.* If the magnetic field is very weak corresponding to a weak field Zeeman effect, the term $ha\mathbf{I}.\mathbf{J}$ in the above is the largest, so that $\mathbf{I}$ and $\mathbf{J}$ are tightly coupled together to form a resultant **F** as in Fig. III. 2 where $\mathbf{F} = \mathbf{I}+\mathbf{J}$. As a consequence, only the components of the electron and nuclear magnetic moments along

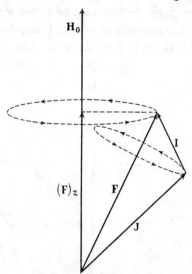

Fig. III. 2. Coupling of nuclear angular momentum **I** and electronic angular momentum **J** in weak magnetic field $\mathbf{H}_0$.

F are effective since the perpendicular components average to zero. The above then becomes approximately

$$\mathscr{H}_{M1} = ha\mathbf{I}.\mathbf{J} - \frac{\mu_J}{J}\mathbf{J}.\frac{\mathbf{F}}{|\mathbf{F}|}\frac{\mathbf{F}}{|\mathbf{F}|}.\mathbf{H}_0 - \frac{\mu_I}{I}\mathbf{I}.\frac{\mathbf{F}}{|\mathbf{F}|}\frac{\mathbf{F}}{|\mathbf{F}|}.\mathbf{H}_0. \quad \text{(III. 89)}$$

But $\quad \mathbf{I}.\mathbf{I} = (\mathbf{F}-\mathbf{J}).(\mathbf{F}-\mathbf{J}) = |\mathbf{F}|^2+|\mathbf{J}|^2-2\mathbf{J}.\mathbf{F}, \quad$ (III. 90)

so the energy for the state specified by quantum numbers F and m, where m is the magnetic quantum number for **F**, is

$$\begin{aligned}W_{M1}(F,m) &= (Fm|\mathscr{H}_{M1}|Fm)\\&= ha(Fm|\mathbf{I}.\mathbf{J}|Fm)-\\&\quad -\frac{\mu_J}{J}\frac{F(F+1)+J(J+1)-I(I+1)}{2F(F+1)}(Fm|\mathbf{F}.\mathbf{H}_0|Fm)-\\&\quad -\frac{\mu_I}{I}\frac{F(F+1)+I(I+1)-J(J+1)}{2F(F+1)}(Fm|\mathbf{F}.\mathbf{H}_0|Fm)\\&= \frac{ha}{2}[F(F+1)-I(I+1)-J(J+1)]-\\&\quad -\left[\frac{\mu_J}{J}\frac{F(F+1)+J(J+1)-I(I+1)}{2F(F+1)}+\right.\\&\quad \left.+\frac{\mu_I}{I}\frac{F(F+1)+I(I+1)-J(J+1)}{2F(F+1)}\right]H_0 m. \quad \text{(III. 91)}\end{aligned}$$

In the low field limit the term proportional to μ_I can often be neglected owing to its small size.

(b) *Strong external field.* In the limit of a very strong magnetic field, corresponding to a Paschen–Back effect, **I** and **J** will each separately precess rapidly about $\mathbf{H}_0$ as in Fig. III. 3, so that F is no longer a good

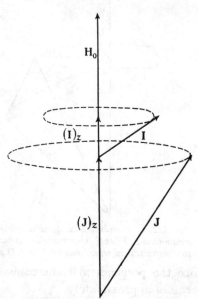

FIG. III. 3. Coupling of nuclear angular momentum **I** and electronic angular momentum **J** in strong magnetic field $\mathbf{H}_0$.

quantum number, but instead the good quantum numbers are the magnetic quantum numbers m_I and m_J of **I** and **J** respectively. Then, from Eq. (III. 88a),

$$W_{M1}(m_I, m_J) = (m_I m_J | \mathcal{H}_{M1} | m_I m_J)$$
$$= ha(m_I m_J | (\mathbf{I})_x (\mathbf{J})_x + (\mathbf{I})_y (\mathbf{J})_y + (\mathbf{I})_z (\mathbf{J})_z | m_I m_J) -$$
$$- \frac{\mu_J}{J} H_0 m_J - \frac{\mu_I}{I} H_0 m_I$$
$$= -\frac{\mu_J}{J} H_0 m_J - \frac{\mu_I}{I} H_0 m_I + ha m_I m_J \qquad \text{(III. 92)}$$

since the diagonal elements of $(\mathbf{I})_x$ and $(\mathbf{I})_y$ are zero.

(c) *Intermediate external field.* The case of intermediate external field is of particular importance in molecular-beam research, but it is more difficult to work out the energy levels since a secular equation must be solved, as in the corresponding fine-structure case (CON 35). The

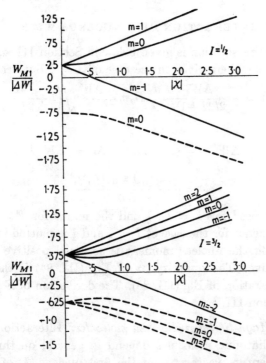

Fig. III. 4a. Variation of the energy with the magnetic field. Nuclear moment assumed positive, $J = \frac{1}{2}$. The dotted lines are the magnetic levels arising from the $F = I - \frac{1}{2}$ state (RAB 36).

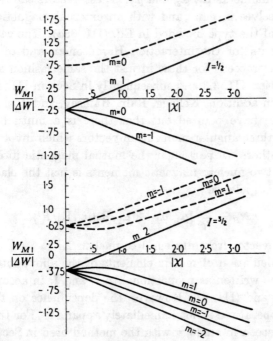

Fig. III. 4b. Same as Fig. 4a but with nuclear moment assumed negative (RAB 36).

procedure for this solution is given below in Section III. 4. The result of the derivation there given in the case of $J = \frac{1}{2}$ but arbitrary I is

$$W_{M1}(F,m) = -\frac{\Delta W}{2(2I+1)} - \frac{\mu_I}{I} H_0 m \pm \frac{\Delta W}{2} \sqrt{\left(1 + \frac{4m}{2I+1} x + x^2\right)},$$

(III. 93)

where

$$\Delta W = \frac{ha}{2}(2I+1), \quad \Delta \nu = |\Delta W|/h,$$

$$x = \frac{(-\mu_J/J + \mu_I/I)H_0}{\Delta W},$$

(III. 94)

and the $+$ is used for $F = I+\frac{1}{2}$ and the minus for $F = I-\frac{1}{2}$. This energy dependence for the case of $I = \frac{1}{2}$ and $\frac{3}{2}$ is plotted in Fig. III. 4. In Fig. III. 4a the nuclear moment is assumed positive and in Fig. III. 4b negative. The quantity $\Delta \nu$ of Eq. (III. 94) is the hyperfine-structure separation of Eq. (III. 88). The derivation of Eq. (III. 93) is given in Section III. 4.

III. 3.4. *Magnetic interactions in molecules.* Interaction energies in rotating $^1\Sigma$ diatomic molecules depend in general on three different angular-momentum vectors: $\mathbf{I}_1$ of the first nucleus, $\mathbf{I}_2$ of the second, and $\mathbf{J}$ of the molecular rotational angular momentum with corresponding magnetic moments $\mathbf{\mu}_1$, $\mathbf{\mu}_2$, and $\mathbf{\mu}_J$. These vectors can then interact among themselves in pairs and with an external magnetic field with interactions of the type included in Eq. (III. 88a). The calculation of matrix elements for the interaction Hamiltonian and of interaction energies then proceeds for these in just the same fashion as discussed above in Section III. 3.3 and subsequently in Section III. 4.

However, in addition, Kellogg, Rabi, Ramsey, and Zacharias (KEL 39a, KEL 40) have pointed out that there is a mutual interaction between the three angular-momentum vectors which involves all three of them together. In particular the mutual magnetic interaction $\mathscr{H}_S$ between the two nuclear magnetic moments is just the classical interaction of two dipoles so

$$\mathscr{H}_S = \frac{1}{R^3}\{\mathbf{\mu}_1 \cdot \mathbf{\mu}_2 - 3(\mathbf{\mu}_1 \cdot \mathbf{R})(\mathbf{\mu}_2 \cdot \mathbf{R})/R^2\},$$

(III. 95)

where $\mathbf{R}$ is a vector from nucleus 1 to nucleus 2.

For the calculation of matrix elements of the above interaction, $\mathbf{\mu}_1$ and $\mathbf{\mu}_2$ can be written as proportional to $\mathbf{I}_1$ and $\mathbf{I}_2$ in accordance with Eqs. (III. 2) and (III. 4). However, the dependence on the angular-momentum operator $\mathbf{J}$ is not immediately apparent. For this reason it is best to proceed in analogy with the method used in Section III. 2.3

to express the quadrupole interaction in terms of angular-momentum operators as in Eq. (III. 43).

Eq. (III. 95) can be rewritten as

$$\mathcal{H}_S = -\frac{(\mu_1/I_1)(\mu_2/I_2)}{R^3} \sum_{ij} [I_{1i} I_{2j}][3 R_i R_j/R^2 - \delta_{ij}], \quad \text{(III. 96)}$$

where the subscripts i and j indicate the different Cartesian components. By the general theorem of Section III. 2.3 in the calculation of matrix elements diagonal in J one may write

$$[3 R_i R_j/R^2 - \delta_{ij}] = C\left[3 \frac{J_i J_j + J_j J_i}{2} - \delta_{ij} J^2\right]. \quad \text{(III. 97)}$$

C can then be evaluated by multiplying the preceding equation by J_i on the left and J_j on the right and summing over i and j. This then gives

$$3(\mathbf{J}\cdot\mathbf{R})^2/R^2 - \mathbf{J}^2 = C[\tfrac{3}{2}(\mathbf{J}^2)^2 + \tfrac{3}{2}\sum_{ij} J_i J_j J_i J_j - (\mathbf{J}^2)^2]. \quad \text{(III. 98)}$$

On the left side one may use the fact that $\mathbf{J}\cdot\mathbf{R}$ vanishes in diatomic molecules since the angular momentum is always perpendicular to the internuclear axis. The right side may be transformed by exactly the same procedures that in Appendix C transform the right side of Eq. (C. 12) to the right side of Eq. (C. 19), so

$$-\mathbf{J}^2 = \tfrac{1}{2} C \mathbf{J}^2[(2J-1)(2J+3)] \quad \text{(III. 99)}$$

and

$$C = -\frac{2}{(2J-1)(2J+3)}. \quad \text{(III. 100)}$$

An alternative evaluation of C can be obtained by taking the $J, m_J = J$ diagonal matrix element of the zz-component of both sides of Eq. (III. 97). The matrix element of the right side is immediately obvious and that of the left side can be evaluated by integration and from the fact that the only θ-dependence of ψ_{JJ} is as $\sin^J\theta$ (CON 35, p. 5). This alternative derivation of course also leads to Eq. (III. 100).

From Eqs. (III. 97) and (III. 100),

$$[3 R_i R_j/R^2 - \delta_{ij}] = -\frac{2}{(2J-1)(2J+3)}\left[3\frac{J_i J_j + J_j J_i}{2} - \delta_{ij}\mathbf{J}^2\right].$$

$$\text{(III. 101)}$$

When this is combined with Eq. (III. 96), there results

$$\mathcal{H}_S = \frac{(\mu_1/I_1)(\mu_2/I_2)}{R^3}\frac{2}{(2J-1)(2J+3)} \times$$
$$\times [\tfrac{3}{2}(\mathbf{I}_1\cdot\mathbf{J})(\mathbf{I}_2\cdot\mathbf{J}) + \tfrac{3}{2}(\mathbf{I}_2\cdot\mathbf{J})(\mathbf{I}_1\cdot\mathbf{J}) - (\mathbf{I}_1\cdot\mathbf{I}_2)\mathbf{J}^2]. \quad \text{(III. 102)}$$

This expression is applicable to both homonuclear and heteronuclear diatomic molecules. However, it is the most convenient expression to use only when the two nuclei are not identical, as in the molecule HD. The matrix elements diagonal in I and J are then obtained directly from Eq. (III. 102) and the known ways in which the angular-momentum components operate on the wave functions (CON 35) as in Appendix C.

If, on the other hand, the two nuclei are identical and with nuclear spin $\frac{1}{2}$ (or spin 1 if J is odd) it is most convenient to use the resultant angular momentum $\mathbf{I}_R = \mathbf{I}_1 + \mathbf{I}_2$ since I_R is a good quantum number in a rotational state of definite J because of the symmetry requirements on the wave function. For this purpose, the combination of Eqs. (III. 96) and (III. 101) can be written in the equivalent form

$$\mathscr{H}_S = \frac{(\mu_1/I_1)(\mu_2/I_2)}{R^3} \frac{2}{3(2J-1)(2J+3)} \times$$

$$\times \sum_{ij} [3I_{1i}I_{2j} - \delta_{ij}\mathbf{I}_1 \cdot \mathbf{I}_2] \left[3\frac{J_iJ_j + J_jJ_i}{2} - \delta_{ij}\mathbf{J}^2\right]. \quad \text{(III. 103)}$$

Then, if I_R is a good quantum number, by the general theorem of Section III. 2.3 we can write

$$[3I_{1i}I_{2j} - \delta_{ij}\mathbf{I}_1 \cdot \mathbf{I}_2] = D\left[3\frac{I_{Ri}I_{Rj} + I_{Rj}I_{Ri}}{2} - \delta_{ij}I_R^2\right]. \quad \text{(III. 104)}$$

D can then be evaluated by multiplying both sides of the above by I_{Ri} on the left and I_{Rj} on the right and summing over i and j. Then, since the operations on the right side of Eq. (III. 104) are identical with those on the right side of Eq. (C. 11) in Appendix C, it can be transformed in the same way to the form of the right side of Eq. (C. 19). Eq. (III. 104) with these modifications then becomes

$$[3(\mathbf{I}_R \cdot \mathbf{I}_1)(\mathbf{I}_2 \cdot \mathbf{I}_R) - (\mathbf{I}_1 \cdot \mathbf{I}_2)\mathbf{I}_R^2] = \tfrac{1}{2} D\, \mathbf{I}_R^2 (2I_R - 1)(2I_R + 3). \quad \text{(III. 105)}$$

But in Eq. (C. 17) of Appendix C it is shown that

$$\mathbf{I}_R \cdot \mathbf{I}_1 = \mathbf{I}_R \cdot \mathbf{I}_2 = \tfrac{1}{2}\mathbf{I}_R^2 \quad \text{(III. 106)}$$

and $\quad \mathbf{I}_1 \cdot \mathbf{I}_2 = \tfrac{1}{2}[(\mathbf{I}_1 + \mathbf{I}_2)^2 - \mathbf{I}_1^2 - \mathbf{I}_2^2] = \tfrac{1}{2}[\mathbf{I}_R^2 - 2\mathbf{I}_1^2] = \tfrac{1}{2}\mathbf{I}_R^2 - \mathbf{I}_1^2. \quad \text{(III. 107)}$

With these values, Eq. (III. 105) immediately yields

$$D = \frac{1}{2}\frac{I_R(I_R+1) + 4I_1(I_1+1)}{(2I_R-1)(2I_R+3)}. \quad \text{(III. 108)}$$

Therefore Eq. (III. 103) may be written as

$$\mathscr{H}_S = \frac{(\mu_1/I_1)(\mu_2/I_2)}{R^3} \frac{I_R(I_R+1) + 4I_1(I_1+1)}{3(2I_R-1)(2I_R+3)(2J-1)(2J+3)} \times$$

$$\times \sum_{ij}\left[3\frac{I_{Ri}I_{Rj} + I_{Rj}I_{Ri}}{2} - \delta_{ij}I_R^2\right]\left[3\frac{J_iJ_j + J_jJ_i}{2} - \delta_{ij}\mathbf{J}^2\right].$$

$$\text{(III. 109)}$$

This is sometimes the most convenient form to use. However, it is often convenient to express it in an alternative form analogous to Eq. (III. 43). For this, the derivation in Appendix C transforming Eq. (C. 1) to Eq. (C. 9) is directly applicable, whence

$$\mathscr{H}_S = \frac{(\mu_1/I_1)(\mu_2/I_2)}{R^3} \frac{I_R(I_R+1)+4I_1(I_1+1)}{(2I_R-1)(2I_R+3)(2J-1)(2J+3)} \times$$

$$\times [3(\mathbf{I}_R \cdot \mathbf{J})^2 + \tfrac{3}{2} \mathbf{I}_R \cdot \mathbf{J} - \mathbf{I}_R^2 \mathbf{J}^2]. \quad \text{(III. 110)}$$

Note that the dependence of this on the operators $\mathbf{I}_R$ and $\mathbf{J}$ is exactly the same as that of Eq. (III. 44b) for the quadrupole interaction of a homonuclear diatomic molecule with two identical nuclei. Consequently, by measurements only in a single homonuclear molecule which possesses both a nuclear quadrupole moment and a spin–spin magnetic interaction it is impossible to distinguish between the spin–spin interaction and the nuclear quadrupole interaction. On the other hand, the spin–spin interaction can often be inferred from a similar measurement in a different molecule, whence the quadrupole interaction can be obtained, as in a comparison of H_2 and D_2. This difficulty does not arise with a molecule such as HD whose nuclei are not identical so the spin–spin and nuclear quadrupole interactions can then be inferred separately.

In addition to the magnetic interactions discussed above, Ramsey (RAM 40, RAM 52b) has pointed out that in precision experiments with molecules account must be taken of the molecular diamagnetic interaction with the external magnetic field and of the dependence of this interaction on molecular orientation. He has also pointed out (RAM 50d, RAM 51c) that a small correction must be made for the magnetic shielding of the nuclei by the electrons of the molecule and for the variation of this shielding with molecular orientation. The detailed discussion of these magnetic interactions and of their theoretical form is given in Chapter VIII.

III. 4. Intermediate coupling

III. 4.1. *Introduction.* In general the nuclear interactions of an atom or molecule in a magnetic field are combinations of several of the interactions described in Sections III. 2 and III. 3. With several such interactions present simultaneously it is not in general possible to select *a priori* a representation for which the energy matrix is diagonal. Often it is possible to select a representation in which the largest matrix elements are diagonal, in which case a standard form of perturbation calculation can be used to make corrections for the effect of the small non-diagonal matrix elements (CON 35). The procedure for such a calculation is given in Section III. 4.3. In many cases, however, the non-diagonal matrix elements are comparable to the diagonal ones, in which case the perturbation procedure is unsatisfactory. In this case the best procedure is to calculate all the matrix elements from the interaction Hamiltonian in any convenient representation. The energy levels of the system may then be obtained by a solution of the corresponding secular equation.

III. 4.2. *Solution of secular equation.* An example of the solution of the secular equation for an intermediate coupling problem is the solution of the intermediate field problem of Section III. 3.3. In this case and if the z-axis is taken in the direction of the external field H_0, the Hamiltonian of the system as given in Eq. (III. 88a) is

$$\mathcal{H} = -dI_z - cJ_z + ha\mathbf{I}\cdot\mathbf{J}$$
$$= -dI_z - cJ_z + haI_zJ_z + \tfrac{1}{2}haI_+J_- + \tfrac{1}{2}haI_-J_+, \quad \text{(III. 111)}$$

where $\quad d = (\mu_I/I)H_0 \quad$ and $\quad c = (\mu_J/J)H_0.$ (III. 112)

As discussed in Section III. 3.3, an Fm representation is most convenient when H_0 is small and an $m_I m_J$ representation is best when H_0 is large. However, in intermediate magnetic fields the matrix is not even approximately diagonal in either representation, so that the secular equation must be solved. As a result it makes little difference which representation is chosen in the calculation of the complete matrix and the secular equation will, of course, be independent of this choice even though the matrices are quite different in form. For the present calculation an $m_I m_J$ representation will be chosen arbitrarily, so the typical matrix element will be $(m_I m_J |\mathcal{H}| m_I' m_J')$; a calculation with an Fm representation has been published elsewhere (MIL 38). The calculation will be limited to atomic states for which $J = \tfrac{1}{2}$ since this greatly simplifies the arithmetic while retaining an example which illustrates the important principles and is in fact the specific example needed in the majority of molecular-beam experiments (BRE 31).

From the commutation rules of Eq. (III. 44), it is easy to show that $F_z = I_z + J_z$ commutes with the Hamiltonian of Eq. (III. 111). Therefore, all matrix elements of $\mathcal{H}$ not diagonal in $m = m_I + m_J$ will necessarily vanish. Therefore, the total-energy matrix will consist of a number of submatrices, each corresponding to a different value of m and with no non-diagonal matrix elements connecting two submatrices of different m. Since by assumption $J = \tfrac{1}{2}$, these submatrices will, at most, be 2×2

corresponding to the two possible states giving the value of m. In an $m_I m_J$ representation these states are $(m_I = m-\frac{1}{2}, m_J = \frac{1}{2})$ and $(m_I = m+\frac{1}{2}, m_J = -\frac{1}{2})$.

Let us first consider the submatrix with $m = I+\frac{1}{2}$. In this case the submatrix is only 1×1, since only one state $(m_I = I, m_J = \frac{1}{2})$ can give the desired m so there are no non-diagonal matrix elements. The same thing is true of the $m = -I-\frac{1}{2}$ submatrix. Since m_I is completely determined by m and m_J, let the energy of the level be designated by W_{m,m_J}. Then, from Eq. (III. 35a),

$$W_{\pm I\pm\frac{1}{2},\pm\frac{1}{2}} = (\pm I, \pm\tfrac{1}{2}|\mathscr{H}|\pm I, \pm\tfrac{1}{2}) = \mp dI \mp \tfrac{1}{2}c + \tfrac{1}{2}haI. \tag{III. 113}$$

However, the other submatrices are 2×2. The secular equation is then of the form

$$\begin{vmatrix} A-W_{m,\pm} & C \\ D & B-W_{m,\pm} \end{vmatrix} = 0, \tag{III. 114}$$

where, from Eq. (III. 35a),

$$\begin{Bmatrix} A \\ B \end{Bmatrix} = (m\pm\tfrac{1}{2}, \mp\tfrac{1}{2}|\mathscr{H}|m\pm\tfrac{1}{2}, \mp\tfrac{1}{2}) = -d(m\pm\tfrac{1}{2})\pm\tfrac{1}{2}c\mp\tfrac{1}{2}ha(m\pm\tfrac{1}{2})$$
and $\tag{III. 115}$

$$C = D = (m\pm\tfrac{1}{2}, \mp\tfrac{1}{2}|\mathscr{H}|m\mp\tfrac{1}{2}, \pm\tfrac{1}{2}) = \tfrac{1}{2}ha(m\pm\tfrac{1}{2}|I_\pm|m\mp\tfrac{1}{2})(\mp\tfrac{1}{2}|J_\mp|\pm\tfrac{1}{2})$$
$$= \tfrac{1}{2}ha\sqrt{\{(I-m+\tfrac{1}{2})(I+m+\tfrac{1}{2})\}}\sqrt{\{(\tfrac{1}{2}+\tfrac{1}{2})(\tfrac{1}{2}+\tfrac{1}{2})\}} = \tfrac{1}{2}ha\sqrt{\{(I+\tfrac{1}{2})^2-m^2\}}. \tag{III. 116}$$

With the above and by straightforward expansions Eq. (III. 114) can be written as
$$a'W_{m,\pm}^2 + b'W_{m,\pm} + c' = 0, \tag{III. 117}$$
where

$$a' = 1,$$
$$b' = 2dm + \tfrac{1}{2}ha,$$
$$c' = [-d(m+\tfrac{1}{2})+\tfrac{1}{2}c-\tfrac{1}{2}ha(m+\tfrac{1}{2})][-d(m-\tfrac{1}{2})-\tfrac{1}{2}c+\tfrac{1}{2}ha(m-\tfrac{1}{2})]-$$
$$-\tfrac{1}{4}h^2a^2[(I+\tfrac{1}{2})^2-m^2]. \tag{III. 118}$$

Therefore, with some straightforward simplification,

$$W_{m,\pm} = \frac{-b'\pm\sqrt{\{(b'^2-4a'c')\}}}{2a'}$$
$$= -\tfrac{1}{4}ha - dm \pm \tfrac{1}{2}\sqrt{\{h^2a^2(I+\tfrac{1}{2})^2+2ha(d-c)m+(d-c)^2\}}. \tag{III. 119}$$

When $H_0 = 0$ so that d and c are also zero the above equation gives W_{m+} as equal to the energy of Eq. (III. 91) when $F = I+\tfrac{1}{2}$ and W_{m-} as the energy for $F = I-\frac{1}{2}$. Hence we can use the designation $W(F,m)$ provided that it is understood that the quantum numbers in this case are those which apply at the $H = 0$ limit and that the upper signs are used for $F = I+\frac{1}{2}$ while the lower ones are used for $F = I-\frac{1}{2}$. The constants a, d, and c can be re-expressed from Eqs. (III. 88) and (III. 112). With these changes and some elementary rearrangements Eq. (III. 119) becomes

$$W(F,m) = -\frac{\Delta W}{2(2I+1)} - \frac{\mu_I}{I} H_0 m \pm \frac{\Delta W}{2} \sqrt{\left(1 + \frac{4m}{2I+1}x + x^2\right)},$$

(III. 120)

where

$$x = \frac{(-\mu_J/J + \mu_I/I)H_0}{\Delta W}$$

(III. 121)

$$= (g_J - g_I)\mu_0 H_0/\Delta W,$$

(III. 121a)

with

$$g_I = -\mu_I/I\mu_0,$$

where μ_0 is the Bohr magneton and a similar definition applies to g_J (see the footnote in Section III. 1). From the nature of the above derivation, the preceding equation would seem to apply for all values of m except $\pm(I+\frac{1}{2})$ for which Eq. (III. 113) is the appropriate expression. However, if the value of Eq. (III. 120) is calculated for the states of $F = I+\frac{1}{2}$ [whence the plus sign is used in Eq. (III. 120)] and $m = \pm(I+\frac{1}{2})$, the square root becomes the square root of a perfect square and the resulting linear equation is just the same as Eq. (III. 113). Hence Eq. (III. 120) may be taken to apply to all states. This is just the relation assumed earlier without proof in Eq. (III. 94). The energy variation is shown graphically in Fig. III. 4 of Section III. 3.3. In Fig. III. 4a the nuclear moment is assumed positive and in Fig. III. 4b negative.

With more complicated Hamiltonians, including such features as quadrupole moment interactions, the procedure is similar except that additional matrix elements must be computed and if the atomic or molecular J is greater than half, the secular equation to be solved in general is of higher degree than quadratic. For example, from Eqs. (III. 44c) and (III. 88a) the Hamiltonian for an atom with both magnetic and quadrupole interactions may be written as

$$\mathcal{H}' = -\left(\frac{\mu_J}{J}\mathbf{J} + \frac{\mu_I}{I}\mathbf{I}\right)\cdot\mathbf{H}_0 + ha\mathbf{I}\cdot\mathbf{J} + hb_1\, 2\mathbf{I}\cdot\mathbf{J}(2\mathbf{I}\cdot\mathbf{J}+1).$$

(III. 122)

Fig. III. 5 illustrates the energy levels of an atom, Ga^{69}, whose nuclear interactions are those of the above Hamiltonian and which is in a $^2P_{\frac{3}{2}}$

atomic state (BEC 48). Other examples of such calculations will appear in later chapters, notably Chapter VIII. In Appendix C, Eqs. (C. 24), (C. 26), and (C. 27), provide useful formulae for the diagonal and non-diagonal matrix elements needed in such calculations.

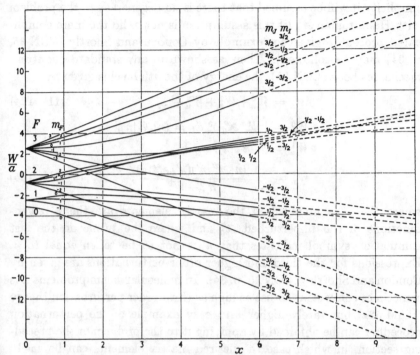

FIG. III. 5. The variation with magnetic field of the energy levels of Ga^{69} in the $^2P_{\frac{1}{2}}$ state. For values of $x = g_J \mu_0 H/ah$ from 0 to 6, the levels have been computed with a Hamiltonian including the magnetic dipole and electric quadrupole interactions. Numerical values for b/a and g_I were taken from Renzetti (REN 40). At $x = 10$, the calculation was made for a high field approximation, and the dotted curves join the two solutions smoothly (BEC 48).

III. 4.3. *Perturbation theory.* Often the energy levels are desired only in regions in which some terms of the Hamiltonian are much larger than others. For example, if H_0 in Eq. (III. 122) is large, the first term may be much larger than the last two. Conversely, for small H_0 the last two terms are large compared with the first. In such cases the desired energy levels can often be obtained more simply by the approximate methods of perturbation theory than from a complete solution of the secular equation. This is especially true if the nature of the Hamiltonian is such that the secular equation is of higher degree than the second, unlike the example of the preceding section.

In this case, one can write

$$\mathcal{H} = \mathcal{H}_0 + \mathcal{H}', \quad \text{(III. 123)}$$

where $\mathcal{H}' \ll \mathcal{H}_0$. In the case of large H_0, for example, $\mathcal{H}_0$ would correspond to the first term of Eq. (III. 122) and $\mathcal{H}'$ to the last two. For simplicity it will be assumed that there is no degeneracy to the problem with Hamiltonian $\mathcal{H}_0$ (if this assumption is not valid the more complicated procedure given for example by Condon and Shortly [CON 35, p. 34] may be followed). Then, as shown in any standard quantum-mechanics book (CON 35), the energy of the nth level is given by

$$W_n = W_n^0 + W_n^1 + W_n^2 + \cdots, \quad \text{(III. 124)}$$

where

$$W_n^0 = \int \psi_n^{0*} \mathcal{H}_0 \psi_n^0 \, d\tau = (n|\mathcal{H}_0|n),$$

$$W_n^1 = (n|\mathcal{H}'|n),$$

$$W_n^2 = \sum{}' \frac{(n|\mathcal{H}'|n')(n'|\mathcal{H}'|n)}{W_n^0 - W_{n'}^0}, \quad \text{(III. 125)}$$

where the $\psi_{n'}^0$'s involved in the matrix elements are the n' eigenfunctions of the unperturbed $\mathcal{H}_0$ and where the prime on the last summation symbol indicates that n' is not to be taken equal to n. Expressions for the third and higher order perturbations are given by Condon and Shortley (CON 35, p. 34). In molecular-beam problems it is sometimes convenient to use as high as third-order perturbation theory (RAM 52a) but rarely higher. An easy example of the perturbation procedure can be obtained by applying it to the problem of the preceding section, in which case the desired matrix elements can be easily obtained from the appropriate parts of Eqs. (III. 115) and (III. 116), with the d and c terms being taken as contributing to $\mathcal{H}_0$ while the a term is part of $\mathcal{H}'$.

IV

NON-RESONANCE MEASUREMENTS OF ATOMIC AND NUCLEAR MAGNETIC MOMENTS

IV. 1. Deflexion and intensity relations

IV. 1.1. *Effective magnetic moment.* Let W be the effective potential energy of an atom or molecule in an external magnetic field H_0. Then in the appropriate cases W is given by Eqs. (III. 76), (III. 120), etc. since W is also equal to the 'spectroscopical energy' of the molecule (ZZA 6). However, from the usual relation between force and potential energy the force on the atom or molecule is given by

$$\mathbf{F} = -\nabla W = -\frac{\partial W}{\partial H_0}\nabla H_0, \qquad (\text{IV. 1})$$

where it is assumed that the only significant dependence of W on position is through the spatial variation of the magnetic field H_0.

For a simple constant magnetic moment independent of H_0, the energy W by Eq. (III. 76) consists of

$$W = -\boldsymbol{\mu}\cdot\mathbf{H}_0 = -\mu_{\text{eff}} H_0, \qquad (\text{IV. 2})$$

where μ_{eff} is the component of $\boldsymbol{\mu}$ along $\mathbf{H}_0$. Then from Eq. (IV. 1),

$$\mathbf{F} = -\frac{\partial W}{\partial H_0}\nabla H_0 = \mu_{\text{eff}}\nabla H_0. \qquad (\text{IV. 3})$$

In the general case where the dependence of W on H_0 is more complicated than in Eq. (IV. 2) it is customary to refer to the factors of Eq. (IV. 1) in the terminology that would apply in the simple constant magnetic moment case. Thus from a comparison of Eqs. (IV. 1) and (IV. 3), the $-\partial W/\partial H_0$ factor is spoken of as the effective component of the magnetic moment in the direction of ∇H_0 or

$$\mu_{\text{eff}} = -\frac{\partial W}{\partial H_0}. \qquad (\text{IV. 4})$$

The comparison of Eqs. (IV. 1) and (IV. 3) shows that the atom with a general energy dependence on H_0 would be deflected in an inhomogeneous magnetic field as if it had a constant magnetic moment whose component along H_0 was given by Eq. (IV. 4).

IV. 1.2. *Deflexion of molecule by magnetic fields.* Consider a molecule of mass m moving with velocity v in a molecular-beam apparatus such as that in Fig. I. 3. When the molecule is acted upon by one of the inhomogeneous magnetic fields, it is accelerated transversely with an acceleration a which, by Eq. (IV. 3), is given by

$$a = F/m = (\mu_{\text{eff}}/m)(\partial H_0/\partial z), \qquad (\text{IV. 5})$$

where the z-axis is taken in the direction of the gradient of H_0.

For many molecular-beam problems it is necessary to calculate the deflexion of the molecule at various places along its path. This is necessary, for example, in calculating whether the molecules will clear the pole faces of the deflecting magnet and in determining the beam intensity to be expected at the detector. The procedures for such calculations were developed in the fundamental paper of Stern (STE 27) on magnetically deflected beams. The sideways displacement of the molecule is of course dependent on the angle θ with which it emerges from the source. However, this angle θ can be determined for a given molecule, velocity, and magnetic-field array if the transverse displacement at the source and at the collimator is taken as zero. This corresponds to the displacement of a beam defined by an infinitely narrow source and detector. However, as will be discussed in the next section, this displacement may also be used to calculate the beam pattern with appreciable source and collimator widths since it is the displacement relative to a straight line passing through the same points at the source and collimator position.

Since the deflexions are small, such approximations as the replacement of $\sin \theta$ by θ are appropriate in the following calculations. Likewise, the number of molecules per unit solid angle may be considered constant for all angles θ considered. The deflexions are calculated at the source (S), the entrance to the A magnet (A'), the exit of the A magnet (A''), the collimator (C), the entrance and exit of the B magnet (B' and B''), and the detector (D). Various distances are designated with the notation of Fig. I. 3. In order that the following calculation may be equally applicable to the molecular-beam electric method, the transverse acceleration in region A will be taken as a_A, in which case a is given by Eq. (IV. 5) if the deflexion is caused by a transverse magnetic field gradient and by the corresponding expression of Chapter XI in the electric case. The transverse displacement relative to a straight line passing through the same positions at the detector and collimator will be designated by $_x s_v$ and the transverse velocity by $_x v_v$, where the first subscript x indicates the position along the beam and second subscript v designates the velocity of the molecule concerned; one or both subscripts will be omitted for simplicity where no clarity is lost thereby.

By definition
$$_Ss = _Cs = 0, \qquad (IV.\ 6)$$
$$_Sv = _{A'}v = v\theta. \qquad (IV.\ 7)$$

Now, the time t_A that the molecule is in region A is l_A/v, etc., so
$$_{A'}s = l_1\theta, \qquad (IV.\ 8)$$
$$_As = _{A'}s + _{A'}vt_A + \tfrac{1}{2}a_A t_A^2$$
$$= \theta(l_1+l_A) + \tfrac{1}{2}a_A l_A^2/v^2, \qquad (IV.\ 9)$$
$$_Av = _Cv = _Bv = _{A'}v + a_A t_A = v\theta + a_A l_A/v, \qquad (IV.\ 10)$$
$$_Cs = 0 = _{A'}s + _{A'}v l_{C'}/v = \theta(l_1+l_A+l_{C'}) + \tfrac{1}{2}a_A l_A(l_A+2l_{C'})/v^2. \qquad (IV.\ 11)$$

Therefore
$$\theta = -\tfrac{1}{2}a_A l_A(l_A+2l_{C'})/(l_1+l_A+l_{C'})v^2, \qquad (IV.\ 12)$$
$$_{B'}s = \theta(l_1+l_A+l_{C'}+l_{C''}) + \tfrac{1}{2}a_A l_A(l_A+2l_{C'}+2l_{C''})/v^2, \qquad (IV.\ 13)$$
$$_Bs = \theta(l_1+l_A+l_{C'}+l_{C''}+l_B) + \tfrac{1}{2}a_A l_A(l_A+2l_{C'}+2l_{C''}+2l_B)/v^2 + \tfrac{1}{2}a_B l_B^2/v^2, \qquad (IV.\ 14)$$
$$_Bv = _Dv = v\theta + a_A l_A/v + a_B l_B/v, \qquad (IV.\ 15)$$
$$_Ds = \theta(l_1+l_A+l_{C'}+l_{C''}+l_B+l_4) +$$
$$+ \tfrac{1}{2}a_A l_A(l_A+2l_{C'}+2l_{C''}+2l_B+2l_4)/v^2 +$$
$$+ \tfrac{1}{2}a_B l_B(l_B+2l_4)/v^2. \qquad (IV.\ 16)$$

Note that, although θ appears in the above formulae, it can for any special case be eliminated with the aid of Eq. (IV. 12).

The most important case of the above formulae is the deflexion at the detector position as in Eq. (IV. 16). In this case, therefore, the elimination of the factor θ from Eq. (IV. 12) will be carried out. If this is done and the first major term of Eq. (IV. 16) is broken into two parts to provide cancellation of the denominator in one part, the resulting expression for the deflexion is

$$_Ds_v = -\tfrac{1}{2}a_A l_A(l_A+2l_{C'})/v^2 - \tfrac{1}{2}a_A l_A(l_A+2l_{C'})(l_{C''}+l_B+l_4)/(l_1+l_A+l_{C'})v^2 +$$
$$+ \tfrac{1}{2}a_A l_A(l_A+2l_{C'}+2l_{C''}+2l_B+2l_4)/v^2 + \tfrac{1}{2}a_B l_B(l_B+2l_4)/v^2. \qquad (IV.\ 17)$$

The first term of the previous equation cancels part of the third term, the remainder of the third term can be multiplied by
$$(l_1+l_A+l_{C'})/(l_1+l_A+l_{C'}),$$
after which some cancellation with the second term is possible, and use may be made of the relation from Fig. I. 3 that $l_5 = l_1+l_A+l_{C'}$ and

$l_6 = l_{C'} + l_B + l_4$. With these simplifications the deflexion at the detector

$$_D s_v = \tfrac{1}{2}(a_A/v^2)l_A(l_A+2l_1)(l_6/l_5) + \tfrac{1}{2}(a_B/v^2)l_B(l_B+2l_4). \qquad \text{(IV. 18)}$$

It should be noted that if only this result were desired it could be most easily obtained by first solving the simple problem that arises when $a_A = 0$ and consequently $\theta = 0$. In this case the second term of Eq. (IV. 18) immediately arises. Then the principle of reversibility of path could be used to infer from this the deflexion that arises when the deflecting magnet is placed before instead of after the collimator. In this way the first term of Eq. (IV. 18) could be found with the l_6/l_5 factor corresponding to the relative geometrical amplification of the source and detector displacements.

An alternative expression to the above is often useful, particularly when dealing with sources and collimators of appreciable width. Let s_0 be the position in the detector plane to which a molecule goes in the absence of the field, let s be the position to which a molecule with the same displacement at the source and collimator goes in the presence of the field so that $s - s_0$ (which will hereafter be called the magnetic deflexion) corresponds to $_D s_v$, and let s_α represent the magnetic deflexion at the detector position of the molecule with the most probable velocity α. Then, from Eq. (IV. 18),

$$s - s_0 = s_\alpha \alpha^2 / v^2, \qquad \text{(IV. 19)}$$

where $\quad s_\alpha = \tfrac{1}{2}(a_A/\alpha^2)l_A(l_A+2l_1)(l_6/l_5) + \tfrac{1}{2}(a_B/\alpha^2)l_B(l_B+2l_4). \qquad \text{(IV. 20)}$

In the case of magnetic deflexions we have from Eqs. (IV. 5)

$$s_\alpha = \frac{\mu_{\text{eff }A}}{2m\alpha^2}\left(\frac{\partial H_0}{\partial z}\right)_A l_A(l_A+2l_1)(l_6/l_5) + \frac{\mu_{\text{eff }B}}{2m\alpha^2}\left(\frac{\partial H_0}{\partial z}\right)_B l_B(l_B+2l_4).$$

$$\text{(IV. 21)}$$

Alternatively, from Eq. (II. 25), if T is the absolute temperature of a source in which α is the most probable velocity,

$$s_\alpha = \frac{\mu_{\text{eff }A}}{4kT}\left(\frac{\partial H_0}{\partial z}\right)_A l_A(l_A+2l_1)(l_6/l_5) + \frac{\mu_{\text{eff }B}}{4kT}\left(\frac{\partial H_0}{\partial z}\right)_B l_B(l_B+2l_4).$$

$$\text{(IV. 22)}$$

IV. 1.3. *Deflected beam shape.* In Section II. 2 the molecular beam shape at the detector was discussed in general when the source and collimator slits provided parallel narrow rectangular apertures for the beam. It was then shown that the beam shape was in general trapezoidal, as shown by the undeflected beam intensity $I_0(s_0)$ in Fig. II. 3 and by Eqs. (II. 21)–(II. 23). The shape of the beam after being deflected by an external magnetic field must now be considered. The procedures for calculating the deflected beam shapes were developed in a fundamental paper by Stern (STE 27). For simplicity in the following discussion it is assumed that the deflecting magnets all follow the collimator, but the results are independent of this assumption as, with the small angle approximation, the collimator is uniformly illuminated by the beam even if it follows a deflecting magnet.

If all the molecules were in the same molecular state and at the same velocity, the calculation would be trivial since the beam shape would then, to the first approximation, be the same as the undeflected shape and would merely be shifted laterally by an amount equal to the magnetic deflexion calculated in the previous paragraph. Most of the complication of the subsequent paragraphs is in allowing adequately

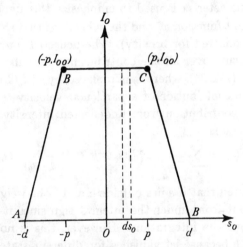

Fig. IV. 1. Beam shape before deflexion with differential element ds_0.

for the smearing effects of the velocity distribution. In the following, s will be the transverse distance from the centre of the beam to the position at which the beam intensity $I(s)$ is being calculated and s_0 for the velocity concerned is the corresponding undeflected position as in the preceding section and measured from the beam centre, so that s and s_0 are related by Eq. (IV. 19). This notation is shown in Fig. IV. 1.

By Eq. (II. 26) the velocity distribution of the molecules in the beam is
$$I(v)\,dv = (2I_0/\alpha^4)v^3 e^{-v^2/\alpha^2}\,dv. \tag{IV. 23}$$
This may be re-expressed in terms of the magnetic deflexion $s-s_0$ by Eq. (IV. 19), so
$$v^2 = s_\alpha \alpha^2/(s-s_0) \tag{IV. 24}$$
and
$$2v\,dv = \{s_\alpha/(s-s_0)^2\}\alpha^2\,ds_0. \tag{IV. 25}$$

Therefore, for a fixed s,

$$dI(s) = I(v)\,dv = \pm I_0 e^{-s_\alpha/(s-s_0)} \frac{s_\alpha^2}{(s-s_0)^3}\,ds_0, \qquad \text{(IV. 26)}$$

where the $\pm$ is introduced to enable ds_0 always to be positive, since in Eq. (IV. 25) ds_0 would be negative for negative s_α. The plus is used for $s_\alpha > 0$ and the minus for $s_\alpha < 0$. If the beam were initially infinitely narrow, Eq. (IV. 26) would give the deflected pattern if all molecules had the same s_α. Allowance will be made below for the initial spread of the beam and for different molecular states.

Let w_i^* be the statistical weight of a particular state whose magnetic deflexion for velocity α is s_α with the direction of the deflexion being from s_0 to s and where w_i^* is taken as zero for the opposite direction of deflexion (the asterisk is used to emphasize this unusual characteristic of the weight function w_i^* and the subscript i that should strictly be added to s_α is omitted for brevity). The deflected intensity $I(s)$ then consists of the sum over all the contributing molecular states i of the integral of Eq. (IV. 26), where I_0 is replaced by $w_i^* I_0(s_0)$ since the integrand gives the contribution of a particular velocity to the intensity. Therefore, the contribution from undeflected intensity $I_0(s_0)$ between $s_0 = a$ and $s_0 = b$ is

$$I(s) = \sum_i \pm \int_a^b w_i^* I_0(s_0) e^{-s_\alpha/(s-s_0)} \frac{s_\alpha^2}{(s-s_0)^3}\,ds_0. \qquad \text{(IV. 27)}$$

It should be noted that if s lies outside a and b the weight function w_i^* can be taken out from under the integral sign since it is then a constant throughout the integration. However, this is not true if s lies between a and b because w_i^* vanishes for different states when $s_0 < s$ and when $s_0 > s$.

It is convenient to introduce a new variable of integration into Eq. (IV. 27). Let

$$x = \frac{s_\alpha}{s-s_0}, \qquad s_0 = s - \frac{s_\alpha}{x}, \qquad ds_0 = \frac{s_\alpha}{x^2}\,dx. \qquad \text{(IV. 28)}$$

Then
$$I(s) = \sum_i \pm \int_{s_\alpha/(s-a)}^{s_\alpha/(s-b)} w_i^* I_0(x) e^{-x} x\,dx. \qquad \text{(IV. 29)}$$

If the sign of s_α is retained, the condition for the vanishing of w_i^* is that it should vanish whenever x is negative.

If $s > b > a$, the preceding equation can be written as

$$I(s) = \sum_i w_i^* \int_{s_\alpha/(s-a)}^{s_\alpha/(s-b)} I_0(x) e^{-x} x\,dx. \qquad \text{(IV. 30)}$$

If $a < s < b$ the above can be written as

$$I(s) = \sum_i w_i^* \int_{s_\alpha/(s-a)}^{\infty} I_0(x)e^{-x}x\,dx - \sum_j w_j^* \int_{\infty}^{s_\alpha/(s-b)} I_0(x)e^{-x}x\,dx. \tag{IV. 31}$$

Note that, for a given molecular state, w_i^* will be zero in one term of the preceding equation but not in the other. Also the signs of the infinite limits need not be correct in the cases when s and $s-s_0$ are opposite in sign since such terms vanish owing to the vanishing of w_i^*.

In many cases the molecular states have symmetrical populations, i.e. the statistical weight of the state with $s_\alpha = |s_\alpha|$ equals that of the state with $s_\alpha = -|s_\alpha|$. For example, this is true of a magnetic moment if all magnetic quantum numbers are equally populated but is not true if some of the states are cut off by collisions with an asymmetrical beam stop. In such symmetrical cases, let w_i' be the statistical weight of the state with positive s_α and be zero for negative s_α so that $\sum w_i' = \frac{1}{2}$. Expressions for w_i' are given in Eqs. (XIII. 11a)–(XIII. 11c). In this case the definitions of w_i^* and w_i' make Eq. (IV. 31) reduce to

$$I(s) = \sum_i w_i' \left[\int_{|s_\alpha|/(s-a)}^{\infty} I_0(x')e^{-x'}x'\,dx' - \int_{\infty}^{-|s_\alpha|/(s-b)} I_0(x'')e^{-x''}x''\,dx'' \right], \tag{IV. 32}$$

where
$$x' = \frac{|s_\alpha|}{s-s_0}, \qquad x'' = \frac{-|s_\alpha|}{s-s_0}. \tag{IV. 33}$$

The $I_0(x')$ is therefore a different function of x' from $I_0(x'')$ since for the latter $|s_\alpha|$ is replaced by $-|s_\alpha|$.

These results may now be applied to the practical case when the undeflected beam is trapezoidal in shape as in Fig. IV. 1. First the case of $s > d$ may be considered. Then from Eqs. (II. 21)–(II. 23)

$$I_0(x) = I_{00}(d+s-s_\alpha/x)/(d-p) \quad \text{for } -d \leqslant s_0 \leqslant -p, \tag{IV. 34}$$

$$I_0(x) = I_{00} \quad \text{for } -p \leqslant s_0 \leqslant p, \tag{IV. 35}$$

$$I_0(x) = I_{00}(d-s+s_\alpha/x)/(d-p) \quad \text{for } p \leqslant s_0 \leqslant d. \tag{IV. 36}$$

From Eqs. (IV. 30) and (IV. 34)–(IV. 36)

$$I(s) = I_{00} \sum_i w_i' \left[\frac{d+s}{d-p} \int_{s_\alpha/(s+d)}^{s_\alpha/(s+p)} e^{-x}\,dx - \frac{s_\alpha}{d-p} \int_{s_\alpha/(s+d)}^{s_\alpha/(s+p)} e^{-x}\,dx + \right.$$
$$\left. + \int_{s_\alpha/(s+p)}^{s_\alpha/(s-p)} e^{-x}\,dx + \frac{d-s}{d-p} \int_{s_\alpha/(s-p)}^{s_\alpha/(s-d)} e^{-x}\,dx + \frac{s_\alpha}{d-p} \int_{s_\alpha/(s-p)}^{s_\alpha/(s-d)} e^{-x}\,dx \right]. \tag{IV. 37}$$

Since the indefinite integral of e^{-x} is $-e^{-x}$ and

$$\int e^{-x} x \, dx = -e^{-x}(x+1), \tag{IV. 39}$$

$$I(s) = I_{00} \sum_i w_i' \left[\frac{d+s}{d-p} \left\{ \left(1 + \frac{s_\alpha}{s+d}\right) e^{-s_\alpha/(s+d)} - \left(1 + \frac{s_\alpha}{s+p}\right) e^{-s_\alpha/(s+p)} \right\} - \right.$$

$$- \frac{s_\alpha}{d-p} \left\{ e^{-s_\alpha/(s+d)} - e^{-s_\alpha/(s+p)} \right\} +$$

$$+ \frac{d-p}{d-p} \left\{ \left(1 + \frac{s_\alpha}{s+p}\right) e^{-s_\alpha/(s+p)} - \left(1 + \frac{s_\alpha}{s-p}\right) e^{-s_\alpha/(s-p)} \right\} +$$

$$+ \frac{d-s}{d-p} \left\{ \left(1 + \frac{s_\alpha}{s-p}\right) e^{-s_\alpha/(s-p)} - \left(1 + \frac{s_\alpha}{s-d}\right) e^{-s_\alpha/(s-d)} \right\} +$$

$$\left. + \frac{s_\alpha}{d-p} \left\{ e^{-s_\alpha/(s-p)} - e^{-s_\alpha/(s-d)} \right\} \right] \tag{IV. 40}$$

$$= I_{00} \sum_i w_i' \frac{1}{d-p} \left[(d+s) e^{-s_\alpha/(s+d)} - \right.$$

$$- \left\{ d+s - s_\alpha \left(1 - \frac{d+s}{s+p} + \frac{d-p}{s+p}\right) - (d-p) \right\} e^{-s_\alpha/(s+p)} +$$

$$+ \left\{ -(d-p) + s_\alpha \left(-\frac{d-p}{s-p} + \frac{d-s}{s-p} + 1\right) + d-s \right\} e^{-s_\alpha/(s-p)} -$$

$$\left. - (d-s) e^{-s_\alpha/(s-d)} \right]. \tag{IV. 41}$$

Since the two factors in parentheses multiplying s_α equal zero as given above, the equation reduces to

$$I(s) = I_{00} \sum_i w_i' \frac{1}{(d-p)} \left[(s+d) e^{-s_\alpha/(s+d)} + (s-d) e^{-s_\alpha/(s-d)} - \right.$$

$$\left. - (s+p) e^{-s_\alpha/(s+p)} - (s-p) e^{-s_\alpha/(s-p)} \right]. \tag{IV. 42}$$

It remains to consider the case when s is less than d. The most interesting case is that for symmetrical populations. Then if s is between $-d$ and $+d$, one or two of the integrals in Eq. (IV. 37) will be split into two parts as in Eq. (IV. 32). However, the indefinite integrals for infinite values of x' or x'' vanish so no contributions from these limits arise. The only other changes from Eq. (IV. 30) are that the s_α in the second part of Eq. (IV. 32) is replaced by $-|s_\alpha|$ both in the integrand and in the limits and that the signs of the terms arising from the upper limits of the integration are reversed. From Eq. (IV. 42) it is apparent that these changes are equivalent to putting absolute values around the s_α and the factors such as $(s-p)$ as in the following equation. It is also clear that the resulting expression is applicable when s is less than $-d$ since the same changes are made in this case. Therefore for all values of s the intensity at the detector is given by the following with the sign of s taken positive:

$$I(s) = I_{00} \sum_i w'_i \frac{1}{(d-p)}[(s+d)e^{-|s_\alpha|/(s+d)} + |s-d|e^{-|s_\alpha|/|s-d|} $$
$$- (s+p)e^{-|s_\alpha|/(s+p)} - |s-p|e^{-|s_\alpha|/|s-p|}]. \quad \text{(IV. 43)}$$

It should be noted, however, that the preceding equation applies only in the case of symmetrical populations and that w'_i is the statistical weight of those states with positive s_α only and vanishes for states of negative s_α so $\sum_i w'_i = \frac{1}{2}$. Expressions for w'_i are given in Eqs. (XIII. 11a)–(XIII. 11c).

From Eqs. (IV. 22) and (IV. 43) the intensity distribution to be expected at the detector can be calculated for any particular configuration of the magnetic fields, etc.

For many experiments—such as the one described in Chapter I—the detector is left at the position of maximum beam for no deflecting fields. The exact calculation of the intensity to be expected in the presence of the deflecting fields is then found by integrating Eq. (IV. 43) over the detector slit. However, a useful approximation can be obtained from the value of $I(s)$ for $s = 0$. From Eq. (IV. 43), this result is

$$I(0) = 2I_{00} \sum_i w'_i \frac{1}{d-p}[de^{-|s_\alpha|/d} - pe^{-|s_\alpha|/p}]. \quad \text{(IV. 44)}$$

It is apparent from this that $I(0)$ equals I_{00} in the limiting case when s_α is zero, as of course it should, provided that use is made of the fact that $\sum_i w'_i = \frac{1}{2}$. Typical of the results that can be calculated numerically from Eq. (IV. 44) are the following for $w_c = w_d = w$ and $l_{sc} = l_{cd}$ so $p = \frac{1}{2}w$ and $d = \frac{3}{2}w$ as in Eq. (II. 20a).

For $I(0) = \frac{1}{2}I_{00}$: $s_\alpha = 3\cdot 2p = 1\cdot 06d = 1\cdot 6w.$ (IV. 45)

For $I(0) = 0\cdot 1 I_{00}$: $s_\alpha = 8\cdot 1p = 2\cdot 7d = 4\cdot 05w.$ (IV. 46)

The beam intensity $I(0)$ as a function of s_α for the preceding case is shown graphically in Fig. IV. 2.

In some cases a more accurate expression than Eq. (IV. 44) is desired for the theoretical intensity when the detector is centred at the undeflected position; in particular allowance is sometimes made for the finite width w_d of the detector. The expression for this intensity is just twice the integral of Eq. (IV. 43) from 0 to $\tfrac{1}{2}w_d$. The integral is obtained most

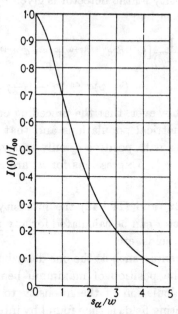

FIG. IV. 2. Dependence of beam intensity at $s = 0$ upon s_α/w for detector of negligible width.

easily if $\tfrac{1}{2}w_d < p$, in which case one can show with the aid of a simple integration by parts that

$$I = 2\int_0^{\tfrac{1}{2}w_d} I(s)\,ds = 2I_{00} \sum_i w_i' w_d \left[1 - \frac{s_\alpha^2}{2w_d(d-p)}\left\{\text{Ei}\left(-\frac{s_\alpha}{d+\tfrac{1}{2}w_d}\right) - \right.\right.$$
$$\left.\left. -\text{Ei}\left(-\frac{s_\alpha}{d-\tfrac{1}{2}w_d}\right) - \text{Ei}\left(-\frac{s_\alpha}{p+\tfrac{1}{2}w_d}\right) + \text{Ei}\left(-\frac{s_\alpha}{p-\tfrac{1}{2}w_d}\right)\right\}\right], \quad \text{(IV. 46}a\text{)}$$

where
$$\text{Ei}(x) = -\int_x^\infty \frac{e^{-t}}{t}\,dt \quad \text{(IV. 46}b\text{)}$$

is the tabulated exponential integral (JAH 45).

Another interesting limiting case of Eq. (IV. 43) is that for a rectangular beam, i.e. for $p = d = a$. At first sight Eq. (IV. 43) is indeterminate in this case, but the limiting value of Eq. (IV. 43) can be found as p approaches d. Then, if the first and third (or second and fourth) terms of Eq. (IV. 43) are combined together with the common factor $1/(d-p)$, the combination is of the form

$$\lim_{\Delta x \to 0} \frac{(x+\Delta x)e^{-|s_\alpha|/(x+\Delta x)} - xe^{-|s_\alpha|/x}}{\Delta x} = \frac{d}{dx} xe^{-|s_\alpha|/x} = \left(1 + \frac{|s_\alpha|}{x}\right) e^{-|s_\alpha|/x}.$$

(IV. 46c)

Therefore, with a rectangular beam of half-width a,

$$I(s) = I_{00} \sum_i w_i' \left[\left(1 + \frac{|s_\alpha|}{s+a}\right) e^{-|s_\alpha|/(s+a)} - \frac{(s-a)}{|s-a|}\left(1 + \frac{|s_\alpha|}{|s-a|}\right) e^{-|s_\alpha|/|s-a|} \right].$$

(IV. 47)

The extra factor $(s-a)/|s-a|$ is introduced to keep the signs correct because of the effect of the absolute-value signs in Eq. (IV. 43). Often

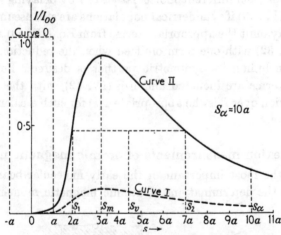

FIG. IV. 3. Deflected beam shape for a rectangular undeflected beam and for a spin of $\frac{1}{2}$. Curve I is for $s_\alpha = 10a$ and is on the same scale as the undeflected curve O. Curve II is the same as curve I except that the ordinate is multiplied by a factor of ten. s_m is the deflexion of the trace maximum while s_v is midway between the two half-maximum positions s_1 and s_2 (FRA 31).

this expression is used instead of Eq. (IV. 43) to obtain approximate beam shapes with trapezoidal beams by approximating the trapezoid with a rectangle of equal height and equal area. A beam shape calculated from Eq. (IV. 47) for a single component of $s_\alpha = 10a$ is shown in Fig. IV. 3. It is apparent from this figure that the maximum of the deflected trace s_m is at a much smaller deflexion than s_α. This can also be seen from Eq. (IV. 47) for $s_m \gg a$ by setting $dI(s)/ds = 0$, whence

$$s_\alpha = \frac{s_m^2 - a^2}{2a} 3\ln\frac{s_m+a}{s_m-a} \approx 3s_m.$$

(IV. 48)

For $s = 0$, Eq. (IV. 47) reduces to

$$I(0) = 2I_{00} \sum_i w'_i \left[1 + \frac{|s_\alpha|}{a}\right] e^{-|s_\alpha|/a}. \tag{IV. 49}$$

A typical application of Eq. (IV. 43) or the approximate Eq. (IV. 47) is in estimating the amount of beam lost by striking a magnet or beam stop at a location other than the detector position. In such a case the s_α at the position of the stop can be calculated from Eqs. (IV. 8)–(IV. 16) or from a similar equation for a different location. With this value of s_α, $I(s)$ may be calculated from Eqs. (IV. 43) or (IV. 47). If this $I(s)$ is then integrated over the values of s blocked by the beam stop, the loss of intensity is obtained. If desired, this can also be done separately for each different molecular state i by delaying the summation in Eq. (IV. 43) if symmetrical populations are still assumed. Otherwise one may omit the appropriate terms from Eq. (IV. 40) as indicated by Eq. (IV. 32) with one term omitted when the reduction in a single state not including its symmetric partner is desired. For $s > d$, of course, all terms are included and Eq. (IV. 42), with the omission of the summation over i, is the appropriate $I(s)$ for such a calculation with a single state.

IV. 2. Deflexion measurements of atomic magnetic moments

Perhaps the most important of the early molecular-beam measurements were the determinations of atomic magnetic moments and the

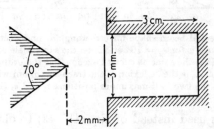

FIG. IV. 4. Stern–Gerlach inhomogeneous magnetic field (GER 24).

direct experimental demonstrations of spatial quantization which these experiments provided. The first of these experiments was that of Stern and Gerlach (STE 21, GER 24). They allowed a beam of silver atoms to pass through an inhomogeneous magnetic field region produced by an iron magnet whose cross-section was as shown in Fig. IV. 4. Had the μ_{eff} of Eqs. (IV. 2) and (IV. 5) been merely the classical component of the atomic magnetic moment, a continuous distribution of values for μ_{eff} would have resulted since all angles of orientation of the magnetic moment would have been possible classically. Consequently, from the second term of Eq. (IV. 22) and from Eq. (IV. 43), a continuous distribution of the intensity at the detector position would have been expected. Instead of this the pattern of the accumulated silver formed a double trace as in Fig. XIV. 10.

This result, however, was completely consistent with quantum theory. According to quantum mechanics (CON 35), if nuclear effects can be omitted, if the magnetic field is sufficiently weak that the Zeeman instead of the Paschen–Back limit applies, and if there is Russell–Saunders coupling,

$$\mu_{\text{eff}} = mg\mu_0, \qquad (\text{IV. 50})$$

where μ_0 is the Bohr magneton $= e\hbar/2mc$, m is the magnetic quantum number of the atom, and g is the Landé g factor (CON 35)

$$g = 1 + \frac{J(J+1)+S(S+1)-L(L+1)}{2J(J+1)}. \qquad (\text{IV. 51})$$

m can take on the $2J+1$ values

$$m = -J, -J+1,\ldots, +J, \qquad (\text{IV. 52})$$

where J is the total angular-momentum quantum number of the atom. The occurrence of a double trace indicates that $J = \tfrac{1}{2}$, as is now known to be the case since the ground state of the silver atom is $^2S_{\frac{1}{2}}$. Actually, at the time of the original Stern–Gerlach experiment, half-integral angular-momentum quantum numbers were unknown, so the experimental results at the same time provided a beautiful confirmation of the then existing quantum ideas of spatial quantization and an anomaly for that theory because of the occurrence of two, instead of the anticipated three, traces.

From the separation of the traces or from the intensity distribution in comparison with Eqs. (IV. 22), (IV. 43), and (IV. 50), the values of mg can be inferred. These values are found to agree with Eq. (IV. 51) for silver, i.e. $g = 2$, corresponding to $L = 0$ and $J = S = \tfrac{1}{2}$.

Since the time of the first Stern and Gerlach experiment many different atoms have been studied by similar techniques (TAY 26, TAY 29, KUR 29, LEU 27, PHI 27, SPE 35, STE 37, WRE 27, FRA 26, SCH 26, MEI 33, MEI 33a). Similar experiments have been performed with paramagnetic molecules such as O_2 (SCH 33). In many of these experiments improved techniques of detection and deflexion such as those described in Chapter XIV were used. However, owing to the velocity distribution, it was difficult to get accurate experimental values for mg, particularly if J were large. In principle this difficulty could be overcome by using a velocity selector to monochromatize the beam, but this often reduces the beam intensity too severely. It was found that, in all cases when there was no reason for the nuclear moment to confuse the result and when the assumed electronic couplings ought to apply, the results were in agreement with the theoretical predictions of the Landé expression of Eq. (IV. 51). A summary of these results has been given by Fraser (FRA 31). However, the effects of the nuclear spins can often be quite large; for an intermediate coupling as in Eq. (III. 93), the nuclear spin affects the orientation of the large electron magnetic moment, so the effects of the nucleus may be much larger than would at first sight be expected from the comparative sizes of the electron and nuclear magnetic moments. The effects of nuclear spins and magnetic moments in magnetic-deflexion experiments will be discussed in the remaining sections of this chapter.

IV. 3. Deflexion measurements of nuclear and rotational magnetic moments in non-paramagnetic molecules

In molecules, such as $^1\Sigma$ molecules, for which there is no resultant electronic magnetic moment, any deflexion in an inhomogeneous magnetic field must be due to (a) the nuclear magnetic moment, (b) the rotational magnetic moment which the molecule acquires by virtue of the rotational angular velocity of the molecule, and (c) any magnetic moment which may be induced diamagnetically in the molecule by the application of the external magnetic field. All of these moments are of the order of nuclear magnetons or less. Since the nuclear magneton is 1/1836 of a Bohr magneton, the deflexion sensitivity must be much greater in these experiments than in the experiments with paramagnetic atoms previously discussed.

By essentially a Stern and Gerlach experiment, of narrower beam and consequently increased sensitivity, Knauer and Stern (KNA 26a, FRA 31) made an approximate measurement of the rotational magnetic moment of the H_2O molecule by the deflexion method.

Of the $^1\Sigma$ molecules in which nuclear effects should be observable, those that have been most carefully studied by the deflexion method are H_2, D_2, and HD. Frisch and Stern (FRI 33c) and Estermann, Stern, and Simpson (EST 33a, EST 33c, EST 37) in a series of experiments measured the magnetic moments of the proton and deuteron by the deflexion of the diatomic hydrogen molecules of a collimated beam which passed through an inhomogeneous magnetic deflexion field of the Stern and Gerlach type. The rotational magnetic moments of these molecules were also determined approximately. Since these experiments were with magnetic moments of nuclear instead of Bohr magneton size, they were necessarily more difficult than those with paramagnetic atoms. Much longer deflecting fields and larger field gradients were needed. These were achieved with an apparatus which is schematically illustrated in Fig. IV. 5.

In experiments with H_2, there is no net nuclear magnetic moment in

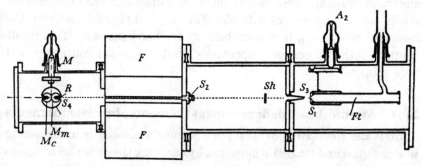

FIG. IV. 5. Proton magnetic moment apparatus of Estermann, Simpson, and Stern. S_1 is the source slit, S_2 the collimating slit, S_3 the foreslit, S_4 the receiver slit, Ft the gas-feeding tube, Sh the shutter, and F the magnet pole pieces (EST 37).

para-H_2 [the form with antisymmetric spin-wave functions or with resultant spin angular momentum $I = 0$ (PAU 35, RAM 53e)]. Consequently, the nuclear magnetic moment in para-H_2 cannot be measured and the measurements must be for ortho-H_2, for which the spin wave function is symmetric in the nuclei and for which $I = 1$. However, the $^1\Sigma$ electronic state is symmetric in the interchange of the two protons and the protons must satisfy Fermi–Dirac statistics. Consequently, the orthohydrogen can exist only in odd rotational states for which the lowest is $J = 1$. Therefore, the H_2 molecules with a resultant nuclear moment also necessarily possess a rotational magnetic moment. The net deflexion is then due to the combination of these two moments. A correction for the contribution of the rotational magnetic moment to the deflexion was then obtained from some auxiliary experiments. In these pure parahydrogen, for which there was no nuclear contribution, was used. From these experiments, the rotational magnetic moment was calculated. This value could then be used to correct for the rotational magnetic-moment contribution in the orthohydrogen experiments. In this way, the nuclear magnetic moments of the proton and deuteron were obtained as well as the rotational magnetic moment of the molecule. The results were in general agreement with the much more accurate later values. The measurement of the proton moment by this method presumed most accurate, however, was $2 \cdot 46 \pm 0 \cdot 08$ nuclear magnetons, which was considerably lower than the present best value $2 \cdot 792743 \pm 0 \cdot 000060$ nuclear magnetons, as discussed in Chapters VI and IX.

Deflexions from pure diamagnetic moments have been studied by Kellogg and Ramsey (KEL 38) in neon and argon, which possess no electronic, nuclear, or rotational magnetic moment. Their experimental results were in agreement with Eqs. (IV. 43) and (IV. 22) and with the assumption that μ_{eff} is due exclusively to diamagnetism. This result supported the theoretical expectation that the nuclei concerned had zero spin.

IV. 4. Atomic-beam deflexion measurements of nuclear moments

Breit and Rabi (BRE 31) first pointed out the possibility of measuring nuclear spins and magnetic moments by studying the deflexion of beams of atoms which have electron moments and are consequently much more easily deflected in an inhomogeneous field (MEI 33, MEI 33a). Data on nuclear spins and moments are obtainable from such studies since, as discussed in detail below, the magnitude of the nuclear moment determines the magnitude of the coupling between the nuclear spin and the electrons; the nuclear magnetic moment thereby also affects the strength of the externally-applied magnetic field which is necessary to produce a given stage of intermediate coupling.

From Eqs. (IV. 22) and (IV. 43), the value μ_{eff} can be obtained from the observed deflexion pattern. However, from Eq. (IV. 4),

$$\mu_{\text{eff}} = -\frac{\partial W}{\partial H_0}, \qquad (IV. 53)$$

where W is the energy of the atom in the magnetic field and is calculated theoretically by the methods discussed in Section III. 4. Typical dependencies of W on H_0 are given in Figs. (III. 4) and (III. 5). In the particular case of atoms whose resultant electronic angular-momentum quantum number J equals $\frac{1}{2}$ but whose nuclear spin is arbitrary, the theoretical dependence of W on H_0 is given in Eq. (III. 120). From this and Eq. (IV. 53),

$$\mu_{\text{eff}} = \frac{\mu_I}{I}m \mp \frac{x+2m/(2I+1)}{2\{1+(4m/(2I+1))x+x^2\}^{\frac{1}{2}}}\left(-\frac{\mu_J}{J}+\frac{\mu_I}{I}\right), \qquad (IV. 54)$$

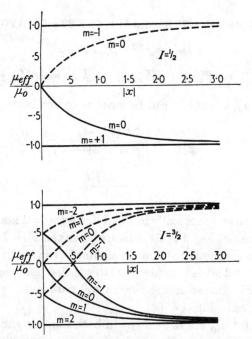

FIG. IV. 6a. Variation of effective magnetic moment with magnetic field. The dotted lines are the moments of the magnetic levels arising from the $F = I - \frac{1}{2}$ state. The nuclear moment is assumed positive (RAB 36).

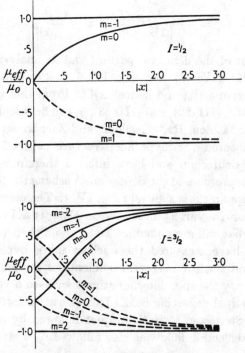

FIG. IV. 6b. Same as Fig. 6a except that the nuclear magnetic moment is assumed negative (RAB 36).

where $\qquad x = (-\mu_J/J + \mu_I/I)H_0/\Delta W \qquad$ (IV. 55)

and $\qquad \Delta W = W^0_{F=I+\frac{1}{2}} - W^0_{F=I-\frac{1}{2}} = h\Delta\nu = ha(I+\frac{1}{2}). \qquad$ (IV. 56)

Since $\mu_I \ll \mu_J$, the above can be written approximately as

$$\mu_{\text{eff}} = \mp \frac{x + 2m/(2I+1)}{2\{1 + (4m/(2I+1))x + x^2\}^{\frac{1}{2}}}\left(-\frac{\mu_J}{J}\right), \qquad \text{(IV. 57)}$$

where $\qquad x = -\dfrac{(\mu_J/J)H_0}{\Delta W}. \qquad$ (IV. 58)

This result is shown graphically in Fig. IV. 6 for $J = \frac{1}{2}$ and $I = \frac{1}{2}$ and $\frac{3}{2}$. The dashed lines of Fig. IV. 6a correspond to the $F = I - \frac{1}{2}$ state provided that the magnetic moment is assumed positive (if it is negative this statement still holds provided that the dashed lines are interchanged in position with corresponding curved full lines and the straight lines are unaltered, as in Fig. IV. 6b).

The curves for μ_{eff} are dependent in both number and shape on the value of the nuclear spin. Consequently, the nuclear spin can be determined from an analysis of the deflexion pattern. Furthermore, the abscissae of the curves are

$$|x| = \frac{-\mu_J/J + \mu_I/I}{|\Delta W|}H_0 \approx -\frac{(\mu_J/J)H_0}{|\Delta W|}. \qquad \text{(IV. 59)}$$

A measurement of the deflexion pattern and its analysis with the aid of Eqs. (IV. 22), (IV. 43), and Fig. IV. 6 at a suitable intermediate field then determines $|x|$ and hence $|\Delta W|$. From this value for $|\Delta W|$ and Eqs. (IV. 56), (III. 82), and (III. 83), μ_I can be calculated.

In the above fashion, Rabi, Kellogg, and Zacharias (RAB 34, RAB 34a) have measured the hyperfine-structure separations in atomic hydrogen and deuterium and have inferred therefrom the magnetic moments of the proton and the deuteron. A schematic diagram of their experimental apparatus is shown in Fig. IV. 7. The magnetic field used was of the two-wire variety described in Chapter XIV. In a related though somewhat different manner Rabi and Cohen (RAB 33, RAB 34b, COH 39) have measured the nuclear spins, hyperfine-structure separations, and nuclear magnetic moments of sodium and caesium. In order to clarify the spin determination they used a velocity selector before the principal deflecting field. This velocity selector consisted of a strong field Stern and Gerlach magnet followed by a selecting slit. The principal deflecting field was also followed by a strong field Stern and Gerlach magnet for the purpose of increasing the intensity by a refocusing action.

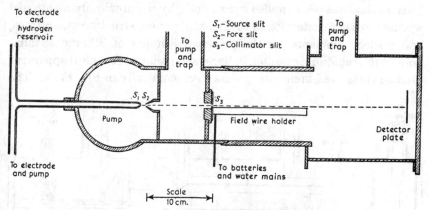

Fig. IV. 7. Apparatus of Rabi, Kellogg, and Zacharias (RAB 34).

IV. 5. Atomic-beam zero-moment method for nuclear moments

A major evil of the atomic-beam deflexion method is that the amount of deflexion suffered by an atom in passing through the inhomogeneous field is dependent on the velocity of the atom, which is different for different atoms as a result of the velocity distribution in the beam. Consequently, the deflexion pattern, instead of being sharp, is smeared, and accurate observations are difficult. Some improvement can be achieved with velocity selection, as discussed at the end of the preceding section, but this is necessarily accompanied by a great loss in beam intensity.

These difficulties have been effectively overcome by the so-called zero-moment method used by Cohen and Millman (COH 34, MIL 35, MIL 36), Fox and Rabi (FOX 35), Manley (MAN 36, MAN 37), and others (MIL 38, HAM 39, REN 40). From the curves of Fig. IV. 6, it can be seen that for a nuclear spin greater than $\frac{1}{2}$ there are one or more values of H for which $\mu_{\text{eff}} = 0$. For these values there will be no deflexion and this will be true for atoms of all velocities. The beam intensity as measured at the undeflected position will therefore rise through large velocity-independent maxima at these values of H_0. The positions of these maxima will also depend on H_0 and not on $\partial H_0/\partial z$, which is measured with more difficulty. From the numbers of such zero-moment maxima and from their relative spacing (supplemented by deflexion-pattern measurements if $I \leqslant \frac{3}{2}$) the nuclear spin is determined.

From Fig. IV. 6 it is apparent that the zero-moment peaks occur for specific values of $|x|$. From these values of $|x|$ and from the values of H_0 at which they occur, $|\Delta W|$ can be calculated from Eq. (IV. 59). This method has been applied successfully by the previously-mentioned workers to a determination of the spins, hyperfine-structure separations, and nuclear moments for the principal isotopes of lithium, sodium, potassium, rubidium, caesium, indium, and gallium. A typical apparatus used in these measurements is shown schematically in Fig. IV. 8. The

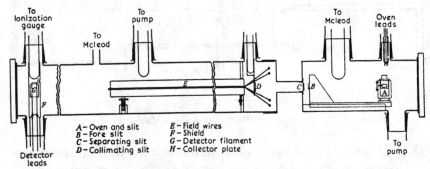

FIG. IV. 8. Millman's apparatus for the study of potassium by the zero-moment method (MIL 35).

two-wire field described in Chapter XIV was used in this apparatus. A typical result with potassium is plotted in Fig. IV. 9 and with indium and a much higher resolution apparatus in Fig. IV. 10.

In all of the preceding atomic-beam methods the nuclear magnetic moment was determined from the hyperfine-structure separation and was therefore uncertain to the extent that the field at the nucleus due to the electrons was not accurately known but was only approximated by formulae similar to Eq. (III. 84). However, Millman, Rabi, and Zacharias (MIL 38) overcame this difficulty in their high-resolution experiments with indium, whose ground electronic state is $^2P_{\frac{1}{2}}$. The success of their method depended on the difference between the exact Eq. (IV. 54) and the approximate Eq. (IV. 57). According to the approximate equation or the resulting Fig. IV. 6, the zero moments for states with the same value of m occur at the same value of $|x|$. However, with the exact Eq. (IV. 54), the zero values of μ_{eff} are at slightly different values of $|x|$ for these states and consequently at slightly different values of H_0. From Eqs. (IV. 54) and (IV. 55), the difference ΔH_0 between the fields for the two zero moments for $J = \frac{1}{2}$ is given approximately by

$$\frac{\Delta H_0}{H_0} = \frac{2I+1}{I}\frac{\mu_I}{\mu_J}\left[1-\left(\frac{2m}{2I+1}\right)^2\right]^{\frac{1}{2}}. \qquad \text{(IV. 60)}$$

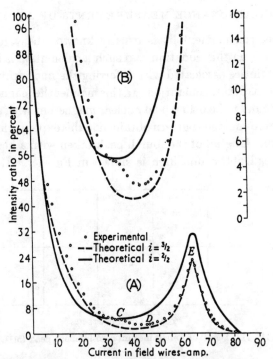

Fig. IV. 9. Variation of beam intensity with field at the position of zero deflexion. 1 ampere equals 1·29 gauss. Curve B is the same as part of curve A but is plotted with an expanded ordinate scale to show the effect of K^{41} (MIL 35).

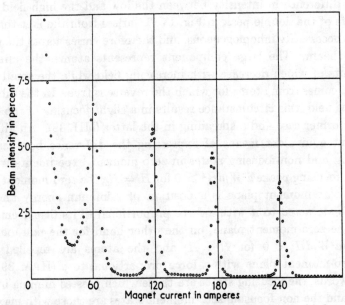

Fig. IV. 10. Plot of indium beam intensity against magnet current with detector fixed at the centre of the beam (MIL 38).

Since the electron moment μ_J is usually known, the value of μ_I can be calculated from this equation and such a calculation is independent of the uncertainties associated with inferring the nuclear moment from the hyperfine-structure separation. In this manner the magnetic moment of In^{115} was found to be $6\cdot40\pm0\cdot20$ nuclear magnetons with a spin of 9/2. The overall zero-moment pattern obtained in this experiment is as shown in Fig. IV. 10. A detail of the fourth peak taken with a beam that was 0·7 mm instead of 2·5 mm high is shown in Fig. IV. 11. The direct

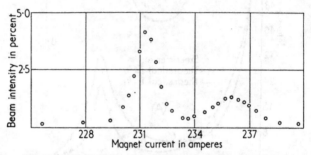

Fig. IV. 11. Detail of fourth indium zero-moment peak (MIL 38).

evaluation of the nuclear magnetic moment depends on the separation of these doublets.

The difference in intensity between the low and the high field components of the double peak in Fig. IV. 11 arises from the fact that the field is necessarily inhomogeneous, and therefore varies across the width of the beam. The large component represents atoms, the effective moments of which decrease with increasing field while the small component comes from atoms for which the reverse is true. In the inhomogeneous field, this circumstance results in a slight focusing of the beam in the former case and a spreading in the latter (MIL 38). An alternative but equivalent means of expressing the difference between the focusing and non-focusing states in zero-moment experiments is that for the focusing peaks $d^2W/dH^2 > 0$ for $H = H_0$, the zero-moment field, so the zero-moment plane is a position of minimum energy and the atom is attracted to it with a force proportional to its displacement y from the zero-moment plane; on the other hand for the non-focusing states $d^2W/dH^2 < 0$ for $H = H_0$ and the atoms are repelled from the zero-moment plane with a force proportional to y (HAM 39). In other words, the focusing states are those which possess minima in Fig. III. 4 and the non-focusing zero-moment states are those with maxima. For $J = \frac{1}{2}$ the focusing component of the doublet occurs at lower H

regardless of the sign of the nuclear moment, as can be inferred from Eq. (IV. 54) if both the explicit and the implicit changes resulting from a reversal of the nuclear sign are made.

In their experiments with indium, Millman, Rabi, and Zacharias (MIL 38) noticed some very small peaks on the early part of the curve of Fig. IV. 10. These they correctly attributed to indium atoms whose electronic states were $^2P_{\frac{3}{2}}$ since about 19 per cent. of the atoms should

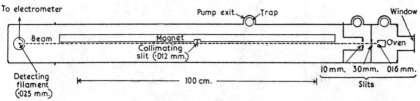

FIG. IV. 12. Schematic top view of indium and gallium zero-moment high-resolution apparatus (HAM 39).

be in this metastable state at the 1,500° K oven temperatures used. Hamilton and Renzetti (HAM 39, REN 40) studied these peaks from the $^2P_{\frac{3}{2}}$ state with a specially-built high-resolution molecular-beam apparatus, shown schematically in Fig. IV. 12, which used the two-wire field described in Chapter XIV. The nature of these peaks is shown in Fig. IV. 13. From an analysis of this result Hamilton and Renzetti found departures from the interval rule since, with $J = \frac{3}{2}$, quadrupole interactions could be observed. In this way they determined both of the constants a and b_1 in Eq. (III. 122). The calculation of the theoretical zero-moment fields is, of course, much more complicated for $J = \frac{3}{2}$ and with quadrupole interactions; these calculations are given by Hamilton and Renzetti (HAM 39, REN 40). The same experimenters also studied the isotopes of gallium in electronic $^2P_{\frac{3}{2}}$ states by the same method (REN 40).

IV. 6. Atomic-beam refocusing method for nuclear moments

Unfortunately the zero-moment method is not applicable for a spin of $\frac{1}{2}$ since from Fig. IV. 6 the only zero moment in this case occurs for $H_0 = 0$, at which there is no deflexion even of a non-zero moment. The zero-moment method is therefore not applicable to the important case of atomic hydrogen. Kellogg, Rabi, and Zacharias (KEL 36), however, devised a refocusing method which shared the zero-moment method's advantage of being velocity-independent, although, unlike the zero-moment method, a difficult determination of $\partial H_0/\partial z$ was still required.

The apparatus used is shown schematically in Fig. IV. 14. An initial

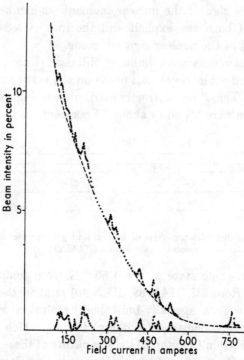

FIG. IV. 13. Indium beam intensity at position of zero deflexion as a function of field current. The small peaks are due to $^2P_{\frac{3}{2}}$ atoms and the large background is due to the first peak of the $^2P_{\frac{1}{2}}$ atoms.

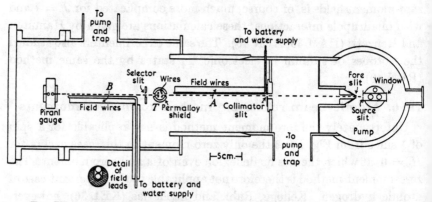

FIG. IV. 14. Diagram of apparatus used for the determination of the sign of the nuclear magnetic moments of hydrogen and deuterium. The permalloy shield and wires were removed from the apparatus for the measurement of the magnitudes of the moments (KEL 36).

deflexion was produced with a weak magnetic field A so μ_{eff}/μ_0 was approximately $\frac{1}{2}$ for the inner states of Fig. IV. 6. The beam was then allowed to pass into a strong field region B with inhomogeneity reversed. From Fig. IV. 6, $\mu_{\text{eff}} = \mu_0$ in this case and hence is known. From the value of the gradient of the second field required exactly to compensate the deflexion by the first field and from the gradient of the first field and the geometrical dimensions of the apparatus, the value of μ_{eff} could be calculated by Eq. (IV. 22). From this and Fig. IV. 6, $|x|$ could be obtained and the value of the H to produce this $|x|$ in the first field could be measured; hence $|\Delta W|$ could be found from Eq. (IV. 59). The value of μ_I could then be inferred from Eq. (III. 82). The wires inside the permalloy shield of Fig. IV. 14 were for the purpose of measuring the sign of the magnetic moment of the proton as discussed in the next section.

IV. 7. Atomic-beam measurements of signs of nuclear moments

From Fig. IV. 6 it is apparent that the sign of the nuclear moment can be determined if it is possible to tell whether μ_{eff} of the state $F = I - \frac{1}{2}$ is positive or negative in strong fields. The identification of which state has $F = I - \frac{1}{2}$ can be made because this state can never have $m = \pm(I+\frac{1}{2})$, whereas the state $F = I+\frac{1}{2}$ can have all possible values of m. Consequently if one of the intermediate states of Fig. IV. 6 (for example the $m = 0$ state) is subjected to a low field whose direction changes rapidly so that non-adiabatic transitions are made to states of other m values but with the selection rule $\Delta F = 0$ being satisfied, the values of $m = \pm(I+\frac{1}{2})$ can be achieved if $F = I+\frac{1}{2}$ but not if $F = I-\frac{1}{2}$. The experimental determination can then be made in an apparatus like that in Fig. IV. 14. In the A field the atoms are deflected by a weak inhomogeneous magnetic field such that $|x| \sim \frac{1}{2}$. The indicated selector slit then is set to pass only those molecules whose μ_{eff} in the A field was, say, positive. The strong inhomogeneous B field is then set to refocus one of these states, say that with $m = 0$. Steady current may then pass through wires inside the permalloy shield to produce a weak magnetic field whose direction as seen by the moving molecule seems to change rapidly. By the processes discussed in greater detail in the next two chapters, this rapid change in direction will induce non-adiabatic transitions between the states of different m values possible for the fixed value of F. If $F = I-\frac{1}{2}$ for the selected state, all of the states arising from the non-adiabatic transitions will have the same value of μ_{eff} in the strong field by Fig. IV. 6 and the refocusing will be as good. On the other hand, if $F = I+\frac{1}{2}$ for the selected state, one of the resulting states will have the opposite sign of μ_{eff} in the B field, and the refocusing will be reduced. Therefore, an experimental determination of whether the transition field current reduces the refocused beam intensity determines which form of Fig. IV. 6 is applicable and hence which sign is correct for the nuclear magnetic moment. The possibility of determining signs of nuclear moments in this manner was first suggested by Rabi (RAB 36).

In this way Kellogg, Rabi, and Zacharias (KEL 36) measured the signs of the magnetic moments of the proton and deuteron. In a similar fashion, Torrey (TOR 37), Gorham (GOR 38), Millman and Zacharias (MIL 37), and Gorham (GOR 38) determined the signs of the nuclear magnetic moments of lithium, potassium, rubidium, and caesium.

IV. 8. Atomic-beam space focusing measurements of nuclear spins

Bennewitz and Paul (ZZA 11) have recently adapted their atomic-beam space focusing techniques to the measurement of nuclear spins and hyperfine structure separations of radioactive isotopes. These focusing techniques are described in Sections V. 5.1 and XIV. 4.7. When suitable weak focusing fields are applied the atoms are in states of intermediate coupling between I and J, and the different orientation states are focused on different concentric rings (ZZA 11). The spins may be determined by counting the number of rings, while $\Delta \nu$ is related to the spacing of the rings.

V
MOLECULAR-BEAM MAGNETIC RESONANCE METHODS

V. 1. Introduction

THE molecular beam magnetic resonance was first introduced by Rabi and his associates (RAB 38, KEL 39) in 1938. An introductory description of the method has been given in Chapter I. However, a number of related experiments were performed prior to 1938. These include experiments such as those described at the end of the preceding chapter in which non-adiabatic transitions were induced to allow the determination of the signs of the nuclear moments.

The first suggestion for the study of non-adiabatic transitions in molecular beams was made by Darwin (DAR 28) in 1928. His interest was in the study of the quantum aspects of the non-adiabatic transitions. In 1931 Phipps and Stern (PHI 31) performed the first experiments of this nature on paramagnetic atoms which passed through weak magnetic fields whose direction varied rapidly in space. Güttinger (GUT 31) and Majorana (MAJ 32) discussed the theory of such molecular beam experiments (the transitions are often called Majorana transitions). Frisch and Segré (FRI 33) continued the experiments with non-adiabatic transitions and found, in agreement with Güttinger's and Majorana's theories, that transitions took place when the rate of change of the direction of the field was large compared with the Larmor frequency $\omega_0 = \gamma_I H_0$, or was comparable to it, and did not otherwise. However, some of the results of Frisch and Segré were not consistent with theoretical expectations when only the effect of the electronic moment was included. Rabi (RAB 36) pointed out that this discrepancy arose from the effects of the nuclear magnetic moment since some of the transitions were performed in such weak fields that weak or intermediate coupling between the nuclei and the electrons prevailed, i.e. the left portions of the curves in Fig. IV. 6 were applicable. The transitions under such circumstances would be quite different from those where the effects of the nuclear spins could be neglected. Rabi showed that the results of Frisch and Segré were consistent with expectations if the effects of the nuclei were included as in Fig. IV. 6. He also pointed out that such non-adiabatic transitions could be used to identify the states and hence to determine the signs of the nuclear magnetic moments as discussed at the end of the preceding chapter. In other papers, Motz (MOT 36), Rabi (RAB 37), and Schwinger (SCH 37a) discussed theoretical means for calculating the transition probability for molecules which passed through a region in which the direction of the field varied rapidly.

In all of the above experiments, however, the direction of the field varied in space and the only time variation arose as the atom passed through the region. Since the atoms possessed a Maxwellian distribution of velocities, the apparent frequencies of the changing field were different for different molecules. Also, the change ordinarily went through only a portion of a cycle so no sharp resonance effects could be expected. The first suggestion that transitions be induced by an oscillatory field from a radiofrequency oscillator was made in 1936 by Gorter (GOR 36). He proposed to detect the transitions by the absorption of the radiofrequency radiation and its reaction on the radiofrequency circuits. Although Purcell, Torrey, and Pound (PUR 46) and Bloch, Hansen, and Packard (BLO 46) achieved success with this method of detection in 1946, Gorter's experiments were unsuccessful in 1936. Rabi (RAB 38, RAB 39, KEL 39a) then proposed the use of such an oscillator driven magnetic field component as the transition inducing field of a molecular beam refocusing experiment. With such a radiofrequency field, the apparent frequency would be almost the same for all molecules regardless of velocity and each molecule would be exposed to a large number of cycles. As a result of these two features sharp resonances are obtained and the transitions occur only in the immediate neighbourhood of certain discrete frequencies. For this reason the technique is called the molecular beam resonance method or molecular-beam radiofrequency spectroscopy.

V. 2. Magnetic resonance method

The basic principles and apparatus for the magnetic resonance method have already been described in Section I. 2 and need not be repeated here. In its original conception it was only for measurements of nuclear magnetic moments (RAB 39, KEL 39a), but it soon became apparent in the experiments of Kellogg, Rabi, Ramsey, and Zacharias (KEL 40) that the method was a general technique for radiofrequency spectroscopy. The nuclear moment measuring application will be discussed in detail in Chapter VI, but for the sake of suitable generality the historical order will be reversed and the resonance method will be discussed in the present chapter in its general radiofrequency spectroscopy form.

As pointed out in Chapter I, the oscillator frequency at which a resonance transition between states of energy W_p and W_q can occur is

$$\nu_0 = (W_q - W_p)/2\pi\hbar. \quad (\text{V. 1})$$

However, in detailed molecular beam considerations, one is concerned not only with the frequency of the resonance but also with the width of the resonance line and the probability of the transition at resonance. For these reasons a calculation of transition probabilities at, and in the neighbourhood of, resonance will be given in the next section. Incidental to this calculation, it will also be shown that the transition probability is a maximum when the oscillator frequency satisfies Eq. (V. 1).

By way of an introduction to the subsequent quantitative calculations, a qualitative physical discussion of the resonance process should be informative. The simplest case is that of a magnetic moment associated with such a large angular momentum that the problem can be treated classically. Consider that such a magnetic moment enters a region of fixed strong homogeneous magnetic field H_0. In this field the magnetic moment will precess about H_0 with the Larmor angular frequency

$$\omega_0 = \frac{\mu_I}{\hbar I} H_0 = \gamma_I H_0 \quad (\text{V. 1}a)$$

as shown in the simple derivation of the precession frequency of a classical top or gyroscope acted upon by a torque. Consider an added weak field H_1 perpendicular to H_0 whose direction is allowed to rotate about an axis parallel to H_0. At any instant the nuclear moment will also tend to precess about H_1 and thereby to change the angle ϕ between μ_I and H_0. However, if the field H_1 rotates at a different rate from that of Eq. (V. 1a) at which the nucleus precesses, the angle ϕ will sometimes tend to increase and sometimes to decrease and on the average will not change. On the other hand, if the rate of rotation of H_1 equals the Larmor frequency, H_1 and the nucleus will precess about H_0 together, so the changes in ϕ will be cumulative and a net change in ϕ will be achieved. This process whereby ϕ changes is related to that which leads to the 'sleeping' of a top. The change in ϕ can be detected by a failure of refocusing in the molecular-beam apparatus, discussed in Chapter I and illustrated schematically in Fig. I. 3. The minimum in refocused beam intensity then occurs when the oscillator angular frequency equals the Larmor frequency of Eq. (V. 1a), which for this case is equivalent to Eq. (V. 1) as shown in Eq. (I. 3). A continuation of the discussion of the magnetic moment case will be given in the next chapter; the remainder of this chapter will be devoted to a discussion of transition probabilities in the general case of radio-frequency spectroscopy.

V. 3. Transition probabilities

V. 3.1. *Transition probability for individual molecules.* In the present section the transition probability between two general states a and b will be calculated for a molecule subject to a suitable oscillatory perturbation. As a result, the calculation is necessarily a formal mathematical one. For a more intuitive and physically interpretable derivation of a special case of the same problem the reader is referred to Chapter VI where the magnetic moment case is derived independently. Theoretical discussions related to molecular beam resonance transition probabilities have been given by Rabi (RAB 37), Schwinger (SCH 37a), Bloch and Siegert (BLO 40), Stevenson (STE 40), Torrey (TOR 41), Ramsey (RAM 50c, RAM 51d), Kruse and Ramsey (KRU 51), Rabi, Ramsey, and Schwinger (RAB 54), and others (CAL 39).

Consider a molecular system which at time $t = 0$ enters a region where it is subjected to an oscillatory perturbation that induces transitions between two molecular energy eigenstates p and q of energies W_p and W_q for neither of which states can there be appreciable transitions to any other state while the system is subjected to the perturbing field. Assume that the perturbation V is of such a form that its matrix elements are

$$V_{pq} = \int \psi_p^* V \psi_q \, d\tau = \hbar b e^{i\omega t}, \quad V_{qp} = \hbar b e^{-i\omega t}, \quad V_{pp} = V_{qq} = 0. \quad (V.\ 2)$$

As will be shown in the next chapter such a perturbation arises, for example, if the transitions of a system containing magnetic moments are induced by a magnetic field rotating with angular velocity ω about the axis of quantization. It should be noted that for many pairs of states p and q, the above perturbation will vanish. In particular, if the perturbation is induced by an oscillatory magnetic field, the b in Eq. (V. 2) vanishes except for allowed magnetic dipole transitions. For an atom in an external magnetic field so strong that the nuclear angular momentum $\mathbf{I}$ and the electronic angular momentum $\mathbf{J}$ are decoupled from each other, the selection rules on the corresponding magnetic quantum numbers are

$$\Delta m_I = 0, \pm 1; \quad \Delta m_J = 0, \pm 1. \quad (V.\ 2a)$$

If, on the other hand, the external field is weak so that the quantum number F of $\mathbf{F} = \mathbf{I}+\mathbf{J}$ is an approximately good quantum number, the selection rules are

$$\Delta F = 0, \pm 1; \quad \Delta m = 0, \pm 1. \tag{V. 2b}$$

If the wave function of the system making transitions between the states p and q is written as

$$\Psi(t) = C_p(t)\psi_p + C_q(t)\psi_q \tag{V. 3}$$

the time-dependent Schrödinger equation is

$$i\hbar \partial \Psi/\partial t = \mathscr{H}\Psi = \mathscr{H}_0\Psi + V\Psi. \tag{V. 4}$$

After multiplication by ψ_p^* or ψ_q^* and integration over all space these reduce with the aid of the orthogonality and normality conditions to

$$i\hbar \dot{C}_p(t) = W_p C_p(t) + \hbar b e^{i\omega t} C_q(t)$$
$$i\hbar \dot{C}_q(t) = \hbar b e^{-i\omega t} C_p(t) + W_q C_q(t). \tag{V. 5}$$

Now assume at time $t = 0$ that

$$C_p(0) = 1, \quad C_q(0) = 0. \tag{V. 6}$$

Then at time t the solution is

$$C_p(t) = (i\cos\Theta \sin\tfrac{1}{2}at + \cos\tfrac{1}{2}at)\exp\{i[\tfrac{1}{2}\omega - (W_p+W_q)/2\hbar]t\}$$
$$C_q(t) = +i\sin\Theta \sin\tfrac{1}{2}at \exp\{i[-\tfrac{1}{2}\omega - (W_p+W_q)/2\hbar]t\} \tag{V. 7}$$

where

$$\cos\Theta = (\omega_0-\omega)/a, \quad \sin\Theta = -2b/a,$$
$$a = [(\omega_0-\omega)^2+(2b)^2]^{\frac{1}{2}}, \quad \omega_0 = (W_q-W_p)/\hbar. \tag{V. 8}$$

The proof of the above is that Eq. (V. 7) clearly satisfies the initial conditions of Eq. (V. 6) and by substitution in Eq. (V. 5) it can be seen directly that the differential equations are satisfied.

The probability $P_{p,q}$ of a transition from state p to state q in view of Eq. (V. 6) is then simply given by

$$P_{p,q} = |C_q(t)|^2 = \sin^2\Theta \sin^2\tfrac{1}{2}at \tag{V. 9}$$

or from Eq. (V. 8)

$$P_{p,q} = \frac{(2b)^2}{(\omega_0-\omega)^2+(2b)^2} \sin^2\{\tfrac{1}{2}[(\omega_0-\omega)^2+(2b)^2]^{\frac{1}{2}}t\}. \tag{V. 10}$$

It can be seen that the first factor in Eq. (V. 10) does indeed reach a maximum of unity when

$$\omega = \omega_0 = (W_q-W_p)/\hbar. \tag{V. 11}$$

Likewise, the frequency full half-width of this factor is $\Delta\omega = 4b$. However, the behaviour when both factors of Eq. (V. 10) are taken into account is less obvious and is dependent on t. This behaviour can be calculated for various choices of t. When $t = \pi/2b$ the transition probability as a function of frequency varies as in the dashed curve of Fig. V. 1. In the relevant molecular beam problems it is not usually Eq. (V. 10) that is desired but its average over the velocities in the beam. The procedures for taking these averages and the results obtain-

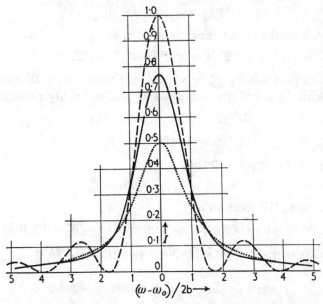

FIG. V. 1. Theoretical line shapes. The intensity I is plotted against $(\omega - \omega_0)/2b$. The dashed curve assumes a uniform velocity and optimum perturbation $(2bl/v = \pi)$. The full curve is averaged over the velocity distribution with optimum perturbation $(2bl/\alpha = 1\cdot200\pi)$. The dotted curve is averaged over the velocity distribution for a large perturbation $(2bl/\alpha \gg \pi)$ (TOR 41).

able therefrom are discussed in the next section. One limiting case can, however, be seen immediately. If $bt \gg \pi$ the effect of the velocity distribution will be merely to replace the $\sin^2$ factor by its average $\langle \sin^2 \rangle = \frac{1}{2}$. Therefore, the transition probability is just $\frac{1}{2}$ times the first factor of Eq. (V. 10) so that resonance frequency and half-width are exactly as discussed at the beginning of this section. The transition probability in this case is plotted as a function of frequency as the dotted curve in Fig. V. 1.

It should be noted that the problem discussed above is not the most general one possible since only two states were assumed to be effective and since the form of perturbation in Eq. (V. 2) is not the most general one possible. For this reason some comment about the significance of the restrictive assumptions is in order.

One of these restrictions is the omission of spontaneous or other transitions induced by any perturbations not included in Eq. (V. 2). For most molecular-beam experiments the omission of spontaneous transitions is a fully valid approximation. The probability A_{pq} of a molecule in state p making a spontaneous transition to a lower state q with the emission of a quantum of frequency $\nu_{pq} = (E_p - E_q)/h$ is (CON 35, p. 93)

$$A_{p,q} = \frac{64\pi^4 \nu_{pq}^3}{3hc^3} |(p|\mathbf{\mu}|q)|^2, \qquad (\text{V. 12})$$

where $\mathbf{\mu}$ is the dipole moment which leads to the transition. In a typical molecular-beam experiment $\mathbf{\mu}$ is the magnetic dipole moment of the nucleus so the order of magnitude of the matrix element is 1 nuclear magneton or 5×10^{-24} erg gauss^{-1} and in a field of 10,000 gauss

$$\nu_{pq} \approx 5 \times 10^{-24} \times 10^4 \times (6 \cdot 6 \times 10^{-27})^{-1} = 8 \times 10^6 \text{ c/s}.$$

Therefore,

$$A_{p,q} \approx \frac{64\pi^4 \times (8 \times 10^6)^3 \times (5 \times 10^{-24})^2}{3 \times 6 \cdot 6 \times 10^{-27} \times (3 \times 10^{10})^3} \approx 1 \cdot 5 \times 10^{-28} \text{ sec}^{-1}.$$

(V. 13)

In other words about 10^{28} seconds is required for a spontaneous transition, which is vastly longer than the 10^{-3} seconds that a molecule moving at 10^5 cm/sec takes to go the length of a typical molecular-beam apparatus. Effects of spontaneous radiation have been considered by Lamb (LAM 50a) in his atomic hydrogen-beam experiments discussed in Chapter XII.

A second restrictive approximation is the omission of possible oscillatory diagonal matrix elements in Eq. (V. 2), i.e. the assumption that $V_{pp} = V_{qq} = 0$ instead of say $\hbar d e^{i\omega t}$. Torrey (TOR 41), however, has considered such diagonal oscillatory matrix elements and has shown that, if $d \ll \omega_0$, the effects of such oscillatory diagonal elements are negligible near resonance.

Another approximation has been the inclusion of only two states p and q above. If these were the only pair of states to which Eq. (V. 11) applied even approximately then this approximation would indeed be valid near the resonance of Eq. (V. 11) since transitions to other states would not be appreciably induced. However, this is not always the case. A particularly important case is that of a magnetic moment of spin greater than $\tfrac{1}{2}$ in an external magnetic field, for which the energy differences of all successive orientation states are the same so if one transition is near resonance they all are. This case is discussed in detail in the next chapter and leads to the so-called Majorana formula (MAJ

32). The invalidity of this assumption also leads in certain cases to the possibility of multiple quanta transitions which have been observed by Kusch (KUS 54a) and which are discussed in greater detail in Section V. 4.3 (RAM 50e, ZZA 46, ZZA 8, ZZA 47).

A final approximation is the assumption that only a single frequency ω is present and that the perturbation is of the form of Eq. (V. 2). Bloch and Siegert (BLO 40) and Stevenson (STE 40) considered the special case of Eq. (V. 2) being replaced by

$$V_{pq} = \hbar b(e^{i\omega t}+e^{-i\omega t}), \quad V_{qp} = \hbar b(e^{-i\omega t}+e^{i\omt}) \qquad (V. 14)$$

since this case is the one which arises when transitions of nuclear moments are induced by an oscillating instead of a rotating magnetic-field component (an oscillating field can be resolved into two fields which rotate in opposite directions at the same frequency as in Eq. (V. 14)). They showed that the effect of the component rotating in the non-resonant direction was essentially negligible near the resonance frequency of the other component; the shift $\omega_r - \omega_0$ of the resonance frequency ω_r due to the presence of the non-resonant oppositely rotating component was shown to be

$$\omega_r - \omega_0 = (2b)^2/4\omega_0. \qquad (V. 15)$$

For typical values of $\omega_0 = 10^7$ cycles sec^{-1} and $2b = 10^4$ cycles sec^{-1} as in a strong field nuclear resonance experiment, the shift in resonance frequency is a negligible 2·5 cycles sec^{-1}. However, for low values of ω_0 this correction may become appreciable.

Ramsey (RAM 50e, KOL 52, ZZA 46) has considered the more general case which also arises frequently in practice of there being present simultaneously two components whose frequencies and perturbations are arbitrary. Let Eq. (V. 2) be replaced by

$$V_{pq} = \hbar b e^{i\omega t}+\hbar b_1 e^{i\omega_1 t}, \quad V_{qp} = \hbar b e^{-i\omega t}+\hbar b_1 e^{-i\omega_1 t}. \qquad (V. 16)$$

Then for $|2b_1| \ll |\omega_0-\omega_1|$ and $|2b| \ll |\omega_0-\omega_1|$, Ramsey (RAM 50e) showed that the transition probability as a function of ω achieves its maximum at $\omega = \omega_r$ where

$$\omega_r-\omega_0 = (\omega_0-\omega_1)[\sqrt{\{1+(2b_1)^2/(\omega_0-\omega_1)^2\}}-1]$$

$$= \frac{(2b_1)^2}{2(\omega_0-\omega_1)}. \qquad (V. 17)$$

It should be noted that this result includes Eq. (V. 15) as a special case applicable when $\omega_1 = -\omega_0$ and $b_1 = b$. It also shows that the effect in shifting the resonance becomes larger as ω_1 approaches ω_0.

V. 3.2. *Transition probability averaged over velocity distribution.*
The transition probability of Eq. (V. 9) applies for a specific time t during which the molecule is exposed to the oscillatory field. However, for molecular beams
$$t = l/v, \qquad (V. 18)$$
where l is the length of the oscillatory field region and v is the molecule's velocity. In an actual beam v is distributed in accordance with the Maxwellian beam velocity distribution of Eq. (II. 26). The transition probability must then be averaged over this distribution. From Eqs. (V. 9) and (II. 26) the average of $P_{p,q}$ is then

$$\langle P_{p,q} \rangle = 2 \int_0^\infty \exp(-y^2) y^3 \sin^2\Theta \sin^2\left(\frac{al}{2\alpha y}\right) dy, \qquad (V. 19)$$

where y is written for v/α.

The numerical work of evaluating Eq. (V. 19) can be much simplified by the use of trigonometric expansions to reduce it to a sum of terms involving only the two integral functions

$$I(x) = \int_0^\infty \exp(-y^2) y^3 \cos(x/y) \, dy, \qquad (V. 20)$$

$$K(x) = \int_0^\infty \exp(-y^2) y^3 \sin(x/y) \, dy. \qquad (V. 21)$$

With the use of the trigonometric identities of Eq. (D. 1) in Appendix D, Eq. (V. 19) becomes
$$\langle P_{p,q} \rangle = \sin^2\Theta[\tfrac{1}{2} - I(al/\alpha)]. \qquad (V. 22)$$

The functions $I(x)$ and $K(x)$ have been calculated by Kruse and Ramsey (KRU 51) and are tabulated in Appendix D, so all numerical integration can be avoided. An alternative tabulation of more restricted applicability has been given by Torrey (TOR 41). The shape of the resonance curve predicted by the above equation is shown as the full curve in Fig. V. 1 when the length of the oscillatory field region and the magnitude of the perturbation are related by

$$2bl/\alpha = 1\cdot 200\pi. \qquad (V. 23)$$

The relation of Eq. (V. 23) was selected because it gives the maximum transition probability at resonance as can be seen from Table D. 1.

From Fig. V. 1 or from the above calculations on which the figure is based, it is apparent that for the conditions of Eq. (V. 23), the full width at half maximum $\Delta \nu$ of the curve for the averaged transition probability is

$$\Delta \nu = \frac{\Delta \omega}{2\pi} = \frac{2b \times 1\cdot 787}{2\pi} = 0\cdot 284(2b). \qquad (V.\ 24)$$

Alternatively, from Eq. (V. 23)

$$\Delta \nu = 1\cdot 072 \alpha/l. \qquad (V.\ 25)$$

Typical experimental resonance curves are shown by the six resonance lines in Fig. I. 5.

V. 4. Magnetic resonance method with separated oscillating fields

V. 4.1. *Introduction.* From Eq. (V. 25) it would appear that an unrestricted increase in the sharpness of the radiofrequency spectral lines could be achieved merely by increasing the length l of the region in which the transitions are induced. However, in practice this unlimited increase in precision cannot be achieved. One limitation, of course, is the loss in beam intensity as the apparatus is lengthened. However, in many cases another limitation often arises first. The molecular energy is often dependent upon the strength of an external magnetic field in a major way, e.g. the energy of a simple magnetic moment is directly proportional to the field strength. It is, however, impossible in practice to achieve completely uniform magnetic fields over great lengths, so the resonance frequencies differ along the length of the beam with the result that the observed resonance pattern is frequently broadened far beyond the theory of Eq. (V. 25); frequently the resultant resonance increases rather than decreases in width as the length l of the transition region is increased.

Ramsey (RAM 49, RAM 50c, RAM 51a, RAM 51d), however, has recently developed a new molecular-beam resonance method which overcomes the effect of field inhomogeneities in adding to the resonance width, and which in addition produces resonances that are 40 per cent. narrower than those of the old method, even when a perfectly homogeneous field is available for the latter. Furthermore, the new method is often more convenient at very high frequencies and reduces Doppler broadening.

The new method arose from the observation that in Rabi's resonance method the oscillatory field was of approximately uniform intensity throughout the regions of the apparatus in which the energy levels of the molecule were investigated. Ramsey pointed out that this was not necessarily the most advantageous method of applying the oscillatory field since resonance curves of a different and often more useful character could be obtained if the amplitude and phase of the oscillating field were varied along the path of the beam. A particular arrangement that is more advantageous in many cases is one in which the oscillatory field is confined to a small region at the beginning and another small region at the end, there being no oscillating field in between.

It is superficially surprising that information can be obtained about the energy levels of a molecule in a region of space without the radiation field inducing the transitions being extended throughout that region. However, by considering the Fourier analysis of an oscillating field on for a time τ, off for a time T, and on again for a time τ, one can see that Ramsey's method of separated oscillating fields produces transitions at frequencies corresponding to the energy levels in the intermediate region in which no oscillating field exists. Alternatively, this can be seen by considering a simple magnetic example as in the following paragraph.

Consider, as at the end of Section V. 2, a magnetic moment associated with such a large angular momentum that the problem can be treated classically. Consider that such a magnetic moment enters a region of strong homogeneous magnetic field H_0 at the entrance and exit ends of which a weak magnetic field perpendicular to H_0 rotates about an axis parallel to H_0. Then, as at the end of Section V. 2, if the angular momentum is initially parallel to the fixed field so that ϕ is equal to zero initially, it is possible to select the magnitude of the rotating field to be such that $\phi = 90°$ at the end of the first oscillating region. In the next region with no oscillating field, the magnetic moment simply precesses with the Larmor frequency appropriate to the magnetic field in that region.

When the magnetic moment enters the second oscillating field region there is again a torque acting to change ϕ. If the frequency of the rotating field is exactly the same as the mean Larmor frequency in the intermediate region, there is no relative phase shift between the angular momentum and the rotating field. Consequently, if the magnitude of the second rotating field and the length of time of its application are equal to those of the first region, the second rotating field has just the same effect as the first one, i.e. increases ϕ by another 90°, making $\phi = 180°$, corresponding to a complete reversal of the direction of the angular momentum. On the other hand, if the field and the Larmor frequencies are slightly different, so that the relative phase angle between the rotating field vector and the precessing angular momentum is changed by 180° while the system is passing through the intermediate region, the second oscillating field has just the opposite effect to the first one, with the result that ϕ is returned to zero. If the Larmor frequency and the rotating field frequency differ by just such an amount that the relative phase shift in the intermediate region is exactly an integral multiple of 360°, ϕ will again be left at 180° just as at exact

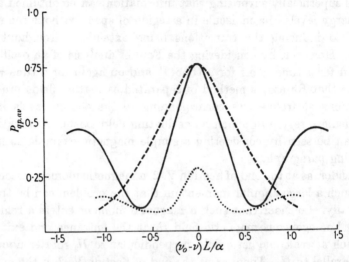

FIG. V. 2. Transition probability averaged over molecular velocity as a function of frequency near resonance. L = distance between oscillating field regions, α = most probable molecule velocity, and ν = oscillator frequency. The full curve is the transition probability for the optimum perturbation ($2bl/\alpha = 0.600\pi$). The dotted curve is for one-third the oscillating field. The dashed curve is for the Rabi method with the same length of C field (RAM 50c).

resonance. However, in a molecular-beam resonance experiment one can easily distinguish between exact resonance and the other cases by the fact that in the former case the condition for no change in the relative phase of the rotating field and of the precessing angular momentum is independent of the molecular velocity, whereas in the other cases the condition for an integral multiple of 360° relative phase shift is velocity dependent since a slower molecule is in the intermediate region longer and so experiences a greater shift than a faster molecule. Consequently, for the molecular beam as a whole, the reorientations are incomplete in all except the resonance cases. Therefore, one would expect a resonance curve similar to that shown in Fig. V. 2, in which the transition probability for a particle of spin $\tfrac{1}{2}$ is plotted as a function of frequency.

V. 4.2. *Transition probability for individual molecule with separated oscillating fields.* The transition probability in the present case can be calculated with procedures very similar to those used in Section V. 3 for Rabi's method. First, a more general solution to Eqs. (V. 5) can be obtained by considering the molecule to enter the oscillatory region at time t_1 with C_p and C_q having the values $C_p(t_1)$ and $C_q(t_1)$. Then, if the molecule remains in that region until time t_1+T, the solution of Eqs. (V. 5) at time t_1+T for this initial condition is

$$C_p(t_1+T) = \{[i\cos\Theta\sin\tfrac{1}{2}aT + \cos\tfrac{1}{2}aT]C_p(t_1) +$$
$$+ [i\sin\Theta\sin\tfrac{1}{2}aT\exp(i\omega t_1)]C_q(t_1)\}\exp\{i[\tfrac{1}{2}\omega-(W_p+W_q)/2\hbar]T\},$$
$$C_q(t_1+T) = \{[i\sin\Theta\sin\tfrac{1}{2}aT\exp(-i\omega t_1)]C_p(t_1) +$$
$$+ [-i\cos\Theta\sin\tfrac{1}{2}aT + \cos\tfrac{1}{2}aT]C_q(t_1)\}\exp\{i[-\tfrac{1}{2}\omega-(W_p+W_q)/2\hbar]T\},$$

(V. 26)

where Θ, ω_0, and a are defined as in Eq. (V. 8). That Eqs. (V. 26) are solutions of Eqs. (V. 5) can be confirmed readily by substitution. Eqs. (V. 26) can also easily be seen to satisfy the above initial conditions. They also reduce to Eqs. (V. 7) obtained earlier in the particular case then applicable of $t_1 = 0$ and initial conditions of Eq. (V. 6). A particular case of Eqs. (V. 26) occurs when $b = 0$, in which case

$$C_p(t_1+T) = \exp[-i(W_p/\hbar)T]C_p(t_1),$$
$$C_q(t_1+T) = \exp[-i(W_q/\hbar)T]C_q(t_1). \qquad (V.\ 27)$$

Now consider a molecule on which the perturbation of Eq. (V. 2) acts while the molecule goes a distance l in time τ, after which it enters a region of length L and duration T in which b is zero, and after that is again acted on by the perturbation for a time τ. To achieve greater generality, corresponding to the experimental impossibility of attaining completely uniform magnetic fields, assume that the energies of the states p and q are not constant in the intermediate region where b is zero, but that the region is divided into a number of sub-regions such that the kth sub-region of duration Δt_k the energies are $W_{p,k}$ and $W_{q,k}$. Then if t is taken to be zero when the first perturbation begins to act, and if
$$C_p(0) = 1, \qquad C_q(0) = 0, \qquad (V.\ 28)$$
successive applications of (V. 26) and (V. 27) yield

$$C_p(\tau) = [i\cos\Theta\sin\tfrac{1}{2}a\tau + \cos\tfrac{1}{2}a\tau]\exp\{i[\tfrac{1}{2}\omega-(W_p+W_q)/2\hbar]\tau\},$$

(V. 29)

$$C_q(\tau) = [i\sin\Theta \sin\tfrac{1}{2}a\tau]\exp\{i[-\tfrac{1}{2}\omega-(W_p+W_q)/2\hbar]\tau\}, \quad (V. 30)$$

$$C_p(\tau+T) = \prod_k [\exp(-iW_{p,k}\Delta t_k/\hbar)]C_p(\tau)$$

$$= \{\exp[-(i/\hbar)\sum_k W_{p,k}\Delta t_k]\}C_p(\tau)$$

$$= [\exp(-i\overline{W}_p T/\hbar)]C_p(\tau), \quad (V. 31)$$

$$C_q(\tau+T) = [\exp(-i\overline{W}_q T/\hbar)]C_q(\tau), \quad (V. 32)$$

$$C_p(2\tau+T) = \{[i\cos\Theta \sin\tfrac{1}{2}a\tau+\cos\tfrac{1}{2}a\tau]C_p(\tau+T)+$$
$$+[i\sin\Theta \sin\tfrac{1}{2}a\tau \exp i\omega(\tau+T)]C_q(\tau+T)\}\times$$
$$\times \exp\{i[\tfrac{1}{2}\omega-(W_p+W_q)/2\hbar]\tau\}, \quad (V. 33)$$

$$C_q(2\tau+T) = \{[i\sin\Theta \sin\tfrac{1}{2}a\tau \exp(-i\omega(\tau+T))]C_p(\tau+T)+$$
$$+[-i\cos\Theta \sin\tfrac{1}{2}a\tau+\cos\tfrac{1}{2}a\tau]C_q(\tau+T)\}\times$$
$$\times \exp\{i[-\tfrac{1}{2}\omega-(W_p+W_q)/2\hbar]\tau\}. \quad (V. 34)$$

Here
$$\overline{W}_p = (1/T)\sum_k W_{p,k}\Delta t_k = (1/L)\sum_k W_{p,k}\Delta L_k$$
$$= \text{space mean value of } W_p, \quad (V. 35)$$

with a similar interpretation for $\overline{W}_q$.

The elimination of the intermediate values of the C's from Eqs. (V. 29)–(V. 34) yields for $C_q(2\tau+T)$,

$$C_q(2\tau+T) = -2i\sin\Theta[\cos\Theta \sin^2\tfrac{1}{2}a\tau \sin\tfrac{1}{2}\lambda T - \tfrac{1}{2}\sin a\tau \cos\tfrac{1}{2}\lambda T]\times$$
$$\times \exp\{-i[(\tfrac{1}{2}\omega+(W_p+W_q)/2\hbar)(2\tau+T)+$$
$$+[(\overline{W}_p-W_p+\overline{W}_q-W_q)/2\hbar]T]\}, \quad (V. 36)$$

where
$$\lambda = [(\overline{W}_q-\overline{W}_p)/\hbar]-\omega. \quad (V. 36a)$$

It follows from Eq. (V. 36) that the probability that the system changes from state p to state q is

$$P_{p,q} = |C_q|^2 = 4\sin^2\Theta \sin^2\tfrac{1}{2}a\tau[\cos\tfrac{1}{2}\lambda T \cos\tfrac{1}{2}a\tau - \cos\Theta \sin\tfrac{1}{2}\lambda T \sin\tfrac{1}{2}a\tau]^2. \quad (V. 37)$$

V. 4.3. *Separated oscillating fields transition probability averaged over velocity distribution.* The above expression for the transition probability can be averaged over the molecular velocity distribution as in Section V. 3.2. Without loss of significant generality, this averaging can be carried out most easily by performing it in only two limiting cases. One of these is for frequencies so close to resonance that $|\omega_0-\omega|$ is much less than $|2b|$, in which case Eq. (V. 37), when integrated (as in Section

V. 3.2) over the molecular velocities with the usual molecular beam weighting factor, becomes

$$\langle P_{p,q}\rangle = 2\int_0^\infty \exp(-y^2)y^3\sin^2\!\left(\frac{2bl}{\alpha y}\right)\cos^2\!\left(\frac{\lambda L}{2\alpha y}\right)dy. \qquad (V.\ 38)$$

The other interesting limit is that for frequencies so far from exact resonance that the quantity $\lambda L/(2\alpha y)$ is sufficiently large that in averaging over the velocities the rapidly varying factor $\sin^2(\lambda L/2\alpha y)$ can be replaced by its average value of $\tfrac{1}{2}$, while $\sin(\lambda L/2\alpha y)$ can be replaced by its average value of zero. In this case we have

$$\langle P_{p,q}\rangle = 4\sin^2\Theta\int_0^\infty \exp(-y^2)y^3\sin^2\!\left(\frac{al}{2\alpha y}\right)\!\left[1-\sin^2\Theta\sin^2\!\left(\frac{al}{2\alpha y}\right)\right]dy.$$

$$(V.\ 39)$$

With the aid of the equations in Appendix D, Eq. (V. 38) becomes

$$\langle P_{p,q}\rangle = \tfrac{1}{4}-\tfrac{1}{4}I\!\left(\frac{\lambda L+4bl}{\alpha}\right)-\tfrac{1}{4}I\!\left(\frac{\lambda L-4bl}{\alpha}\right)+\tfrac{1}{2}I(\lambda L/\alpha)-\tfrac{1}{2}I(4bl/\alpha),$$

$$(V.\ 40)$$

while Eq. (V. 39) becomes

$$\langle P_{p,q}\rangle = \sin^2\Theta(1-\tfrac{3}{4}\sin^2\Theta)-\tfrac{1}{2}\sin^2 2\Theta\, I(al/\alpha)-\tfrac{1}{2}\sin^4\Theta\, I(al/2\alpha). \qquad (V.\ 41)$$

From Eq. (V. 40) and Appendix D it may be seen that the transition probability is a maximum for $\lambda = 0$ or for

$$\omega = (\overline{W}_q-\overline{W}_p)/\hbar. \qquad (V.\ 41a)$$

Both Eqs. (V. 40) and (V. 41) depend on the value assumed for the experimentally arbitrary parameter $2bl/\alpha$. Corresponding to Eq. (V. 23), the maximum transition probability at exact resonance in the present case is achieved for

$$2bl/\alpha = 0\cdot 600\pi. \qquad (V.\ 42)$$

For this choice of $2b$, the average transition probability in the immediate vicinity of resonance as given by Eq. (V. 40) has been plotted as the full line in Fig. V. 2 where the abscissa is $(\nu_0-\nu)L/\alpha$ and $\nu = \omega/2\pi$. In contrast to Eq. (V. 25) the full resonance width at half intensity is given by

$$\Delta\nu = 0\cdot 65\alpha/L. \qquad (V.\ 42a)$$

For comparative purposes the theoretical transition probability with the Rabi method and a perfectly uniform field is plotted with dashed lines on the same curve. A sharper though weaker resonance can be obtained by using a smaller value of b than in Eq. (V. 42) since the slower molecules are then relatively more effective and these are in the homogeneous field region for a longer time. This is shown by the dotted curve in Fig. V. 2 which is for

$$2bl/\alpha = 0.200\pi. \qquad (V.\ 43)$$

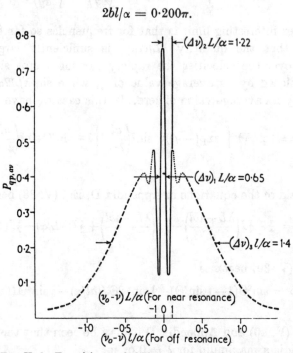

FIG. V. 3. Transition probability as a function of frequency. The full line is calculated near resonance, the dashed line is calculated off resonance, and the dotted line is interpolated between (RAM 50c).

As an indication of the general appearance of the transition probability over the entire resonance region, the transition probabilities from Eqs. (V. 40) and (V. 41) are plotted on the same curve in Fig. V. 3. This can be compared with the experimental result in Fig. V. 4. In a typical apparatus, such as that in Table XIII. I, the resonance width is 300 c/s or 10^{-8} cm^{-1}.

With the method of separated oscillating fields several interesting possibilities arise when more than two states have to be considered. In particular, consider three possible states p, q, and r such that in the end regions the oscillatory field can induce transitions between p and q and between q and r, i.e. such that the difference between $(W_q - W_p)/h$ and $(W_r - W_q)/h$ is appreciable but less than the width of the resonance in the end field alone which is the broad background resonance of Fig. V. 2. In such a case Ramsey (RAM 50e, SIL 50), has shown theoretically that there should be three resonance interference patterns centred at the frequencies

$$\nu_{p,q} = (W_q - W_p)/h, \qquad \nu_{q,r} = (W_r - W_q)/h \qquad (V. 43a)$$

and
$$\nu_{r,p} = (W_r - W_p)/2h. \qquad (V. 43b)$$

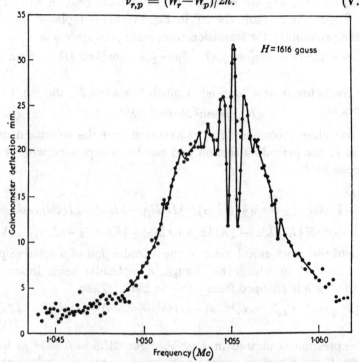

Fig. V. 4. Experimental transition probability for the central resonance of D_2 as a function of frequency with separated oscillating fields method (KOL 50a, KOL 52).

The last of these is something of a novelty and is at a frequency that corresponds to a two-quantum absorption. These multiple quantum transitions are distinct from those discussed in Section X. 4.3 below for the reasons there given. When more than three states are involved, the results are a generalization of the above with such resonance frequencies as $(W_s - W_p)/3h$, etc., being possible. Kusch (KUS 54a, ZZA 47), in experiments with a Rabi single oscillatory field, has observed such multiple transitions experimentally. Similar multiple quantum radio-frequency transitions have recently been observed with techniques of optical spectroscopy by Brossel, Cagnac, and Kastler (ZZA 7, ZZA 8).

V. 4.4. *Phase shifts in the molecular-beam method of separated oscillating* *fields.* In all of the preceding sections it has been assumed implicitly that the entrance and exit oscillating fields were of the same phase. Ramsey and Silsbee (RAM·51d) have pointed out that it is often advantageous to introduce a relative phase shift between the two fields. If the second oscillating field leads the first by a phase angle δ the Eqs. (V. 33) and (V. 34) are modified by the replacement of $\omega(\tau+T)$ by $\omega(\tau+T)+\delta$. As a result the λT in Eq. (V. 37) is replaced by $\lambda T-\delta$ and the probability for transition from state p to state q is

$$_\delta P_{p,q} = 4\sin^2\Theta \sin^2 \tfrac{1}{2}a\tau[\cos\tfrac{1}{2}(\lambda T-\delta)\cos\tfrac{1}{2}a\tau - \cos\Theta\sin\tfrac{1}{2}(\lambda T-\delta)\sin\tfrac{1}{2}a\tau]^2.$$
(V. 44)

Near resonance, where $|\omega_0-\omega|$ is much less than $2b$, the Eq. (V. 44) reduces to

$$_\delta P_{p,q} = \sin^2 2b\tau \cos^2 \tfrac{1}{2}(\lambda T-\delta).$$ (V. 44a)

This transition probability can be averaged over the velocity distribution as in the preceding section and can be re-expressed with the aid of Appendix D as

$$\langle _\delta P_{p,q}\rangle$$
$$= \tfrac{1}{4}-\tfrac{1}{2}I(4bl/\alpha)+\cos\delta[\tfrac{1}{2}I(\lambda L/\alpha)-\tfrac{1}{4}I(4bl/\alpha+\lambda L/\alpha)-\tfrac{1}{4}I(4bl/\alpha-\lambda L/\alpha)]+$$
$$+\sin\delta[\tfrac{1}{2}K(\lambda L/\alpha)-\tfrac{1}{4}K(4bl/\alpha+\lambda L/\alpha)+\tfrac{1}{4}K(4bl/\alpha-\lambda L/\alpha)]. \quad \text{(V. 45)}$$

One of the most useful cases of the introduction of a relative phase shift δ is that in which the change in molecular-beam intensity is measured as δ is changed from 0 to π radians. Then

$$\langle _0 P_{p,q}\rangle - \langle _\pi P_{p,q}\rangle = I(\lambda L/\alpha)-\tfrac{1}{2}I(4bl/\alpha+\lambda L/\alpha)-\tfrac{1}{2}I(4bl/\alpha-\lambda L/\alpha).$$
(V. 46)

This expression is plotted in Fig. V. 5 for $2bl/\alpha = 0.600\pi$ as in Eq. (V. 42). For comparison purposes an experimental curve of the change in beam intensity for the nuclear resonance of ortho-D_2 is shown in Fig. V. 6.

The change in the ordinate between maxima and minima in Fig. V. 5 is just double that in Fig. V. 2. Therefore the phase-shifting method provides the equivalent of an effective doubling of beam intensity over mere observations of the change in beam intensity as the oscillator is successively turned off and on. Furthermore, a mere reversal of phase tends to give less radiofrequency pick-up in other parts of the apparatus than when the oscillator is turned off and on.

An alternative case of considerable interest is that in which δ is changed from $\tfrac{1}{2}\pi$ to $-\tfrac{1}{2}\pi$. Then Eq. (V. 45) becomes

$$\langle _{\tfrac{1}{2}\pi} P_{p,q}\rangle - \langle _{-\tfrac{1}{2}\pi} P_{p,q}\rangle = K(\lambda L/\alpha)-\tfrac{1}{2}K(4bl/\alpha+\lambda L/\alpha)+\tfrac{1}{2}K(4bl/\alpha-\lambda L/\alpha).$$
(V. 47)

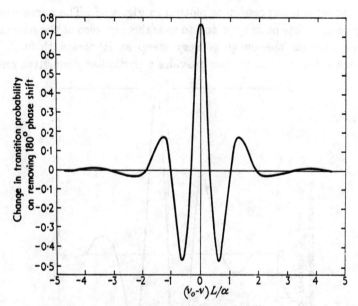

Fig. V. 5. Theoretical change in transition probability on removing 180° phase shift (RAM 51d).

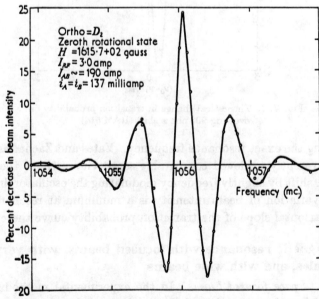

Fig. V. 6. Experimental change in beam intensity on removing 180° phase shift with ortho-D_2 in zeroth rotational state (RAM 51d).

The graph of this expression is plotted in Fig. V. 7. The curve can be seen to be of a dispersion type and to pass through zero at the resonance frequency. Since the curve is very steep as it passes through the resonance value, this technique provides a particularly sensitive means

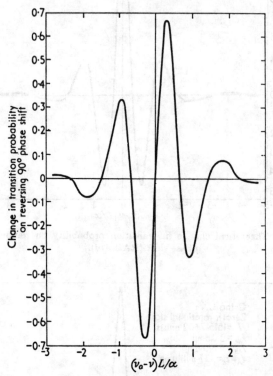

Fig. V. 7. Theoretical change in transition probability on reversing 90° phase shift (RAM 51d).

for locating the exact resonance frequency. Yates and Zacharias (BRO 54) have recently achieved comparable sensitivity without the use of any phase shift by slightly frequency modulating the oscillator, in which case the variation of beam intensity is a minimum at resonance due to the horizontal slope of the transition probability curve there.

V. 5. Magnetic resonance with focused beams, with very slow molecules, and with wide beams

V. 5.1. *Space focused beams.* In the experimental methods so far described there has sometimes been a velocity refocusing of the beam so that a magnetically spread out beam is refocused, the refocusing being independent of velocity. However, the beam intensity has been no greater than it would have been in the absence of any magnetic deflexion at all, i.e. it has never been greater than that predicted by Eq. (II. 15) which assumes a simple diminution of intensity with distance L as $1/L^2$.

Vauthier (VAU 49), Friedburg and Paul (FRI 51), and Korsunskii and Fogel (KOR 51) have suggested the use of inhomogeneous magnetic fields to produce some genuine focusing of the neutral particles in a molecular beam.

The focusing magnetic fields used by Friedburg and Paul (FRI 51) are described in detail in Section XIV. 4.7 and are illustrated schematically in Fig. XIV. 32. As shown in Eq. (XIV. 14) these fields are such that the restoring force F on an atom with a constant magnetic moment is given by

$$F = -D\mathbf{r}_1, \qquad (V. 48)$$

where $\mathbf{r}_1$ is the projection of the position vector $\mathbf{r}$ of the atom on a plane perpendicular to the dominant beam direction, which is taken as the z axis. In such a force field the atom undergoes simple harmonic oscillations with angular frequency $\omega = \sqrt{(D/M)}$. This oscillatory motion is extended along the z axis because of the velocity component v_z. All atoms which originate from a source point $x_0 y_0 z_0$ diverging with arbitrary velocity components v_x and v_y will converge after half a period at the point $-x_0$, $-y_0$, $z_0+(\pi/\omega)v_z$. For a single velocity, then, a sharp real image of the source is formed. As discussed in Section XIV. 4.7, with practical values of magnetic fields and with a monoenergetic beam, it is possible at 20 cm from the source to concentrate into a point the same intensity that would otherwise be distributed over a circular disk 1·4 cm in diameter. Unfortunately, however, molecular beams are not monoenergetic so there is a large chromatic aberration. Nevertheless, even with the chromatic aberration the focusing provides a considerable improvement in beam intensity.

Bennewitz and Paul (ZZA 11) have recently modified their method to apply to atoms whose magnetic moments depend on the strength of the external magnetic field. In this way they have measured nuclear spins.

Hamilton, Lemonick, Pipkin, and Reynolds (HAM 53) have partially overcome the effects of chromatic aberration by the use of one converging and one diverging lens in series. The optical analogue to their system is shown in Fig. V. 8. They found that with the use of two lenses that deviated the atoms equal amounts in opposite directions much of the chromatic aberration was overcome. Actually, only one type of lens was constructed, but it was divergent or convergent according to the orientation of the atomic magnetic moment. Consequently, only atoms which undergo the desired transition in the C field are unobstructed by the various stops in the apparatus. The arrangement of stops used is shown schematically in Fig. V. 9; typical atom paths are shown

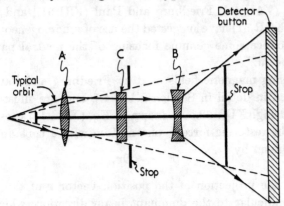

FIG. V. 8. Optical analogue to Hamilton's focusing atomic-beam apparatus (HAM 53).

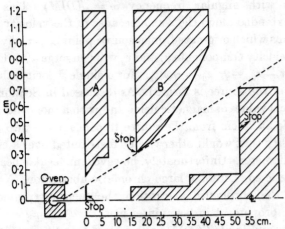

FIG. V. 9. Stops used in focusing atomic-beam apparatus. The diagram is distorted by having a much larger vertical than horizontal scale (HAM 53).

schematically on a distorted scale in Fig. V. 10. Iron magnets were used to produce the focusing fields as shown in the end view of Fig. V. 11. The apparatus has been used to measure spins of radioactive isotopes and the atoms that successfully traversed the maze of stops were detected by the radioactivity accumulated on the detector. When a resonance was attained with the oscillatory magnetic field the amount of collected radioactivity increased.

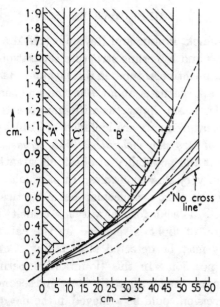

FIG. V. 10. Typical atomic paths in focusing atomic-beam apparatus. The diagram is distorted by having a much larger vertical than horizontal scale (HAM 53).

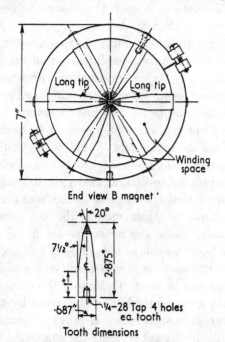

FIG. V. 11. End view of iron focusing magnet. The marks on the pole tips indicate the varying aperture with distance along the magnet (HAM 53).

Hamilton, Lemonick, Pipkin, and Reynolds (HAM 53, BRO 54) have measured the spins and hyperfine structure separations of Cu^{64}, Ag^{111}, Au^{199}, and other radioactive isotopes in this fashion. Hamilton estimates that his focusing apparatus is about 100 times as sensitive as a conventional molecular-beam apparatus with non-directive sources. The results of these measurements are included in Tables VI. 1 and IX. 1.

V. 5.2. *Beams of very slow molecules. Wide beams.* Zacharias (ZAC 53) has pointed out the possibility of obtaining greatly increased precision by the use of very slow molecules. In particular he has proposed the possibility of shooting a molecular beam vertically and then of observing only those molecules of such very low velocity ($\sim$ 600 cm sec^{-1}) that their direction of flight is reversed by the earth's gravitational field, so that they may be detected after falling back approximately to their original position. In this 'fountain' experiment the atoms would be in the apparatus for approximately one second with the result that very high precision could be achieved in the measurement according to Eq. (V. 42a). Zacharias (ZAC 53) has proposed using Ramsey's method of separated oscillatory fields in this experiment with the first field acting on the beam as it is on the way up while the second acts on the falling atoms on the way down.

The obvious objection to the proposed experiment is that the desired very slow molecules form only a small fraction of the number of molecules emerging thermally from a source at any reasonable temperature so that a large loss of beam intensity should be expected. Zacharias has proposed to compensate for this loss in several ways. One is by the use of space focusing beams as discussed in Sections V. 5.1 and XIV. 4.7. The focusing techniques should be particularly well suited to the proposed experiment since the bad effects of chromatic aberration in the focusing should be diminished by the limited range of atomic velocity distributions that are effective in the 'fountain' experiment.

Zacharias also hopes to regain some of the lost intensity by the use of very wide beams. Since the molecules are going so slowly, large deflexions are easily obtainable so that magnetic deflexions of the order of centimetres become possible. Consequently very wide beams of comparable dimensions can be used. Ordinarily the gain from the use of wide beams would be considerably compensated by the necessity for the use of a lower source pressure. However, Zacharias proposes to overcome this difficulty by subdividing the source slit into many small parallel channels by the use either of stacked hypodermic needles or stacked crimped foils so that each individual channel is of smaller dimensions. As discussed in Section XIV. 2.2 such sources provide greater directivity and consequent greater economy of source material. At the time of writing, the 'fountain' experiment is under construction but not yet tested. It is, however, a promising new development in high precision measurements. As discussed in greater detail in Section IX. 11 such an experiment may provide a frequency standard accurate to one part in 10^{13} which should be sufficient to detect changes of the relativistic gravitational red shift over the surface of the earth.

Collision alignment techniques for wide beams are discussed on page 168 and in Appendix G (ZZA 25).

V. 6. Distortions of molecular-beam resonances

In precision molecular-beam studies one of the most important problems is the measurement of the frequency of a resonance. However, although it is ordinarily easy to determine the frequency at which a peak occurs, it is not so easy to be certain that the experimental peak is exactly at the Bohr frequency $(E_n - E_m)/h$ for the levels concerned. In other words, distortions of the resonances may lead to a false result. In the present section various distortions of nuclear resonances will be described. These distortions can ordinarily be overcome with sufficient care, but they must be taken into consideration in the design of precision experiments if misleading results are to be avoided. Although some of the distortions that occur with the single oscillatory field method do not occur with the separated oscillatory field method, almost all of the distortions with the latter method occur with the former. To avoid repetition, however, distortions that occur in both methods will be described only for the separated oscillatory field case since at least qualitative generalizations to the corresponding distortions in the single oscillatory field case are apparent. However, from this one should not infer that there are more distortion difficulties in the separated oscillatory field case; on the contrary there are fewer and they are more easily analysed.

One of the earliest observed distortions was the Millman effect (MIL 39), which is a resonance asymmetry that arises if the direction of the oscillatory field varies along the length of the oscillatory field region. This asymmetry is discussed in detail in Section VI. 4 so it will not be described in the present section. The Millman effect can ordinarily be avoided by the use of suitably designed oscillatory field loops as discussed in Section XIV. 5.2. It is also less likely to occur with the separated oscillatory field method.

When the wavelength of the oscillatory radiation becomes comparable to the length of the oscillatory field, a molecular-beam resonance can become distorted by Doppler effect. This can be seen from the following discussion if $(\Delta \nu)_D$ represents the Doppler shift for a molecule of velocity v while $(\Delta \nu)_R$ represents the natural radiation width of the resonance. Then by the usual Doppler effect discussion (HEI 53),

$$(\Delta \nu)_D \approx \nu \frac{v}{c}. \qquad \text{(V. 49)}$$

On the other hand, from Eq. (V. 25),

$$(\Delta \nu)_R \approx v/l, \qquad \text{(V. 50)}$$

so
$$(\Delta \nu)_D/(\Delta \nu)_R \approx l\nu/c = l/\lambda. \qquad \text{(V. 51)}$$

This shows that the Doppler distortion becomes serious as the wavelength of the radiation approaches the length of the oscillatory field. An alternative explanation of the Doppler effect is to say that if there are standing waves in the oscillatory field region, the motion of the molecule along the standing wave pattern will create an apparent additional oscillatory frequency which must be added to or subtracted from the applied frequency to obtain the effective frequency. A related explanation is to say that the standing waves in the oscillatory field region are due to a superposition of two running waves, one moving in the same direction as the beam and one in the opposite direction. Then to the moving molecule the wave going in the same direction will appear to have a lower frequency and that in the opposite direction a higher frequency (DAL 53). If both of these waves are of equal amplitude the Doppler effect will produce a broadening of the resonance and not a shift in its position. On the other hand if one running wave is of larger amplitude than the other the resonance will be asymmetrically distorted. Doppler effect can be most easily overcome by the use of Ramsey's separated oscillatory field method with each oscillatory field region being short compared with a wavelength (RAM 50c).

With the single oscillatory field method variations in the value of H_0, the presumed uniform and constant magnetic field, can cause great distortions to the resonance. If the field changes are so great that the resonance condition occurs at only a few places along the field, the apparatus will operate as an accidental, multiple, and somewhat random separated oscillatory field apparatus; numerous spurious interference peaks can then occur. In this way several sharp isolated resonance lines have been produced where only one should occur, and in some cases a single sharp resonance has been produced at a position far away from that appropriate to the average magnetic field. Fig. V. 12 illustrates spurious resonances produced by non-uniformities of H_0. It should be noted that the mere fact that a line is apparently of theoretical narrowness is not proof that it is undistorted by field inhomogeneities since it may happen that only one of the interference peaks is sufficiently strong to be easily observable.

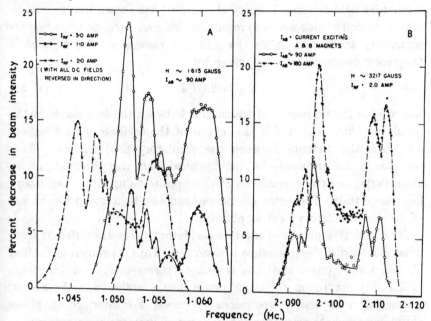

Fig. V. 12. Illustration of multiple peaks produced by non-uniformities of the constant field H_0 with the single oscillatory field method. The transitions are for nuclear reorientations of the deuteron in ortho-D_2 in the zeroth rotational state, for which only a single resonance frequency should be observed, and for which only a single one is observed with Ramsey's separated oscillatory field method (PHI 50).

With the separated oscillatory field method the above difficulty is alleviated. However, it must be remembered in this case, as pointed out by Ramsey (RAM 50c) and shown in Eq. (V. 41a), that the resonance occurs for

$$\nu = [\overline{W}_q(H) - \overline{W}_p(H)]/h \qquad (V.\ 52)$$

and not for $[W_q(\bar{H}) - W_p(\bar{H})]/h$. For energy levels linearly dependent on H, such as interactions of constant magnetic moments, these are the same but for energy levels not linearly dependent on H this is not the case.

As discussed in Section V. 4.3, Ramsey (RAM 50e, SIL 50) has discussed theoretically and Kusch (KUS 54a) has observed experimentally resonance at frequencies such as $(W_r - W_p)/2h$. Erroneous results would be obtained if such a multiple resonance were confused with the normal single quantum resonance. As discussed in Section V. 4.3 these multiple transitions occur both with the single and the separated oscillatory field methods (ZZA 47).

When a greatly excessive oscillatory field is used in the separated oscillatory field method, Ramsey (RAM 50e) has shown that two additional velocity independent resonance minima are produced symmetrically above and below the principal resonance maxima with a frequency displacement $(\Delta \nu)_E$ given by

$$\Delta \nu_E = 2\nu_T l/L, \qquad (V.\ 53)$$

where ν_T is the frequency of transitions between the two states in the oscillatory field region, l is the length of the oscillatory field region, and L is the distance between the oscillatory field regions. These resonances can ordinarily be distinguished from the principal one by their shape and the dependence of their positions upon the oscillatory field amplitude. The extra resonances can also be avoided by the use of optimal oscillatory field amplitudes.

If the relative phases of the two oscillatory fields are different from what is intended the resonance is distorted. Thus, as shown in Section V. 4.4, for 0° phase shift the resonance corresponds to a transition probability maximum, for a 180° phase shift a minimum, and for a 90° phase shift a dispersion-type resonance curve. For intermediate phase shifts intermediate resonance curves occur. Therefore, if an extraneous phase shift is present but unknown the resonance frequency will be incorrectly determined. Consequently, in precision measurements the relative phases must be carefully measured. This problem is equally important in the Rabi single oscillatory field method as pointed out by Ramsey (RAM 49, RAM 50c), since different parts of a relatively long oscillatory field loop may easily oscillate in different phases thereby leading to peculiarly shaped resonance curves. Spurious results have been obtained in this way with a single oscillatory field (DAL 53, KIN 54, JAC 54). Woodgate and Hellworth (ZZA 12) have noted both theoretically and experimentally that when atomic magnetic σ-transitions ($\Delta m = 0$) are observed with a single oscillatory field primarily designed for π-transitions (as Figs. I. 6 and XIV. 34), it is particularly easy for 180° phase shifts to occur between the beginning and the end of the oscillatory field region since the necessary component of the radiofrequency field parallel to H_0 can arise from the curvature of the radiofrequency field lines and may be oppositely directed at the two ends.

If the values of H_0 in the transition regions of the separated oscillatory field method are far different from the average value of H_0 in the intermediate region, Ramsey (RAM 50e) has shown that the resonance is distorted asymmetrically and the position of the resonance maximum is shifted. If $(\Delta\omega)_1$ is the difference between the resonance frequency in the oscillatory region and in the average intermediate region, the shift $\Delta\omega_0$ of the observed resonance is approximately

$$\Delta\omega_0 \sim \frac{l}{L}(\Delta\omega)_1. \qquad (V.\ 54)$$

Consequently, this distortion can be eliminated by the use of sufficiently short oscillatory regions, which has other advantages as well, including greater ease in locating the resonance.

If the two oscillatory fields in the separated field method have different amplitudes, Ramsey has shown (RAM 50f) that the shape of the curve is altered but only in a symmetrical manner. In fact, for suitable choices of unequal oscillatory field values the resonance can be changed from a transition probability maximum to a minimum. This can be seen qualitatively by the fact that if the first oscillatory field reoriented most of the nuclei about 270° while the second reoriented them 90°, the exact resonance frequency, at which the two reorientations would combine to give 360°, would correspond to a transition probability minimum. Since this distortion is symmetric and has no dispersion curve character it fortunately leads to no error in the determination of the resonance frequency. The shape of a single oscillatory field resonance curve is also modified if the amplitude or direction of the oscillatory field varies throughout the oscillatory field region (RAM 49, RAM 50e).

In Eq. (V. 17) above it was shown that if two oscillatory fields of different frequencies were present simultaneously, the one whose perturbation was $2b_1$ would perturb the resonance curve of the other so that the resonance maximum instead of occurring at the proper value ω_0 would occur at the angular frequency ω_r where

$$\omega_r = \omega_0 + \frac{(2b_1)^2}{2(\omega_0 - \omega_1)}. \qquad (V.\ 55)$$

There are several ways in which such extraneous oscillatory fields can occur in practice. One is by virtue of the fact that oscillatory instead of rotating fields are ordinarily used and, as discussed in Section V. 3.1, the oppositely rotating component corresponds to a frequency $\omega_1 = -\omega_0$. This effect was first considered by Bloch and Siegert (BLO 40) and was shown ordinarily to be of negligible importance, as can be seen from Eq. (V. 55). However, in practice, there are other sources of oscillatory fields, whose frequencies ω are much closer to ω_0, and whose effects from Eq. (V. 55) may thereby be much greater. For example, two independent oscillatory fields are sometimes applied simultaneously in order that two different transitions can be compared. Alternatively, a second frequency is sometimes unintentionally present when the desired frequency is produced by frequency multiplication from a crystal-controlled low-frequency oscillator and when the undesired frequencies are not adequately filtered out. In either of these cases the magnitude of the displacement of the desired resonance by the additional oscillatory field should be estimated from Eq. (V. 55) to eliminate possible misleading results. An extraneous oscillatory frequency can also be produced accidentally if an iron magnet is used at low magnetic fields. Owing to imperfections in the iron, the direction of the field may then change along the molecular path and to the moving molecule this will appear to be an oscillatory field. The magnitude of this effect in any case can then be estimated with the aid of Eq. (V. 55) (RAM 50e ZZA 46).

VI
NUCLEAR AND ROTATIONAL MAGNETIC MOMENTS

VI. 1. Resonance measurements of nuclear magnetic moments

THE present chapter will be concerned with resonance measurements of nuclear and rotational magnetic moments from their interactions with external magnetic fields H_0. Resonance measurements of atomic hyperfine structure and of paramagnetic atoms will be deferred to Chapter IX even though these experiments yield nuclear moment information. Therefore, it will be assumed throughout the remainder of this chapter that all the atoms considered are non-paramagnetic; this assumption is valid, for example, in $^1\Sigma$ diatomic molecules.

In the preceding chapter the positions, widths, and shapes of the molecular-beam resonances were calculated both for Rabi's and for Ramsey's methods. The corresponding nuclear magnetic moment expressions can be obtained immediately as special cases of these general formulae for nuclei of spin $\frac{1}{2}$, which thereby can have only two possible orientation states (magnetic moments with spins greater than $\frac{1}{2}$ are discussed in Section VI. 3 below). The energies of the two possible orientation states of a nucleus of spin $\frac{1}{2}$ and gyromagnetic ratio γ_I in a magnetic field H_0 are from Eqs. (I. 2) and (III. 3)

$$W_{\pm\frac{1}{2}} = \pm\tfrac{1}{2}\gamma_I \hbar H_0. \tag{VI. 1}$$

From Eq. (V. 11), if state p is taken as that with $m_I = +\tfrac{1}{2}$, the resonance frequency is at

$$\omega_0 = \gamma_I H_0 = (\mu_I/I)H_0/\hbar, \tag{VI. 2}$$

i.e. at the Larmor frequency.

In order that the expressions for the resonance shapes of the previous chapter may apply to magnetic moments, the value of b in Eq. (V. 2) must be found. Assume that the nucleus is in a fixed external magnetic field H_0 with a weak component H_1 that is perpendicular to H_0 and rotating about it with angular velocity $\boldsymbol{\omega} = -\omega\mathbf{k}$. Then if the z-axis is along H_0 the resultant magnetic field is

$$\mathbf{H} = H_0\mathbf{k} + H_1\cos\omega t\,\mathbf{i} - H_1\sin\omega t\,\mathbf{j}, \tag{VI. 3}$$

and as in Eq. (III. 35b)

$$\mathcal{H} = -\mu_I \cdot \mathbf{H} = -\gamma_I \hbar \mathbf{I} \cdot \mathbf{H}$$
$$= -\gamma_I \hbar I_z H_0 - \tfrac{1}{2}\gamma_I \hbar I_+ H_1 e^{i\omega t} - \tfrac{1}{2}\gamma_I \hbar I_- H_1 e^{-i\omega t}. \quad \text{(VI. 4)}$$

But from Eq. (V. 2)

$$\hbar b e^{i\omega t} = V_{pq} = (\tfrac{1}{2}|\mathcal{H}|-\tfrac{1}{2}) = -\tfrac{1}{2}\gamma_I \hbar H_1 e^{i\omega t}(\tfrac{1}{2}|I_+|-\tfrac{1}{2})$$
$$= -\tfrac{1}{2}\gamma_I \hbar H_1 e^{i\omega t}, \quad \text{(VI. 5)}$$

where the last step follows from Eq. (III. 35a). Therefore

$$-2b = \gamma_I H_1 = (\mu_I/I)H_1/\hbar = \omega_0 H_1/H_0. \quad \text{(VI. 6)}$$

With these values for ω_0 and $2b$ all the line shape formulae of the preceding chapter are fully applicable.

The above calculations show that the refocused beam intensity in a molecular beam resonance experiment as described in Chapter I will pass through a minimum of intensity at the Larmor frequency of Eq. (VI. 2) with a resonance shape given by Eqs. (VI. 6) and by Eqs. (V. 22), (V. 40), or (V. 45). Of course, if the deflecting magnets are set for refocusing when a transition does occur (so-called 'flop-in' experiments), the beam intensity minimum will be replaced by a maximum as described in Section I. 2.

In actual practice an oscillatory field is often used instead of a rotating field. However, an oscillatory field can always be resolved into two rotating fields of equal amplitudes and frequencies that rotate in opposite directions. If H_1' is the amplitude of the oscillatory field and H_1 the amplitude of one of the rotating components,

$$H_1 = \tfrac{1}{2} H_1' \quad \text{(VI. 7)}$$

since the two rotating components add together to give the maximum oscillatory amplitude. Only the component rotating in the correct direction to give resonance is really effective, so that all the previous results apply in the oscillatory case provided H_1 is determined by Eq. (VI. 7). The extraneous component introduces a slight shift in the resonance frequency, but this is ordinarily negligible as shown in Eq. (V. 15).

VI. 2. Interpretation with rotating coordinate system

VI. 2.1. *Introduction.* Although the transition probabilities for nuclear moments have been derived above as special cases of the more general calculations in the preceding chapter they can also be calculated directly. Since the methods for the latter calculations are easily interpretable physically and are often of great help in providing a simplified analysis of resonance experiments, they will be given in this section. These methods utilize a transition to a rotating frame of reference in which the discussion of the problem is much simplified. In fact on the rotating coordinate system the problem often becomes so simple that it can be solved merely by inspection. The extensive and explicit use of the rotating coordinate system procedures in nuclear resonance problems was introduced by Bloch, Ramsey, Rabi, and Schwinger; it has been described in detail by Rabi, Ramsey, and Schwinger (RAB 54).

The rotating coordinate system method is equally applicable to classical and quantum mechanical systems. Because of the great simplicity and extensive applicability of the classical description it will be given as well as the quantum mechanical one.

VI. 2.2. *Classical theory.* Consider a system consisting of one or more nuclei or atoms all of which have the same constant gyromagnetic ratio γ_I. Then the equation of motion of the system in a stationary coordinate system is

$$\hbar \frac{d\mathbf{I}}{dt} = \boldsymbol{\mu} \times \mathbf{H} = \gamma_I \hbar \mathbf{I} \times \mathbf{H}. \qquad (\text{VI. 8})$$

But if $\partial/\partial t$ represents differentiation with respect to a coordinate system that is rotating with angular velocity $\boldsymbol{\omega}$,

$$\frac{d\mathbf{I}}{dt} = \frac{\partial \mathbf{I}}{\partial t} + \boldsymbol{\omega} \times \mathbf{I}, \qquad (\text{VI. 9})$$

where $\mathbf{I}$ on both sides of the equation is the angular momentum as measured by the stationary observer, but the $\partial \mathbf{I}/\partial t$ represents how a rotating observer would find the stationary observer's $\mathbf{I}$ to vary as a function of time. By rearrangement Eq. (VI. 9) becomes

$$\hbar \frac{\partial \mathbf{I}}{\partial t} = \gamma_I \hbar \mathbf{I} \times (\mathbf{H} + \boldsymbol{\omega}/\gamma_I) = \gamma_I \hbar \mathbf{I} \times \mathbf{H}_{er}, \qquad (\text{VI. 10})$$

where $\mathbf{H}_{er}$ is the effective field in the rotating coordinate system and is defined by
$$\mathbf{H}_{er} = \mathbf{H} + \boldsymbol{\omega}/\gamma_I. \qquad (\text{VI. 11})$$

In other words, the effect of the rotation of the coordinate system is merely to change the effective field by the added term $\boldsymbol{\omega}/\gamma_I$.

This result can readily be applied to interpret the effect of the rotating magnetic fields used in molecular beam magnetic resonance experiments. In most of these there is a constant field $\mathbf{H}_0$ about which another (usually much weaker) field $\mathbf{H}_1$ perpendicular to $\mathbf{H}_0$ rotates with angular velocity $-\omega$. However, from the point of view of a coordinate system rotating with $\mathbf{H}_1$, none of the magnetic fields are changing as a function of time. Therefore, the axes of the rotating coordinate system can be selected so that

$$\mathbf{H}_0 = H_0 \mathbf{k}, \qquad \mathbf{H}_1 = H_1 \mathbf{i}, \qquad \boldsymbol{\omega} = -\omega \mathbf{k}. \qquad (\text{VI. 12})$$

Then on the rotating coordinate system

$$\mathbf{H}_{er} = (H_0 - \omega/\gamma_I)\mathbf{k} + H_1 \mathbf{i}, \qquad (\text{VI. 13})$$

as is shown schematically in Fig. VI. 1. Since this field is constant in

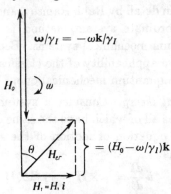

Fig. VI. 1. Effective magnetic field in the rotating coordinate system (RAB 54).

time the solution to the motion of the system is much simpler in the rotating coordinate system than in the stationary system. From Eq. (VI. 13) it follows that the magnitude of the effective magnetic field is

$$|H_{er}| = \sqrt{\{(H_0 - \omega/\gamma_I)^2 + H_1^2\}} = |a/\gamma_I|, \qquad (\text{VI. 14})$$

where

$$a = \sqrt{\{(\omega_0 - \omega)^2 + (\gamma_I H_1)^2\}} = \sqrt{\{(\omega_0 - \omega)^2 + (\omega_0 H_1/H_0)^2\}}, \qquad (\text{VI. 15})$$

with ω_0 by definition being $\gamma_I H_0$. Likewise the angle Θ of $\mathbf{H}_{er}$ relative to $\mathbf{H}_0$ is given by the following, and is consequently similar to the Θ of Eq. (V. 8):

$$\cos \Theta = (\omega_0 - \omega)/a, \qquad \sin \Theta = \gamma_I H_1/a = (\omega_0 H_1/H_0)/a. \qquad (\text{VI. 16})$$

From this it is apparent that when $\omega = \omega_0$, $\Theta = 90°$ and a magnetic moment initially parallel to $\mathbf{H}_0$ can precess about $\mathbf{H}_{er}$ until it becomes antiparallel to $\mathbf{H}_0$. In other words, such a moment can have its orientation relative to $\mathbf{H}_0$ changed most completely when $\omega = \omega_0$, so ω_0 can be considered as the resonance frequency of the system.

If one next goes to a second rotating coordinate system which rotates about $\mathbf{H}_{er}$ with a suitable angular velocity, the effective field $\mathbf{H}_{err}$ in the doubly rotating system can be reduced to zero. In this doubly rotating coordinate system the problems become trivial since there is then no magnetic field and consequently no change in the orientation of I. If **a** is the angular velocity about $\mathbf{H}_{er}$ which reduces $\mathbf{H}_{err}$ to zero, **a** must be determined by

$$0 = \mathbf{H}_{err} = \mathbf{H}_{er} + \frac{\mathbf{a}}{\gamma_I}. \qquad (VI.\ 17)$$

Hence, from (VI. 14) if $\boldsymbol{\alpha}$ is a unit vector parallel to $\mathbf{H}_{er}$

$$\mathbf{a} = -\gamma_I \mathbf{H}_{er} = -a\boldsymbol{\alpha}, \qquad (VI.\ 18)$$

where a is the quantity previously defined in Eq. (VI. 15).

As is shown later, the above considerations apply to a quantum mechanical as well as to a classical system. Consequently, either classically or quantum mechanically on the doubly rotating coordinate system with the two rotational angular velocities $\boldsymbol{\omega}$ and $\mathbf{a}$, there is no effective resultant field and the state of the system remains constant in time.

The above can be directly applied to various problems arising in nuclear resonance experiments. As a first application the method can be used to demonstrate the criterion for the rate of change of a field to be 'adiabatic', i.e. to be such that a nuclear moment preserves its magnetic quantum number (classically its angle) relative to the field as the field is moved. Let the field be rotated with angular velocity $-\omega_1$. Then for this problem H_0 is zero and H_1 is the full field H. On the rotating coordinate system

$$\mathbf{H}_{er} = (-\omega_1/\gamma_I)\mathbf{k} + H\mathbf{i}. \qquad (VI.\ 19)$$

The nuclei will then preserve their orientation relative to H provided $\mathbf{H}_{er}$ is approximately equal to $H\mathbf{i}$ or that

$$|\omega_1| \ll |H\gamma_I|, \qquad (VI.\ 20)$$

which can be written alternatively as

$$|\dot{\mathbf{H}} \times \mathbf{H}|/H^2 \ll |\gamma_I H|. \qquad (VI.\ 21)$$

The use of the rotating coordinate system also provides a simple pictorial interpretation of the transition process in the molecular beam magnetic-resonance method. A singly rotating coordinate system rotating with the oscillator frequency $-\omega$ can be used throughout. Prior to the molecule reaching the rotating field region H_1 equals zero and $\mathbf{H}_{er}$ has the value $(H_0 - \omega/\gamma_I)\mathbf{k}$. As the molecule enters the transition region where the rotating field is being established $\mathbf{H}_{er}$ changes. Conditions are usually such that near resonance the condition of Eq. (VI. 21) applied to $\mathbf{H}_{er}$ is violated. Consequently the transition under such circumstances is not adiabatic and the nuclear moments do not follow

H_{er} as H_1 is established. After H_1 achieves its full value, H_{er} is $(H_0-\omega/\gamma_I)\mathbf{k}+H_1\mathbf{i}$ and the nuclei precess about this effective field. When the molecules leave the rotating field region H_{er} again changes too rapidly for the nuclei to follow, and they are left with the orientation relative to the z axis to which they have precessed in the region of the rotating field. At exact resonance this precession is about a field H_1 which is perpendicular to the original direction of the field, and consequently the change of orientation can be large.

The qualitative analysis of the preceding paragraph can also be expressed quantitatively. Assume the I is initially parallel to H_0. Then in the rotating coordinate system I will precess about H_{er} with the precession angle Θ and at an angular velocity a. If α is the angle between H_0 and I, the simple geometry of the above precession is such that after a time interval t_2-t_1

$$\cos\alpha = \cos^2\Theta + \sin^2\Theta \cos a(t_2-t_1)$$

$$= 1-2\sin^2\Theta \sin^2\frac{a(t_2-t_1)}{2}. \qquad (\text{VI. 22})$$

On the other hand, as shown in the next section, the quantum mechanical angular momentum operators satisfy the same Eq. (VI. 8). Since this equation is linear, the expectation values satisfy the same equation. Therefore, a correspondence between the classical and quantum mechanical solutions can be established by requiring them to agree on their predicted average z component and on the total probability. For a spin of $\frac{1}{2}$ these requirements are

$$P_{\frac{1}{2},\frac{1}{2}} - P_{\frac{1}{2},-\frac{1}{2}} = \cos\alpha,$$

Therefore, $\qquad P_{\frac{1}{2},\frac{1}{2}} + P_{\frac{1}{2},-\frac{1}{2}} = 1. \qquad (\text{VI. 23})$

$$P_{\frac{1}{2},-\frac{1}{2}} = \frac{1-\cos\alpha}{2} = \sin^2\Theta \sin^2\frac{a(t_2-t_1)}{2}$$

$$= \frac{(\omega_0 H_1/H_0)^2}{(\omega_0-\omega)^2+(\omega_0 H_1/H_0)^2}\sin^2\frac{a(t_2-t_1)}{2}. \qquad (\text{VI. 24})$$

This result is in complete agreement with the combination of Eqs. (V. 10) and (VI. 6).

Likewise the rotating coordinate system analysis procedure is applicable to Ramsey's resonance method with separated oscillating fields (RAM 50c). In this case the description through the first oscillating field is just the same as in the preceding paragraph. After leaving the oscillating field, with this method, the nucleus enters an intermediate region where there is only H_0, so the magnitude of H_1 is zero. Relative to the singly rotating coordinate system, the nucleus in this region precesses about $(H_0 - \omega/\gamma_I)\mathbf{k}$ until it reaches the second oscillating field. As a result of this precession the nucleus will in general have a different orientation relative to $\mathbf{H}_{er}$ in the second rotating field region from its orientation in the first. On the other hand, if the average value of $H_0 - \omega/\gamma_I$ in the intermediate region is zero the orientation of the nucleus relative to $\mathbf{H}_{er}$ is the same in the second oscillating field region as in the first. This will be true regardless of the velocity of the molecule. However, if the average of $H_0 - \omega/\gamma_I$ has any value other than zero, the orientation of the nucleus relative to $\mathbf{H}_{er}$ in the second field is in general different from that in the first and the magnitude of the difference will depend upon the velocity of the molecule. When the combined effects of the two rotating field regions are averaged over the velocity of the molecules, the transition probability is therefore a maximum for ω equal to the average value of $\gamma_I H_0$ in the intermediate region.

VI. 2.3. *Quantum theory.* Relation (VI. 11) derived above can be proved equally as well by quantum mechanical as by classical means. As indicated in the previous section, one procedure is simply to use the fact that Eq. (VI. 8) applies to the quantum mechanical angular momentum operators. Eq. (VI. 8) can indeed easily be proved (RAB 54) to follow from the standard operator relation

$$(\hbar/i)\mathbf{dI}/dt = [\mathscr{H}, \mathbf{I}] = \mathscr{H}\mathbf{I} - \mathbf{I}\mathscr{H}, \quad (\text{VI. 25})$$

where the Hamiltonian $\mathscr{H}$ is taken as

$$\mathscr{H} = -\gamma_I \hbar \mathbf{I}.\mathbf{H}. \quad (\text{VI. 26})$$

Alternatively, from a wave mechanical point of view, the Schrödinger equation for the problem relative to a stationary coordinate system is

$$i\hbar\dot{\Psi} = \mathscr{H}\Psi = -\gamma\hbar\mathbf{I}.\mathbf{H}\Psi. \quad (\text{VI. 27})$$

However, in quantum mechanics the finite rotation operator (KEM 37, pp. 247, 307, and 532) for the coordinates to be rotated through an angle $\mathscr{S}$ about an axis along which the system's angular momentum $\mathbf{I}$ has the component $I_{\mathscr{S}}$ is the unitary operator $\exp(i\mathscr{S}I_{\mathscr{S}})$. Let Ψ and $\mathbf{H}$ be the wave function and field relative to a stationary coordinate system whereas Ψ_r and $\mathbf{H}_r$ are the same quantities relative to coordinates rotating with angular velocity $\boldsymbol{\omega}$. These quantities are related by the unitary transformation so that

$$\Psi = \exp(-i\boldsymbol{\omega}.\mathbf{I}t)\Psi_r, \quad (\text{VI. 28})$$

$$\mathbf{I}.\mathbf{H}_r = \exp(i\boldsymbol{\omega}.\mathbf{I}t)\mathbf{I}.\mathbf{H}\exp(-i\boldsymbol{\omega}.\mathbf{I}t). \quad (\text{VI. 29})$$

If Eq. (VI. 28) is substituted in Eq. (VI. 27) and if Eq. (VI. 29) is used to simplify the resulting expression, one obtains immediately

$$i\hbar\dot{\Psi}_r = -\gamma\hbar\mathbf{I}.(\mathbf{H}_r+\boldsymbol{\omega}/\gamma)\Psi_r = -\gamma\hbar\mathbf{I}.\mathbf{H}_{er}\Psi_r, \qquad (\text{VI. 30})$$

where $\mathbf{H}_{er}$ is given by Eq. (VI. 11), with the understanding that H is to be expressed relative to the rotating coordinate system. This result justifies the application of the previous discussions to quantum mechanical as well as classical systems.

The probabilities for transition of the system from a state of one magnetic quantum number to another can readily be calculated with the above procedure. Consider the case discussed earlier in which the magnetic fields are given by Eq. (VI. 12) and assume that up to time t_1 the magnitude of H_1 is zero after which it is H_1 until time t_2. As use will be made of the stationary, singly, and doubly rotating coordinate systems previously discussed, wave functions relative to these three systems will be designated as $\Psi(t)$, $\Psi_r(t)$, and $\Psi_{rr}(t)$, respectively. Since $\mathbf{H}_{err}$ for the doubly rotating coordinates is zero,

$$\Psi_{rr}(t_2) = \Psi_{rr}(t_1) = \Psi_r(t_1). \qquad (\text{VI. 31})$$

However, as between times t_1 and t_2 the doubly rotating system has rotated through an angle $-a(t_2-t_1)$ relative to the singly rotating system one must use the previous finite rotation operator to relate $\Psi_{rr}(t_2)$ and $\Psi_r(t_2)$, with the result that

$$\Psi_r(t_2) = \exp(ia[t_2-t_1]\boldsymbol{\alpha}.\mathbf{I})\Psi_{rr}(t_2). \qquad (\text{VI. 32})$$

Hence from Eq. (VI. 31)

$$\Psi_r(t_2) = \exp(ia[t_2-t_1]\boldsymbol{\alpha}.\mathbf{I})\Psi_r(t_1). \qquad (\text{VI. 33})$$

In a similar fashion, this can be reduced to a non-rotating coordinate system with the result

$$\Psi(t_2) = \exp(-i\omega t_2\mathbf{k}.\mathbf{I})\exp(ia[t_2-t_1]\boldsymbol{\alpha}.\mathbf{I})\exp(i\omega t_1\mathbf{k}.\mathbf{I})\Psi(t_1). \qquad (\text{VI. 34})$$

It should be noted, however, that because the components of I do not commute one cannot perform all the operations appropriate to exponentials of ordinary numbers; instead, the exponentials may be taken as defined by their series expansion.

From the above, the transition probability from a state m to a state m' can be calculated by taking $\Psi(t_1) = \Psi_m$, in which case

$$P_{m,m'} = |(\Psi_{m'}, \Psi(t_2)|^2$$
$$= |(m'|\exp(i\omega t_2\mathbf{k}.\mathbf{I})\exp(ia[t_2-t_1]\boldsymbol{\alpha}.\mathbf{I})\exp(-i\omega t_1\mathbf{k}.\mathbf{I})|m)|^2$$
$$= |(m'|\exp(ia[t_2-t_1]\boldsymbol{\alpha}.\mathbf{I})|m)|^2. \qquad (\text{VI. 35})$$

The last simplifying step is a consequence of Ψ'_m and $\Psi'_{m'}$ being eigenfunctions of k.I.

In general, the numerical evaluation of Eq. (VI. 35) is somewhat cumbersome because of the non-commutativity of the terms in the exponent. However, a series expansion of the exponential may be used. In the special case of spin $\frac{1}{2}$, it becomes much simplified, for then $I = \frac{1}{2}\sigma$, where σ is the Pauli spin operator. Since, for the Pauli spin operator, $(\alpha.\sigma)^2$ equals one, the series expansion of the above exponential together with the series expansion for sine and cosine merely give

$$\exp(i\tfrac{1}{2}a[t_2-t_1]\alpha.\sigma)$$
$$= \cos\frac{a(t_2-t_1)}{2} + i\alpha.\sigma\sin\frac{a(t_2-t_1)}{2}$$
$$= \cos\frac{a(t_2-t_1)}{2} + i(\sigma_x\sin\Theta + \sigma_z\cos\Theta)\sin\frac{a(t_2-t_1)}{2}. \quad \text{(VI. 36)}$$

Therefore $\quad P_{\frac{1}{2},-\frac{1}{2}} = \sin^2\Theta \sin^2\frac{a(t_2-t_1)}{2}, \quad$ (VI. 37)

which is the desired transition probability that is applicable to the conventional molecular beam resonance method. It should be noted that the above agrees with the classically derived expression in Eq. (VI. 24). The averaging over the velocity distribution is, of course, the same as in Chapter V.

The above procedure may also be used to calculate the transition probability applicable to the molecular beam resonance method with separated oscillating fields. In this case, Eq. (VI. 34) can be applied separately to the first oscillating field region, to the intermediate region, and to the second oscillating field region. The resulting three equations can then be combined to express the final state of the system in terms of the initial state and from this the transition probabilities can be calculated as in Eq. (VI. 35).

VI. 3. Spins greater than one-half

All of the transition probability expressions previously derived have been limited to spin $\frac{1}{2}$. In Chapter V this arose from the limitation to cases with only two relevant states and in the preceding section a spin of $\frac{1}{2}$ was assumed to simplify the final evaluation. From Eq. (I. 2) it is apparent that for a magnetic moment of spin one or greater interacting with a magnetic field the resonance frequencies corresponding to all the possible $\Delta m_I = \pm 1$ are the same.

The problem of transitions with spin greater than $\frac{1}{2}$ was first solved by Majorana (MAJ 32) by group theoretical methods. He showed that the transition probabilities $P_{m,m'}$ for a general spin I could be related directly to the probability $P_{\frac{1}{2},-\frac{1}{2}}$ calculated above for the transition of a nucleus of spin $\frac{1}{2}$ provided the two nuclei were assumed to have the same gyromagnetic ratio. Bloch and Rabi (BLO 45) provided a simpler derivation of the Majorana formula by synthesizing a moment of high spin from a number of spin $\frac{1}{2}$ constituents. A still different derivation has been given by Hammermesh (HAM 47) from Eq. (VI. 35) but without restricting the spin to $\frac{1}{2}$ in the next step as was done in the above discussion. Schwinger (SCH 52, ZXZ 54) and Ramsey (RAM 50e) further simplified the Bloch and Rabi procedure. Such a derivation of the Majorana formula is given in Appendix E (ZZA 47).

The result of the derivation in Appendix E is that the probability $P_{m,m'}$ for the moment starting with magnetic quantum number m to end with m' after the perturbation is given by

$$P_{m,m'} = (I-m)!\,(I+m)!\,(I-m')!\,(I+m')!\,(\sin\tfrac{1}{2}\alpha)^{4I} \times$$

$$\times \left[\sum_r \frac{(-1)^r(\cot\tfrac{1}{2}\alpha)^{m+m'+2r}}{(I-m-r)!\,(I-m'-r)!\,(m+m'+r)!\,r!}\right]^2, \quad \text{(VI. 38)}$$

where
$$\sin^2\tfrac{1}{2}\alpha = P_{\frac{1}{2},-\frac{1}{2}}, \quad \text{(VI. 39)}$$

with $P_{\frac{1}{2},-\frac{1}{2}}$ being the transition probability for a nucleus of spin $\frac{1}{2}$ and the same gyromagnetic ratio. The value of $P_{\frac{1}{2},-\frac{1}{2}}$ and hence of $\sin^2\tfrac{1}{2}\alpha$ in Eq. (VI. 39) can be found in the appropriate cases from Eqs. (V. 10), (V. 37), (V. 44), (VI. 6), or (VI. 24). The summation index r is allowed to range in the above over all values for which no argument of a factorial becomes negative. The above equation is frequently written in a superficially different form with the summation index $\lambda = I-m'-r$. However, the form of Eq. (VI. 38) possesses the advantage of demonstrating explicitly the symmetry in m and m'. From Eq. (VI. 38) and the restrictions on r it is apparent that

$$P_{m,m'} = P_{m',m} = P_{-m,-m'}. \quad \text{(VI. 40)}$$

In using the Majorana formula to calculate the average probability over the molecular velocity distribution, one must be careful to calculate $P_{m,m'}$ first at a specific velocity and then average over the velocity distribution to obtain $\langle P_{m,m'}\rangle$. An incorrect result will be obtained if $\langle P_{\frac{1}{2},-\frac{1}{2}}\rangle$ is used in Eq. (VI. 39).

Since Eq. (VI. 38) is not very convenient, it is often useful to consider the special case which is the probability for spin I of a transition from $m = I$ to $m' = I-1$. In this case it is easy to see (RAM 50e) that

$$P_{I,I-1} = \frac{(2I)!}{(2I-1)!}(1-\sin^2\tfrac{1}{2}\alpha)^{2I-1}\sin^2\tfrac{1}{2}\alpha. \quad \text{(VI. 41)}$$

This can be differentiated with respect to α and equated to zero to obtain the conditions for which $P_{I,I-1}$ is a maximum. The result is that the maximum is for

$$\sin \tfrac{1}{2}\alpha = \sqrt{(P_{\frac{1}{2},-\frac{1}{2}})} = \sqrt{\left(\frac{1}{2I}\right)}. \qquad (VI.\ 42)$$

The physical reason that $P_{I,I-1}$ does not continue to increase with increasing $P_{\frac{1}{2},-\frac{1}{2}}$ is, of course, that for too large values of the latter too many transitions take place from $m = I-1$ to $m' = I-2$, etc. With the above value for $\sin \tfrac{1}{2}\alpha$, Eq. (VI. 41) gives for its maximum possible value

$$(P_{I,I-1})_{\max} = \left(1 - \frac{1}{2I}\right)^{2I-1} \qquad (I > \tfrac{1}{2}),$$

$$(P_{\frac{1}{2},-\frac{1}{2}})_{\max} = 1 \qquad (I = \tfrac{1}{2}). \qquad (VI.\ 43)$$

The form which Eq. (VI. 38) acquires in a specific case can easily be worked out, say for a spin of $\tfrac{3}{2}$. If c is written for $\cos\tfrac{1}{2}\alpha$ and s for $\sin\tfrac{1}{2}\alpha$, the result is

$$P_{\frac{3}{2},\frac{3}{2}} = c^6, \quad P_{\frac{3}{2},\frac{1}{2}} = 3s^2c^4, \quad P_{\frac{3}{2},-\frac{1}{2}} = 3s^4c^2, \quad P_{\frac{3}{2},-\frac{3}{2}} = s^6,$$

$$P_{\frac{1}{2},\frac{1}{2}} = 4c^2s^4 - 4c^4s^2 + c^6, \quad P_{\frac{1}{2},-\frac{1}{2}} = 4c^4s^2 - 4c^2s^4 + s^6. \qquad (VI.\ 44)$$

All other values of $P_{m,m'}$ are determined from the above and Eq. (VI. 40). If $\sin\tfrac{1}{2}\alpha$ is selected in accordance with Eq. (VI. 42) to maximize $P_{\frac{3}{2},\frac{1}{2}}$, the numerical values of $P_{m,m'}$ are

$$P_{\frac{3}{2},\frac{3}{2}} = 0\cdot 296, \quad P_{\frac{3}{2},\frac{1}{2}} = 0\cdot 444, \quad P_{\frac{3}{2},-\frac{1}{2}} = 0\cdot 222, \quad P_{\frac{3}{2},-\frac{3}{2}} = 0\cdot 037,$$

$$P_{\frac{1}{2},\frac{1}{2}} = 0, \quad P_{\frac{1}{2},-\frac{1}{2}} = 0\cdot 333. \qquad (VI.\ 45)$$

VI. 4. Signs of nuclear moments

It is apparent from the above discussion—Eq. (VI. 24) for example—that the sign of a nuclear magnetic moment can easily be determined in the molecular-beam magnetic-resonance method. All that is required is the use of a rotating instead of an oscillatory field, in which case, by a determination of the direction of rotation (sign of ω) that gives resonance, the sign of ω_0 and hence of γ_I is determined. The signs are in fact most easily kept clear by noting that the transition field must precess in the same direction as a classical magnetic moment with the same sign as the nuclear moment concerned. The sign of the neutron magnetic moment, for example, has been determined with a rotating field (ROG 49).

Millman (MIL 39), however, has pointed out that it is often not necessary to introduce explicitly a rotating field since the design of the transition field wires is often such as to introduce a slight rotating component accidentally. The loop carrying the oscillatory current is often in the form shown in Fig. VI. 2. As the beam approaches the radiofrequency loop and is at position D the oscillatory field is approximately in the direction of the beam, but after the molecules are inside the loop as at F the direction of the oscillatory field is approximately perpendicular to the beam; in other words, the direction of the oscillatory field is rotated approximately ninety degrees as the molecule enters

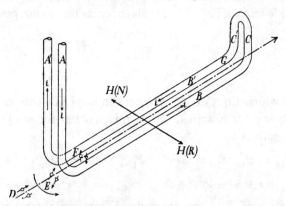

Fig. VI. 2. Schematic perspective diagram of the two-wire oscillating field. The large arrows indicate the direction of the constant field and the small arrows the oscillatory fields. $H(N)$ indicates the magnetic field in its normal direction and $H(R)$ in the reverse direction (MIL 39).

the loop. For the loop of Fig. VI. 2 it is apparent that the field rotates another ninety degrees in the same direction as the molecule leaves the loop. If the transition region on entering the loop is of length Δl, the apparent increment of angular frequency due to the rotation will be $v\pi/2\Delta l$; incremental frequencies of 2×10^5 radians sec^{-1} are often found in practice. If the basic oscillatory field is resolved into two oppositely rotating components, the above incremental frequency will increase the magnitude of the frequency for the component rotating in the same direction as the rotation from the loop geometry and will decrease the magnitude of the other component. However, depending on the sign of the nuclear magnetic moment, only one of these rotating components is effective in producing transitions. Therefore, transitions are induced by two oscillatory frequencies, one in the main part of the loop at the basic oscillator frequency and one in the end regions at a frequency that is either higher or lower than that of oscillator depending on the sign of the nuclear moment. These two effective frequencies give rise to an asymmetry of the resonance pattern with the direction of the asymmetry depending on the sign of the nuclear magnetic moment.

Experimental curves demonstrating the Millman effect are shown in Fig. VI. 3. The curve indicated by N corresponds to the normal direc-

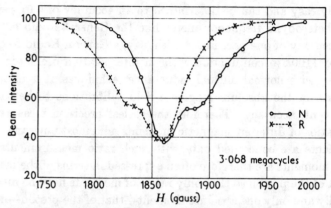

FIG. VI. 3. Pair of Li^7 curves from Li_2 molecules. N refers to the normal direction of the constant field and R the reverse. N corresponds to the positive Z direction of a right-handed system of axes where the direction of the positive molecular velocity determines the positive X axis and the downward direction determines the positive Y axis (MIL 39).

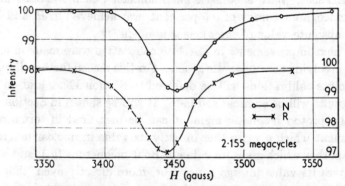

FIG. VI. 4. Pair of Li^6 resonance curves from LiCl molecules. N and R have the same meaning as in Fig. VI. 3 (MIL 39).

tion of the uniform magnetic field and R to the reversed direction showing that the direction of the asymmetry is reversed when the direction of the nuclear precession is reversed. Often the asymmetry is much less but still detectable as in Fig. VI. 4.

It is, of course, possible to design the radiofrequency current loops in such a way that Millman effect is avoided. Such designs, as illustrated in Chapter XIV, are often of value in precision measurements, though they sacrifice information on signs of nuclear moments.

VI. 5. Absolute values of nuclear moments

All magnetic resonance measurements of nuclear moments depend on a precision measurement of ω_0 and of H_0, from which γ_I can be determined from Eq. (VI. 2). Absolute frequency measurements are relatively easy and can be determined to an accuracy of 0·01 per cent. with a heterodyne frequency meter like the General Radio 620A and to an accuracy of one part in 10^8 with, say, a General Radio frequency standard 1100A in conjunction with an interpolation oscillator 1107A. However, with normal mutual inductance techniques, it is difficult to be certain of the absolute magnetic field calibrations to better than 0·5 per cent. accuracy. Therefore the easiest precision measurements are of ratios of different magnetic moments, and most nuclear moment experiments are concerned only with such ratio measurements. The nuclear moments are therefore often expressed in terms of the magnetic moment of the proton so that only ratios of moments need be measured ordinarily and only one absolute moment—that of the proton—need be measured in order for all the moments to be on an absolute scale.

The earliest absolute calibrations in nuclear resonance experiments were by Rabi, Millman, Kusch, Zacharias, Kellogg, and Ramsey (RAB 39, KEL 39a). With a ballistic galvanometer and a standard mutual inductance, an accuracy of 0·5 per cent. was achieved in H_0 and hence in the absolute value of the nuclear moment.

The next improvements in absolute calibration were made in experiments involving paramagnetic atoms. For this reason further discussion of absolute calibrations will be deferred to Section IX. 2 and only the final results will be summarized here. It will be shown in Section IX. 2 that the proton magnetic moment can be measured in terms of one experiment which gives rather directly its value in nuclear magnetons (μ_N), another which gives its value in Bohr magnetons (μ_0), and a third that gives its value in ergs gauss^{-1} or, more directly even, that gives the gyromagnetic ratio γ_p in sec^{-1} gauss^{-1}. The results of these experiments when a shielding correction (see Section VI. 7) is made, as indicated by subscript c, are

$$\mu_{p,c} = (2{\cdot}792743 \pm 0{\cdot}000060)\mu_N$$

$$\mu_{p,c} = (0{\cdot}00152101 \pm 0{\cdot}00000002)\mu_0$$

$$\gamma_{p,c} = 2\mu_p/\hbar = (26753{\cdot}05 \pm 0{\cdot}60) \text{ sec}^{-1} \text{ gauss}^{-1}, \qquad \text{(VI. 46)}$$

where μ_N is the nuclear magneton and μ_0 the Bohr magneton. The experiments (BLO 50, HIP 49, LIN 50, JEF 51, TAU 49, GAR 49, THO 49) which lead to these values will be described in Chapter IX.

With these values only ratios of moments need be measured since absolute values can then be obtained from the above. In the tables of nuclear moments it will be assumed that the magnetic moment of the proton is exactly 2·792743 nuclear magnetons.

VI. 6. Nuclear magnetic moment measurements

The above molecular-beam resonance techniques have been used to measure a number of nuclear magnetic moments and their signs in many

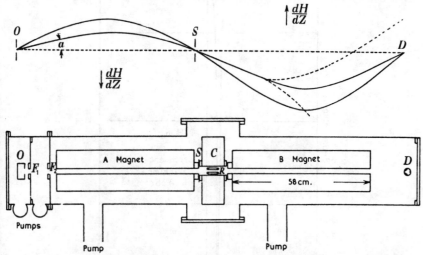

FIG. VI. 5. Schematic diagram of a molecular-beam apparatus and of some molecular paths. The two full curves in the upper part of the figure indicate the paths of two molecules having different moments and velocities and whose moments do not change during passage through the apparatus. The dotted curves indicate possible changes in path if the component of magnetic moment is changed. The motion in the z direction is greatly exaggerated (KEL 46).

experiments performed by Rabi, Millman, Kusch, Zacharias, Kellogg, Ramsey, and others (RAB 39, KEL 39a, KUS 39, KUS 39a, KUS 39b, MIL 39a, MIL 39b, HAY 41, MIL 41, BRO 47, KUS 49c, RAM 52c, BRA 52). A typical apparatus used in these measurements is shown schematically in Fig. VI. 5.

A typical fully resolved resonance is that obtained with HD and shown in Fig. VI. 6. The sharp central minimum arises from HD in the zeroth rotational state. For this state none of the molecular interactions of Chapter VIII contribute and the nuclear interaction with the external field is all that remains. The line is therefore just at the Larmor frequency and the width is the theoretical width of Eq. (VI. 24). The small subsidiary maxima are due to the $J = 1$ state and are displaced from the Larmor frequency as a result of the various molecular interactions discussed in Chapter VIII. With the separated oscillatory field such a resolved resonance is of the form shown in Fig. V. 6.

On the other hand, the nuclear moment resonances often have the appearance of Fig. VI. 7. The width of this resonance is much greater

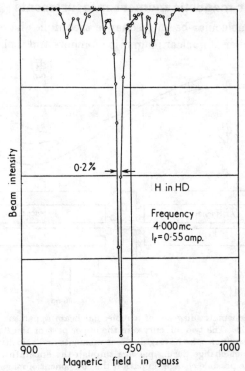

FIG. VI. 6. Resonance curve for HD molecules. The largest resonance corresponds to the transition of the proton spin for the zero rotational state (KEL 39a).

than either the theoretical widths of Eq. (VI. 24) or the width to be expected from possible field inhomogeneities. The explanation of the width is that there are nuclear quadrupole and spin rotational interactions in the molecules as discussed in Chapters III and VIII. These give rise to slight displacements of the individual resonances about the Larmor frequency as shown in Chapter VIII. However, a large number of rotational states are excited in many experiments; for NaBr, as an example, an average value of rotational quantum number $J = 40$ is to be expected at the temperature of the molecular-beam oven. Since, for each of these, $2J+1$ different orientation states can occur, there will be so many closely spaced resonances that a resolution into separate lines is impossible and only an average resonance pattern that is much broadened as in Fig. VI. 7 will be observed. The results of nuclear moment measurements by molecular-beam methods are given in the tables of Section VI. 9.2.

Since nuclear quadrupole interactions in a molecule are often comparable to the interaction with the external magnetic field, a means of distinguishing that the observed resonance is indeed of the form of Eq. (VI. 2) is needed. This is ordinarily done by varying the magnetic field

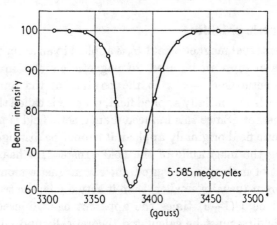

FIG. VI. 7. Resonance curve of the Li⁷ nucleus observed in LiCl (RAB 39).

by about a factor of two or so and noting whether the frequency changes proportionally with H_0 as implied in Eq. (VI. 2). As a further check, and as a means of identifying the nucleus, the moment is usually measured in more than one molecule containing the same nucleus.

Ordinarily the preceding tests are adequate. However, Ramsey (RAM 48) has pointed out that these tests are not sufficient in certain cases. In particular, if the molecule possesses a quadrupole interaction that is large compared with the magnetic interaction with an external field, a resonance will occur under certain circumstances at just double the Larmor frequency and with a field and molecule dependence similar in many ways to a resonance at the Larmor frequency. However, in such a case, Ramsey (RAM 48) has shown that a double Larmor frequency resonance can be identified by its shape, its intensity, and its variation with H_0. The predicted twice Larmor frequency resonances have been observed experimentally by Coté and Kusch (COT 53) at low magnetic fields with alkali halides. Coté and Kusch have also extended Ramsey's theory to include the effect of spin-rotational magnetic interaction as in Chapter VIII on the shape of the resonance. The theories of Ramsey and of Coté and Kusch are discussed further in Section VIII. 5.

A source of error that cannot be eliminated by the preceding tests is the effect of magnetic shielding, since the resonance frequency for a magnetically shielded nucleus varies proportionally with H_0 just as for an unshielded one. This correction will be discussed in the next section.

VI. 7. Magnetic shielding

When an external magnetic field H_0 is applied to an atom or molecule, the electrons acquire an induced diamagnetic circulation which produces a magnetic field $-\sigma H_0$ at the position of the nucleus which partially cancels the initially applied field, where σ is called the magnetic shielding constant. Since this induced diamagnetic field is proportional to the magnetic field originally applied, it cannot be distinguished from it by varying the magnitude of the field. Instead, it has the experimental effect of making a measured nuclear magnetic moment appear slightly different (usually smaller) than it actually is; the true magnetic moment will be $1/(1-\sigma)$ times the apparent one. Consequently the magnetic shielding must be calculated theoretically and allowed for in inferring the magnitude of the nuclear moment from the externally applied field and the observed resonance frequency.

The first calculations of magnetic shielding were made by Lamb (LAM 41) who showed that for atoms only

$$\sigma = \frac{e^2}{2mc^2} \int \frac{(x^2+y^2)\rho}{r^3}\, d\tau = \frac{e^2}{3mc^2} \int \frac{\rho}{r}\, dr \qquad \text{(VI. 47)}$$

$$= \frac{e^2}{3mc^2} \Big\langle \sum_k 1/r_k \Big\rangle.$$

where ρ is the density of electrons. The result is restricted to atoms only because of the implicit assumption that the nuclear electrostatic field is spherically symmetric. The preceding equation follows almost immediately from Larmor's theorem and Ampère's law of electromagnetism. Values for this correction have been tabulated by Lamb (LAM 41) with Thomas–Fermi, and in a few cases Hartree wave functions, and by Dickinson (DIC 50a) with Hartree and Hartree–Fock functions. Although Lamb's theory applies directly only to atoms, most measurements are made with molecules. Since no better theory was available for a long time, all corrections in molecules were made according to Lamb's theory, the correction being made only for the atom concerned.

However, it was pointed out later by Ramsey (RAM 50a, RAM 50d), that with the increased accuracy of nuclear moment measurements the error from this approximate calculation of the diamagnetic correction was appreciable. For very light molecules, like hydrogen in particular, the error in Lamb's value of the diamagnetic correction is comparable to the magnitude of the correction itself. Ramsey (RAM 50d, RAM 51, RAM 52e) showed that the average over all molecular orientations of the magnetic shielding correction could be expressed as

$$\sigma = \sigma^L + \sigma^{HF}, \qquad (\text{VI. 48})$$

where
$$\sigma^L = \frac{e^2}{3mc^2}\int \frac{\rho}{r}\,d\tau = \frac{e^2}{3mc^2}\Big\langle \sum_k 1/r_k \Big\rangle \qquad (\text{VI. 49})$$

and
$$\sigma^{HF} = -\tfrac{4}{3}\,\text{Re} \sum_{n'\lambda'}{}' \frac{\big(n\lambda\big|\sum_k \mathbf{m}_k^0/r_k^3\big|n'\lambda'\big)\cdot\big(n'\lambda'\big|\sum_k \mathbf{m}_k^0\big|n\lambda\big)}{E_{n'}-E_n}. \qquad (\text{VI. 50})$$

The first integral is to be taken over the entire molecule and the origin of r and r_k is the nucleus for which the shielding is desired. Likewise $\sum_k$ is over all electrons in the molecule. Re denotes the real part only of the subsequent expression and the prime on the $\sum'_{n'\lambda'}$ indicates that the state $n' = n$ is not included in the summation. $\big(n\lambda\big|\sum_k \mathbf{m}_k^0\big|n'\lambda'\big)$ is the matrix element between the molecular electronic ground state n and the electronic excited state n' (with the molecular axis in an orientation expressed by λ) of the orbital magnetic moment operator $\mathbf{m}_k^0$ for the kth electron whose z component for example is

$$m_{zk}^0 = -(e\hbar/2mci)\left[x_k \frac{\partial}{\partial y_k} - y_k \frac{\partial}{\partial x_k}\right]. \qquad (\text{VI. 51})$$

The σ^{HF} of Eq. (VI. 50) corresponds to the second-order paramagnetism (often called the high frequency term) which enters into the theory of the ordinary diamagnetic susceptibility of molecules as developed by Van Vleck (VAN 32, p. 275).

Since the σ^{HF} depends on the excited electronic states of the molecule it is very difficult to evaluate. However, if $E_n - E_0$ is replaced by the suitable average energy ΔE of the excited states it can be extracted from the summation and the summation evaluated with the aid of the closure property (SCH 49) to give

$$\sigma^{HF} = -\frac{4}{3\Delta E}\left(n\left|\sum_{jk} \mathbf{m}_j^0 \cdot \frac{\mathbf{m}_k^0}{r_k^3}\right|n\right). \quad (\text{VI. 52})$$

Nevertheless, even this is difficult to evaluate.

However, Ramsey (RAM 50d, RAM 52f, RAM 53a) has shown that for linear molecules σ^{HF} in Eq. (VI. 50), even before the preceding approximation, can be evaluated from experimental data on the spin rotational magnetic interaction discussed in Chapter VIII. In particular if c_W is the spin rotational interaction constant of Chapter VIII for the nucleus considered and if effects of zero point vibration and centrifugal stretching are temporarily neglected, Ramsey showed that

$$\sigma^{HF} = -\frac{e^2}{6mc^2\mu_N}\left[\sum_i \frac{2Z_i\mu_N}{R_i} - \frac{2\pi(\mu'R^2)c_W}{M\gamma_I}\right], \quad (\text{VI. 53})$$

where $(\mu'R^2)$ is the moment of inertia of the molecule, R_i is the distance of the nucleus with charge Z_i from the nucleus where the shielding is desired, and M is the mass of the proton. Later, Ramsey (RAM 52f, RAM 53a, HAR 53) showed that an evaluation of $\langle\sigma\rangle$ was also possible with a suitable allowance for molecular vibration and rotation. The effects of vibration and centrifugal stretching are discussed in further detail in Section VIII. 7.

The close relation between spin-rotational interaction and magnetic interaction might appear at first sight to be a surprising accident. However, it can be seen to be quite reasonable from the following point of view. Larmor's theorem states essentially that the application of a magnetic field **H** produces a similar effect on the electrons as a rotation of angular velocity

$$\omega = e\mathbf{H}/2mc. \quad (\text{VI. 54})$$

The electronic contribution to the spin-rotational interaction is determined by the magnetic field at the nucleus resulting from the electronic motion which is a consequence of the rotation at angular velocity ω. On the other hand, the magnetic shielding depends on the magnetic field at the nucleus resulting from the electronic motion which is a consequence of the application of an external field **H**. Since ω and **H** are related by Eq. (VI. 54), the shielding and spin-rotational interaction constants can be related. This argument can indeed be made quantitative and Eq. (VI. 53) can be derived in this way (RAM 50d).

With the value of Harrick, Barnes, Bray, and Ramsey (HAR 53) for c_{HW} and with Newell's theoretical value (NEW 50a, SMA 51) for the first term, Eq. (VI. 51) yields for the average value of σ in the first rotational state of H_2

$$\sigma^L = (3\cdot 21 \pm 0\cdot 02) \times 10^{-5},$$

$$\sigma^{HF} = -(0\cdot 59 \pm 0\cdot 03) \times 10^{-5},$$

$$\sigma = (2\cdot 62 \pm 0\cdot 04) \times 10^{-5}. \qquad (VI.\ 55)$$

In a similar fashion the σ for D_2 has been found (RAM 52f, RAM 53a, SMA 51, RAM 52c, and HAR 53) to have almost the same value. It should be noted that Eq. (VI. 53) comes directly from Eq. (VI. 50) without the approximations involved in Eq. (VI. 52). As an alternative to the above evaluation for H_2, Hylleraas and Skavlem (HYL 50) and Ishiguro and Koide (ISH 54) have attempted to calculate the last term in Eq. (VI. 50) from approximate hydrogen wave functions, but this procedure should be much less accurate because of the approximations involved and is best interpreted as a theoretical calculation of the experimentally measurable spin-rotational magnetic interaction constant.

Because of the difficulty of evaluating the diamagnetic correction in this manner for molecules other than hydrogen, it is still customary to approximate the magnetic shielding in heavy molecules by the correction for the single atom containing the nucleus concerned. This procedure is relatively fairly accurate for nuclei of large Z, since the innermost electrons are most effective in producing magnetic shielding owing to the $1/r$ factor in Eq. (VI. 47). The validity of this approximation has been considered theoretically by Ramsey (RAM 52f).

From Eq. (VI. 49) and (VI. 50) one would expect to find differences in the magnetic shielding in different molecules. Such effects have indeed been discovered in independent nuclear paramagnetic resonance experiments by Knight (KNI 49), Dickinson (DIC 50), and Proctor and Yu (PRO 50). Such shifts of resonance frequency in different chemical compounds have sometimes been called the chemical effect. As examples, shifts of 0·015 per cent. and 0·05 per cent. have been found in different compounds of nitrogen. Similarly shifts of a few parts in a million have been found in H_2, water, and mineral oil so the σ in these should be taken as $2\cdot 62 \times 10^{-5}$, $2\cdot 56 \times 10^{-5}$, and $2\cdot 78 \times 10^{-5}$ respectively (THO 50, GUT 51).

In molecular beam experiments where the molecules are not subject to frequent collisions, the shielding can be observed in different orientation states of the molecule and it will not necessarily be the same in different states. For such cases, Ramsey (RAM 51c, RAM 52a) has shown that the shielding constant $\sigma(J)$ of a linear molecule should depend on the rotational angular momentum operator $\mathbf{J}$ as

$$\sigma(\mathbf{J}) = \tfrac{2}{3}\sigma_\sigma + \tfrac{1}{3}\sigma_\pi + (\sigma_\sigma - \sigma_\pi)\frac{2}{3(2J-1)(2J+3)}[3(\mathbf{J}\cdot\mathbf{H})^2/H^2 - \mathbf{J}^2],$$
(VI. 56)

where σ_σ is the magnetic shielding of the nucleus when the field $\mathbf{H}$ is perpendicular to the internuclear axis and σ_π is the shielding parallel to the internuclear axis. The above dependence on the operator $\mathbf{J}$ can be proved by averaging over the molecular orientations of the m_J state concerned as in Eq. (III. 45) ff. or by proceeding analogously to Eq. (III. 97) (RAM 51c). If x_{0k}, y_{0k}, z_{0k} are Cartesian coordinates fixed in the molecule of the kth electron with the z_0 axis parallel to the internuclear line,

$$\sigma_\pi = (e^2/2mc^2)\Big(n\Big|\sum_k (r_k^2 - z_{0k}^2)/r_k^3\Big|n\Big),$$

$$\sigma_\sigma = \tfrac{1}{2}(3\sigma - \sigma_\pi),$$
(VI. 57)

where σ is given by Eq. (VI. 48). If σ_{J,m_J} is the value of the shielding for the rotational state J with magnetic quantum number m_J, the above gives for example with linear molecules

$$\sigma_{1,\pm 1} - \sigma_{1,0} = \frac{e^2}{10mc^2}\Big\langle \sum_k (3z_{0k}^2 - r_k^2)/r_k^3 \Big\rangle - \tfrac{3}{5}\sigma^{HF}.$$
(VI. 58)

Since σ^{HF} can be evaluated from Eq. (VI. 53), a measurement of the above difference in shielding yields an experimental value of the quadrupole moment of the electron distribution weighted with a $1/r_k^3$ factor. So far the molecular beam experiments have not been quite accurate enough for such a determination, but with some of the new techniques now under development, as discussed in Chapters V and XIV, such a measurement should soon be possible.

VI. 8. Molecular rotational magnetic moments

VI. 8.1. *Experimental measurements of rotational magnetic moments.* Even non-paramagnetic molecules like those in $^1\Sigma$ states can acquire a small magnetic moment by virtue of the rotation of the molecule. The order of magnitude to be expected for such a rotational moment can be seen from the easily calculated result that two protons rotating about their common centre of mass with an angular momentum $\hbar$ will possess a magnetic moment of one nuclear magneton. The presence of the negative electrons and neutral nuclear constituents in general reduce the magnetic moment below one nuclear magneton.

The first measurements of rotational magnetic moments by the magnetic resonance method were those by Ramsey (RAM 40) on H_2, HD, and D_2. A typical resonance curve is shown in Fig. VI. 8; the multiplicity of resonances is due to the molecular interactions discussed in detail in Chapter VIII. He found the magnetic moments of these molecules to be 0.8787 ± 0.0070, 0.6601 ± 0.0050, and 0.4406 ± 0.0030 nuclear magnetons, respectively, with all of the moments positive. To within the precision of the experiment these values are in the ratio of 4:3:2,

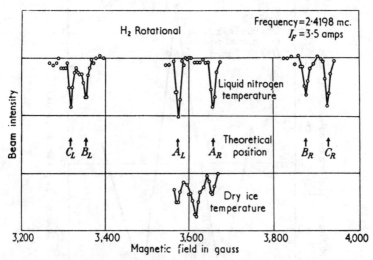

FIG. VI. 8. The radiofrequency spectra of H_2 corresponding to transitions in which the rotational angular momentum changes its orientation. The oscillator frequency is 2.4198 megacycles and the oscillating field is about 10 gauss (oscillating current 3.5 amps) (RAM 40).

as would be expected from the relative rotational angular velocities and as discussed in greater detail in Section VI. 8.2. Likewise the rotational magnetic moment in the second rotational state was measured for H_2 and it was found to be just twice that for the first rotational state within the experimental error, as indicated by the location of the additional peak on the lower curve of Fig. VI. 8.

More recently the rotational magnetic moments μ_J of H_2 and D_2 have been remeasured with much greater accuracy with the separated oscillatory field method by Harrick and Ramsey (HAR 52) and by Barnes, Bray, and Ramsey (BAR 54). A typical resonance curve for one transition in their experiment is shown in Fig. VI. 9. If $^H\langle\mu_J\rangle_1$ indicates the average value of the rotational magnetic moment μ_J for the H_2 molecule in the zeroth vibrational and first rotational state, their results may be expressed as

$${}^H_0\langle\mu_J/J\rangle_1 = 0{\cdot}88291\pm0{\cdot}0007 \text{ nucl. magns.} \qquad (\text{VI. 59})$$

$${}^D_0\langle\mu_J/J\rangle_1 = 0{\cdot}442884\pm0{\cdot}000052 \text{ nucl. magns.} \qquad (\text{VI. 60})$$

$${}^H_0\langle\mu_J/J\rangle_2 = 0{\cdot}882265\pm0{\cdot}000035 \text{ nucl. magns.} \qquad (\text{VI. 61})$$

$${}^D_0\langle\mu_J/J\rangle_1(M_d/M_p)/{}^H_0\langle\mu_J/J\rangle_1 = 1{\cdot}00274\pm0{\cdot}00011 \qquad (\text{VI. 62})$$

$${}^H\langle\mu_J/J\rangle_2/{}^H_0\langle\mu_J/J\rangle_1 = 0{\cdot}99927\pm0{\cdot}00009. \qquad (\text{VI. 63})$$

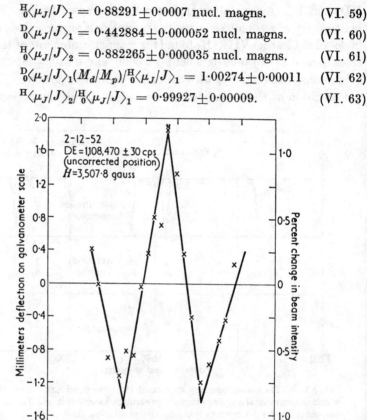

Fig. VI. 9. Typical D_2 rotational resonance for $I = J = 1$ (BAR 54).

The above results agree with Ramsey's (RAM 40) earlier values to well within the experimental error of the earlier experiments. However, the much greater accuracy of the new experiments provides proof that the magnetic moment of H_2 in the second rotational state is not exactly double that in the first and that the rotational magnetic moments of H_2 and D_2 are not inversely proportional to the nuclear masses as would be expected on the simplest theory. For example, Eqs. (VI. 62) and (VI. 63) would equal unity if the simple theory applied. As discussed in the next section and in Chapter VIII, the failures of the simple theory can be attributed to the effects of zero point vibrations and centrifugal stretchings in the molecules. See also Appendix G.

VI. 8.2. Theory of rotational magnetic moments. An approximate theory of the rotational magnetic moment of H_2 was developed by Wick (WIC 33, WIC 33a, BRO 41). Ramsey extended this theory to diatomic molecules in general (RAM 40) and later to general linear molecules (RAM 52f). The extension of the theories to non-linear polyatomic molecules has been given by Eshbach and Strandberg (ESH 52) and Schwartz (SCH 53). They have shown that in polyatomic molecules the rotational gyromagnetic ratio is, in general, a tensor of the second rank. Ramsey (RAM 52f) has extended the theories to include the effects of molecular vibrations and centrifugal stretchings on rotational magnetic moments.

Superficially it might appear that the rotational magnetic moment could be calculated by supposing that the electronic charge distribution rotates rigidly with the molecule. However, such an assumption leads to much too large an electronic contribution. This is well illustrated by the facts that such a calculation would predict a rotational magnetic moment of $-2 \cdot 72$ nuclear magnetons for the H_2 molecule, which is far different both in sign and magnitude from the above experimental result of $+0 \cdot 88291$ nuclear magnetons which is fairly close to the maximum possible positive value of 1 nuclear magneton that would occur if the electrons made no contribution whatsoever. This result indicates that there is a large backward slippage of the electrons, as is confirmed in the discussion of spin-rotational magnetic interactions in Chapter VIII.

Just as it was shown in Section VI. 7 that there was a close relation between magnetic shielding and spin-rotational magnetic interaction, so here there is for the same reason a similar relation between molecular diamagnetic susceptibility and rotational magnetic moment; the rotational angular velocity of the molecule and the magnetic field of the corresponding susceptibility are related by Eq. (VI. 54). As first shown by Van Vleck (VAN 32, p. 275), the diamagnetic susceptibility of a single linear molecule is given by

$$\chi_M/N_0 = \xi = -\frac{e^2}{6mc^2}\left\langle \sum_k r_k^2 \right\rangle + \xi^{HF}, \qquad \text{(VI. 64)}$$

where ξ^{HF} is the high-frequency term (VAN 32, p. 275) of the diamagnetic susceptibility and is given by

$$\xi^{HF} = \frac{2}{3}\sum_{n'\lambda'}{}' \frac{\left(n\lambda\left|\sum_k \mathbf{m}_k^0\right|n'\lambda'\right).\left(n'\lambda'\left|\sum_k \mathbf{m}_k^0\right|n\lambda\right)}{E_{n'}-E_n} \qquad \text{(VI. 65)}$$

where the quantities concerned are defined as in Eq. (VI. 50). The close similarity of the above to Eqs. (VI. 48) and (VI. 50) is apparent. The choice of the origin of coordinates in the present case, however, is arbitrary (VAN 32, p. 276).

The relation of Wick (WIC 33) and Ramsey (RAM 40, RAM 52f) mentioned above between the rotational magnetic moment μ_J and the diamagnetic susceptibility of a diatomic molecule in the absence of vibrational and centrifugal effects is

$$\xi^{HF} = \frac{e^2 R^2}{12mc^2}\left\{\frac{2Z_1 Z_2}{Z_1+Z_2} + 2(Z_1+Z_2)\frac{D^2-d^2}{R^2} - \frac{2\mu'\mu_J}{MJ\mu_N}\right\}, \quad \text{(VI. 66)}$$

where Z_1 and Z_2 are the two nuclear charges, R is the internuclear spacing, D is the distance from the centre of mass to the centroid of nuclear charge, d is the distance from the centre of mass to the centroid of the electron charge distribution, μ' is the reduced mass of the molecule, and M is the mass of the proton. For molecules with no permanent dipole moment, like H_2, $d = D$ and the second term in the brackets vanishes. Eq. (VI. 66) is analogous to Eq. (VI. 53) in the magnetic shielding case. The corrections to Eq. (VI. 66) for molecular vibration and centrifugal stretching have been given by Ramsey (RAM 52f) and will be discussed further in Section VIII. 6.

The measured values of μ_J can be used in the preceding equation to determine the high frequency terms in the diamagnetic susceptibility with the result (including vibrational and centrifugal corrections)

$$^{D}_{0}\langle\chi^{HF}\rangle_1 = ^{D}_{0}\langle N_0 \xi^{HF}\rangle_1 = (0{\cdot}0978 \pm 0{\cdot}0003) \times 10^{-16} \text{ ergs gauss}^{-2} \text{ mole}^{-1},$$

$$^{H}_{0}\langle\chi^{HF}\rangle_1 = (0{\cdot}1036 \pm 0{\cdot}0005) \times 10^{-16} \text{ ergs gauss}^{-2} \text{ mole}^{-1},$$

$$^{H}_{0}\langle\chi^{HF}\rangle_2 = (0{\cdot}1049 \pm 0{\cdot}0008) \times 10^{-16} \text{ ergs gauss}^{-2} \text{ mole}^{-1}. \quad \text{(VI. 67)}$$

From these values, from the measurements of the ordinary diamagnetic susceptibility (HAV 33), and from Eq. (VI. 64) the value of $\langle \sum_k r_k^2 \rangle$ can be found. In this way Ramsey (RAM 53c) found that the mean square distance $\langle r^2 \rangle$ of an electron from the mid-point of the H_2 molecule is

$$\langle r^2 \rangle = \tfrac{1}{2}\langle \sum_k r_k^2 \rangle = (0{\cdot}7258 \pm 0{\cdot}0022) \times 10^{-16} \text{ cm}^2. \quad \text{(VI. 68)}$$

VI. 9. Results of nuclear moment measurements

VI. 9.1. Introduction.
Although the methods discussed in this chapter form some of the principal means for the measurement of nuclear magnetic moments, other molecular beam experiments which yield

nuclear magnetic-moment data are discussed in Chapters III, IV, VII, IX, X, and XII. However, the best magnetic moment results by all of these methods as well as by non-molecular beam experiments will be included in the tables of the present chapter. Likewise, for tabular compactness and convenience the values of the nuclear spins and nuclear electric quadrupole moments will be included in the same tables, even though the methods of measuring these quantities are not discussed in the present chapter but are instead distributed throughout Chapters III, IV, VIII, IX, X, and XI. A few magnetic octupole moments have so far been measured, and these have been included in the tables; the octupole results are discussed in Section IX. 9.

VI. 9.2. *Nuclear moment tables.* The nuclear moment values that have so far been measured for nuclei in their ground states are listed in Table VI. I. The best known values are listed, regardless of whether the measurements were by molecular-beam experiments or by other nuclear moment techniques (RAM 53e). However, explicit references to original papers are given only for the molecular-beam experiments; for non-molecular beam measurements, references are merely given to various printed tables of nuclear moments which in turn list the original publications.

All the magnetic moments in Table VI. I are based on the assumption that the magnetic moment of the proton is exactly 2·792743, the most probable value at the time the table was prepared. The magnetic shielding corrections used in obtaining Table VI. I are listed in Table VI. II. Extensive use has been made of the tables of Ramsey (RAM 53e) and Walchli (WAL 53) in the preparation of these tables.

VI. 10. Significance of nuclear spin and nuclear magnetic moment results

VI. 10.1. *Introduction.* Although the present chapter is primarily on nuclear magnetic moments, the theories of nuclear spins and magnetic moments are so closely related that the significance of both spin and magnetic moment results will be discussed in the remainder of the chapter. A further advantage of discussing nuclear spins at this place is that discussions of the methods of spin measurements are distributed throughout a number of different chapters, including Chapters III, IV, VIII, IX, and X; consequently no single chapter is especially appropriate for the discussion of the significance of the spin results. On the other hand, nuclear electric quadrupole moments will not be discussed

TABLE VI. I

Nuclear Moments

The quantities and units of this table are defined as in the text. The asterisk (*) after a mass number indicates a radioactive isotope and the superscript m indicates a metastable isomer. All magnetic moments are relative to an assumed magnetic moment for the proton of 2·792743 and are corrected for magnetic shielding with the shielding constants of Table VI. II. Moment values marked with an asterisk (*) were added in proof and were not given consistent shielding and absolute value adjustments. As the listed experimental errors include no allowance for the error in the fundamental absolute proton moment calibration, all magnetic moment errors should be increased to allow for the 0·005 per cent. error in that calibration if the error in the absolute value of a moment is desired. Almost all values of Q are subject to a large uncertainty due to lack of knowledge of q. Quantities in parentheses following a number indicate the uncertainty of the last printed numeral of the preceding number. Only moments of stable nuclei and nuclear ground states are listed in the table except in a few cases that are designated with an asterisk (*) or a superscript m. Moment values calculated from theory or about which there is considerable uncertainty are enclosed in brackets [ ]. The present best value of the nuclear magneton μ_N is $5 \cdot 05038(36) \times 10^{-24}$ erg gauss^{-1} (DUM 53). Explicit references are given only to molecular-beam papers; otherwise the reference symbol ZZ is to the tables of Ramsey (RAM 53c), Walchli (WAL 53), and Fuller and Cohen (ZZB 7) where the original papers are listed. In some cases data published after 1 December 1954 are not included in this table, although many results up to May 1964 are included.

Z	Atom	A	I	μ (μ_N)	Q (10^{-24} cm^2)	Ω (10^{-24} μ_N cm^2)	References
0	n	1*	½	−1·913139(45)	..	..	ALV 40, ARN 47, BLO 48, ROG 49, COR 53, COR 54, ZZ
1	H	1	½	+2·792743(0)	..	..	KEL 39a, MIL 41, TAU 49, THO 49, GAR 49, HIP 49, BRA 52, SMA 51, RAM 52c, JEF 51, THO 50, RAM 50d, KOL 52, ZZ
		2	1	+0·8574073(2)	+0·00282(2)	..	BLO 50, KEL 39a, KOL 52, BRA 52, KEL 39, KEL 40, HAR 53, NOR 40, NEW 50, ZZ, ZZB 11
		3*	½	+2·97884(1)	..	..	ZZ
2	He	3	½	−2·127544(7)	..	..	ZZ
		4	0	..	..	..	ZZ
3	Li	6	1	+0·822008(22)	+0·023(2) × $Q(Li^7)$	..	MAN 37, KUS 49a, KUS 49b, KUS 49c, LOG 52, KUS 49, KUS 40, ZZ
		7	3/2	+3·256310(85)	[−0·04]	..	FOX 35, MIL 41, KUS 49a, KUS 49, KUS 40, KUS 49b, LOG 52, KUS 49c, MIL 37, STE 53, ZZ
4	Be	9	3/2	−1·17737(41)	±0·03	..	KUS 39, ZZ
5	B	10	3	+1·80081(49)	+0·06(4) 2·084(2)$Q(B^{11})$	..	MIL 39a, ZZ
		11	3/2	+2·68852(4)	+0·0355(2)	..	MIL 39a, ZZA 13, ZZ
6	C	12	0	..	..	..	ZZ
		13	½	+0·702381(2)	..	..	HAY 41, ZZ
		14*	0	..	..	..	ZZ
7	N	14	1	+0·40371(6)	+0·01	..	KUS 39a, ZZ
		15	½	−0·28313(12)	..	..	ZAC 40, ZZ

TABLE VI. I (cont.)

Z	Atom	A	I	μ (μ_N)	Q (10^{-24} cm^2)	Ω (10^{-24} μ_N cm^2)	References
8	O	16	0	..	..	..	ZZ
		17	$\frac{5}{2}$	$-1\cdot89370(9)$	$-0\cdot026$	..	ZZ
		18	0	..	$\vert<4\times10^{-3}\vert$	..	ZZ
9	F	18*	[1]	$[+0\cdot8(1)]$	..	..	ZZ
		19	$\frac{1}{2}$	$+2\cdot62850(5)$	..	..	MIL 41, ZZ
10	Ne	20	[0]	~ 0	..	..	KEL 38, ZZ
		21	$\frac{3}{2}$	$-0\cdot6618$	0·09	..	ZZ
		22	[0]	~ 0	..	..	KEL 38, ZZ
11	Na	22*	3	$+1\cdot7469(22)$	..	..	SMI 51, DAV 48a, DAV 48b, DAV 49a, DAV 49c, ZZ
		23	$\frac{3}{2}$	$+2\cdot21753(10)$	$+0\cdot10(6)$	..	KUS 39a, LOG 52, MIL 40, ZZA 23, MIL 41, ZZ
		24*	4	$+1\cdot688(5)$	..	..	BEL 53, ZZ
12	Mg	24	[0]	~ 0	..	..	ZZ
		25	$\frac{5}{2}$	$-0\cdot85532(14)$	$+0\cdot22$	..	ZZ
		26	[0]	~ 0	..	..	ZZ
13	Al	26*	[5]	$[+2\cdot8(2)]$	..	..	ZZ
		27	$\frac{5}{2}$	$+3\cdot641421(30)$	$+0\cdot155(3)$	..	LEW 48, LEW 49, KOS 52, LEW 53, MIL 39b, ZZ
14	Si	28	[0]	..	~ 0	..	ZZ
		29	$\frac{1}{2}$	$[-]0\cdot55525(11)$	~ 0	..	ZZ
		30	[0]	..	~ 0	..	ZZ
15	P	30*	[1]	$[+0\cdot6(1)]$	..	..	ZZ
		31	$\frac{1}{2}$	$+1\cdot13162(31)$	..	..	ZZ
16	S	32	0	..	..	..	ZZ
		33	$\frac{3}{2}$	$+0\cdot64342(13)$	$-0\cdot050$	..	ZZ
		34	[0]	..	$\vert<2\times10^{-3}\vert$	..	ZZ
		35*	$\frac{3}{2}$	$[+1\cdot0(1)]$	$+0\cdot035$	..	ZZ
		36	[0]	..	$\vert<0\cdot01\vert$	..	ZZ
17	Cl	34*	3	$[+1\cdot4(4)]$	..	..	ZZ
		35	$\frac{3}{2}$	$+0\cdot821808(71)$	$-0\cdot0782(2)$	$-0\cdot019$	JAC 51, DAV 48, DAV 49b, LIV 51, KUS 39b, LOG 52, SHR 40, ZZ
		36*	2	$+1\cdot28538(6)$	$0\cdot0172(4)$	..	ZZ
		37	$\frac{3}{2}$	$+0\cdot68409(6)$	$-0\cdot0616(2)$	$-0\cdot015$	JAC 51, DAV 48, DAV 49b, LIV 51, KUS 39b, LOG 52, SHR 40, ZZ
18	A	40	[0]	~ 0	..	..	KEL 38
19	K	38	[3]	$[+1\cdot4(4)]$	..	..	ZZ
		39	$\frac{3}{2}$	$+0\cdot39146(7)$	$+0\cdot09$	..	KUS 39a, KUS 40, LOG 52, MIL 35, ZZ
		40*	4	$-1\cdot2981(4)$	$-0\cdot07$	..	ZAC 42, DAV 49a, DAV 49c, ZZ
		41	$\frac{3}{2}$	$+0\cdot215173(84)$	$1\cdot220(2)Q(K^{39})$	..	KUS 40, MIL 35, LEE 52, ZZ
		42*	2	$-1\cdot137(5)$	..	..	BEL 53, ZZ
20	Ca	40	[0]	~ 0	..	..	ZZ
		43	$\frac{7}{2}$	$-1\cdot317201(12)$	..	..	ZZ
21	Sc	45	$\frac{7}{2}$	$+4\cdot75631(12)$	$-0\cdot22$	..	ZZ
22	Ti	47	$\frac{5}{2}$	$-0\cdot788130(84)$	..	..	ZZ
		49	$\frac{7}{2}$	$-1\cdot10377(12)$	..	..	ZZ
23	V	50	6	$+3\cdot34702(94)$* for NaVO$_3$	..	..	ZZ
		51	$\frac{7}{2}$	$+5\cdot1470(57)$* for NaVO$_3$	$+0\cdot27$	..	ZZ
24	Cr	53	$\frac{3}{2}$	$-0\cdot474391(42)$	$-0\cdot03$	..	BRI 53, ZZ
25	Mn	55	$\frac{5}{2}$	$+3\cdot46766(14)$	$+0\cdot4$	..	ZZ
26	Fe	57	$\frac{1}{2}$	$+0\cdot0905$	..	..	ZZ
27	Co	57*	$\frac{7}{2}$	$4\cdot65(20)$	..	..	ZZ
		58*	2	$+3\cdot5(3)$	..	..	ZZ
		59	$\frac{7}{2}$	$+4\cdot6488(5)$	$+0\cdot4$	..	ZZ
		60*	5	$+3\cdot5(5)$	..	..	ZZ
28	Ni	61	$[\frac{3}{2}]$	$\pm 0\cdot75$	..	..	ZZ
29	Cu	63	$\frac{3}{2}$	$+2\cdot22664(17)$	$-0\cdot157(5)$ $1\cdot084(7)Q(Cu^{65})$	..	ZZ
		64*	1	$\pm 0\cdot216(4)$	..	..	HAM 53, BRO 54, ZZ
		65	$\frac{3}{2}$	$+2\cdot38473(45)$	$-0\cdot145(5)$	..	ZZ
30	Zn	64	[0]	~ 0	..	..	ZZ
		66	[0]	~ 0	..	..	ZZ
		67	$\frac{5}{2}$	$+0\cdot87571(10)$	$+0\cdot17$	..	ZZ

TABLE VI. 1 (cont.)

Z	Atom	A	I	μ (μ_N)	Q (10^{-24} cm^2)	Ω (10^{-24} μ_N cm^2)	References
		68	[0]	~0			ZZ
31	Ga	68*	1				ZZ
		69	$\frac{3}{2}$	+2·01605(51)	+1·19	0·14	BEC 48, KUS 48, KUS 50a, LOG 51, FOL 50, REN 40, KOS 52, DAL 54, SCH 54, ZZ
		71	$\frac{3}{2}$	+2·56158(26)	+0·119(3)	0·18	BEC 48, KUS 48, KUS 50a, LOG 51, FOL 50, REN 40, KOS 52, DAL 54, SCH 54, ZZ
32	Ge	70	[0]		$\|<7\times10^{-3}\|$		ZZ
		72	[0]		$\|<7\times10^{-3}\|$		ZZ
		73	$\frac{9}{2}$	−0·87914(12)	−0·21(10)		ZZ
		74	[0]		$\|<7\times10^{-3}\|$		ZZ
		76	[0]		$\|<7\times10^{-3}\|$		ZZ
33	As	75	$\frac{3}{2}$	+1·43896(16)	+0·3(2)		ZZ
34	Se	74	[0]				ZZ
		76	[0]		$\|<2\times10^{-3}\|$		ZZ
		77	$\frac{1}{2}$	+0·534058(14)	$\|<2\times10^{-3}\|$		ZZ
		78	[0]	~0	$\|<2\times10^{-3}\|$		ZZ
		79*	$\frac{7}{2}$	−1·018(15)	0·7		ZZ
		80	0		$\|<2\times10^{-3}\|$		ZZ
		82	[0]	~0			ZZ
35	Br	79	$\frac{3}{2}$	+2·10555(30)	+0·33(2) 1·1973(6) × Q(Br^{81})		BRO 47, LEE 52, BRO 54, ZZ
		81	$\frac{3}{2}$	+2·26958(3)	+0·28(2)		BRO 47, BRO 54, ZZ
36	Kr	82	[0]	~0			ZZ
		83	$\frac{9}{2}$	−0·969	+0·23	−0·18	ZZ
		84	[0]	~0			ZZ
		86	[0]	~0			ZZ
37	Rb	81*	$\frac{3}{2}$	+2·00(6)			NIE 54, ZZA 14, ZZ
		85	$\frac{5}{2}$	+1·35268(11)	+2·0669(5) × Q(Rb^{87})		KUS 39b, LOG 52, MIL 40, HUG 50, TRI 54, MIL 36, MIL 37, ZZA 24, ZZ
		86*	2	[−]1·67(40)			BEL 51, BEL 53, ZZ
		87	$\frac{3}{2}$	+2·750529(38)	+0·14(1)		KUS 39b, LOG 52, MIL 40, MIL 36, MIL 37, ZZA 24, ZZ
38	Sr	86	[0]				ZZ
		87	$\frac{9}{2}$	−1·09302(13)	+0·36		ZZ
		88	[0]	~0			ZZ
39	Y	89	$\frac{1}{2}$	−0·137314(29)			ZZ
40	Zr	91	$\frac{5}{2}$	−1·303			ZZ
41	Nb	93	$\frac{9}{2}$	+6·16713(35)	−0·22		ZZ
42	Mo	92	[0]	~0			ZZ
		94	[0]	~0			ZZ
		95	$\frac{5}{2}$	−0·9135			ZZ
		96	[0]	~0			ZZ
		97	$\frac{5}{2}$	−0·9327			ZZ
		98	[0]	~0			ZZ
		100	[0]	~0			ZZ
43	Tc	99	$\frac{9}{2}$	+5·68048(35)	0·34(17)		ZZ
44	Ru	99	$\frac{5}{2}$	−0·63			ZZ
		101	$\frac{5}{2}$	1·09(3)μ(Ru^{99})			ZZ
45	Rh	102	$\frac{1}{2}$	[−]0·11			ZZ
		103	$\frac{1}{2}$	−0·088512(?)			ZZ
46	Pd	105	$\frac{5}{2}$	−0·6015			ZZ
		111*	$\frac{5}{2}$				ZZ
47	Ag	107	$\frac{1}{2}$	−0·113556(14)			WES 53, ZZ
		109	$\frac{1}{2}$	−0·13053(19)			WES 53 ZZ
		111*	$\frac{1}{2}$	±0·146(2)*			HAM 53, BRO 54, ZZ
48	Cd	110	[0]	~0			ZZ
		111	$\frac{1}{2}$	−0·59499(8)			ZZ
		112	[0]	~0			ZZ
		113	$\frac{1}{2}$	−0·62243(8)			ZZ
		114	[0]	~0			ZZ
		116	[0]	~0			ZZ
49	In	111*	$\frac{9}{2}$	+5·53	+0·85		ZZ

Table VI. 1 (cont.)

Z	Atom	A	I	μ (μ_N)	Q (10^{-24} cm^2)	Ω (10^{-24} μ_N cm^2)	References
		113	$\frac{9}{2}$	+5·52317(54)	0·820	+0·57	HAR 42, MAN 50, KOS 52, ZZ
		114*	1	+4·7	..	..	ZZ
		115	$\frac{9}{2}$	+5·53441(66)	0·834	+0·56	HAR 42, MAN 50, MIL 38, KOS 52, SCH 54, KUS 54, ZZ
50	Sn	111*	$\frac{7}{2}$	..	..	..	ZZ
		115	$\frac{1}{2}$	−0·917798(76)	..	..	ZZ
		116	[0]	∼0	..	..	ZZ
		117	$\frac{1}{2}$	−0·99990(19)	..	..	ZZ
		118	[0]	∼0	..	..	ZZ
		119	$\frac{1}{2}$	−1·04611(84)	..	..	ZZ
		120	[0]	∼0	..	..	ZZ
51	Sb	121	$\frac{5}{2}$	+3·35892(19)	−0·29	..	ZZ
		123	$\frac{7}{2}$	+2·54653(3)	1·2751(2) × $Q(Sb^{121})$	..	ZZ
52	Te	124*	[3]	[+3·2(2)]	..	..	ZZ
		123	$\frac{1}{2}$	−0·73587(23)	..	..	ZZ
		125	$\frac{1}{2}$	−0·88716(26)	..	..	ZZ
		126	[0]	∼0	..	..	ZZ
		128	[0]	∼0	..	..	ZZ
		130	[0]	∼0	..	..	ZZ
53	I	126*	2	..	..	..	ZZ
		127	$\frac{5}{2}$	+2·80897(23)	−0·79	0·17(3)	JAC 54, SCH 54, ZZ
		129*	$\frac{7}{2}$	+2·617266(12)	−0·55 0·701213(15) × $Q(I^{127})$	..	ZZ
		131*	$\frac{7}{2}$	..	..	..	ZZ
54	Xe	129	$\frac{1}{2}$	−0·776786(53)	..	..	ZZ
		131	$\frac{3}{2}$	+0·690635(85)	−0·12	+0·048	ZZ
		132	[0]	∼0	..	..	ZZ
		134	[0]	∼0	..	..	ZZ
		136	[0]	∼0	..	..	ZZ
55	Cs	131*	$\frac{5}{2}$	+3·48(4)	..	..	BEL 53, ZZ
		133	$\frac{7}{2}$	+2·57887(30)	−0·003(2)	..	KUS 39a, LOG 52, MIL 37, MIL 40, ZZ
		134*	[4]	+2·95(1)	..	..	BEL 53, JAC 52, ZZA 15, ZZA 16, ZZ
		134^m*	8	+1·10(1)	..	..	COH 54, BRO 54, ZZ
		135*	$\frac{7}{2}$	+2·7382(19)	+0·049	..	DAV 49a, DAV 49c, ZZ
		137*	$\frac{7}{2}$	+2·8502(25)	+0·050	..	DAV 49, DAV 49a, DAV 49c, ZZ
56	Ba	134	[0]	∼0	..	..	ZZ
		135	$\frac{3}{2}$	+0·8355(24)	+0·18	..	HAY 41, ZZ
		136	[0]	∼0	..	..	ZZ
		137	$\frac{3}{2}$	+0·9324(27)	+0·28	..	HAY 41, ZZ
		138	[0]	∼0	..	..	ZZ
57	La	138	5	+3·707	±0·8	..	ZZ
		139	$\frac{7}{2}$	+2·77807(61)	+0·22	..	ZZ
58	Ce*	141	[$\frac{7}{2}$]	±0·9	..	..	ZZ
59	Pr	141	$\frac{5}{2}$	+4·5	−0·054	..	LEW 53a, ZZ
60	Nd	143	$\frac{7}{2}$	−1·0(2) 1·6083(2) $\mu(Nd^{145})$	−0·6	..	ZZ
		145	$\frac{7}{2}$	−0·71	−0·3	..	ZZ
61	Pm	147	[$\frac{7}{2}$]	±3·2	±0·7	..	ZZ
62	Sm	147	$\frac{7}{2}$	−0·78(37)	..	..	ZZ
		149	$\frac{7}{2}$	−0·65(23)	..	..	ZZ
63	Eu	151	$\frac{5}{2}$	+3·464 2·235(30) × $\mu(Eu^{153})$	+0·95	..	ZZ
		153	$\frac{5}{2}$	+1·530	+2·5	..	ZZ
64	Gd	155	$\frac{3}{2}$	−0·27	+1·3	..	ZZ
		157	[$\frac{3}{2}$,$\frac{7}{2}$,$\frac{5}{2}$]	−0·36	+1·5	..	ZZ
65	Tb	159	$\frac{3}{2}$	1·5(4)	..	..	ZZ
66	Dy	161	$\frac{5}{2}$	±0·42	±1·1	..	ZZ
		163	$\frac{5}{2}$	±0·58	±1·3	..	ZZ
67	Ho	165	$\frac{7}{2}$	+4·1	+3·0	..	ZZ
68	Er	167	$\frac{7}{2}$	−0·56	+2·8	..	ZZ

Table VI. 1 (cont.)

Z	Atom	A	I	μ (μ_N)	Q (10^{-24} cm^2)	Ω ($10^{-24} \mu_N$ cm^2)	References
69	Tm	169	$\frac{1}{2}$	−0·229			ZZ
70	Yb	171	$\frac{1}{2}$	+0·493			ZZ
		173	$\frac{5}{2}$	−0·678	+3·0		ZZ
71	Lu	175	$\frac{7}{2}$	+2·9(5)	+5·9		ZZ
		176*	7	+3·18	+7(1)		ZZ
72	Hf	177	$\frac{7}{2}$	+0·61	+3		ZZ
		178	[0]	∼0			ZZ
		179	$\frac{9}{2}$	−0·47	+3		ZZ
		180	[0]	∼0			ZZ
73	Ta	181	$\frac{7}{2}$	+2·1	+4·2		ZZ
74	W	182	[0]				ZZ
		183	$\frac{1}{2}$	+0·11846(13)			ZZ
		184	[0]				ZZ
		186	[0]				ZZ
75	Re	185	$\frac{5}{2}$	+3·17156(34)	+2·8		ZZ
		187	$\frac{5}{2}$	+3·17591(34)	+2·6		ZZ
76	Os	187	$\frac{1}{2}$	+0·067			ZZ
		189	$\frac{3}{2}$	+0·655914(78)	+0·8		ZZ
77	Ir	191	$\frac{3}{2}$	+0·2(1)	+1·2(7)		ZZ
		193	$\frac{3}{2}$	+0·17(3)	+1·0(5)		ZZ
78	Pt	194	[0]	∼0			ZZ
		195	$\frac{1}{2}$	+0·60596(21)			ZZ
		196	[0]	∼0			ZZ
79	Au	197	$\frac{3}{2}$	+0·14(2)	0·56(10)	+0·0112	WES 53, ZZ
		198*	2	±0·50(4)			ZZA 56
		199*	$\frac{3}{2}$	±0·24(2)			ZZA 56
80	Hg	198	[0]	∼0			ZZ
		199	$\frac{1}{2}$	+0·504117(41)			ZZ
		200	[0]	∼0			ZZ
		201	$\frac{3}{2}$	−0·5567	+0·5	−0·13	ZZ
		202	[0]	∼0			ZZ
		204	[0]	∼0			ZZ
81	Tl	203	$\frac{1}{2}$	+1·6116(14)			ZZ
		205	$\frac{1}{2}$	+1·62734(42)			ZZ
82	Pb	204	[0]	∼0			ZZ
		206	[0]	∼0			ZZ
		207	$\frac{1}{2}$	+0·58943(14)			ZZ
		208	[0]	∼0			ZZ
83	Bi	209	$\frac{9}{2}$	+4·07970(81)	−0·4		ZZ
		210* (RaE)	1	±0·0442	±0·13		SMI 54, PHY 52, ZZ
		214* (RaC)	2				
84	Po	209*	$\frac{1}{2}$				ZZ
89	Ac	227*	$\frac{3}{2}$	+1·1	+1·7		ZZ
91	Pa	231*	$\frac{3}{2}$	±1·98			ZZ
92	U	233*	$\frac{5}{2}$	+0·54	+3·5		ZZ
		235*	$\frac{7}{2}$	−0·35	+4·1		ZZ
93	Np	237*	$\frac{5}{2}$	6·0(25)	∼15		ZZ
		239*	$\frac{5}{2}$				ZZ
94	Pu	239*	$\frac{1}{2}$	+0·21			ZZ
		241*	$\frac{5}{2}$	−0·73	+5·6		ZZ
					≠ 0		ZZ
95	Am	241*	$\frac{5}{2}$	+1·4	+4·9		ZZ
		243*	$\frac{5}{2}$	+1·4	+4·9		ZZ

Table VI. II

Magnetic Shielding Corrections

This table lists the magnetic shielding constants σ used in the calculation of Table VI. I. The effective magnetic field at the nucleus is $(1-\sigma)H_0$, where H_0 is the applied magnetic field. The value of σ for $Z = 1$ comes from the combination of Ramsey's shielding theory (RAM 50d, RAM 52f, RAM 53a) and the experiments of Harrick, Barnes, Bray, Ramsey, and Thomas (HAR 53, THO 50). Values for other Z's are taken from Dickinson's calculations (DIC 50a). Owing to chemical shifts, the shielding constants actually vary from one compound to another, but the values below were used for normal corrections.

Z	σ	Z	σ	Z	σ
1	0·0000278*	32	0·00273	63	0·00693
2	0·000060	33	0·00285	64	0·00709
3	0·000101	34	0·00296	65	0·00724
4	0·000149	35	0·00308	66	0·00740
5	0·000199	36	0·00321	67	0·00756
6	0·000261	37	0·00333	68	0·00772
7	0·000325	38	0·00345	69	0·00788
8	0·000395	39	0·00358	70	0·00804
9	0·000464	40	0·00371	71	0·00820
10	0·000547	41	0·00384	72	0·00837
11	0·000629	42	0·00397	73	0·00853
12	0·000710	43	0·00411	74	0·00869
13	0·000795	44	0·00425	75	0·00885
14	0·000881	45	0·00438	76	0·00901
15	0·000970	46	0·00452	77	0·00917
16	0·00106	47	0·00465	78	0·00933
17	0·00115	48	0·00478	79	0·00949
18	0·00124	49	0·00491	80	0·00965
19	0·00133	50	0·00504	81	0·00982
20	0·00142	51	0·00517	82	0·00998
21	0·00151	52	0·00531	83	0·0101
22	0·00161	53	0·00545	84	0·0103
23	0·00171	54	0·00559	85	0·0105
24	0·00181	55	0·00573	86	0·0106
25	0·00191	56	0·00587	87	0·0108
26	0·00202	57	0·00602	88	0·0110
27	0·00214	58	0·00616	89	0·0111
28	0·00226	59	0·00631	90	0·0113
29	0·00238	60	0·00647	91	0·0115
30	0·00249	61	0·00662	92	0·0116
31	0·00261	62	0·00678		

* The value of 0·0000278 applies to mineral oil. For H_2 σ should be 0·0000262 and for H_2O 0·0000256.

in the remainder of the present chapter since Chapter XI is devoted to quadrupole moments. Magnetic octupole results are discussed in Section IX. 9.

Some of the most interesting nuclear magnetic moment results have been the anomalies in comparisons of the ratios of nuclear magnetic moments of two isotopes with the ratios of the hyperfine structure separations of the same two isotopes. These hyperfine structure anomalies, however, are discussed and interpreted in Chapter IX instead of the present chapter. Likewise, the important results on the anomalous electron magnetic moment are discussed in Chapter IX instead of in the present chapter.

Since the present book is primarily concerned with methods of molecular beam measurements and not with the theory of nuclear moments, the interpretations discussed in the remainder of the chapter will be described only briefly. For more detailed theoretical discussions the reader should read the original papers referred to below, the author's *Nuclear Moments* (RAM 53e), or the standard books on nuclear theory (BLA 52, SAC 53, ROS 48, ZZA 18).

VI. 10.2. *Relations between nuclear statistics, spins, and mass numbers.* A nucleus whose mass number A is odd is found experimentally to satisfy Fermi–Dirac statistics and a nucleus whose mass number is even satisfies Bose–Einstein statistics. This is reasonable since the proton and neutron each separately satisfies Fermi–Dirac statistics; hence an interchange of two identical nuclei each of which has n such nucleons will multiply the wave function by $(-1)^n$ as can be seen by interchanges of the nucleons one at a time. Hence, if n is even, there is no change in the sign of the wave function, whereas if n is odd the sign is reversed.

From Table VI. 1 it is apparent that the nuclear spin is half-integral if the mass number A is odd and integral if the mass number is even. A combination of this result with that in the preceding paragraph implies that Fermi–Dirac statistics is associated with half integral spins while Bose–Einstein statistics and even spins are associated. Pauli (PAU 40) from general field theory considerations has shown that this association must occur for all elementary particles. Pauli's theorem is reasonable since a particle with spin $\frac{1}{2}$ requires a Dirac type of equation with negative energy states and only with Fermi–Dirac statistics can these negative energy states be filled so that particles in positive energy states may be excluded from them.

VI. 10.3. *Nucleon magnetic moments.* The proton magnetic moment is not exactly one nuclear magneton and the neutron magnetic moment is not exactly zero, as would be expected from the simplest considerations based on Dirac's electron theory. It is, however, always possible to add an additional magnetic moment (PAU 32) to the one that arises naturally in Dirac's theory. Attempts have been made (ZXW 54, MAR 52) with partial success to account for the anomalous nucleon moments on the basis of the meson theory of nuclear forces, since the resultant magnetic moment is contributed to by the magnetic moment and currents of the mesons which have a finite probability of existence within the range of forces of the heavy particle.

VI. 10.4. *Deuteron magnetic moment.* The magnetic moment of the proton plus that of the neutron is close to the moment of the deuteron, but not exactly equal to it; the sum is too high by 0.0225 ± 0.003 nuclear magnetons. In this case, then, the nucleon moments are nearly but not exactly additive. That there should be a departure from simple additivity was predicted by Rarita and Schwinger (RAR 41) on the basis of their theory of the deuteron electric quadrupole moment. The theory of the deuteron, especially as concerns its electric quadrupole moment is discussed in Section XI. 5.1. As shown there, the deuteron in its ground state is in a mixture of $^3S_1$ and $^3D_1$ states. A departure from simple additivity of the proton and neutron magnetic moments in the deuteron then arises from the orbital magnetic moment of the $^3D_1$ state.

The deuteron magnetic moment, including the effect of the D state, may be calculated in a straightforward fashion (RAR 41, SCH 46, BLA 52) analogous to the procedure for calculating the Landé g factor of atoms (CON 35) or to Eq. (III. 91). Since only the components of magnetic moment parallel to **I** are effective and since the magnetic moment by definition is the effective component of μ_I along the z axis when $m_I = I$, the magnetic moment is, in general,

$$\mu = \mu_N \sum_i (g_{li}\mathbf{l}_i + g_{si}\mathbf{s}_i) \cdot \mathbf{I} \cdot \mathbf{k}/\mathbf{I} \cdot \mathbf{I}, \qquad (\text{VI. 69})$$

where g_{li} is the orbital g factor for the ith nucleon with orbital angular momentum $\mathbf{l}_i$ and spin angular momentum $\mathbf{s}_i$. Then for the proton in the deuteron $\mathbf{l}_p = \tfrac{1}{2}\mathbf{L}$ with $g_{lp} = 1$ while for the neutron $g_{ln} = 0$. Therefore, for the deuteron

$$\begin{aligned}\mu_d &= \mu_N[\tfrac{1}{2}\mathbf{L} + g_{sp}\mathbf{s}_p + g_{sn}\mathbf{s}_n] \cdot \mathbf{I} \cdot \mathbf{k}/\mathbf{I} \cdot \mathbf{I} \\ &= \mu_N[\tfrac{1}{2}(g_{sp}+g_{sn})(\mathbf{s}_p+\mathbf{s}_n) + \tfrac{1}{2}(g_{sp}-g_{sn})(\mathbf{s}_p-\mathbf{s}_n) + \tfrac{1}{2}\mathbf{L}] \cdot \mathbf{I} \cdot \mathbf{k}/\mathbf{I} \cdot \mathbf{I}. \end{aligned}$$
$$(\text{VI. 70})$$

But $(\mathbf{s}_p-\mathbf{s}_n)$ has a zero expectation value in triplet states and in general $\mathbf{s}_p+\mathbf{s}_n = \mathbf{S}$ so

$$\mu_d = \mu_N[\tfrac{1}{2}(g_{sp}+g_{sn})\mathbf{S}+\tfrac{1}{2}\mathbf{L}].\mathbf{II}.\mathbf{k}/\mathbf{I}.\mathbf{I}. \qquad (\text{VI. 71})$$

But $\mathbf{I} = \mathbf{S}+\mathbf{L}$ so

$$\mu_d = \mu_N[\tfrac{1}{2}(g_{sp}+g_{sn})\mathbf{I}-\tfrac{1}{2}(g_{sp}+g_{sn}-1)\mathbf{L}].\mathbf{II}.\mathbf{k}/\mathbf{I}.\mathbf{I}. \qquad (\text{VI. 72})$$

Then as $\mathbf{I}.\mathbf{k} = m = I$ and by the analogue to Eq. (III. 90)

$$\mu_d = \mu_N\left[\tfrac{1}{2}(g_{sp}+g_{sn})I-\tfrac{1}{2}(g_{sp}+g_{sn}-1)\frac{I(I+1)+L(L+1)-S(S+1)}{2(I+1)}\right].$$
(VI. 73)

Now for either a $^3S_1$ or a $^3D_1$ state $I = S = 1$ so $I(I+1) = S(S+1) = 2$. On the other hand if P_D is the fraction of D state as in Section XI. 5.1,

$$\langle L(L+1)\rangle = 0(1-P_D)+2(2+1)P_D = 6P_D. \qquad (\text{VI. 74})$$

Therefore, $\quad \mu_d = \mu_p+\mu_n-\tfrac{3}{2}(\mu_p+\mu_n-\tfrac{1}{2}\mu_N)P_D. \qquad (\text{VI. 75})$

The experimental values of μ_p, μ_n, and μ_d when substituted into the above equation give about 4 per cent. for P_D, which is consistent with the deuteron quadrupole moment as discussed in Section XI. 5.1.

In the above calculation, relativistic effects have been neglected, although these are appreciable. Relativistic and other corrections have been discussed by Sachs (SAC 47), Breit (BRE 47), Primakoff (PRI 47), and others (ZXX 54). These effects also make contributions to Eq. (VI. 75) of several per cent., so the best that can be inferred from the experimental magnetic moments is that P_D lies between about 2 and 6 per cent.

VI. 10.5. H^3 *and* He^3 *magnetic moments.* The magnetic moments of H^3 and He^3 are not exactly equal to the moments of the proton and neutron; such equality would occur with central forces and an S ground state for then in $_1H^3$ as an example the two neutron moments would just cancel, leaving the proton moment. Sachs and Schwinger (SAC 46a) and Anderson (AND 49) found that although the existence of tensor forces easily accounts for an orbital contribution to the moment and a consequent non-additivity of the intrinsic moments, it is necessary to make very artificial assumptions to account for the observed results on this basis alone. A more natural explanation offered by Villars (VIL 48) and further discussed by Austern, Sachs, and others (AUS 51 SAC 53, VIL 48) is that, in addition to the above orbital contribution a magnetic moment component arises from the exchange currents of the mesons which, on the meson theory of forces, exchange back and forth between the particles to give rise to the nuclear forces.

VI. 10.6. *Systematics of nuclear spins and magnetic moments.* Except for Eu, all measured isotopes whose constitutions differ by just two neutrons and which have the same spin have magnetic moments which are the same to within 30 per cent., and in eleven such cases the agreement is within 12 per cent. The magnetic moment ratio $\text{In}^{115}/\text{In}^{113}$ = 1·00224 while the electrical quadrupole moment ratio for the same two isotopes is 1·0146. It is as if the two added neutrons chiefly cancelled each other's effect. This tendency of two successive neutrons just to cancel each other lends support to the various shell theories of nuclear structure that are discussed in the next section.

Most nuclei with numbers of protons equal to the number of neutrons and with each of these numbers odd have a spin of 1 like the deuteron. At one time it was thought that all such nuclei would have a spin of 1. Now, however, as shown in the tables, it is known that B^{10} and Na^{22} depart from this rule.

It has been pointed out by Schmidt (SCH 37b) that almost all nuclear moments of nuclei with odd mass number fall between two limits set by very simple (too simple) considerations. These assumptions correspond to assuming that the entire nuclear spin and magnetic moment arise from the extra nucleon of the type for which there is an odd number. Then one limit corresponds to the spin and orbital angular momentum of that nucleon being parallel while for the other limit they are antiparallel.

The Schmidt lines on the above assumptions may easily be evaluated theoretically. From Eq. (VI. 69) when only a single particle contributes to the magnetic moment and angular momentum and from the analogue to Eq. (III. 90),

$$\mu = \mu_N(g_L \mathbf{L} + g_S \mathbf{S}) \cdot \mathbf{I} \cdot \mathbf{k}/\mathbf{I} \cdot \mathbf{I}$$

$$= g_L \mu_N \frac{I(I+1)+L(L+1)-S(S+1)}{2(I+1)} +$$

$$+ g_S \mu_N \frac{I(I+1)+S(S+1)-L(L+1)}{2(I+1)} \quad \text{(VI. 76)}$$

since $\mathbf{I} \cdot \mathbf{k} = m_I = I$ for the state used in defining μ. Since $S = \tfrac{1}{2}$ and $I = L \pm \tfrac{1}{2}$, one can eliminate S and L in Eq. (VI. 76) to obtain the following relations:

(a) for $I = L+\tfrac{1}{2}$,
$$\mu/\mu_N = (I-\tfrac{1}{2})g_L + \tfrac{1}{2}g_S \quad \text{(VI. 77)}$$

and (b) for $I = L-\tfrac{1}{2}$,
$$\mu/\mu_N = \frac{I^2+\tfrac{3}{2}I}{I+1} g_L - \tfrac{1}{2} \frac{I}{I+1} g_S. \quad \text{(VI. 78)}$$

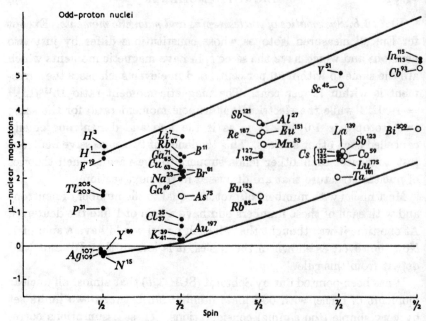

Fig. VI. 10. Magnetic moment and Schmidt lines as a function of nuclear spin for nuclei with an odd number of protons and an even number of neutrons (RAM 53e).

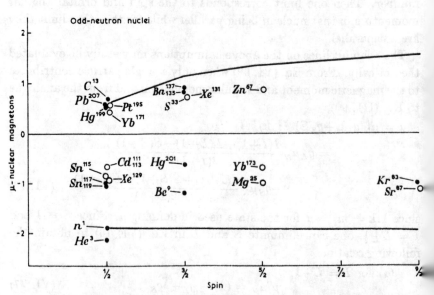

Fig. VI. 11. Magnetic moment and Schmidt lines as a function of nuclear spin for nuclei with an odd number of neutrons and an even number of protons (RAM 53e).

For odd proton nuclei $g_L = 1$ and $g_S = 5\cdot58$. For odd neutron nuclei $g_L = 0$ and $g_S = -3\cdot83$. With these values Eqs. (VI. 77) and (VI. 78) give the two Schmidt lines for each case.

The two Schmidt lines from the above formulae are plotted in Fig. VI. 10 for odd proton nuclei and in Fig. VI. 11 for odd neutron nuclei. The experimental magnetic moments are plotted in the same figures. They indeed fall well between the Schmidt limits. Theoretical reasons for the Schmidt lines being significant are discussed in the next section on nuclear shell models.

Inglis (ING 38) has pointed out that if the orbital g factor is assumed to be $\tfrac{1}{8}$ for a neutron instead of 0 and $\tfrac{7}{8}$ for a proton instead of 1, the above lines are altered in such a way that they go through the two main groups of nuclei.

Margenau and Wigner (MAR 40a, ROS 48) have calculated limiting curves analogous to the Schmidt curves but on the basis of a liquid drop model of the nucleus. Their results, however, are in less good agreement with experiment than the simple Schmidt lines.

The experimental significance of the Schmidt lines provides considerable support to the nuclear shell models discussed in the next section According to the simplest of these models, as shown in the next section, the nuclear magnetic moments should fall on, rather than between, the Schmidt lines.

Causes of departure from the Schmidt lines have been extensively discussed. Foldy (ZXW 54), Townes (ZXW 54), and others have considered the deviations from the Schmidt lines which can arise from a less simplified coupling for the nucleons. Schawlow and Townes (SCH 51a, ZZA 19) have pointed out that the departures for odd A nuclei show the similarity between neutron states and proton states in nuclei. If the deviations from the Schmidt lines were due to admixtures of the two states with the same $I(L+\tfrac{1}{2}$ and $L-\tfrac{1}{2})$ and if neutron and proton states were similar, corresponding odd neutron and odd proton nuclei (with the odd number of neutrons in one equal to the odd number of protons in the other) should have magnetic moments which lie the same fractional distance between the $L+\tfrac{1}{2}$ and $L-\tfrac{1}{2}$ Schmidt limits. Schawlow and Townes (SCH 51a, ZZA 19) point out that this is approximately true experimentally. Miyazawa (MIY 51), de Shalit (SHA 51), and Bloch (BLO 51) suggest that the departures from the Schmidt lines are due at least in part to a change in the intrinsic magnetic moment of the nucleon when it is in a nucleus. Such a change might be expected in a meson theory of the anomalous nucleon moment since the Pauli exclusion principle will inhibit, say, the dissociation of a proton into a neutron and a meson in dense nuclear matter. Bohr and Mottelson (BOH 53, BOH 51) have modified the shell model to provide coupling to the collective motions of the nucleus as a whole; these modifications are discussed in Section XI. 5.2 and among other things they account for departures from the Schmidt lines.

The spin measurements of members of various radioactive chains have been of special value to the theories of the radioactivity concerned. For detailed discussions of the relations between spin measurements and radioactivity the reader is referred to recent books on nuclear theory (BLA 52, SAC 53) and beta radioactivity (PHY 52). The spin measurements of K^{40} and Na^{22}, for example, show that their β decays are accompanied by large spin changes which confirms the explanation of their long radioactive half-lives as being due to the forbiddenness of the transition under the angular momentum selection rules. A very recent experiment is the molecular beam determination by Smith (SMI 54, PHY 52) that RaE has a spin of 1 and not 0, as previously assumed; this result has eliminated what was formerly considered to be definite evidence for a pseudoscalar interaction (SMI 54, PHY 52, PET 52).

VI. 10.7. *Nuclear models.* For the lighter nuclei, Flowers, Inglis, Bethe, Feenberg, Wigner, Sachs, and others (ZZA 17, ING 38, SAC 53, MAR 40a, ZXW 54) have achieved considerable success by interpreting the nuclear spins and magnetic moments in terms of independent particle or Hartree models with various types of coupling. These theories are not described here since they have been summarized in books on theoretical nuclear physics (SAC 53, ROS 48, BLA 52, ZZA 18).

Sachs (SAC 46) has shown that particularly simple relationships hold for mirror or conjugate nuclei (isobaric nuclei such that the number of neutrons in the first equals the number of protons in the second). These relationships arise from assumptions that there is charge independence of nuclear forces, that the nuclei are sufficiently light for the asymmetry from the Coulomb force to be neglected, and that the nuclear moment is the resultant of the spin and orbital magnetic moments of the constituent nucleons. Magnetic moment data are available for such a mirror pair in H^3 and He^3. The discussion of any self-conjugate nucleus, i.e. a nucleus for which the number of neutrons equals the number of protons, is also simplified for the same reasons as are the comparisons of mirror nuclei (ROS 48, SAC 53).

Although for many nuclear physics calculations—especially at high nuclear excitation energies—a model of the nucleus analogous to a liquid drop seems most suitable (WEI 50), there is nevertheless also strong empirical evidence for some form of shell structure in nuclei. Much of this evidence depends on nuclear spin and magnetic moment data; some of the spin and moment evidence has already been given in the preceding section while the remainder depends on a more detailed development of specific shell models and will be given near the end of the present section. From other branches of nuclear physics, there is also now abundant evidence that certain special numbers of neutrons or of protons in a nucleus form particularly stable configurations even in heavy nuclei as was first pointed out by Elsasser (ZXW 54), Mayer (MAY 48), Jensen, Haxel, and Suess (JEN 48), and others (ZXW 54). The so-called magic numbers for which there is special stability are 2, 8, 20, 28, 50, 82, and 126. Thus the elements with unusually large numbers of isotopes and with the largest difference of mass number between the lightest and heaviest isotope are those with 20, 50, or 82 protons. Similar evidence comes from relative abundances, neutron binding energies, neutron absorption cross-sections, etc., as discussed in greater detail in Jensen and Mayer's recent book (ZZA 18), in the author's *Nuclear Moments* (RAM 53e, p. 137), and elsewhere (BLA 52, SAC 53). Nuclear electrical quadrupole moment results can also be correlated with the magic numbers, as discussed in Chapter XI and shown in Fig. XI. 2.

It is somewhat surprising that a shell structure should exist in heavy nuclei because the interaction between neighbouring particles should be very strong indeed. The shell structure may be made possible, however, by the neighbouring potentials blending together to form a roughly uniform potential and by virtue of the nuclear system being highly degenerate so that the Pauli exclusion principle reduces the probability of nucleon collisions (WEI 50).

The high stability of a system with 2, 8, or 20 like nucleons can easily be accounted for if the nucleons are assumed to move approximately independently in a square potential well. For example, 20 like nucleons would correspond to two $1s$, six $1p$, ten $1d$, and two $2s$ like nucleons. However, the high stability of 28, 50, 82, and 126 does not follow immediately from such a model. Feenberg (FEE 49), Nordheim (NOR 49), Mayer (MAY 48), and Jensen, Haxel, and Suess (JEN 48) have shown that the mean potential can be modified in various ways to produce the magic numbers. Mayer (MAY 48) and Jensen, Haxel, and Suess (JEN 48, HAX 48) have accomplished this by a single-particle model with an assumed strong spin-orbital coupling.

How a suitable spin-orbital coupling can accomplish this can be seen with the aid of Fig. VI. 12. For an independent particle model in which each nucleon is assumed to move in the mean potential of the

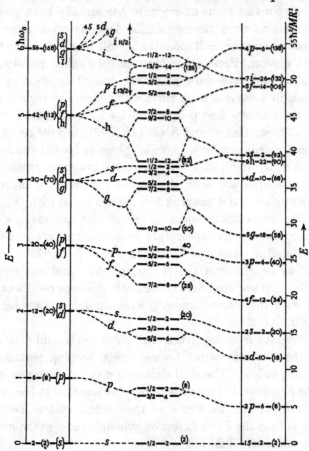

FIG. VI. 12. Nuclear shells. The energy levels of an infinitely deep rectangular well are shown on the right and those of a simple harmonic oscillator potential on the left. The numbers parentheses are the accumulated number of like nucleons in that shell or lower. The levels in the centre are at energies intermediate between those of the square well potential and the simple harmonic oscillator potential. In addition a strong spin-orbital interaction is assumed for these levels (JEN 48).

nucleus unaffected by the location of other nucleons, one can calculate the allowed energies of the nucleon for different assumed mean potentials inside the nucleus. For two forms of this mean potential the calculations are particularly simple. One of these is a simple three-dimensional square well potential of infinite depth for which the radial eigenfunctions are Bessel functions (SCH 49). The energy eigenvalues for such a system can be calculated readily (SCH 49, JEN 48) and are plotted on the right side of Fig. VI. 12 along with the designation of the state, the number $2(2l+1)$ of like nucleons that can exist in each shell according to the Pauli principle, and the accumulated number of particles in that shell or lower. The accumulated numbers are in parentheses. Another potential for which calculations are simple (SCH 49, JEN 48) is the simple harmonic oscillator potential, for which the allowed energies and corresponding numbers are plotted on the left side of Fig. VI. 12. The shape of the mean nuclear potential well is probably somewhat intermediate between the two models, as is indicated by the interconnexion of the corresponding states of the two models by dotted lines. The above postulates alone are not adequate to give all the magic numbers. Mayer (MAY 48) and Jensen, Haxel, and Suess (JEN 48, HAX 48) pointed out independently that the magic numbers would be produced if one assumed the existence of a strong spin-orbital interaction such that the larger j is lower in energy. From Fig. VI. 12 it is immediately apparent that such a splitting of levels of different j can give the magic numbers.

With the aid of suitable rules for the coupling of the different nucleon angular momenta, one can calculate nuclear spins and magnetic moments from the above model. If either the number of neutrons or number of protons is even, Mayer (MAY 48) and Jensen, Haxel, and Suess (JEN 48) give the coupling rule that an even number of like nucleons in a shell of the angular momenta just compensate, whereas an odd number of like nucleons in a shell add up to give the angular momentum of a single nucleon in the shell. Thus for In^{113} there are 64 neutrons and 49 protons, so that from Fig. VI. 12 all shells are closed except for a $g_{9/2}$ shell; therefore the spin should be 9/2 as it is experimentally. For both the number of neutrons and the number of protons being odd, Nordheim (NOR 49) proposes the coupling rule that, if the spin and orbital angular momenta for one kind of a nucleon are parallel while they are antiparallel for the other kind of nucleon, the nuclear spin is ordinarily small, i.e. close to the difference in the two j's of the two kinds of nucleons; on the other hand the resultant spin is ordinarily close to the sum of the two j's if the spins and orbital angular momenta of the two types of nucleons are similarly coupled (both parallel or both antiparallel). These coupling rules are not completely inviolate.

For nuclei whose total number of nucleons A is odd, the above model then predicts that the magnetic moments should fall on the Schmidt lines calculated in Eqs. (VI. 77) and (VI. 78) and shown in Figs. VI.10

and VI. 11. As shown in the figures the observed moments are closely correlated with the Schmidt lines, but they fall between the lines rather than on them. Various reasons that have been suggested for the observed moments not falling on the lines are discussed at the end of the preceding section. Extensive applications of nuclear shell principles to specific nuclei have been given in various review articles (KLI 52, GOL 52, BOH 53).

Bohr and Mottleson (BOH 53, BOH 51) have modified the nuclear shell models to provide coupling to the collective motions of the nucleus as a whole. These modifications are discussed in Section XI. 5.2 since they are of particular value in accounting for the observed nuclear quadrupole moments. When there is strong surface coupling between the particle motion and the collective motion of the nucleus, the equivalent of the Schmidt lines are brought closer together in better agreement with experiment (BOH 53). Detailed calculations of theoretical nuclear magnetic moments have been made by Bohr and Mottleson (BOH 53); their results agree better with experiment than do the simple shell theories.

VII

NEUTRON BEAM MAGNETIC RESONANCE

VII. 1. Introduction

A NEUTRON beam is essentially a special and particularly simple case of a neutral molecular beam. As a result, a complete discussion of neutron-beam experiments both at high and low energies should strictly be included within the present work. On the other hand, the experimental techniques and results are ordinarily very different from those with molecular beams and the subject is such a vast one that its inclusion would double the size of the book. For this reason all high-energy neutron experiments and almost all of those at thermal energies will be omitted. For discussions of the omitted experiments the reader is referred to the excellent books on various topics of nuclear physics that have recently appeared (HUG 53a, SEG 53, BLA 52).

The experiments on the neutron magnetic moment are, however, so closely related to molecular beam resonance experiments both in the nature of the results and in the resonance techniques employed that a discussion of them is included. However, even in these experiments no detailed discussion will be given of the techniques for producing thermal neutron beams or of detecting them.

Since all neutron moment experiments have utilized polarized neutron beams, the discussion of the resonance experiments will be preceded by a discussion of the means for producing polarized neutron beams.

VII. 2. Polarized neutron beams

VII. 2.1. *Neutron beams.*

In the early neutron magnetic moment experiments the source of neutrons was a beryllium target bombarded by high-energy deuterons from a cyclotron. In more recent experiments the source of neutrons has been the fission process in a chain-reacting pile. In each of these cases the neutrons as they first emerge from the nucleus are at kinetic energies of several MeV. However, they can easily be slowed down to low energies by elastic collisions with the nuclei of a moderator which consists of light nuclei for which the energy transfer in an elastic collision is relatively large. H, D, Be, and C are the most common moderating materials. Although hydrogen absorbs the most energy from the neutron in a single collision because of the approximate mass equality of neutrons and protons, it also has the largest neutron capture cross-section, so the other materials often provide the most useful moderators. A 2-MeV neutron after approximately 18 collisions with H and 115 collisions with C will be slowed to thermal energies (0·025 eV).

The neutrons in such a moderator eventually diffuse to the surface and emerge to make up the neutron beams of the resonance experiment. If it is desired, the mean energy of these neutrons can be further reduced by passing them through a BeO or C filter (HUG 53a) which scatters out those neutrons whose wavelengths are less than 4·5 Å (BeO) or 6·5 Å (C). Such filtering, however, is not ordinarily needed in neutron magnetic resonance experiments since other means are often available for eliminating the unwanted fast neutrons. The physical mechanism of the filtering process is explained in the next section in terms of Eq. (VII. 4).

The slow neutrons which emerge from either the moderator or the filter can easily be collimated into a beam by thin sheets of Cd. The neutron beam may be detected with a BF_3 gas-filled proportional counter (SEG 53). Enriched B^{10} is preferable in the counter since the $_5B^{10}(n\alpha)\,_3Li^7$ reaction is the one which makes such counters effective as a result of its large cross-section ($3{,}990 \times 10^{-24}$ cm^2 at the thermal neutron velocity of 2,200 metres sec^{-1}).

VII. 2.2. *Polarization of neutron beams.* Four different techniques have been employed to polarize neutron beams. One of these is polarization by transmission, which was first suggested theoretically by Bloch (BLO 36, BLO 37). He pointed out that a neutron beam passing through a magnetized ferromagnetic substance will become partially polarized as a result of the interference between the nuclear scattering and the magnetic scattering caused by the magnetic interaction of the neutron magnetic moment with the atomic magnetic moment. Thus if ψ_t is the neutron wave function for the total scattering from a ferromagnetic microcrystal at the Bragg angle, ψ_n for the nuclear scattering, and ψ_m for the magnetic scattering,

$$|\psi_t|^2 = |\psi_n + \psi_m|^2 = |\psi_n|^2 + \psi_n^* \psi_m + \psi_n \psi_m^* + |\psi_m|^2. \qquad \text{(VII. 1)}$$

The last term is ordinarily negligible. However, depending on the relative orientation of the nuclear moment and the electron moments, the interference terms above will either add to or subtract from the dominant nuclear scattering. Hence, if the electron moments of a material are to a considerable extent aligned, as is the case in strongly saturated iron, and if a neutron beam is allowed to pass through the magnetically saturated material, the emerging beam will be polarized since neutrons of one spin orientation will be scattered more than the other.

Bloch's original theory of magnetic scattering has been improved by Schwinger (SCH 37, BLO 37) and by Halpern, Johnson, Holstein, and Hammermesh (HAL 41). The complete theoretical calculation of the polarization is quite complicated as it requires detailed knowledge of the distribution of the magnetic scattering caused by the orbital electrons (form factor), as well as the calculation of somewhat complicated crystalline effects in the iron. The net result of these theories is that the total cross-section σ for neutrons in completely saturated iron is given by
$$\sigma = \sigma_0 \pm p, \qquad (\text{VII. 2})$$

where σ_0 is the cross-section for unmagnetized iron, p is a complicated average of the interference between nuclear and magnetic scattering, whose sign depends on the orientation of the neutron magnetic field relative to the magnetic induction.

Polarization by transmission has been studied experimentally by a number of observers including Powers (POW 38); Bloch, Hammermesh, and Staub (BLO 43); Bloch, Condit, and Staub (BLO 46a); and Hughes, Wallace, and Holtzmann (HUG 48), and others (ZXX 54, HUG 53a). Ordinarily either the above quantity p or the neutron polarization P defined by
$$P = \frac{n_+ - n_-}{n_+ + n_-}, \qquad (\text{VII. 3})$$

has been measured in these experiments. In the preceding equation n_+ is the number of neutrons with spins parallel to the field and n_- is the number antiparallel.

One of the most significant implications of both the theoretical and the experimental work is the necessity of a high degree of magnetic saturation. This is needed because neutrons already polarized in the initial section of the iron are depolarized when they later encounter magnetic domains that are not completely aligned with the magnetizing field, since the neutrons in such a domain will precess magnetically about an axis not parallel to the magnetizing field. Saturation to within 0·1 per cent. of complete saturation is necessary in order to approach the full polarization effect in hot rolled steel 4 cm long. With magnetizing fields of 10,000 oersted the transmission of neutrons through a single block 4 cm long is increased by as much as 22 per cent. when the field is applied.

The production of polarized neutrons by transmission involves a serious loss of intensity because of the large thickness of iron involved. In addition this method suffers from the inconvenience of very strong magnetizing fields. A further disadvantage is that neutrons of very long wavelength ($\lambda > 4\cdot04$ Å with iron) cannot be polarized in this way since the iron acts as a pass filter for $\lambda > 2d_m$, where d_m is the maximum d for the crystal in the Bragg relation

$$n\lambda = 2d \sin\theta \qquad (\text{VII. 4})$$

and θ is the glancing angle. The neutrons are passed freely for such λ's because there is no possible orientation of the polycrystal for which the Bragg relation can be satisfied in such a case, so strong Bragg reflections are not possible. This failure to polarize very slow neutrons is particularly unfortunate since it is just these neutrons which give the greatest precision in magnetic resonance experiments.

A second method of polarization involving neutron diffraction from magnetite (Fe_3O_4) has been devised by Shull, Wollan, Koehler, and Fermi (SHU 51). The magnitude of the Bragg reflection from a single magnetized ferromagnetic crystal depends on both the nuclear and the magnetic scattering amplitudes with interference between them being possible. Shull, Wollan, and Koehler found for Bragg scattering from the (220) plane of magnetized magnetite that almost complete destructive interference was achieved for one neutron spin orientation while a large reflected amplitude could be achieved for the other. Approximately 100 per cent. polarization was achieved experimentally; the Bragg reflected polarized neutrons are also monochromatic, which can be both an advantage and a disadvantage owing to loss of intensity. This method, like the preceding one, suffers from its inability to polarize long wavelength neutrons owing to the long wavelength cut off by possible Bragg reflections.

A third method of polarizing neutron beams has been to collimate the beam well, then to deflect it in an inhomogeneous magnetic field similar to those used with molecules in the preceding chapters, and finally to select the desired part of the beam. This method has been used by Sherwood and others (SHE 54).

For neutron-beam magnetic-resonance experiments, by far the most useful of the currently available polarization techniques is that of total external reflection by magnetized ferromagnetic mirrors. When a neutron passes through a ferromagnetic domain where the absorption is small compared with the scattering, its effective wavelength is altered, i.e. the domain has a net index of refraction. As with an optical index of refraction the physical basis of the wavelength change is the vector addition of the amplitudes of the forward scattered waves and the incident wave. The index of refraction of course depends on the magnitudes of both the nuclear and the magnetic interactions. Halpern, Hammermesh, and Johnson (HAL 41), Akheiser and Pomeranchuck (AKH 48), Eckstein (ECK 49), and Halpern (HAL 49) have indeed shown for a single domain that

$$n^2 = 1 - \lambda^2 Na/\pi \pm \mu_n B_s/E, \qquad (VII.\ 5)$$

where n is the index of refraction, N the number of nuclei per cm^3, a the amplitude for coherent nuclear scattering (RAM 53g), B_s the magnetic induction (assumed at the saturation value inside any single domain), μ_n the neutron magnetic moment, and E the neutron energy. A simplified derivation of part of this formula is given in a book by Fermi (FER 50, p. 201).

When the index of refraction is less than unity a total external reflection can be achieved just as total internal reflection occurs in optics. The critical glancing angle θ_c below which there is total reflection is determined as in optics by

$$\cos \theta_c = n \sin \tfrac{1}{2}\pi = n. \qquad (VII.\ 6)$$

Even though a domain size is approximately 10^{-3} cm in diameter and the neutron wavelength is 10^{-8} cm, Snyder (HUG 53a, HUG 51) has pointed out that in reflection at the glancing angles of critical reflection experiments the neutron scattering is coherent over an appreciable number of domains; therefore the average n and consequently the average B should be used in Eqs. (VII. 5) and (VII. 6). Consequently

$$\cos \theta_c = 1 - \theta_c^2 = 1 - \lambda^2 Na/\pi \pm \mu_n B/E,$$

so
$$\theta_c = \sqrt{(\lambda^2 Na/\pi \mp \mu_n B/E)}. \qquad (VII.\ 7)$$

Alternatively, this may be written as

$$\theta_c = \sqrt{(\lambda^2 N/\pi)}\sqrt{(a \mp a_m)}, \qquad (VII.\ 8)$$

where a_m is the effective magnetic scattering amplitude

$$a_m = \frac{\mu_n B \pi}{E \lambda^2 N} = \frac{2\pi \mu_n B M_n}{h^2 N}, \qquad (VII.\ 9)$$

M_n being the mass of the neutron.

For cobalt the nuclear scattering amplitude is $3 \cdot 78 \times 10^{-13}$ cm and the saturated magnetic scattering amplitude is $\pm 4 \cdot 61 \times 10^{-13}$ cm. From Eq. (VII. 8), therefore, if B is greater than 82 per cent. of its saturation value one neutron orientation will be totally reflected and the other will not. Since the neutrons not totally reflected are reflected to a negligible extent, virtually 100 per cent. polarization should be produced in this way.

The polarization of neutrons by total external reflection has been studied experimentally by Hughes and Burgy (HUG 51). They measured the polarization by a double reflection effect as illustrated in Fig. VII. 1. They indeed found that 100 per cent. polarization could be achieved

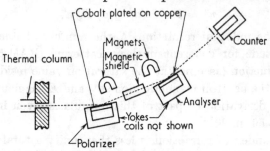

Fig. VII. 1. Apparatus for the production of neutron polarization and for measurement of the polarization by double reflection (HUG 51).

with a high enough B. They found, however, that special care had to be taken to avoid reorientation transitions (the Majorana transitions of Section V. 1) in the region between the two magnetic mirrors where the direction of the field might change rapidly compared with the Larmor frequency. The neutrons that pass closest to the mirror edges are particularly susceptible to such Majorana transitions. Also a magnetically shielded region had to be provided so that the neutron spins would not merely follow the field as it reversed between one mirror and the next, i.e. this principal change in field direction had to take place rapidly compared with the Larmor frequency.

One difficulty with cobalt mirrors is that they are difficult to magnetize; this is particularly true if the cobalt is hot forged in such a way as to orient the hexagonal crystals in a direction perpendicular to the direction of magnetization (COR 54). Corngold, Cohen, Ramsey, and Hughes (COR 53, COR 54) have found that an alloy of 93 per cent. cobalt and 7 per cent. iron could be much more easily magnetized since it forms a face-centred phase at room temperature. Such an alloy easily gave a 90 per cent. polarization and proved to be very effective as a polarizer and analyser in a neutron resonance experiment. The mirrors used in total reflection experiments are ordinarily ground and polished to optical flatness.

VII. 3. Neutron magnetic moment

The earliest resonance measurement of the neutron magnetic moment was that of Alvarez and Bloch (ALV 40). A schematic diagram of their apparatus is shown in Fig. VII. 2. Their neutrons were produced by deuteron bombardment of Be; they were polarized and analysed by passing them through magnetized pieces of Swedish iron about 4 cm thick. With these pieces of iron the neutron transmission increased about 6 per cent. on application of the magnetic field. A Rabi type

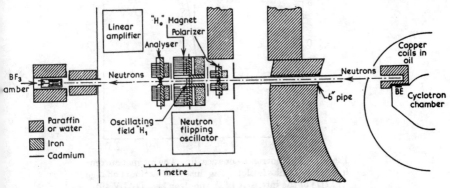

FIG. VII. 2. Plan of apparatus for Alvarez and Bloch neutron magnetic moment measurement (ALV 40).

single-oscillating field was used to induce the transitions. A typical neutron resonance curve is shown in Fig. VII. 3. The magnetic field was calibrated in terms of the magnetic field and frequency inside a cyclotron tuned to accelerate protons. The result of this experiment was $\mu_n = 1.93 \pm 0.02$ nuclear magnetons.

The next neutron magnetic moment experiment was that of Arnold and Roberts (ARN 47). They used neutrons from a chain-reacting pile and polarized them by transmission through a magnetized Armco iron block. They also used a single oscillatory field region. Their uniform magnetic field was calibrated in terms of the proton resonance frequency in a nuclear resonance absorption experiment (PUR 46, BLO 46, RAM 53e, p. 54). Their result was $\mu_n = -1.9103 \pm 0.0012$ nuclear magnetons.

At about the same time a similar experiment was carried out by Bloch, Nicodemus, and Staub (BLO 48). Their apparatus is shown schematically in Fig. VII. 4. Their neutrons originated from deuteron bombardment of beryllium in a cyclotron and were polarized by transmission through magnetized iron. A single oscillatory field region was used. Some increase in precision was achieved by using a less than

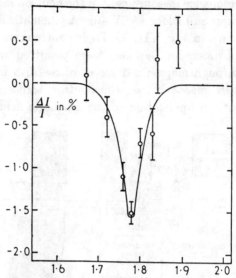

Fig. VII. 3. Neutron resonance dip. The magnet current in arbitrary units is plotted against the fractional change ($\Delta I/I$) of the intensity of the neutron beam (ALV 40).

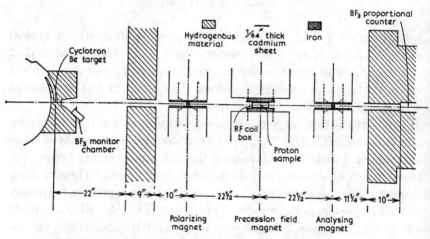

Fig. VII. 4. Experimental arrangement for neutron magnetic moment experiment of Bloch, Nicodemus, and Staub (BLO 48).

normal oscillatory current in which case the resonances were somewhat narrowed since the slower neutrons were then relatively more effective. A typical neutron resonance is shown in Fig. VII. 5. The uniform magnetic field region was calibrated in terms of the Larmor frequency of the proton moment as measured by the nuclear induction method (BLO 46) which was developed explicitly for the neutron moment experiment. The result of this experiment was $\mu_n/\mu_p = -0\cdot685001 \pm 0\cdot000030$

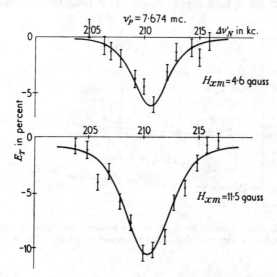

FIG. VII. 5. Neutron resonances observed at 7·674 Mc.
$E_T = (I_{on}-I_{off})/(I_{off}-I_{Cd})$ (BLO 48).

with no shielding correction for the proton in H_2O. Rogers and Staub (ROG 49) have used a rotating instead of an oscillating magnetic field to induce the transitions and have thereby determined the sign of the magnetic moment directly by observing the sense of the Larmor precession (see Section VI. 4). The neutron moment was negative, as was to be expected from the approximate additivity of the neutron moment and the proton moment to give the experimental deuteron moment within 1 per cent. Further discussion is given in Appendix G.

The most recent and most precise measurement of the neutron magnetic moment is that of Corngold, Cohen, and Ramsey (COR 53, COR 54). Their experiment employed many techniques not available to their predecessors. A schematic view of the apparatus is shown in Fig. VII. 6. The source of neutrons was the fission process in a pile with the neutrons being slowed down by a graphite moderator. When desired a BeO filter (HUG 53) could be inserted to reduce the number of high-energy neutrons and hence to give narrower lines; the filter, however, was not ordinarily used since the high-energy neutrons could also be eliminated by a suitable choice of angles for the polarizing mirrors.

The neutrons were polarized and analysed by total reflection from mirrors of a 93 per cent. cobalt 7 per cent. iron alloy as discussed in the previous section. A 90 per cent. polarization as defined by Eq. (VII. 3) was obtained on a single reflection. The reflection process had the

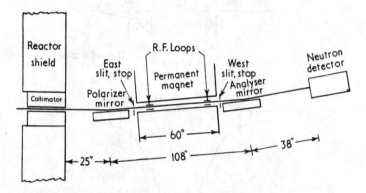

FIG. VII. 6. Experimental arrangement for neutron magnetic moment experiment of Corngold, Cohen, and Ramsey (COR 53, COR 54).

advantage of also greatly reducing the average energy of the neutrons since, from Eq. (VII. 8), only neutrons with a λ greater than a certain minimum value will be totally reflected. This is illustrated in Fig. VII. 7, where the theoretical beam velocity distribution for room temperature is drawn and where the cut-off spectrum is indicated for a neutron glancing angle of 0·005 radians. From this it is apparent that the mean velocity of the neutrons is reduced by about a factor of four, which leads to a very desirable fourfold increase in the sharpness of the observed resonance.

The uniform field was 60 inches long with a gap of 0·7 inches. Permanent Alnico V magnets supplied both the uniform field and the fields for the polarizing and analysing mirrors. A typical cross-section of the C magnet is shown in Fig. VII. 8. The use of permanent magnets simplified the problem of the constancy of the field in time, though moderately uniform thermal drifts of 0·2 gauss per day were found at 8,500 gauss. The magnetic fields were hand shimmed for maximum uniformity as indicated by proton magnetic resonance measurements.

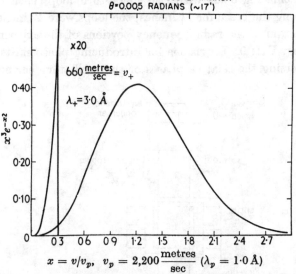

Fig. VII. 7. Neutron velocity spectrum. The curve extending beyond 1·5 is the theoretical neutron velocity distribution before reflection at a 0·005 radian glancing angle from a magnetized iron-cobalt mirror. The reflection cuts off the velocity of even the favourably polarized neutrons at $x = v/v_p = 0·3$ as shown by the curve at the left, whose vertical scale is multiplied by a factor of 20 (COR 53, COR 54).

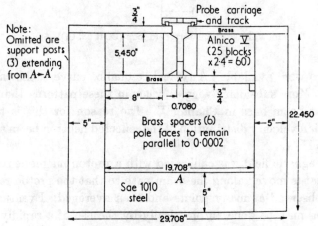

Fig. VII. 8. Cross-section of the 60 in. long permanent magnet (COR 53, COR 54)

The resonance transitions were induced by Ramsey's method of separated oscillating fields with two single-turn loops each of which extended along the beam for 1⅛ inches; the loops were 42 inches apart. A block diagram of the radiofrequency portions of the experiment are shown in Fig. VII. 9. Provisions for introducing phase shifts and for quickly reversing the relative phases of the oscillatory current in the

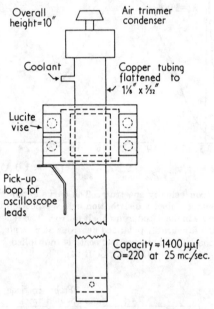

FIG. VII. 9. Oscillatory field loop (COR 53, COR 54).

two loops were available. A typical resonance curve is shown in Fig. VII. 10. More subsidiary wiggles occur in these patterns than in the molecular beam ones in Chapter V. The reason for this is the non-Maxwellian velocity distribution of the reflected neutron beam as shown in Fig. VII. 7.

The magnetic field was calibrated with a proton magnetic resonance probe that is moved along the beam path so that the proton resonance can be observed at many points and then averaged. Eventually this averaging may be done automatically by the use of a rapidly flowing stream of water or of hydrogen gas and a nuclear induction technique analogous to the use of separated oscillating fields in molecular beam experiments (PUR 52, COR 54, DIC 54, ZZA 61).

The results of this experiment are as follows where the subscript u, w

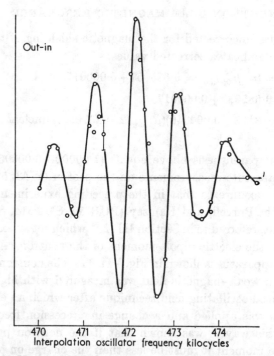

Fig. VII. 10. Neutron magnetic resonance curve. The intensity with loops 180° out of phase minus the intensity in phase is plotted as a function of frequency. Additional subsidiary maxima occur as a result of the non-Maxwellian velocity distribution shown in Fig. VII. 7 (COR 53, COR 54).

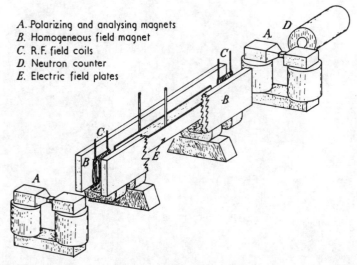

A. Polarizing and analysing magnets
B. Homogeneous field magnet
C. R.F. field coils
D. Neutron counter
E. Electric field plates

Fig. VII. 11. Schematic diagram of apparatus for setting experimental upper limit to neutron electric dipole moment (SMI 51a, SMI 51b).

indicates values uncorrected for the magnetic shielding of the protons in water and c indicates corrected values:

$$(\mu_n/\mu_p)_{u,w} = (\nu_{0n}/\nu_{0p})_{u,w} = 0{\cdot}685057 \pm 0{\cdot}000017,$$

$$(\mu_n/\mu_p)_c = 0{\cdot}685039 \pm 0{\cdot}000017,$$

$$\mu_n = -(1{\cdot}913138 \pm 0{\cdot}000045)(\mu_{p,u\cdot w}/2{\cdot}7926700...) \text{ nuclear magnetons},$$

(VII. 10)

where the last parenthesis above equals $(1{\cdot}00000 \pm 0{\cdot}00005)$ with the present best absolute measurements on the proton (WAL 53).

A similar procedure to that in the preceding experiment has been used by Smith, Purcell, and Ramsey (PUR 50, SMI 51a, SMI 51b) in the experiment, referred to in Section III. 2.2, which set an experimental upper limit to the electric dipole moment of the neutron. A schematic view of their apparatus is shown in Fig. VII. 11. The neutron precession frequency in a weak magnetic field was measured with high precision by the separated oscillating field technique after which an electric field of 215 kV/cm was applied and a change in precession frequency was sought. In this way it was shown that if the neutron possessed an electric dipole moment it must be less than the charge on the electron multiplied by a distance $D = 5 \times 10^{-20}$ cm.

VIII
NUCLEAR AND MOLECULAR INTERACTIONS IN FREE MOLECULES

VIII. 1. Introduction

IN the two preceding chapters, consideration has been limited to cases where the only significant interaction energy of the molecule was that between the external magnetic field and the nuclear and rotational magnetic moments. However, other interactions such as those of nuclear quadrupole moments are often of comparable importance. The theory and experimental results of measurements of the various interactions which occur in non-paramagnetic molecules will be discussed in this chapter. Since all of the molecular-beam resonance measurements of this nature so far have been made on $^1\Sigma$ diatomic molecules, the discussion will be concerned primarily with such molecules; theoretical discussions of polyatomic molecules and of other states including paramagnetic ones have been given by Frosch and Foley (FRO 52), Herzberg (HER 45, HER 50a), and others (TOW 54, VAN 51). Some of the experimental data of this nature have been obtained by the molecular-beam electric-resonance method discussed in Chapter X. Where results obtained by that method are particularly relevant they will be discussed in the present chapter. However, the present chapter will emphasize primarily the molecular-beam magnetic-resonance experiments, Chapter X being reserved for the detailed discussion of the electric-resonance results.

Historically, in the earliest molecular-beam magnetic-resonance experiments only the interaction of the nuclear moment with the external magnetic field was anticipated to be of importance. The interactions which make up the bulk of the present chapter were added only after experiments showed the existence of structure and excessive breadth in the resonance patterns. Although historically the experiments have thereby preceded the theory at most stages, the historical order will be reversed for the sake of clarity in the present chapter and the theory of each interaction will be discussed in advance of the experimental data.

VIII. 2. Nuclear and rotational magnetic moments

In a strong external magnetic field, the largest interactions are often those of the nuclear magnetic moments and of the molecular rotational magnetic moments with the external magnetic field. The effects of these interactions when they occur essentially alone have already been discussed in Chapter VI. However, at present these interactions must be assumed to make up only a part of the total molecular Hamiltonian. By Sections III. 3.2, VI. 7, and VI. 8, these interactions make the following contributions to the Hamiltonian in a $^1\Sigma$ diatomic molecule

$$\mathcal{H}_{1H} = -[1-\sigma_1(\mathbf{J})]\boldsymbol{\mu}_1 \cdot \mathbf{H} = -[1-\sigma_1(\mathbf{J})]a_1 h\mathbf{I}_1 \cdot \mathbf{H}/H, \quad \text{(VIII. 1)}$$

$$\mathcal{H}_{2H} = -[1-\sigma_2(\mathbf{J})]a_2 h\mathbf{I}_2 \cdot \mathbf{H}/H, \quad \text{(VIII. 3)}$$

$$\mathcal{H}_{JH} = -[1-\sigma_J(\mathbf{J})]\boldsymbol{\mu}_J \cdot \mathbf{H} = -[1-\sigma_J(\mathbf{J})]b_J h\mathbf{J} \cdot \mathbf{H}/H, \quad \text{(VIII. 4)}$$

where μ_J is the rotational magnetic moment and

$$a_1 = \mu_{I1} H/I_1 h = 2\pi\gamma_{I1} H, \quad \text{(VIII. 5)}$$

$$a_2 = 2\pi\gamma_{I2} H, \quad \text{(VIII. 6)}$$

$$b_J = \mu_J H/Jh = 2\pi\gamma_J H. \quad \text{(VIII. 7)}$$

In the above $\sigma_1(\mathbf{J})$ and $\sigma_2(\mathbf{J})$ are given by Eq. (VI. 56).

If the diatomic molecule is a homonuclear molecule like H_2, it is usually most convenient to express the results in terms of the resultant angular momentum
$$\mathbf{I}_R = \mathbf{I}_1 + \mathbf{I}_2 \quad \text{(VIII. 8)}$$
since I_R is a good quantum number in a rotational state of definite J because of the symmetry requirements on the wave function with identical nuclei. In this case $a_1 = a_2 = a_I$ and

$$\mathcal{H}_{IH} = -[1-\sigma_I(\mathbf{J})]a_I h\mathbf{I}_R \cdot \mathbf{H}/H. \quad \text{(VIII. 9)}$$

Values of the nuclear magnetic moments are listed in Chapter XV. The values of the rotational magnetic moments of H_2, D_2, and HD are listed in Section VIII. 10.

Further discussion of the nuclear and rotational magnetic moment interactions is given in Chapter VI. Values of the nuclear magnetic moments are listed in that chapter and of the rotational magnetic moments so far measured by molecular beams in Section VI. 8.

VIII. 3. Nuclear spin-spin magnetic interactions

VIII. 3.1. *Direct spin-spin magnetic interaction.* The first experimental evidence for a spin-spin magnetic interaction in molecules was found in the experiments of Kellogg, Rabi, Ramsey, and Zacharias on H_2 molecules (KEL 39a, RAM 40) as shown in Fig. I. 5. They found that their results could best be interpreted by including a term in the molecular Hamiltonian analogous to the interaction between two classical magnetic moments a distance R apart, where R is the internuclear separation. The form of this interaction was discussed in Chapter III and is given in general by the $\mathcal{H}_S$ of Eq. (III. 95). In the case when the rotational angular momentum quantum number J is a good quantum number (as it usually is), the interaction $\mathcal{H}_S$ can be rewritten as in Eq. (III. 102) or, with a slight change of notation, as

$$\mathcal{H}_S = \frac{5d'_M h}{(2J-1)(2J+3)}[\tfrac{3}{2}(\mathbf{I}_1\cdot\mathbf{J})(\mathbf{I}_2\cdot\mathbf{J})+\tfrac{3}{2}(\mathbf{I}_2\cdot\mathbf{J})(\mathbf{I}_1\cdot\mathbf{J})-(\mathbf{I}_1\cdot\mathbf{I}_2)\mathbf{J}^2],$$
(VIII. 10)

where $\qquad d'_m = \frac{2}{5}\frac{(\mu_1/I_1)(\mu_2/I_2)}{h}\langle 1/R^3\rangle.$ (VIII. 11)

If the molecule is homonuclear, the interaction can be rewritten as Eq. (III. 110), which with a slight change in notation becomes

$$\mathcal{H}_S = \frac{5d_M h}{(2J-1)(2J+3)}[3(\mathbf{I}_R\cdot\mathbf{J})^2+\tfrac{3}{2}\mathbf{I}_R\cdot\mathbf{J}-\mathbf{I}_R^2\mathbf{J}^2],\quad \text{(VIII. 12)}$$

where
$$d_M = \frac{(\mu_1/I_1)^2}{5h}\langle 1/R^3\rangle\frac{I_R(I_R+1)+4I_1(I_1+1)}{(2I_R-1)(2I_R+3)}$$

$$= \tfrac{1}{2}d'_M\frac{I_R(I_R+1)+4I_1(I_1+1)}{(2I_R-1)(2I_R+3)}.\qquad\text{(VIII. 13)}$$

Kolsky, Phipps, Ramsey, Silsbee, and their associates (KOL 50, KOL 50b, KOL 52, HAR 53) have measured the radiofrequency spectrum of H_2 with high precision as shown in Fig. VIII. 1. They find that the best fit is obtained for a value

$${}^H_0\langle d_M\rangle_1 = 57{,}671\pm 24 \text{ c/s},\qquad \text{(VIII. 14)}$$

where the subscripts and superscripts around the brackets indicate average values for the molecule H_2 in the zeroth vibrational and first rotational state. Ramsey (RAM 52d, RAM 52f) has compared this with the theoretical value to be expected from Eq. (VIII. 13) and the known value of μ_I. In the calculation of ${}^H_0\langle 1/R^3\rangle_1$ suitable allowance must be made for zero point vibration and centrifugal stretching as discussed in Section VIII. 7. The theoretical and experimental values (RAM 52d, RAM 52f, HER 50) agree to within the 0·07 per cent. accuracy of the theoretical calculation. This indicates that, if there is any long range non-magnetic nuclear tensor interaction [interaction of the form of Eq. (III. 95)] between two protons 0·74 Å apart, the interaction potential must be less than 10^{-19} MeV (RAM 52d, RAM 52f). Alternatively, if it is assumed that the entire interaction is magnetic, the measurement supplies a precision value for ${}^H_0\langle 1/R^3\rangle_1$ as listed in the tables of Section VIII. 10.

It is of interest to note that $\mathcal{H}_S$ averages to zero over all molecular orientations. This can be seen either by averaging Eq. (III. 95) over all directions of **R** or the equations of this section equally over all molecular orientation states. For this reason, an interaction of the form of $\mathcal{H}_S$ cannot be observed in nuclear paramagnetic resonance experiments of the Purcell (PUR 46) and Bloch (BLO 46) type with gases or liquids such that frequent collisions provide an averaging over all orientation states.

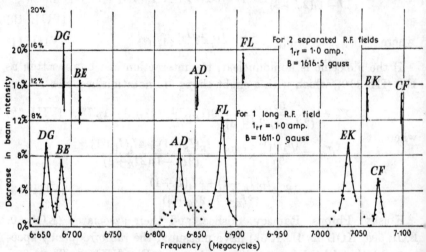

FIG. VIII. 1. Radiofrequency spectrum of ortho-H_2 for transitions in which $\Delta m_I = \pm 1$ and $\Delta m_J = 0$. The upper curve is with Ramsey's method of separated oscillating fields while the lower curve is by Rabi's method in the same apparatus. The relative displacement between the two curves is due merely to the difference in the magnitude of the uniform fields (KOL 52).

VIII. 3.2. *Electron-coupled interactions between nuclear spins in molecules.* Despite the conclusion of the preceding paragraph that no interaction of the form $\mathcal{H}_S$ could be observed in nuclear paramagnetic resonance experiments, a spin-spin structure has been observed in some high resolution experiments of this type by Gutowsky, McCall, Slichter, and McNeil (GUT 51a) and independently by Hahn and Maxwell (HAH 51). They found that their observations could be accounted for if an interaction of the form $h\, \delta\mathbf{I}_1 \cdot \mathbf{I}_2$ is assumed to exist between the two nuclei. If magnetic shielding of the nuclear spin-spin interaction by the orbital motion of the electron is included, the form of Eq. (III. 95) is modified in such a way that a residual spin-spin interaction of the desired form is obtained. However, several attempts to account for the experimental results in this way led to interactions that were much too weak (GUT 51a, MCN 51, HAH 51, DRE 52).

The solution to this difficulty was proposed by Ramsey and Purcell

(RAM 52b, RAM 53b), who suggested a mechanism which would give rise to an $\mathbf{I}_1.\mathbf{I}_2$ interaction of the magnitude observed. This mechanism is the magnetic interaction between each nucleus and the electron spin of its own atom together with the exchange coupling of the electron spins with each other. The mechanism corresponds to the fact that the magnetic interaction of one nucleus with the electron of its atom will make the electron of that atom tend to lie more frequently antiparallel to the nuclear spin than parallel to it. On the other hand, the two electron spins in the singlet state must be antiparallel to each other so that the electron of the other atom will tend to lie more frequently parallel to the spin of the first nucleus. However, the electron of that atom interacts magnetically with the second nucleus. The combination of these interactions therefore provides a spin interaction between the two nuclei. In terms of perturbation theory the proposed mechanism corresponds to a second-order perturbation by the higher electronic triplet states of the molecule, the perturbing interaction being the magnetic interaction of each nucleus with the electron spins. The reason why electron spin effects enter here, whereas they are omitted in ordinary magnetic-shielding calculations, is that here the magnetic fields from both nuclei vary over the molecule, whereas in the ordinary magnetic-shielding case one of the perturbing fields is the externally applied one, which is uniform over the molecule and hence affects both electron spins alike.

Detailed calculations including both this contribution and that of the orbital motions of the electrons have been made by Ramsey (RAM 53b). He has shown that in general the electron coupled nuclear spin-spin interaction in a diatomic molecule is of the form

$$\mathcal{H}_{Se} = h\delta\mathbf{I}_1.\mathbf{I}_2 + h\sum_{ij}\mathfrak{b}_{ij}I_{1i}I_{2j}, \qquad \text{(VIII. 15)}$$

where $\mathfrak{b}_{ij}$ is the ij Cartesian component of a traceless tensor of the second rank (not to be confused with the Kroeneker δ_{ij}).

Only when frequent collisions average the molecular orientation equally over all directions does the interaction reduce to the simpler $h\delta\mathbf{I}_1.\mathbf{I}_2$. Hence in the molecular beam case the full expression should be retained although the tensor term in a case like HD is probably only about a tenth as large as the simpler $h\delta\mathbf{I}_1.\mathbf{I}_2$ term.

The general theoretical expressions for δ and $\mathfrak{b}$ are quite complicated and are given in the literature (RAM 53d). However, in the case of HD over 90 per cent. of the interaction is of the form

$$\delta_{\mathrm{HD}} = (64\mu_0^2 h\gamma_\mathrm{H}\gamma_\mathrm{D}/9\Delta)|\psi|^2_{a\mathrm{H}b\mathrm{D}}, \qquad \text{(VIII. 16)}$$

where Δ is a suitable (RAM 53d) mean energy of the excited electronic states of the HD molecule, $|\psi|^2_{aHbD}$ is the probability density for one electron to be on the proton and the other on the deuteron. The other quantities have their earlier meanings.

Following the suggestion of Ramsey and Purcell (RAM 52b), measurements of δ in HD have been made by Smaller (SMA 52), Carr and Purcell (CAR 52a), and Wimett (WIM 53) by methods of nuclear paramagnetic resonance with the result that δ is $43\pm0\cdot 5$ c/s. So far this has not been

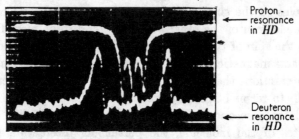

FIG. VIII. 2. Proton and deuteron nuclear paramagnetic resonances with a volume of gaseous HD. The multiplicity of the resonances is presumably due to electron-coupled nuclear spin interaction (RAM 52b, RAM 53d). The scales for the two curves are indicated by the fact that the two deuteron resonances are 43 c/s apart, whereas the outermost proton resonances are each 43 c/s from the central one (WIM 53).

clearly observed in molecular beam experiments but it should be soon since its magnitude is just comparable to the present experimental error in the beam resonance experiments. Fig. VIII. 2 shows nuclear paramagnetic resonance patterns for HD with the triple proton resonance and the double deuteron one. Eq. (VIII. 16) can be made to agree with the experimental value of δ in HD provided Δ is assumed to have the reasonable value of 1·4 Rydbergs.

VIII. 4. Spin-rotational magnetic interactions

The first experimental evidence for a molecular spin-rotational magnetic interaction was found by Kellogg, Rabi, Ramsey, and Zacharias (KEL 39a, RAM 40) in their molecular-beam study of the radiofrequency spectrum of H_2. To account for their experimental results as shown in Fig. I. 5, they found that they had to add to the molecular Hamiltonian of a $^1\Sigma$ diatomic molecule the two terms

$$\mathscr{H}_{1J} = -c_1 h\mathbf{I}_1.\mathbf{J},$$
$$\mathscr{H}_{2J} = -c_2 h\mathbf{I}_2.\mathbf{J}. \qquad \text{(VIII. 17)}$$

These terms could be interpreted as the interaction of the nuclear magnetic moment μ_i with the effective magnetic field $H'_i J$ due to the molecular rotation. In this case

$$c_i = \mu_i H'_i/I_i h. \qquad \text{(VIII. 18)}$$

It should be noted that some confusion in notation exists in this subject,

TABLE VIII. I

Spin-Rotation Interaction Constants in Diatomic Molecules

The interaction constant $c_i = -c'_i/h$ is defined by Eq. (VIII. 17). The letter M under method indicates a measurement by the magnetic resonance method while the letter E indicates the electric resonance method. Results in cycles per second.

Molecule	Nucleus (i)	$c'_i/h = -c_i$ (c/s)	Method	References
H_2	H^1	$-113,904 \pm 30$	M	KEL 39a, RAM 40, KOL 52, HAR 52
HD	H^1	$-87,200 \pm 2,000$	M	KEL 39a, RAM 40
HD	D^2	$-13,400 \pm 340$	M	KEL 39a, RAM 40
D_2	D^2	$-8,788 \pm 40$	M	KEL 39a, RAM 40, KOL 52, HAR 52
LiF	F^{19}	$\pm 19,000 \pm 8,000$	M	NIE 47, WIC 48
Li^6F	F^{19}	$+37,300 \pm 700$	E	SWA 52, ZZA 22
RbF	F^{19}	$\pm 9,100 \pm 900$	M	ZZA 21
$Rb^{85}F$	F^{19}	$\pm 11,000 \pm 300$	E	HUG 50
$Rb^{87}F$	F^{19}	$\pm 14,000 \pm 400$	E	HUG 50
CsF	F^{19}	$+16,000 \pm 2,000$	E	TRI 48
CsF	F^{19}	$\pm 12,000 \pm 600$	M	NIE 47, ZZA 21
NaF	Na^{23}	$\sim \pm 1,000$	M	ZEI 52, COT 53
NaCl	Na^{23}	± 370 (or -160)	M	ZEI 52, COT 53
NaBr	Na^{23}	$\sim \pm 670$	M	NIE 47, COT 53
NaI	Na^{23}	$\sim \pm 690$(or-230)	M	NIE 47, COT 53
TlCl	Cl^{35}	$\pm 1,400 \pm 100$	M	ZEI 52
TlCl	Cl^{35}	$+1,200 \pm 200$	E	CAR 52
TlCl	Cl^{37}	$\pm 1,100 \pm 100$	M	ZEI 52
TlCl	Cl^{37}	$+1,000 \pm 200$	E	CAR 52
KBr	Br^{79}	$\sim \pm 210$	M	COT 53
KBr	Br^{81}	$\sim \pm 1,300$	M	COT 53
RbF	Rb^{87}	$\pm 1,100 \pm 100$	M	BOL 52
RbCl	Rb^{87}	$\pm 1,000 \pm 100$	M, E	BOL 52, TRI 54
CsF	Cs^{133}	$0 \pm 1,000$	E	TRI 48
TlCl	Tl	$+73,000 \pm 2,000$	E	CAR 52

and some authors (ZEI 52, SWA 52) use c_i to designate the quantity called $-hc_i$ above. To prevent confusion their c_i will be called c'_i so

$$c'_i = -hc_i. \qquad \text{(VIII. 18a)}$$

For homonuclear diatomic molecules the above interactions may be combined as in the preceding sections of this chapter to

$$\mathcal{H}_{IJ} = -ch\mathbf{I}_R \cdot \mathbf{J}. \qquad \text{(VIII. 19)}$$

From the precision measurements of the radiofrequency spectrum of H_2 and D_2 by Kolsky, Phipps, Ramsey, Silsbee, and their associates (KOL 52, HAR 53) as shown in Fig. VIII. 1, values of the spin-rotational interaction constant c have been derived, as in Table VIII. I.

Molecules with large moments of inertia like LiF cannot be studied as easily as H_2 because the large number of rotational states excited leads to a number of closely spaced and unresolvable radiofrequency spectral lines. However, Nierenberg and Ramsey (NIE 47) have pointed out that values of the spin rotational interaction constant c can be inferred from the shape of the resultant resonance pattern which is

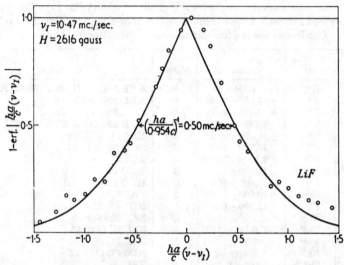

FIG. VIII. 3. Comparison of theory and experiment for the radiofrequency spectrum of LiF. The experimental points are plotted as reduction in beam intensity (normalized to unity for the maximum) versus the difference in frequency from the Larmor frequency normalized by the factor $ha/c = 1·91 \times 10^6$ sec. $a^2 = h^2/8\pi^2 IkT = \pi/4J^2$ and c is the parameter of Eq. (VIII. 17). The full line is the calculated line shape for a total orientation dependent Hamiltonian equal to $-\mu_N g_I \mathbf{I}\cdot\mathbf{H} - c\mathbf{I}\cdot\mathbf{J}$ (NIE 47).

a statistical superposition of the individual resonances for all the rotational states that are excited. Some of their theoretical analysis is summarized in Section VIII. 5. With their statistical theory and experimental results as shown in Fig. VIII. 3, they measured c in LiF.

These methods have subsequently been extended to TlCl, RbF, RbCl NaF, and NaCl by Zeiger, Bolef, and Rabi (ZEI 52, BOL 52, ZEI 50). They have found an actual splitting of the zero magnetic field nuclear quadrupole resonance as shown in Fig. VIII. 4. With their analysis which is an extension of that of Nierenberg and Ramsey (NIE 47) and is outlined in Section VIII. 5, c can be determined from the separation of the peaks, which is approximately $2cJ$.

Also, by methods that will be discussed in Chapter X, the molecular beam electric-resonance method has been applied to the determination of c by Trischka (TRI 48) with CsF, by Hughes and Grabner (HUG 50, HUG 50a) with RbF, by Swartz and Trischka (SWA 52) with LiF, and by Carlson, Lee, Fabricand, and Rabi (CAR 52, LEE 52) with TlCl, KCl, and KBr. The values obtained for the spin-rotational magnetic

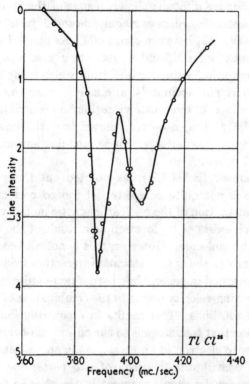

FIG. VIII. 4. Zero field spectrum of Cl^{35} in $TlCl^{35}$. Line intensity (decrease in beam at the detector) is expressed in cm deflexion on a galvanometer scale. Peak intensity corresponds to a 0·5 per cent. decrease in total beam. The radiofrequency current was 12 amperes. The width of the line at half intensity of each component is 100 kc, considerably greater than the 3 kc resolution half-width of the apparatus for this molecule (ZEI 52).

interaction parameter c in all of these experiments are listed in Table VIII. I. See also Appendix G.

The theory of spin-rotational magnetic interaction has been developed by Wick (WIC 48), Foley (FOL 47a), Brooks (BRO 41), and Ramsey (RAM 50d, RAM 52e, RAM 52f, RAM 53a). One consequence of these theories (RAM 50d) is the close relationship between the spin-rotational magnetic interaction and the high frequency terms σ^{HF} in the magnetic shielding constant. The justification of this relation has already been given in Section VI. 7 above, the relationship itself being given in Eq. (VI. 53). Since this equation expresses the spin rotation interaction constant c in terms of σ^{HF} and since the theory of σ^{HF} is given in Eq. (VI. 50), the combination of the two equations provides the theoretical expression for c. As mentioned in Section VI. 7, numerical calculations with Eq. (VI. 50) are difficult; the chief uses of the theory have been in indicating reasonable orders of magnitude and in justifying Ramsey's relation of Eq. (VI. 53) between c and σ^{HF}. As pointed out in Section VI. 7, the values of c for H_2 and D_2 have been used with Eq. (VI. 53) to provide an empirical value for σ^{HF} from which σ can be calculated.

Eq. (VI. 53) is not accurately applicable to any actual molecule because of the effects of molecular vibration and centrifugal stretching. Ramsey (RAM 52f) has, however, shown how the equation can be modified to include these effects. These modifications are discussed in Section VIII. 7 below.

Recently Ramsey (RAM 53a) has pointed out that the theory as outlined above is not quite complete. It indeed correctly allows for the magnetic interaction of the nuclear magnetic moment and the magnetic field which exists at its location as a result of the motion of the remainder of the molecule. However, this is not all because there are also contributions to the spin-rotational interaction from the fact that the nucleus concerned in general has both a non-vanishing acceleration and a non-vanishing velocity owing to the combination of the molecular vibrations and rotations. This results in a contribution to spin-rotational interaction that is analogous to the entire spin-orbital interaction of a single valence electron in an atom. As in spin-orbital interactions in atoms, this contribution consists of two parts: one is due to the motion of the magnetic dipole moment in the electric field that gives rise to the nuclear acceleration and the other is the relativistic Thomas precession. The acceleration and the electric field are of course simply related by Newton's law

$$M_N d\mathbf{v}/dt = Ze\mathbf{E}, \qquad (\text{VIII. 20})$$

where M_N is the mass of the nucleus concerned.

By a calculation analogous to that in the atomic spin orbital case Ramsey (RAM 53a) showed that the combination of the simple term and the Thomas term gave an acceleration contribution c_A to the spin-rotational magnetic-interaction constant c that was given by

$$\begin{aligned}{}_v^x\langle c_A \rangle_J &= (\gamma/2\pi)[1-(Ze/2M_N c\gamma)]\mathbf{I}.(\mathbf{E}\times\mathbf{v}/c)/\mathbf{I}.\mathbf{J} \\ &= (\gamma/2\pi)(\hbar/M_N c)[1-Ze/2M_N c\gamma]{}_v^x\langle E/R\rangle_J. \qquad (\text{VIII. 21})\end{aligned}$$

It is of interest to note that, if $\gamma = Ze/M_N c$ (as it is for an electron), the square bracket reduces to $\frac{1}{2}$ which is the well-known Thomas factor. The indicated average has been carried out for the vibrational state $v = 0$ and rotational state $J = 1$ for molecular H_2 and D_2 with the result (RAM 53a)

$$^H_0\langle c_A\rangle_1 = 1,059 \text{ c/s},$$
$$^D_0\langle c_A\rangle_1 = 30.6 \text{ c/s}. \qquad \text{(VIII. 22)}$$

The experimental quantity c that comes from Eq. (VIII. 19) is then the combination of the above quantity c_A and the c_W to which the preceding theory and Eq. (VI. 53) apply where

$$c = c_A + c_W. \qquad \text{(VIII. 23)}$$

In deriving values of the shielding constant σ as above, the experimental c must be reduced to c_W with (VIII. 23) before Eq. (VI. 53) is used. Ordinarily this correction is negligible but in the case of H_2 it is large in comparison with the experimental error.

VIII. 5. Nuclear electrical quadrupole interaction

The first and most important nuclear electrical quadrupole interaction observed in polyatomic molecules was the quadrupole moment

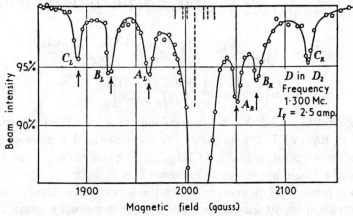

FIG. VIII. 5. Radiofrequency spectrum of D_2 at 80° K arising from transitions of the resultant nuclear spin. Dotted lines indicate the spectrum predicted on the assumption of no deuteron electrical quadrupole moment. Arrows indicate spectrum predicted on assumption of suitable quadrupole moment (KEL 40).

of the deuteron discovered by Kellogg, Rabi, Ramsey, and Zacharias (KEL 39, KEL 40) in molecular D_2 and HD. They found the radiofrequency spectrum to be as in Fig. VIII. 5. On the other hand, from the spin-spin and spin-rotational magnetic interactions of the two preceding sections and from a calibration of the relevant parameters with the H_2 results, the predicted positions of the resonances would be as shown by the dashed lines. They assumed that the discrepancy was due to a deuteron quadrupole moment and found in this way that complete agreement between experiment and theory could be achieved.

From Eq. (III. 43) and (III. 49) with a slight change in nomenclature, it is apparent that the nuclear quadrupole interaction in a diatomic molecule can be written as

$$\mathcal{H}_Q = \mathcal{H}_{Q1} + \mathcal{H}_{Q2}, \qquad (\text{VIII. 24})$$

where $\mathcal{H}_{Q1} = \dfrac{5d'_{Q1} h}{(2J-1)(2J+3)} [3(\mathbf{I}_1 . \mathbf{J})^2 + \tfrac{3}{2}\mathbf{I}_1 . \mathbf{J} - \mathbf{I}_1^2 \mathbf{J}^2], \qquad (\text{VIII. 25})$

$\mathcal{H}_{Q2}$ is similarly expressed with subscript 2 replacing subscript 1, and

$$d'_{Q1} = -\dfrac{eqQ_1}{10hI_1(2I_1-1)}. \qquad (\text{VIII. 26})$$

For homonuclear diatomic molecules this can be re-expressed with the aid of Eq. (III. 44b) as

$$\mathcal{H}_Q = \dfrac{5d_Q h}{(2J-1)(2J+3)} [3(\mathbf{I}_R . \mathbf{J})^2 + \tfrac{3}{2}(\mathbf{I}_R . \mathbf{J}) - \mathbf{I}_R^2 \mathbf{J}^2], \qquad (\text{VIII. 27})$$

where

$$d_Q = -\dfrac{eqQ_1}{10hI_1(2I_1-1)} \left[1 - \dfrac{I_R(I_R+1) + 4I_1(I_1+1)}{(2I_R-1)(2I_R+3)} \right]$$

$$= d'_{Q1} \left[1 - \dfrac{I_R(I_R+1) + 4I_1(I_1+1)}{(2I_R-1)(2I_R+3)} \right]. \qquad (\text{VIII. 28})$$

It should be noted that the dependence of $\mathcal{H}_Q$ on the operators $\mathbf{I}_R$ and $\mathbf{J}$ in Eq. (VIII. 27) is exactly the same as the dependence of $\mathcal{H}_S$ on the same operators in Eq. (VIII. 12). For this reason, in measurements of a single homonuclear molecule like D_2 with a single value of I_R, no experimental distinction can be made between nuclear electrical quadrupole moment interactions and spin-spin magnetic interactions. Only the combination of the two are observed with the partial Hamiltonian

$$\mathcal{H}_{QS} = \mathcal{H}_Q + \mathcal{H}_S = \dfrac{5dh}{(2J-1)(2J+3)} [3(\mathbf{I}_R . \mathbf{J})^2 + \tfrac{3}{2}(\mathbf{I}_R . \mathbf{J}) - \mathbf{I}_R^2 \mathbf{J}^2],$$

(VIII. 29)

where $d = d_M + d_Q.$ \qquad (VIII. 30)

However, from measurements of d_M in a molecule like H_2 for which d_Q vanishes the value of d_M for a structurally similar molecule like D_2

can be inferred from Eq. (VIII. 13); with this and the measured value of d in D_2 the value of d_Q can be obtained from Eq. (VIII. 30). In this calculation, of course, allowance must be made for the differences in $\langle 1/R^3 \rangle$ in Eq. (VIII. 13) because of the different amplitudes of zero point vibration and centrifugal stretching in the two molecules. Procedures for making this allowance are discussed in Section VIII. 7. It should be noted that in a heteronuclear molecule like HD, the same formal similarity between Eqs. (VIII. 25) and (VIII. 10) does not occur, so the value of d'_{Q1} can be obtained by measurements on the single molecule HD.

Kolsky, Phipps, Ramsey, Silsbee, and their associates (KOL 50a, KOL 51, KOL 52, PHI 50, SIL 50, HAR 53) have made precision measurements on the radiofrequency spectrum of D_2 and find

$$^D_0\langle d \rangle_1 = 25{,}237 \pm 10 \text{ c/s.} \qquad \text{(VIII. 31)}$$

A typical resonance spectral line on which this is based is shown in Fig. VIII. 6. The apparatus they used is the one shown in Figs. I. 2, I. 3, and I. 4. From this eqQ_D can be obtained with the aid of Eq. (VIII. 28). This result and its implications are given in the tables of Section VIII. 10 and Chapter XI.

With heavy diatomic molecules like Na_2^{23}, too many rotational states should be excited for separately resolved lines to be observed for each possible transition. Nevertheless, Kusch, Millman, and Rabi (KUS 39a) observed a gross structure to the resonance with Na_2^{23} and K_2^{39} as in Fig. VIII. 7. They suggested that nuclear quadrupole interactions would give rise to displacements of resonances from the Larmor frequency, as in the earlier D_2 experiments. Although such displaced resonances might not be separately resolvable owing to the large number of rotational states excited, they might combine statistically to give a quadrupole structure of the form observed.

The theory of the magnetic resonance spectra to be expected with heavy heteronuclear diatomic molecules one of whose nuclei possesses an electrical quadrupole moment has been developed by Feld and Lamb (FEL 45). They first calculated the energies of the specific states of the molecule from a Hamiltonian containing terms like those of Eqs. (VIII. 1) and (VIII. 25). Such energies could easily be calculated in either the limits of very strong or very weak magnetic fields since only diagonal matrix elements are required in these limits and these can be determined from Eqs. (C. 24), (C. 26), and (C. 27) of Appendix C. The allowed transition energies w between pairs of these states were then

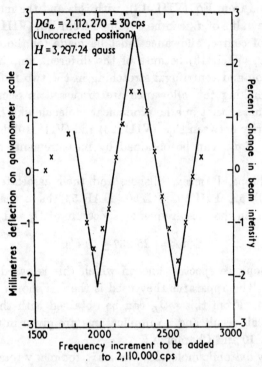

Fig. VIII. 6. D_2 nuclear transition in high magnetic field (HAR 53).

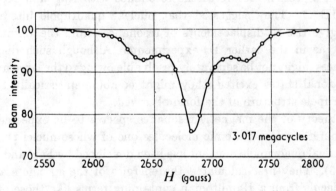

Fig. VIII. 7. Resonance curve for Na^{23} observed in Na_2 with an oscillating current of about 3 amperes (KUS 39a).

found. In general w depends on J and m_J but in the limit of large J, to which they limited their consideration, the approximate dependence is only on m_J/J or on
$$z \equiv m_J/J. \qquad \text{(VIII. 32)}$$
For example, with $I = 1$ and in the limit of large H, the transition energies $w(m_I, m_I')$ for transitions with $\Delta m_I = -1$, $\Delta m_J = 0$ can easily be seen to be in this approximation

$$w(1, 0) = (\mu_I/I)H - \tfrac{3}{16}eqQ(3z^2-1),$$
$$w(0, -1) = (\mu_I/I)H + \tfrac{3}{16}eqQ(3z^2-1). \qquad \text{(VIII. 33)}$$

In general, for each possible nuclear transition one may write
$$w(m_I, m_I') = f(z). \qquad \text{(VIII. 33a)}$$

For one of these transitions and for any energy interval Δw there corresponds an interval
$$\Delta z = \Delta w \frac{dz}{dw}. \qquad \text{(VIII. 34)}$$

Since the number of values of z or of m_J within the interval Δw, and hence the number of transitions within the interval, is proportional to Δz, one may define a 'density of transitions' D given by
$$D(z) = \Delta z/\Delta w = 1/(dw/dz) = 1/f'(z). \qquad \text{(VIII. 35)}$$

The reciprocal of the slope of the w versus z curve will thus be proportional to the number of states involved in transitions having a given energy difference; hence, the depth of the resonance minimum varies with this quantity. The product of $D(z)$ by the transition probability $P(z)$ is a measure of the relative strength of the transition, as a function of the parameter z. $D(z)P(z)$ is therefore called the 'shape function' $S(z)$ for a transition. However, from Eq. (VIII. 33), or its analogue, z can be expressed as a function of w and this can be substituted in $S(z)$ to express the shape function as a function of w, $S(w)$. Clearly then, $S(w)$ indicates the shape of the line as a function of frequency.

This can readily be applied to the above example. In the strong field case $P(z)$ is independent of molecular orientation so $P(z)$ can be taken as independent of z; therefore, except for a constant factor, S equals $D(w)$. From Eqs. (VIII. 35) and (VIII. 33) for the transition
$$m_I = 1 \to m_I = 0$$
$$S = k'D(w) = k\{-[w-(\mu_I/I)H]/\tfrac{3}{16}eqQ+1\}^{-\frac{1}{2}}, \qquad \text{(VIII. 36)}$$
where from Eq. (VIII. 33) w is limited to the range between
$$(\mu_I/I)H - \tfrac{3}{8}eqQ \quad \text{and} \quad (\mu_I/I)H + \tfrac{3}{16}eqQ$$
since z must lie between $+1$ and -1. It is apparent that this has the

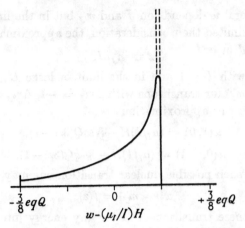

Fig. VIII. 8. Theoretical shape of resonance line for $I = 1$, corresponding to the transitions $m_I = 1 \to m_I = 0$, $\Delta m_J = 0$ in strong magnetic field as a function of the transition energy. The ordinate unit is arbitrary, as is the height at which the line is cut off. The calculation of shape includes only first-order perturbation terms (FEL 45).

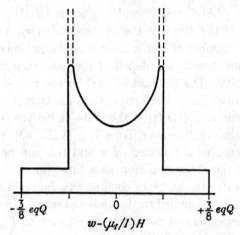

Fig. VIII. 9. Theoretical shape of the resonance line for $I = 1$, corresponding to the transitions $\Delta m_I = \pm 1$, $\Delta m_J = 0$ in strong magnetic field as a function of the transition energy. Otherwise the description of Fig. VIII. 8 applies (FEL 45).

shape shown in Fig. VIII. 8. On the other hand, it is easy to see that the other state of Eq. (VIII. 33) produces a curve that is just the mirror image of the preceding one. Consequently, the two states together should lead to a resonance as shown in Fig. VIII. 9. Of course, the infinite divergence is a theoretical idealization and would not occur in practice because of the radiation line breadth of Eq. (V. 24), because of the simultaneous occurrence of spin-rotational interactions as in the preceding section, and because of the effects of zero point vibration.

With spin $\frac{3}{2}$ the idealized line shape is predicted by a similar analysis to be as in Fig. VIII. 10. This is similar to the nature of the experimental results in Fig. VIII. 7. Feld and Lamb (FEL 45) also extended

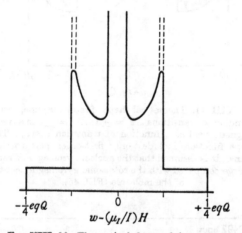

FIG. VIII. 10. Theoretical shape of the resonance line for $I = \frac{3}{2}$, corresponding to the transitions $\Delta m_I = \pm 1$, $\Delta m_J = 0$ in a strong magnetic field as a function of transition energy. Otherwise the description of Fig. VIII. 8 applies (FEL 45).

their theory to the case of weak magnetic fields. They predicted an idealized resonance as shown in Fig. VIII. 11 centred at a frequency determined by the quadrupole interaction. This idealized resonance would shrink to zero width for $H = 0$. Although the calculations of Feld and Lamb were limited to heteronuclear molecules, Foley (FOL 47) extended the theory to homonuclear molecules and showed that resonances of similar shape were to be expected.

The first extensive experimental study of the quadrupole structure of the resonance spectra of heavy diatomic molecules was that of Nierenberg, Ramsey, and Brody (NIE 46, NIE 47) with NaBr, NaCl, and NaI. Typical experimental resonances which they obtained are shown in Figs. VIII. 12, VIII. 13, and VIII. 14. The first of these is at high magnetic fields, the second in weak magnetic fields, and the third in an essentially zero field. These experiments provided the first observations of radiofrequency spectra of molecules in the weak and

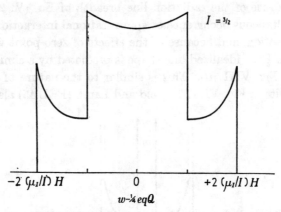

FIG. VIII. 11. Theoretical shape of resonance line, corresponding to transitions $\Delta F = \pm 1$, $\Delta m = \pm 1$ in a weak magnetic field as a function of transition energy. The shape function includes only first-order perturbation terms. It is assumed that the nuclear gyromagnetic ratio is large compared with the rotational gyromagnetic ratio of the molecule (FEL 45).

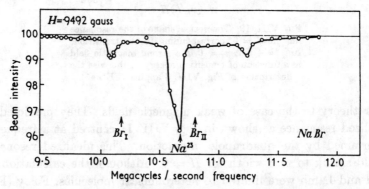

FIG. VIII. 12. Radiofrequency spectrum of NaBr at 9,492 gauss. The middle arrow indicates the position at which the Na^{23} resonance would occur if there were no orientation dependent interactions. The additional arrows indicate the resonance positions for the two bromine isotopes that are not observed in this spectrum (NIE 47).

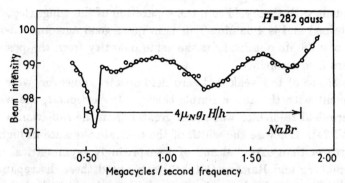

FIG. VIII. 13. Radiofrequency spectrum of NaBr in a field of 282 gauss. The spectrum is broadened approximately proportionately to the field and is approximately centred about the zero-field resonance frequency of Fig. VIII. 14. Three minima are observed in agreement with theory and with separations approximately in agreement with the theoretical Fig. VIII. 11 (NIE 47). The sharp peak on the left is probably due to the $J+\tfrac{1}{2} \to J-\tfrac{1}{2}$ transitions of page 227 (COT 53).

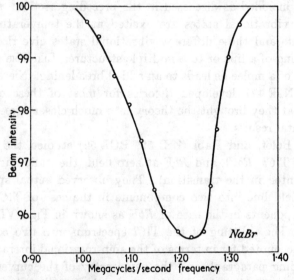

FIG. VIII. 14. Zero-field radiofrequency spectrum of NaBr. The radiofrequency current is 3 amps. This spectrum is essentially due to the interaction of the nucleus with the molecule only. Note that the position of the minimum is almost exactly at a frequency equal to the difference between the frequencies of the side minima of Fig. VIII. 12, as expected theoretically. The width is much larger than the natural width (NIE 47).

zero field limits. The general nature of the results at high fields is in agreement with theory. From the separation of the wings, $|eqQ/h|$ can be obtained and is 4·68 Mc. This is in quite good agreement with the value of 4·88 Mc obtained for the same quantity from the position of the zero field resonance.

The shape of the weak and zero field spectra, however, were not in agreement with the above simple theory. In particular, the width of the zero field resonance was much greater than the radiation width of Eq. (V. 24). Likewise the width of the central resonance in high fields was greater than expected and of a surprisingly asymmetrical nature.

Nierenberg and Ramsey (NIE 47) attributed these discrepancies to the following. (a) Higher order perturbations should be included since the magnetic interaction in the high-field experiments is not infinitely great in comparison to the quadrupole interaction. (b) J is not actually infinite and terms in $1/J$ should be included; these would provide a broadening effect. (c) A spin-rotational interaction of the form $-c\mathbf{I}.\mathbf{J}$ should be included as discussed in the preceding section. (d) Several molecular vibrational states are excited at the temperatures of the experiments and these different vibrational states give rise either to a broadening of a line or to a multiple structure. Likewise centrifugal expansion of a molecule leads to an added broadening. Nierenberg and Ramsey (NIE 47) developed theories for most of these effects and showed that they brought the theory into much closer agreement with experimental results.

Zeiger, Bolef, and Rabi (ZEI 52, BOL 52) studied the resonance spectra of TlCl, RbCl, and RbF at zero field (the italic nucleus is the one reoriented in the transition). They observed actual splittings of the zero field line into two components in the case of TlCl and into seven components in the case of RbF as shown in Figs. VIII. 15 and VIII. 16. The splitting of the TlCl spectrum into two components could be accounted for in terms of the spin-rotational interaction (c) of the preceding paragraph and the asymmetry of the curves could be attributed to the vibrational and centrifugal effects listed as (d) above. Fig. VIII. 17 is a comparison between the experiment and the theory of Zeiger, Bolef, and Rabi when the parameters of the theory were most favourably selected. In the case of RbF, where Rb^{85} has a spin of $\tfrac{3}{2}$, the effects of the finiteness of J, as in (b) of the preceding paragraph, are much more important and the gross splitting is due to the combination of this and the spin-rotational interaction (c). The widths and shapes of the resonances are contributed to by the vibrational and centrifugal

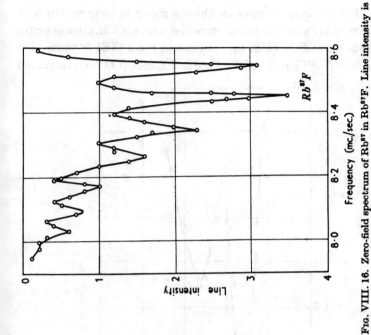

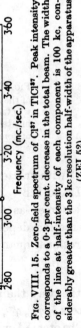

Fig. VIII. 15. Zero-field spectrum of Cl^{37} in $TlCl^{37}$. Peak intensity corresponds to a 0·3 per cent. decrease in the total beam. The width of the line at half-intensity of each component is 100 kc, considerably greater than the 3 kc resolution half-width of the apparatus (ZEI 52).

Fig. VIII. 16. Zero-field spectrum of Rb^{87} in $Rb^{87}F$. Line intensity is expressed in centimetres deflexion on a galvanometer scale. Peak intensity corresponds to a 0·2 per cent. change in the total beam. Radiofrequency current was 15 amps. The widths of the lines at half-intensity are considerably greater than the 6 kc resolution half-widths expected for this molecule (BOL 52).

effects (d). The theoretical curve for the resonance to be expected with the most favourably selected parameters and with six vibrational states included is shown in Fig. VIII. 18. Zeiger and Bolef (ZEI 52) extended their analysis to $Na^{23}Cl$ and $Na^{23}F$. Coté and Kusch (COT 53) restudied

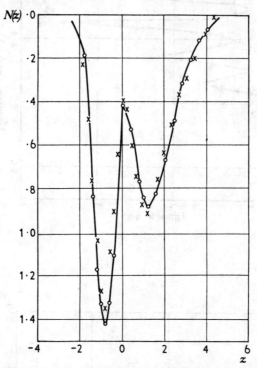

Fig. VIII. 17. Envelope of the theoretical resonance curves of the first five vibrational states for TlCl. $\gamma \equiv (\bar{J}/2c_a)eq^{(J)}Q$ was taken to be 0·30 for the best agreement with the experimental resonance curves of $TlCl^{35}$ and $TlCl^{37}$. For purposes of comparison, the experimental points of the $TlCl^{37}$ resonance in Fig. VIII. 15 are indicated here by crosses (ZEI 52).

NaCl with greater resolution and also made observations on NaF, NaBr, NaI, and KBr. Zeiger and Bolef, in their analysis, found it convenient to assume—as they showed to be reasonable—that the quadrupole interaction depended on the rotational quantum number J and the vibrational quantum number v as

$$eqQ = e\bar{q}Q + eq^{(J)}QJ(J+1) + eq^{(v)}Qv. \qquad \text{(VIII. 37)}$$

In this way the analysis of their experiment gave values of the parameters $\bar{q}$, $q^{(J)}$, $q^{(v)}$, and c. The results for the first three of these are given in Chapter XI while the values of c are given in Section VIII. 4.

Logan, Coté, and Kusch (LOG 52, KUS 49, KUS 49b, KUS 53) have measured the signs of the nuclear quadrupole interactions in molecular-beam magnetic-resonance experiments with heavy atoms. They inserted an obstacle in the beam path where the beam was considerably

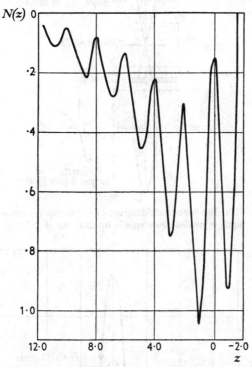

Fig. VIII. 18. Envelope of the theoretical resonance curves corresponding to the first six vibrational states of $Rb^{87}F$, obtained by selecting $\gamma \equiv (\bar{J}/2c_a)eq^{(J)}Q = 0.30$. This theoretical curve is to be compared with Fig. VIII. 16 (BOL 52).

deflected as shown in Fig. VIII. 19. In this way they could attenuate the beam due to one sign of components of μ_I without affecting appreciably the components of opposite sign. From Eq. (VIII. 36) or from Figs. VIII. 8 and VIII. 9, it can be seen that for one sign of quadrupole interaction this will reduce the wing on one side of the high field resonance curve while for the opposite sign the other wing would be reduced. This is shown clearly in the experimental curves of Fig. VIII. 20. From these observations, they inferred the signs of a number of quadrupole interactions. Their results are included in the tables of Chapter XI.

Van Vleck (KEL 40), Foley (FOL 47a), and Ramsey (RAM 53) have

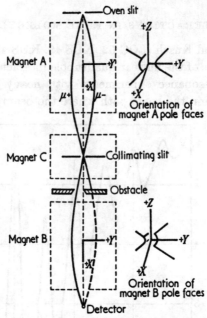

Fig. VIII. 19. Schematic diagram of apparatus for measuring signs of nuclear quadrupole interactions (LOG 52).

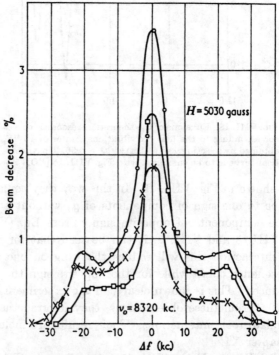

Fig. VIII. 20. Nuclear resonance spectrum of Li[7] in LiI. The curve with circles is obtained with full beam, the curve with squares with molecules of positive total moment removed from the beam, and the curve with crosses with molecules of negative total moment removed from the beam (LOG 52).

pointed out that second-order magnetic interactions between the nucleus and either the orbital or spin magnetic field of the electrons can give rise to pseudo-quadrupole effect in molecules, i.e. to a contribution to the apparent quadrupole moment interaction in a molecule, which is not due to a nuclear quadrupole moment at all but to a second-order magnetic interaction between the nucleus and the electrons. Ramsey

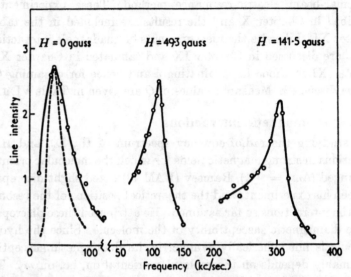

Fig. VIII. 21. Nuclear resonance spectra $(J+\frac{1}{2}) \rightarrow (J-\frac{1}{2})$ of Na²³ in NaCl at several magnetic fields. The dotted curve is a theoretical curve using the values $b/h \equiv eqQ/8h = -709$ kc and $c'/h = -0.16$ kc (COT 53).

(RAM 53) showed that in D_2 the largest contribution was due to the electron spin magnetic moment, and even this gave rise to an apparent quadrupole interaction of only -2.4 c/s in contrast to the experimental value of $224{,}992 \pm 100$ c/s in the same molecule.

As mentioned at the end of Section VI. 6, Ramsey predicted (RAM 48) theoretically a resonance at approximately twice the nuclear Larmor frequency when the quadrupole interaction was large compared with the magnetic interaction. This resonance corresponds to transitions analogous to Eq. (VIII. 33) in which F goes from $J+\frac{1}{2}$ to $J-\frac{1}{2}$ since the quadrupole interaction to first order cancels in such a transition. This resonance has been observed experimentally by Coté and Kusch (COT 53) as shown in Fig. VIII. 21. They have also extended Ramsey's theory to include the effect of a spin-rotational interaction as in Section VIII. 4 in modifying the shape of the resonance. Although their experimental results with NaF, NaCl, NaBr, NaI, and KBr are in general agreement with theory, the values of the spin-rotational interaction constant c to yield the best agreement are rather different from the values of c inferred from other transitions. For this reason the values of c obtained in this way are placed in parentheses or marked as approximate in the tables of Section VIII. 4.

A number of quadrupole interactions have been measured with the molecular-beam electric-resonance method. These experiments are described in Chapter X and the results are included in the tables of Chapter XI. Likewise the measurements of quadrupole interactions in atoms are discussed in Chapter IX and tabulated in Chapter XI. In Chapter XI methods for evaluating q and hence for obtaining Q will also be discussed. Measured values of Q are given in Table VI. I.

VIII. 6. Diamagnetic interactions

In studying the radiofrequency spectrum of H_2, D_2, and HD for transitions in strong magnetic fields for which the molecular orientation is changed ($\Delta m_J = \pm 1$), Ramsey (RAM 40) noted slight discrepancies between his experiments and the theoretical positions of the resonances with the interactions so far assumed. He attributed these discrepancies to the diamagnetic susceptibility of the molecule. Since the hydrogen molecule is not spherically symmetric, its diamagnetic susceptibility presumably depends on the molecular orientation, i.e. on m_J. Therefore, in a transition with a change of m_J, a contribution to the energy difference and hence the resonance frequency arises from the dependence of the diamagnetic susceptibility upon orientation.

In later theoretical discussions, Ramsey developed the theory of this interaction and the implications of the parameters measurable thereby (RAM 50b, RAM 52a). He showed that if ξ_σ is the molecular diamagnetic susceptibility for a magnetic field applied perpendicular to the internuclear line of a diatomic molecule and if ξ_π is the susceptibility parallel to the internuclear axis, the added terms to the molecular Hamiltonian from diamagnetic susceptibility could be written in the form

$$\mathscr{H}_D = -\frac{5fh}{3(2J-1)(2J+3)}\{3(\mathbf{J}\cdot\mathbf{H})^2/H^2 - \mathbf{J}^2\} - gh, \quad \text{(VIII. 38)}$$

where

$$f = (\xi_\sigma - \xi_\pi)H^2/5h, \quad \text{(VIII. 39)}$$

$$g = (\tfrac{1}{3}\xi_\sigma + \tfrac{1}{6}\xi_\pi)H^2/h. \quad \text{(VIII. 40)}$$

The form of Eq. (VIII. 38) is essentially the same as that of Eq. (VI. 56) and for the same reasons. Eq. (VIII. 38) can be proved by averaging the diamagnetic susceptibility energy over the molecular orientations for the state concerned as in Eq. (III. 45) ff. or by proceeding analogously to Eq. (III. 97) (RAM 52a, RAM 51c). From Eq. (VIII. 38) it can easily be seen that, in a molecular state with $J = 1$,

$$\xi_{\pm 1} - \xi_0 = 2hf/H^2 = \tfrac{2}{5}(\xi_\sigma - \xi_\pi), \qquad \text{(VIII. 41)}$$

where $\xi_{\pm 1}$ represents the diamagnetic susceptibility for $m_J = \pm 1$ and ξ_0 represents the susceptibility for $m_J = 0$.

The theory of diamagnetism has already been discussed in Section VI. 8.2. However, in that section the expression given in Eq. (VI. 64) had already been averaged over all orientations. Prior to the averaging, the diamagnetic susceptibilities ξ_σ and ξ_π can be written (VAN 32, p. 275) as

$$\xi_\sigma = -\frac{e^2}{4mc^2} \left\langle \sum_k (x_{0k}^2 + z_{0k}^2) \right\rangle + \tfrac{3}{2}\xi^{HF},$$

$$\xi_\pi = -\frac{e^2}{4mc^2} \left\langle \sum_k (x_{0k}^2 + y_{0k}^2) \right\rangle, \qquad \text{(VIII. 42)}$$

where $x_0 y_0 z_0$ form a Cartesian system fixed in the molecule with the z_0 axis along the internuclear line as in Section VI. 7. The consistency of Eqs. (VIII. 42) and (VI. 64) can be seen by the fact that with these expressions

$$\xi = \tfrac{2}{3}\xi_\sigma + \tfrac{1}{3}\xi_\pi, \qquad \text{(VIII. 43)}$$

as should be the case since there are two coordinate axes perpendicular to the molecular axis and one parallel to it.

From Eq. (VIII. 42) and from the equality of $\langle x_{0k}^2 \rangle$ and $\langle y_{0k}^2 \rangle$ by symmetry

$$5hf/H^2 = \xi_\sigma - \xi_\pi = -\frac{e^2}{8mc^2} \left\langle \sum_k (3z_{0k}^2 - r_k^2) \right\rangle + \tfrac{3}{2}\xi^{HF}. \qquad \text{(VIII. 44)}$$

This relation was first derived by Ramsey (RAM 50b) who pointed out that it provided an experimental means for determining the quadrupole moment Q_e of the electron distribution in a diatomic molecule since

$$Q_e = \left\langle \sum_k (3z_{0k}^2 - r_k^2) \right\rangle = N\langle 3z_0^2 - r^2 \rangle = -\frac{8mc^2}{e^2}[(\xi_\sigma - \xi_\pi) - \tfrac{3}{2}\xi^{HF}]$$

$$= -\frac{8mc^2}{e^2}[5hf/H^2 - \tfrac{3}{2}\xi^{HF}], \qquad \text{(VIII. 45)}$$

where N is the number of electrons in the molecule. The quantity $\xi_\sigma - \xi_\pi$ in the above can be obtained from the experimental measurement of f in Eq. (VIII. 38) above and ξ^{HF} can be found from the rotational magnetic moment with the aid of Eq. (VI. 66).

Harrick and Ramsey (HAR 52) and Barnes, Bray, and Ramsey (BAR 54) have measured f/H^2 in H_2 and D_2 and have found

$$^H_0\langle f/H^2\rangle_1 = -(27{\cdot}6\pm1{\cdot}5)\times 10^{-6}\text{ c/s/gauss}^{-2},$$

$$^D_0\langle f/H^2\rangle_1 = -(26{\cdot}2\pm3{\cdot}0)\times 10^{-6}\text{ c/s/gauss}^{-2}. \qquad \text{(VIII. 46)}$$

From these values of $\xi_{\pm 1}-\xi_0$, $\xi_\sigma-\xi_\pi$ and Q_e could be inferred from the preceding equations, with the results given in the tables of Section VIII. 10. From the value of Q_e, from Eq. (VIII. 45) with $N = 2$, from Eq. (VI. 68), and from the symmetry result that $\langle x_0^2\rangle = \langle y_0^2\rangle$, the values of $\langle x_0^2\rangle$ and $\langle z_0^2\rangle$ can be found separately. In this way Ramsey (RAM 53c) found that relative to the centre of the H_2 molecule the principle second moments of the electron distribution are

$$\langle x_0^2\rangle = \langle y_0^2\rangle = (0{\cdot}2144\pm 0{\cdot}0015)\times 10^{-16}\text{ cm}^2,$$

$$\langle z_0^2\rangle = (0{\cdot}2969\pm 0{\cdot}0022)\times 10^{-16}\text{ cm.}^2 \qquad \text{(VIII. 47)}$$

VIII. 7. Effects of molecular vibration

Most of the molecular interactions discussed in the preceding sections depend either directly or indirectly on the internuclear spacing. This spacing varies with time if the molecule is vibrating. Even when the molecule is in its lowest vibrational state there is some zero-point vibration. In addition, molecules in different rotational states will possess different mean internuclear spacings as a result of centrifugal stretching. Consequently, the various interactions must be averaged over the vibration and centrifugal stretching. Thus in Eq. (VIII. 11) the average value $^x_v\langle 1/R^3\rangle_J$ must be used, where the symbol $^x_v\langle\ \rangle_J$ indicates the expectation value for molecule x in vibrational state v and rotational state J (x is replaced by H for H_2, by D for D_2, by HD for HD, etc.). A typical important vibrational effect is that $\langle 1/R^3\rangle$ cannot be replaced by $\langle 1/R^2\rangle^{\frac{3}{2}}$.

Allowance for zero-point vibration in magnetic-resonance experiments was first made in connexion with the nuclear quadrupole interactions by Nordsieck (NOR 40), Newell (NEW 50), Ramsey (KOL 51, RAM 52f), and Zeiger and Bolef (ZEI 52). The effect of vibration and centrifugal stretching in such cases is to replace the q in Eq. (VIII. 26) ff. by $^x_v\langle q\rangle_J$. Eq. (VIII.37) provides a semi-empirical means for allowing for vibrational and centrifugal effects (ZEI 52). In the case of H_2 the parameters for the J and v dependence of q have been calculated theoretically (NOR 40, NEW 50).

These procedures have been generalized by Ramsey (RAM 52f) and extended to the other interactions in the molecule. Ordinarily all the theoretical equations of the preceding sections need to be modified by merely taking expectation values ($_v^x\langle \ \rangle_J$) of both sides of the equation. Most of the quantities whose expectation values are desired are either of the form R^n or can be reduced to sums of this form by an expansion in a series $\sum_q a_q(R-R_e)^q$ about R_e. For the zeroth vibrational state of a diatomic molecule, Ramsey (RAM 52f) has shown

$$_0^x\langle (R/R_e)^n \rangle_J = 1 + (B_e/\omega_e)[(aR_e)C_1\tfrac{3}{2} + C_2] +$$
$$+ (B_e/\omega_e)^2[(aR_e)^3 C_1 \tfrac{13}{12} + (aR_e)^2 C_2 \tfrac{15}{4} + (aR_e)C_3 \tfrac{11}{2} + 3C_4 + 4C_1 J(J+1)],$$
(VIII. 48)

where the symbols have their usual molecular significance as in Section X. 2.2 (HER 50a) and where $C_1 = n$, $C_2 = C_1(n-1)/2$, $C_3 = C_2(n-2)/3$, and $C_4 = C_3(n-3)/4$. The quantity a is the asymmetry parameter (also called β) of the Morse potential (MOR 29). If the molecule is in a low vibrational state it is most advantageous to use the Morse potential that best fits the actual potential at its bottom rather than the one that gives the correct dissociation energy. Thus, the potential can be fitted to the leading terms of the Dunham (DUN 32, NEW 50) power series expansion of the potential, in which case

$$aR_e = -a_1,$$
(VIII. 49)

where a_1 is the second coefficient of the Dunham series of Eq. (X. 17) in Section X. 2. Alternatively, it can be determined from the conventional molecular parameters α_e, etc., of Chapter X and the standard literature (HER 50a) by

$$aR_e = 1 + \alpha_e \omega_e/6B_e^2.$$
(VIII. 50)

For higher vibrational states, approximately the correct result can be obtained by including only the first two terms of Eq. (VIII. 48) and by multiplying the second one by $2(v+\tfrac{1}{2})$ (ZEI 52). Eq. (VIII. 48) is plotted graphically (RAM 52f) as a function of n in Fig. VIII. 22, where the parameters are selected to have their best values for H_2 (HER 50, HER 50a) including $aR_e = 1\cdot608$ and $R_e = 0\cdot74166$ Å. The ratio of $\langle (R/R_e)^n \rangle$ for D_2 to that for H_2 is plotted in Fig. VIII. 23.

Although in most cases the desired equations including vibrational effects can be obtained by merely applying expectation brackets to the previously derived equations, this is not always true. For example, if Eqs. (VI. 53) and (VI. 66) are solved for c_W and μ_J respectively and if the expectation values are taken, the results will involve $\langle \sigma^{HF}/R^2 \rangle$ and $\langle \xi^{HF} R^2 \rangle$. However, it is $\langle \sigma^{HF} \rangle$ that is desired in shielding calculations and $\langle \xi^{HF} \rangle$ in diamagnetic theory. To go from the former to the latter,

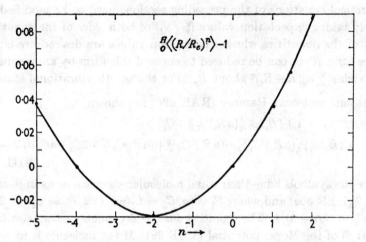

Fig. VIII. 22. $^H_0\langle(R/R_e)^n\rangle_1 - 1$ as a function of n (RAM 52f).

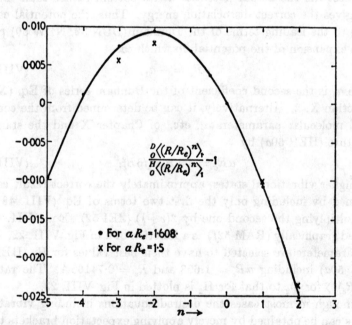

Fig. VIII. 23. $^D_0\langle(R/R_e)^n\rangle_1 / ^H_0\langle(R/R_D)^n\rangle_1 - 1$ as a function of n (RAM 52f).

the dependences of ξ^{HF} and σ^{HF} on R are needed. These can be assumed to be of the form
$$\xi^{HF} = F_e(R/R_e)^l,$$
$$\sigma^{HF} = G_e(R/R_e)^m. \qquad (\text{VIII. 51})$$

With these assumptions, Eqs. (VI. 53) and (VI. 66) are replaced by

$${}^x_a\langle\xi^{HF}\rangle_v = (e^2/12mc^2)R_e^2\,{}^x_v\langle(R/R_e)^l\rangle_J\,{}^{x'}_v\langle(R/R_e)^{l-2}\rangle_{J'}^{-1} \times$$
$$\times \{2Z_1Z_2/(Z_1+Z_2)+2(Z_1+Z_2)\,{}^{x'}_v\langle(D^2-d^2)/R^2\rangle_{J'}-$$
$$-(2\mu'/M\mu_N)\,{}^{x'}_v\langle\mu_J/J\rangle_{J'}\} \qquad (\text{VIII. 52})$$

and for diatomic molecules

$${}^x_v\langle\sigma^{HF}\rangle_J = -(e^2/6mc^2\mu_N)R_e^2\,{}^x_v\langle(R/R_e)^m\rangle_J\,{}^{x'}_v\langle(R/R_e)^{m-2}\rangle_{J'} \times$$
$$\times \{2Z\mu_N R_e^{-3}\,{}^{x'}_v\langle(R/R_e)^{-3}\rangle_{J'}-(2\pi\mu'/M\gamma_N)\,{}^{x'}_v\langle c_W\rangle_{J'}\}. \qquad (\text{VIII. 53})$$

At first sight this would appear to be no improvement since l and m are not known. However, Ramsey (RAM 52f, RAM 52c) has pointed out that it is sometimes possible to determine μ_J or c_W in two similar molecules which differ only in rotational state or in isotopic mass. If one attributes the differences only to changes of reduced mass, rotational quantum number, centrifugal stretching, and zero-point vibration, the values of l or m can be determined. With these values, $\langle\xi^{HF}\rangle$ and $\langle\sigma^{HF}\rangle$ can be found from the preceding equations.

Barnes, Bray, and Ramsey (RAM 52b, BAR 54, HAR 53) have measured various molecular interaction constants in molecular H_2 and D_2 with sufficient accuracy for the vibration and centrifugal effects to be important. In this way they have found for these molecules the values of l and m. The numerical values previously quoted and listed in Section VIII. 10 for $\langle\sigma^{HF}\rangle$ and $\langle\xi^{HF}\rangle$ are based on the above theory and these values of l and m.

Zeiger and Bolef (ZEI 52, BOL 52) have studied quadrupole interactions by the molecular-beam magnetic-resonance method as discussed in detail in Section VIII. 5. They find best agreement between theory and experiment if they assume a vibrational and centrifugal dependence of the quadrupole interaction as in Eq. (VIII. 37). Their numerical results are given in the tables of Chapter XI. Hughes and Grabner (HUG 50, HUG 50a) and Lee, Carlson, Fabricand, and Rabi (LEE 52) have also studied the dependence of q on v and J by the electric-resonance method. Their results are discussed in Chapters X and XI.

VIII. 8. Combined Hamiltonian

In the preceding sections we have considered the different interactions one at a time. In actual practice these occur simultaneously. Therefore, the Hamiltonian for a $^1\Sigma$ diatomic molecule in a magnetic field is the sum of the terms of the above type. Thus, for a heteronuclear diatomic molecule

$$\mathscr{H} = \mathscr{H}_{1H}+\mathscr{H}_{2H}+\mathscr{H}_{JH}+\mathscr{H}_{S}+\mathscr{H}_{Se}+\mathscr{H}_{1J}+\mathscr{H}_{2J}+\mathscr{H}_{Q1}+\mathscr{H}_{Q2}+\mathscr{H}_{D},$$
(VIII. 54)

while for a homonuclear diatomic molecule

$$\mathscr{H} = \mathscr{H}_{IH}+\mathscr{H}_{JH}+\mathscr{H}_{S}+\mathscr{H}_{Se}+\mathscr{H}_{IJ}+\mathscr{H}_{Q}+\mathscr{H}_{D}, \quad \text{(VIII. 55)}$$

where the quantities are defined in previous sections.

When this is written out in full in the homonuclear case it becomes (RAM 52a)

$$\mathscr{H} = -[1-\sigma_I(\mathbf{J})]a_I h\mathbf{I}_R.\mathbf{H}/H - [1-\sigma_J(\mathbf{J})]bh\mathbf{J}.\mathbf{H}/H -$$

$$-ch\mathbf{I}.\mathbf{J} + \frac{5dh}{(2J-1)(2J+3)}[3(\mathbf{I}.\mathbf{J})^2 + \tfrac{3}{2}\mathbf{I}.\mathbf{J} - \mathbf{I}^2\mathbf{J}^2] -$$

$$-\frac{5fh}{3(2J-1)(2J+3)}[3(\mathbf{J}.\mathbf{H})^2/H^2 - \mathbf{J}^2] - gh + \mathscr{H}_{Se}$$
(VIII. 56)

with the notation of the earlier parts of the chapter.

VIII. 9. Matrix elements and intermediate couplings

Any theoretical calculation of the energies corresponding to the above Hamiltonian must be preceded by a calculation of the matrix elements. The procedures for such calculations have been given in Chapter III and in Appendix C. By a straightforward application of the formulae given there one can, for example, easily see that the non-vanishing matrix elements of $\mathscr{H}$ for homonuclear diatomic $^1\Sigma$ molecules with $I = J = 1$ is as shown in Table VIII. II in an $m_I m_J$ representation and in the literature for an Fm representation (RAM 52a).

With these matrix elements the perturbation theory of the energy in weak and strong magnetic fields can be developed as in Section III. 4.3. Expressions for the energies in both of these limits to third order perturbation theory have been given by Ramsey (RAM 52a).

For intermediate values of the magnetic field the secular equation must be solved as in Section III. 4. The secular equation in this case, however, is more complicated in that it involves a cubic equation. It has been solved numerically by Ramsey (RAM 52a) for the parameters most suitable for H_2 ($a = 4,258H$ c/s, $b = 671 \cdot 7H$, $c = 113,904$, $d = 57,671$) and for D_2 ($a = 653 \cdot 6H$ c/s, $b = 336 \cdot 8H$, $c = 8,788$, $d = 25,237$). The energy levels vary as a function of the magnetic field

Table VIII. II
Non-vanishing Matrix Elements of $\mathscr{H}$ in $m_I m_J$ Representation
The quantities a, b, c, etc., are defined by Eq. (VIII. 56).

m_I	m_J	m_I'	m_J'	$(m_I m_J|\mathscr{H}|m_I' m_J')/h$
±1	±1	±1	±1	$\mp(1-\sigma_{i1})a \mp (1-\sigma_{J1})b - c + (1/2)d - (1/3)f - g$
±1	0	±1	0	$\mp(1-\sigma_{i0})a - d + (2/3)f - g$
0	±1	0	±1	$\mp(1-\sigma_{J1})b - d - (1/3)f - g$
±1	∓1	±1	∓1	$\mp(1-\sigma_{i1})a \pm (1-\sigma_{J1})b + c + (1/2)d - (1/3)f - g$
0	0	0	0	$2d + (2/3)f - g$
±1	0	0	±1	$-c + (3/2)d$
0	±1	±1	0	$-c + (3/2)d$
0	0	±1	∓1	$-c - (3/2)d$
±1	∓1	0	0	$-c - (3/2)d$
∓1	±1	±1	∓1	$3d$

Table VIII. III
Experimental Results and Derived Quantities for H_2
The symbols used in this table are defined in this chapter. The original experimental data and its analysis are contained in the reports of Kolsky, Phipps, Ramsey, and Silsbee (KOL 52) and Barnes, Bray, Harrick, and Ramsey (HAR 52, HAR 53, BRA 52, BAR 54, RAM 2f, RAM 53a).

Quantity	Result
c_H	$113{,}904 \pm 30$ c/s
d_H	$57{,}671 \pm 24$ c/s
b_H'/ν_D	1.029766 ± 0.000040
b_H''/ν_D	1.029014 ± 0.000030
H_H'	26.752 ± 0.007 gauss
H_H''	33.862 ± 0.015 gauss
H_{eH}'	-21.739 ± 0.021 gauss
${}_0^H\langle \sigma^{HF} \rangle_1$	$(-0.59 \pm 0.03) \times 10^{-5}$
${}_0^H\langle \bar{\sigma} \rangle_1$	$(2.62 \pm 0.04) \times 10^{-5}$
${}_0^H\langle K(R) \rangle_1^{\text{exp}}$	$(4.7750 \pm 0.0010) \times 10^{-22}$ ergs
${}_0^H\langle K(R) \rangle_1^{\text{theor}}$	$(4.7738 \pm 0.0030) \times 10^{-22}$ ergs
${}_0^H\langle R^{-3} \rangle_1^{-1/3}$	$(0.74677 \pm 0.00010) \times 10^{-8}$ cm
${}_0^H\langle \mu_J/J \rangle_1$	0.882910 ± 0.000080 nucl. magn.
${}_0^H\langle \mu_J/J \rangle_2$	0.882265 ± 0.000035 nucl. magn.
${}_0^H\langle \mu_J/J \rangle_2 / {}_0^H\langle \mu_J/J \rangle_1$	0.99927 ± 0.00009
$d\ln \xi^{HF}/d\ln R = l$	3.80 ± 0.12
${}_0^H\langle \xi^{HF} \rangle_1$	$(1.719 \pm 0.009) \times 10^{-31}$ ergs gauss^{-2} molecule^{-1}
${}_0^H\langle f/H^2 \rangle_1$	$-(27.6 \pm 1.5) \times 10^{-6}$ c/s gauss^{-2}
${}_0^H\langle \xi_{\pm 1} - \xi_0 \rangle_1$	$-(3.66 \pm 0.20) \times 10^{-31}$ ergs gauss^{-2} molecule^{-1}
${}_0^H\langle Q_e \rangle_1^{\text{expt}}$	$(0.333 \pm 0.019) \times 10^{-16}$ cm^2
${}_0^H\langle Q_e \rangle_1^{\text{theor}}$	$(0.345 \pm 0.010) \times 10^{-16}$ cm^2
$\langle r^2 \rangle$	$(0.7258 \pm 0.0022) \times 10^{-16}$ cm^2
$\langle x_0^2 \rangle = \langle y_0^2 \rangle$	$(0.2144 \pm 0.0015) \times 10^{-16}$ cm^2
$\langle z_0^2 \rangle$	$(0.2969 \pm 0.0022) \times 10^{-16}$ cm^2

as shown in Figs. VIII. 24 and VIII. 25. These energy levels may be differenced in accordance with the selection rules of Eqs. (V. 2a) and (V. 2b) to yield the transition frequencies. Typical of these is Fig. VIII. 26 which shows, for H_2, the resonances that are allowed in strong magnetic fields.

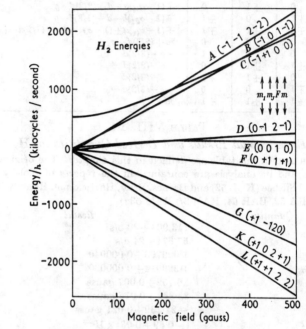

FIG. VIII. 24. Energies of H_2 states as functions of magnetic field (RAM 52a).

In the cases of H_2 and D_2, the radiofrequency spectrum has been studied with precision by Kolsky, Phipps, Ramsey, and Silsbee (KOL 52, RAM 51b) in strong and weak magnetic fields and by Barnes, Bray, Harrick, and Ramsey (HAR 53) in intermediate fields. The results have all been found to be consistent with the Hamiltonian for unique values of the few parameters given above.

For heavier molecules essentially the same procedures can be followed. However, with larger values of I and J the solutions are in general more involved and perturbation theory in either the low or high field limit is ordinarily most suitable. As an added simplification it is often most convenient to discuss the limit of $J \gg 1$. In this case the calculations are considerably simplified since the rotational angular momentum may be treated classically. Examples of such calculations were given in Section VIII. 5, above.

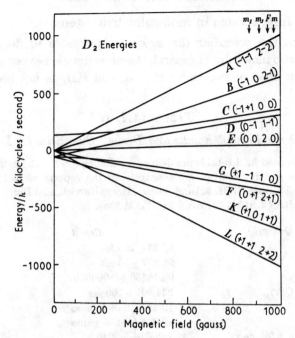

FIG. VIII. 25. Energies of D_2 states as functions of magnetic field (RAM 52a).

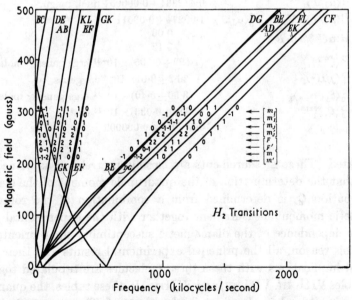

FIG. VIII. 26. Frequencies of H_2 transitions that are allowed in the strong field limit (RAM 52a).

VIII. 10. Interactions in molecular hydrogens

Although the preceding discussion is applicable to diatomic non-paramagnetic molecules in general, the most detailed experiments have so far been done on molecular H_2, D_2, and HD, as has been already

TABLE VIII. IV
Experimental Results and Derived Quantities for D_2

The symbols used in this table are defined in this chapter. The original experimental data and its analysis are contained in the reports of Kolsky, Phipps, Ramsey, and Silsbee (KOL 52) and Barnes, Bray, Harrick, and Ramsey (HAR 52, HAR 53, BRA 52, BAR 54, RAM 52f, RAM 53a).

Quantity	Result
c_D	$8{,}773 \pm 25$ c/s
d_D	$25{,}237 \pm 10$ c/s
b'_D/ν_D	0.516550 ± 0.000016
$^D_0\langle eqQ/h \rangle_1$	$224{,}992 \pm 100$ c/s
H'_D	13.445 ± 0.060 gauss
S_D	19.306 ± 0.008 gauss
$M_d c_D \gamma_p / M_p c_H \gamma_d$	1.0068 ± 0.0040
$^D_0\langle \sigma \rangle_1 / ^H_0\langle \sigma \rangle_1$	1.005 ± 0.003
Q	$(2.738 \pm 0.014) \times 10^{-27}$ cm^2
$^D_0\langle \mu_J/J \rangle_1$	0.442884 ± 0.000051 nucl. magn.
$^D_0\langle \mu_J/J \rangle_1 (M_d/M_p)/^H_0\langle \mu_J/J \rangle$	1.00274 ± 0.00011
$d \ln \xi^{HF}/d \ln R$	$3.72 \pm ^{0.06}_{0.12}$
$^D_0\langle \xi^{HF} \rangle_1$	$(1.622 \pm 0.005) \times 10^{-31}$ ergs gauss^{-2} molecule^{-1}
$^D_0\langle f/H^2 \rangle_1$	$-(26.2 \pm 3.0) \times 10^{-6}$ c/s gauss^{-2}
$^D_0\langle \xi_{\pm 1} - \xi_0 \rangle_1$	$-(3.50 \pm 0.40) \times 10^{-31}$ ergs gauss^{-2} molecule^{-1}
$^D_0\langle Q_e \rangle^{\text{expt}}_1$	$(0.318 \pm 0.030) \times 10^{-16}$ cm^2
μ_d/μ_p	0.307010 ± 0.000009

indicated. These measurements are closely interrelated as witness the fact that the determination of the quadrupole moment of the electron distribution Q_e is determined from a combination of the rotational magnetic moment measurement together with the experimental value of the dependence of the diamagnetic susceptibility upon orientation. For this reason, all the principal experimental results and their theoretical implications with these three molecules are tabulated together in Tables VIII. III, VIII. IV, and VIII. V. In these tables, the quantities listed are defined as elsewhere in this chapter. The purpose in listing $^D_0\langle \mu_J/J \rangle_1 (M_d/M_p)/^H_0\langle \mu_J/J \rangle_1$, $^H_0\langle \mu_J/J \rangle_2 / ^H_0\langle \mu_J/J \rangle_1$, and $M_d c_D \gamma_p / M_p c_H \gamma$.

is that on the simplest theories these quantities should be exactly 1. The departure from 1 is attributable to the effects of zero-point vibration and centrifugal stretching.

TABLE VIII. V

Experimental Results for HD

The original experimental data are contained in the reports of Kellogg, Rabi, Ramsey, and Zacharias (KEL 39a, KEL 40, RAM 40). Since these authors assumed the proton moment was 2·785 nuclear magnetons instead of the present best value 2·79275, a small correction factor of 1·0028, which is the ratio of these two absolute calibrations, has been appropriately applied to give the figures in the second column below. H' is related to c_i by Eq. (VIII. 18).

Quantity	Reported value
H'_P	20·5±0·2 gauss
$c_{p'}$	87,000±2,000 c/s
H'_D	20·2±0·5 gauss
c_d	13,400±340 c/s
$^{HD}_0\langle \mu_p/R^3 \rangle_1$	33·8±1·0 gauss
$^{HD}_0\langle \mu_d/R^3 \rangle_1$	10·6±0·2 gauss
$^{HD}_0\langle -5e^2 q_J Q/4\mu_d \rangle_1$	87·2±1·0 gauss
$^{HD}_0\langle eqQ/h \rangle_1$	227,000±4,000 c/s
μ_J	0·6619±0·0050 nuclear magnetons
$^{HD}_0\langle \xi_{HF} \rangle_1$	$(1·54\pm0·12)\times 10^{-31}$ ergs gauss^{-2} molecule^{-1}
$^{HD}_0\langle \xi_{\pm 1}-\xi_0 \rangle_1$	$-(4·6\pm3·5)\times 10^{-31}$ ergs gauss^{-2} molecule^{-1}

VIII. 11. Molecular polymerization

It has been suspected for a long time (RAM 48) that when such molecules as NaCl emerged from a molecular-beam source some of the molecules might be polymerized to $(NaCl)_n$. However, conclusive experimental evidence for dimerization has now been obtained by Ochs, Coté, Kusch, and Miller (OCH 53, KUS 53a, BRO 54) and some evidence for trimers has also been obtained.

Ochs, Coté, and Kusch (OCH 53) carefully studied the $\Delta m_I = \pm 1$, $\Delta m_J = 0$ transition for both Na and Cl in NaCl in high magnetic fields. If there were no polymerization the theoretical treatment of Nierenberg and Ramsey (NIE 47) should apply as discussed in Section VIII. 5. However, they usually found that the central resonance of the NaCl curve corresponding to Fig. VIII. 12 for NaBr was split into three subsidiary resonance maxima instead of the two predicted by Nierenberg and Ramsey (NIE 47). Ochs, Coté, and Kusch also could find no other internal interactions of a diatomic molecule to account for the third resonance. They therefore suggested that the additional resonance was due to the existence of dimers $(NaCl)_2$. This suggestion was supported by the fact that on one occasion an oven condition was found which produced a resonance in agreement with the theory of Nierenberg and Ramsey, whereas in other experiments with different oven conditions the above disagreement occurred. As a confirmation of the dimer interpretation Ochs, Coté, and Kusch raised the temperature of the oven slits about 100° C while the temperature of the melt in the main body of the oven was kept approximately the same. They found in this case that the peak which did not fit the Nierenberg–Ramsey theory was diminished in intensity at the higher temperatures, as would be reasonable since fewer dimers should then occur. Ochs, Coté, and Kusch (OCH 53) have found similar evidence for the dimerization of NaI and NaF.

In addition to the above evidence for the dimerization of NaF, Kusch (KUS 53a) has found independent evidence for this molecule from its weak field spectrum. As discussed in Section VIII. 5, for a large quadrupole interaction and a very weak external magnetic field Ramsey (RAM 48) showed theoretically that with diatomic molecules the resonance whose frequency is approximately proportional to the external magnetic field should be at twice the Larmor frequency of the nucleus concerned instead of at the Larmor frequency. Coté and Kusch indeed found such a double Larmor frequency resonance but, in addition, with NaF they also found a resonance at the Larmor frequency which was difficult to interpret with any simple diatomic molecule theory. Kusch (KUS 53a) has suggested that the resonance at the Larmor frequency is due to molecular dimers. From his results he estimates that at his oven temperature of 950° C approximately 10 per cent. of the beam consists of dimers or higher polymers.

Miller and Kusch (BRO 54) with a grooved rotating drum velocity selector have accurately studied the velocity distribution of the molecules emerging from a molecular-beam oven under different circumstances. From the analysis of their experimental results they have inferred the degree of polymerization of various molecular beams. They have found (BRO 54) approximately no dimerization of CaCl and CsBr 8 per cent. dimerization with RbCl and RbI, 13 to 15 per cent. dimerization of KCl and KI, 20 per cent. dimerization for NaF, 24 per cent. for NaCl and NaI, and 60 per cent. for LiCl. With LiCl Miller and Kusch also find evidence for trimers.

IX
ATOMIC MOMENTS AND HYPERFINE STRUCTURES

IX. 1. Introduction

ALTHOUGH the magnetic resonance method was originally developed for the study of nuclear magnetic moments in non-paramagnetic molecules, it was applied by Kusch, Millman, and Rabi (KUS 40) to the study of paramagnetic atoms in 1940. They used the same apparatus shown in Fig. VI. 5 that had been applied earlier to the study of molecules. However, because of the much larger magnetic moments of the paramagnetic atoms, the deflecting magnets were operated at fields of 50 to 100 gauss instead of their designed rating of 12,000 gauss since such fields gave adequate deflecting power in the long magnets whose ratio of gradient to field was about 8. Likewise, a much weaker oscillatory field could be used since the moment was about 2,000 times as big. On the other hand, this advantage was compensated by the fact that much higher frequencies had to be used, since the hyperfine structure separation of K^{39} is 462 Mc, for example.

The theoretical variation of the atomic energy levels with external magnetic field has already been calculated in Chapter III. The original experiments of Kusch, Millman, and Rabi were with atoms whose $J = \frac{1}{2}$, so Eq. (III. 120) and Fig. III. 4 apply.

One important way in which these experiments differ from most of those with molecules is that the effective magnetic moment of the atom in the deflecting field is dependent on the strength of the deflecting field as can be seen from Eq. (IV. 54) and Fig. IV. 6. As a result, added versatility in the experiments can frequently be achieved by the use of suitably chosen values for the deflecting fields. Two states which might have approximately the same μ_{eff} at high fields might have quite different values at a suitable intermediate field so that a transition between the two states could be detected in the latter circumstance though not in the former.

Kusch, Millman, and Rabi (KUS 40, MIL 40) studied the radiofrequency spectra of Li^6, Li^7, K^{39}, K^{41}, Na^{23}, Rb^{85}, and Cs^{133} in this fashion. They studied these atoms at fields between 0·05 and 4,000 gauss, i.e. from the ordinary Zeeman region for the hyperfine structure to the complete Paschen–Back region. They observed a number of different transitions consistent with the selection rules of Eq. (V. 2a) that in strong fields

$$\Delta m_I = 0, \pm 1; \quad \Delta m_J = 0, \pm 1 \tag{IX. 1}$$

and in weak fields $\quad \Delta F = 0, \pm 1; \quad \Delta m = 0, \pm 1.$ (IX. 2)

The transitions with $\Delta m = 0$ occur only if there is some oscillatory component parallel to the field. With a simple hairpin a weak component of this nature ordinarily occurs automatically. However, as discussed in Chapter XIV the RF field can be designed either to accentuate or limit this component, as is desired. The transitions in which m is changed by ± 1 are often called π transitions while those for which m is unchanged are called σ transitions. This is opposite to the usual assignment in electric dipole transitions since magnetic dipole transitions are the ones that occur here. The π lines are induced by components of the oscillatory field perpendicular to H_0 while the σ lines are produced by components parallel to H_0. The selection rules above may be violated when the multiple quantum transitions discussed in Chapter V occur.

Typical of the results of Kusch, Millman, and Rabi is the resonance curve of Fig. IX. 1 at such a weak oscillating field that only the $\Delta m = \pm 1$ resonances are observed, and the curve of Fig. IX. 2 at a stronger oscillating field for which the $\Delta m = 0$ transition occurs as well. They found that all of their results were in agreement with Eq. (III. 120). From their results they were able to calculate the atomic hyperfine structure separations of the atoms concerned. These values along with those of other atoms are given in a table in the next section.

The hyperfine structure separations can be calculated either from the low field transitions with $|\Delta F| = 1$ or from high field transitions with $\Delta m_I = \pm 1, \Delta m_J = 0$ since from Eq. (III. 120) the transition frequency in such a case for $J = \frac{1}{2}$ and at high fields is

$$\nu_{m_I, m_I-1} = \pm \Delta W/(2I+1)h + g_I \mu_0 H_0/h, \tag{IX. 3}$$

where g_I is the ordinary Landé g factor for the nucleus in terms of the Bohr magneton as in Eq. (III. 121) (see the footnote to Section III. 1) These transitions have the advantage of frequently being at a lower frequency than the $|\Delta F| = 1$ transitions in low fields; for example with $I = \frac{3}{2}$ the Eq. (IX. 3) resonance is approximately at $\Delta W/4h$ instead of $\Delta W/h$. In the early experiments this was particularly necessary since it removed the need for extremely high frequency oscillators which were not easily available then. The field dependence of these transitions at slightly lower fields can be calculated easily from Eq. (III. 120) and shown to be as in Fig. IX. 3 in the case of Li[7].

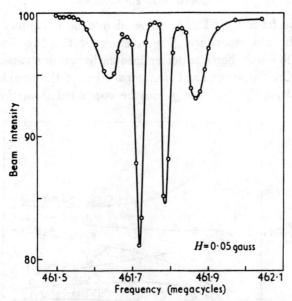

FIG. IX. 1. The Zeeman pattern of the line $|\Delta F| = 1$ of K^{39} observed for very low amplitude of oscillating field. Only π components appear (KUS 40).

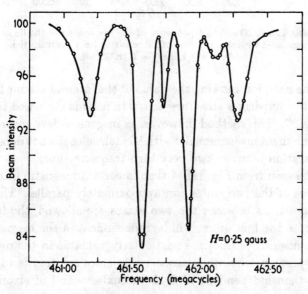

FIG. IX. 2. The Zeeman pattern of the line $|\Delta F| = 1$ of K^{39} with sufficiently large oscillating field to show both σ and π components (KUS 40).

Note that from Eq. (IX. 3) there should be a doublet structure to the high field resonance. The average of the two lines is just $\Delta W/(2I+1)h$ which therefore determines the hyperfine structure separation ΔW. On the other hand, the separation of the doublet by Eq. (IX. 3) is $2g_I \mu_0 H_0/h$. Hence g_I can be computed from this directly

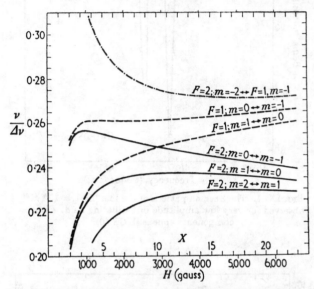

FIG. IX. 3. The field dependence of the frequencies of the lines associated with the spectrum of Li[7] resulting from the transitions $\Delta m_I = \pm 1$, $\Delta m_J = 0$ (KUS 40).

without the need for knowing the value of the internal atomic field that enters in the hyperfine structure separation, as is discussed further in Section IX. 7. This method, however, is in general less accurate than magnetic moment measurements with $^1\Sigma$ molecules since it depends on a small separation between two very high frequency lines.

It can be seen from Fig. III. 4 that in certain magnetic field region the energies of the two states run approximately parallel. This mean that if transitions between the two states are allowed, the transition frequency, to the first order, will be independent of the magnetic field Such field insensitive lines are particularly valuable in the determination of hyperfine structure separations $\Delta \nu$ since the results to the first order are then independent of field fluctuations and of errors in field calibration.

If sufficient detection sensitivity is available, the deflecting fields of an atomic apparatus may be made sufficiently long that adequate deflexions for resonance observations can be achieved at the low deflecting fields corresponding to intermediate coupling. In such cases it is apparent from Figs. III. 4 and IV. 6 that any allowed transition can be observed. On the other hand if detection sensitivity is low, the beam intensity can be greatly increased by the use of short fields with such low deflecting power that transitions can be detected only in strong deflecting fields and only if the magnetic moments of the two states differ by approximately a Bohr magneton. Such transitions in the strong field limit must then correspond to $\Delta m_J = \pm 1$, i.e. they must be 'across the diagram' transitions in Figs. III. 4 and IV. 6. It is apparent from the figures that under these conditions only a fraction of the possible transitions can be observed. Thus, for $J = \frac{1}{2}$ only one of the many possible $\Delta F = 0$, $\Delta m_F = \pm 1$ transitions can be observed.

Since the pioneering experiments of Kusch, Millman, and Rabi, many magnetic resonance studies of atoms have been completed (KUS 51). In any one of these experiments the radiofrequency spectrum could be observed under a wide variety of circumstances; consequently, many different results could be obtained from the same experiment, such as values for μ_I, I, $\Delta\nu$, etc. This makes the division of the experiments into categories for clarity of discussion difficult since each experiment may simultaneously belong to several categories. However, most of the experiments have had a primary objective with other results being of secondary importance; this forms the basis of the division of the remainder of the chapter into sections. These sections in order are (1) Introduction; (2) Absolute scale for nuclear moments; (3) Nuclear spins from atomic hyperfine structure; (4) Atomic hyperfine structure separations for $J = \frac{1}{2}$; (5) Atomic magnetic moments and the anomalous electron moment; (6) Hyperfine structure of atomic hydrogen; (7) Direct nuclear moment measurements with atoms; (8) Quadrupole interactions; (9) Magnetic octupole interactions; (10) Anomalous hyperfine structure and magnetic moment ratios; (11) Frequency standards and atomic clocks; and (12) Atomic-beam resonance method for excited states. The experiments on absolute values of nuclear moments are discussed first because of the dependence of many of the other experiments upon the absolute calibrations of the magnetic fields.

IX. 2. Absolute scale for nuclear moments. The fundamental constants

As discussed in Section VI. 5, most nuclear and atomic magnetic moment measurements involve chiefly the measurement of a frequency and a magnetic field. Although high-precision frequency measurements are easy, high-precision absolute magnetic field calibrations are very difficult. However, once a single magnetic moment has been measured absolutely its resonance frequency can easily be used to calibrate any other field absolutely. Therefore, the present section will be concerned with the experiments to fix the magnetic moment of one nucleus—the proton—in absolute terms.

The first experiment of this nature was that of Millman and Kusch (MIL 41) using the apparatus and method described in the preceding section. They calibrated their magnetic field in terms of the spin magnetic moment of the electron, which they assumed to be exactly one Bohr magneton. This calibration is possible in an atomic experiment since, from Eq. (III. 121), the effectiveness of an external magnetic field in producing a given stage of intermediate coupling in Fig. III. 4 depends on

$$x = (g_J - g_I)\mu_0 H_0/2\pi\hbar\Delta\nu, \qquad (IX. 4)$$

where g_J and g_I are the ordinary Landé g factors as in Eqs. (III. 121a) and (IX. 3) and where the second term in the numerator is only a small correction easily determined with adequate precision by the measured value of g_I with approximately calibrated fields. The value of $\Delta\nu$ can be found by observations of field insensitive lines as in the preceding section. Observations of the experimental spectrum then determine experimentally the value of x from which the value of H_0 can be inferred if μ_J is assumed to be exactly one Bohr magneton. The field sensitive transitions chosen by Kusch and Millman for their field calibrations in an (F, m) representation were $(2, -2) \leftrightarrow (1, -1)$ of Na^{23}, $(3, -3) \leftrightarrow (2, -2)$ of Rb^{85}, and $(4, -2) \leftrightarrow (4, -1)$ of Cs^{133}. The absolute calibrations obtained in this way were found to be consistent with each other. Nuclear magnetic moments of the proton in KOH and other nuclei were then measured in the calibrated field to provide what were thought to be absolute values for the nuclear moments. In these terms, the proton magnetic moment was found to be $2 \cdot 7896 \pm 0 \cdot 0008$ nuclear magnetons.

This absolute calibration was retained with confidence until the hyperfine structure separation of atomic hydrogen was measured in molecular-beam resonance experiments by Nafe, Nelson, and Rabi (NAF 48) and by Nagle, Julian, and Zacharias (NAG 47). These experiments are described in detail in Sections IX. 5 and IX. 6. Their result, as later improved by Prodell and Kusch (PRO 50, ZZA 28) and Wittke and Dicke (ZZA 29), was

$$\Delta\nu_H = 1,420 \cdot 40577 \pm 0 \cdot 00005 \text{ Mc.}$$

On the other hand, from the above Kusch and Millman value for μ_p and from Eq. (III. 82) with the reduced mass correction there suggested, one would expect $\Delta\nu_H = 1{,}416{\cdot}97\pm0{\cdot}54$ Mc. Breit (BRE 47a) suggested that this discrepancy might be due to the electron possessing a magnetic moment that was slightly different from one Bohr magneton as implied by Eq. (III. 82). Ramsey (BRE 47a) pointed out that such an assumed change in the electron moment would also modify the value of the proton magnetic moment since the magnetic field in the proton experiment of Millman and Kusch was calibrated in terms of the electron-magnetic moment. Both of these effects were then included by Breit (BRE 47a) in considering the atomic hyperfine structure separation.

Kusch, Foley, Taub, and Mann (KUS 48, FOL 48, KUS 49a, MAN 50a), as described in greater detail in the next section, confirmed the anomalous electron spin magnetic moment by comparing the values of the atomic gyromagnetic ratio g_J of In and Ga in the $^2P_{\frac{3}{2}}$ with the values of g_J for the same atoms in $^2P_{\frac{1}{2}}$ states and with the g_J values of Na and other alkalis in a $^2S_{\frac{1}{2}}$ state. The electron spin moment in such experiments can be measured because in a $^2P_{\frac{1}{2}}$ state the resultant electron magnetic moment is the difference between an orbital and a spin contribution whereas in a $^2P_{\frac{3}{2}}$ state it is the sum of the two contributions; in a $^2S_{\frac{1}{2}}$ state it is just the electron spin by itself. The orbital magnetic moment in these experiments was assumed to be exactly one Bohr magneton. Schwinger (SCH 48) showed that such an anomalous electron spin moment should be expected on the basis of a suitable relativistic quantum electrodynamics owing to processes of virtual emission and reabsorption both of photons and of electron-positron pairs. The most accurate measurement of the electron magnetic moment is that due to Koenig, Prodell, and Kusch (KOE 51) using the magnetic field standardization of Gardner and Purcell (GAR 49) described in the next paragraph. Koenig, Prodell, and Kusch measured the magnetic moment of atomic hydrogen in a $^2S_{\frac{1}{2}}$ state which has a purer ground state than the indium, sodium, etc., of the earlier experiments; details of the experiment will be given in Section IX. 5. Their experimental result for the magnetic moment of the electron, after a suitable relativistic correction (KOE 51) was $(1{\cdot}001146\pm0{\cdot}000012)$ Bohr magnetons. This is in excellent agreement with the theoretical result of $1{\cdot}0011454$ Bohr magnetons obtained by Karplus and Kroll (KAR 52) in an extension of Schwinger's theory to the fourth order.

Since by the preceding paragraph the electron moment cannot be assumed to have the simple value of one Bohr magneton, it is not suitable as the absolute standard in terms of which fields and nuclear moments are calibrated. The first experiment to calibrate magnetic fields and nuclear moments in terms of a more suitable standard was that of Taub and Kusch (TAU 49) who measured the proton resonance frequency of NaOH in a magnetic field calibrated from the atomic g_J values of the ground states of Cs^{133} and In^{115} which had been determined in the experiments of the preceding paragraph by Kusch, Foley, and Taub (KUS 48, FOL 48, KUS 49a). In this way the proton magnetic moment could be expressed in terms of the orbital magnetic moment of the electron which was assumed to be one Bohr magneton with the result that $\mu_p/\mu_0 = 0.00152106 \pm 0.00000008$. Subsequently Gardner and Purcell (GAR 49) performed a hydrogen resonance absorption experiment in a field calibrated by the cyclotron frequency for electrons with the direct result that

$$\mu_{p,c} = (0.00152101 \pm 0.00000002)\mu_0, \quad \text{(IX. 5)}$$

where the subscript c indicates that the value has been corrected for diamagnetic shielding (RAM 50d) and u indicates uncorrected values. The calibration of Eq. (IX. 5) at the time of writing remains the standard in magnetic field calibrations when atomic or other magnetic moments are to be expressed in Bohr magnetons.

Magnetic moments may be expressed in terms of the nuclear magneton as a result of absolute calibrations due to Hipple, Sommer, and Thomas (HIP 49) and to Bloch and Jeffries (BLO 50, JEF 51). In these experiments the proton nuclear induction resonance was measured in a field calibrated in terms of the cyclotron frequency of the proton. The results were consistent within experimental error; those of Hipple, Sommer, and Thomas (HIP 49) had the least error. From these experiments the best value (WAL 53) in mineral oil without diamagnetic corrections is $\mu_{p,u} = (2.792670 \pm 0.000060)\mu_N$. This can be corrected for magnetic shielding as shown by Ramsey (RAM 50d, HAR 53) and Thomas (THO 50) with the shielding constant $\sigma = 2.78 \times 10^{-5}$ for mineral oil, which is based on a shielding constant in H_2 of 2.62×10^{-5}. The corrected value then is (WAL 53)

$$\mu_{p,c} = (2.792743 \pm 0.000060)\mu_N. \quad \text{(IX. 6)}$$

Yet another calibration has been provided by Thomas, Driscoll, and Hipple (THO 49) who measured the gyromagnetic ratio γ_p of the proton in a magnetic field that was calibrated at the National Bureau of Standards in terms of the force on a wire carrying a known current.

Their uncorrected result was $\gamma_{p,u} = 2\mu_p/\hbar = 26{,}752 \cdot 3 \pm 0 \cdot 6 \text{ sec}^{-1} \text{ gauss}^{-1}$ in mineral oil. With the above diamagnetic correction

$$\gamma_{p,c} = 26{,}753 \cdot 05 \pm 0 \cdot 60 \text{ sec}^{-1} \text{ gauss}^{-1}. \quad \text{(IX. 7)}$$

A wealth of fundamental data is provided by the precision values in Eqs. (IX. 5), (IX. 6), and (IX. 7) together with the above value for the electron spin magnetic moment and the molecular beam fine-structure measurements on atomic hydrogen and deuterium discussed in Chapter XII. One important application of these data has been to combine them with other measurements to obtain much more accurate values of the fundamental constants. As a simple illustration, the ratio of Eq. (IX. 6) to Eq. (IX. 5) provides the ratio M/m of the proton to the electron mass directly. Most of the other fundamental constants depend upon several of the measured data and in a more complex way. The analysis of the new data to obtain values of the fundamental constants has been described in detail by DuMond and Cohen (DUM 50, DUM 53). Greatly improved values of the fundamental constants have been obtained in this way. The best present values of the fundamental constants are listed in Appendix A.

IX. 3. Nuclear spins from atomic hyperfine structure

One of the important physical quantities that can be obtained from magnetic resonance studies of atoms is the spin of a nucleus. This can be obtained from hyperfine structure observations in many ways since many of the features of hyperfine structure patterns, e.g. the number of lines, depend on the nuclear spin. However, for the reasons discussed near the end of Section IX.1, the most convenient means is often to study transitions between the states of $m = -(I-\tfrac{1}{2})$ and $m = -(I+\tfrac{1}{2})$ for $F = I+\tfrac{1}{2}$ as in Fig. III. 4a and Eq. (III. 93). In this case, at low magnetic fields where Eq. (III. 91) applies approximately, the frequency of the transition is

$$\nu = (\tfrac{1}{2}g_J + Ig_I)\mu_0 H_0/(I+\tfrac{1}{2})h \approx \tfrac{1}{2}g_J \mu_0 H_0/(I+\tfrac{1}{2})h. \quad \text{(IX. 8)}$$

Since the above is independent of $\Delta\nu$ and since μ_J is ordinarily known from the nature of the atomic state or from other observations, the value of I can be inferred directly.

In experiments on isotopes of low relative abundance it has usually been convenient to adjust the deflecting fields so that the desired transition is refocused ('flop-in' experiments) since there is then less background intensity than if everything except the resonance was refocused. Also short deflecting fields and 'across the diagram' transitions may be used as in Section IX. 1. Additional reduction of background in such a case can be achieved by the insertion of an obstacle wire as in Fig. I. 6 to block out atoms of such small moment or high velocity that they are not appreciably deflected. Some special alignment problems of 'flop-in' experiments are discussed in Section XIV. 8.2. The use of a mass spectrometer to analyse the ions which leave the surface ioniza-

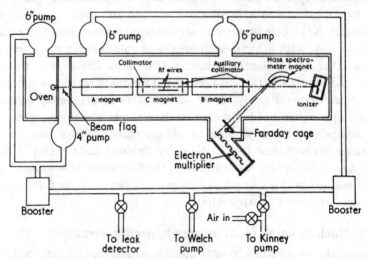

FIG. IX. 4. Schematic diagram to illustrate the atomic-beam apparatus used to measure K^{40} (EIS 52).

tion detector reduces unwanted background still further; typical apparatuses constructed in this fashion are illustrated schematically in Figs. I. 6 and IX. 4. Electron multipliers are frequently used to increase the detection sensitivity as in the previous diagrams. The use of mass spectrometers and electron multipliers in conjunction with molecular beam experiments is described in greater detail in Chapter XIV. In this manner radioactive isotopes of alkalis have been detected. However, Bellamy and Smith (SMI 51, BEL 51, BEL 53) and others (HAM 53, COH 54, BRO 54) have also made measurements on radioactive isotopes by using the radioactivity as a means of detection. Where neither a surface ionization detector nor a Pirani gauge was suitable, as with gold, Wessel and Lew (WES 53) have introduced an electron bombardment ionizer. These alternative detection means used in hyperfine structure measurements are described in Chapter XIV.

Values of I have been measured for a number of isotopes of low abundance including K^{40*} by Zacharias (ZAC 42); Na^{22*} by Davis (DAV 48a, DAV 48b, DAV 49c), and Nagle (NAG 49); Cs^{134*} by

Jaccarino, Bederson, and Stroke (JAC 52); Na^{24}*, K^{42}*, Rb^{86}*, Cs^{131}*, Cs^{134}*, and RaE* by Bellamy and Smith (SMI 51, BEL 53, SMI 54); Ag^{111} and Cu^{64} by Hamilton, Lemonick, Pipkin, and Reynolds; Cs^{134m}* by Cohen, Gilbert, and Wexler; and Rb^{81}* by Nierenberg, Hobson, and Silsbee. The results of these spin measurements are included in the Table VI. 1.

IX. 4. Atomic hyperfine structure separations for $J = \frac{1}{2}$

The methods of Section IX. 1 have been applied to measurements of a number of atomic hyperfine structures. In the cases where either I or J equals $\frac{1}{2}$ the theory of the energy levels is particularly simple and is as given in Eq. (III. 93) and Fig. III. 4; in such cases no complications from nuclear electric quadrupole interactions can arise, as was discussed in Chapter III. The values of $\Delta \nu = \Delta W/h$ can then be obtained by analysis of the resonance pattern at various values of the magnetic field as discussed in Section IX. 1.

The use of field insensitive lines to determine $\Delta \nu$ is particularly effective since the results are then independent of the field calibration. Such field insensitive lines may often be found at either low or high fields as in Section IX. 1. However, Kusch and Taub (KUS 49a) have pointed out a particularly effective means for observing such lines in intermediate fields. If the frequency of the transitions is plotted as a function of H as in Figs. IX. 3 and IX. 5, it is seen that certain of the frequencies pass through a maximum. If this maximum frequency is measured, $\Delta \nu$ can be obtained from Eq. (III. 93) without a direct measurement of x. In particular, the frequency of a line arising from the transition $(I \pm \frac{1}{2}, m) \leftrightarrow (I \pm \frac{1}{2}, m-1)$ is given by

$$f = \frac{\Delta \nu}{2}\{[1+4mx/(2I+1)+x^2]^{\frac{1}{2}} - [1+4(m-1)x/(2I+1)+x^2]^{\frac{1}{2}}\} + $$
$$\pm g_I \mu_0 H_0/h, \quad \text{(IX. 9)}$$

where x is given by Eq. (IX. 4). All the lines appear as components of doublets of frequency separation $2(\mu_I/Ih)H_0$ except when the magnetic level $m = \pm(I+\frac{1}{2})$ is involved in the transition. The mean frequency f_m is then as given as in Eq. (IX. 9) without the last term. Kusch and Taub (KUS 49a) introduce the quantities x_1 and x_2 defined by

$$x_1 = -2m/(2I+1) \quad \text{and} \quad x_2 = -2(m-1)/(2I+1).$$

In terms of these it is easy to show from Eq. (IX. 9) that f_m has its maximum for

$$x = (x_1 x_2 + 1)/(x_1 + x_2) + [(x_1 x_2 + 1)^2/(x_1 + x_2)^2 - 1]^{\frac{1}{2}}. \quad \text{(IX. 10)}$$

The theoretical value of $f_m/\Delta\nu$ may then be determined by Eqs. (IX. 9) and (IX. 10). It can be shown that a maximum occurs only when $-(I-\tfrac{3}{2}) \leqslant m \leqslant 0$, $-(I-\tfrac{1}{2}) \leqslant m-1 \leqslant -1$. In the case of Na^{23}, for example, where $I = \tfrac{3}{2}$ the only line for which a maximum occurs is the

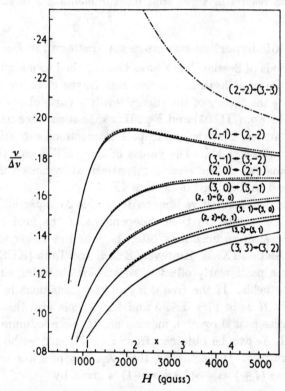

Fig. IX. 5. The field dependence of the frequencies of the lines of Rb^{85} associated with the transitions $\Delta m_I = \pm 1$, $\Delta m_J = 0$ (MIL 40).

line $(F, 0) \leftrightarrow (F, -1)$. For Cs where $I = \tfrac{7}{2}$, a maximum will occur for the lines $(F, -2) \leftrightarrow (F, -3)$, $(F, -1) \leftrightarrow (F, -2)$, and $(F, 0) \leftrightarrow (F, -1)$.

From an observation of the maximum mean frequency of the doublet it is thus possible to determine $\Delta\nu$ with the same precision as that with which the frequency is measured. This method of determining $\Delta\nu$ has a marked advantage over other methods since in the region of field at which the frequency is a maximum, the line frequencies are almost entirely field independent. Accordingly small inhomogeneities in the magnetic field will not affect the width and shape of the lines. A significant measurement of the line frequency may be made to a precision limited only by the accuracy of the frequency measuring equipment and by the natural width of the lines; the possibility of systematic error arising from asymmetrical broadening is avoided. It is not necessary to determine the frequency of the line at the precise maximum since $f/\Delta\nu$ varies slowly with field in the neighbourhood of the maximum, so corrections can be made from Eq. (IX. 9) with approximate values for H obtained from the doublet separation.

Isotopes of very low abundance have been measured by Zacharias (ZAC 42, DAV 49a), Davis (DAV 48b, DAV 49, DAV 49a), Nagle (DAV 49a), Bellamy (SMI 51, BEL 51, BEL 53), Smith (SMI 51, BEL 53),

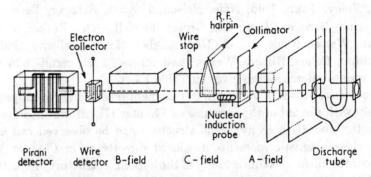

FIG. IX. 6. Diagram of apparatus used to study He in its metastable triplet state (HUG 53).

Wessel (WES 53), Lew (WES 53), and others (EIS 52, ZXX 54) with the various mass spectrometer, electron multiplier, and radioactivity techniques described in the previous section. In the experiments of Weinreich, Grosof, and Hughes (WEI 53) the hyperfine structure of the metastable $^3S_1$ state of He^3 was studied with atoms excited by an electrical discharge in the source. A schematic diagram of their apparatus is shown in Fig. IX. 6. Although $J = 1$ in this state the interpretation of the experiment is similar to those with $J = \frac{1}{2}$ since $I = \frac{1}{2}$ whence only two F states are possible.

It should be noted, as pointed out in Chapter III, that the value of $\Delta\nu$ deduced from the Breit–Rabi formula, Eq. (III. 93), will not always be the true hyperfine structure separation. Non-vanishing matrix elements of the electron-nucleus interaction operator which are not diagonal in J lead to a perturbation of the levels of the same F. Thus with a $P_{\frac{1}{2}}$ state, as in Berman's (BER 52, SCH 54) studies of Tl, the $F = 1$ level will be perturbed by the $P_{\frac{3}{2}}$ level. Thus if $\Delta\nu'$ is the true hyperfine structure separation and $\Delta\nu$ that deduced from the Breit–Rabi formula, then, as shown by Berman (BER 52, SCH 54),

$$\Delta\nu' = \Delta\nu - \frac{|(J=\tfrac{1}{2}|H'|J=\tfrac{3}{2})|^2}{W_{\frac{3}{2}}-W_{\frac{1}{2}}}, \qquad \text{(IX. 11)}$$

where H' is the operator connecting the fine structure levels. The above correction has been calculated by Frosch (BER 52, SCH 54) for a number of elements and shown to be 15·2 kc in Tl^{205}, for example, which is negligible in contrast to the $(21{,}311{\cdot}48\pm 0{\cdot}19)$ Mc for its $\Delta\nu$.

Hyperfine structure separations in a number of atoms have been measured by the above methods by Kusch, Millman, Rabi, Zacharias, Lew, Foley, Taub, Feld, Nafe, Nelson, Hughes, Eisinger, Bederson, Jaccarino, Berman, Weinreich, Grosof, Prodell, Ochs, Becker, Mann, Logan, Wessel, Nagle, Julian, Davis, Zabel, Hardy, Bellamy, Smith, Hamilton, Cohen, Gilbert, Wexler, and others. Their results and the appropriate references are given in Table IX. 1.

From these values of $\Delta\nu$ the nuclear moments can be inferred approximately with the aid of the formulae of Chapter III. However, in most cases the nuclei whose hyperfine structure can be observed can also have their magnetic moments measured directly as in Chapter VI, without the theoretical uncertainty of the hyperfine structure interaction coefficient. Consequently the measurements of hyperfine structure separations are now used but rarely to determine nuclear magnetic moments of abundant isotopes. However, in many cases with isotopes of low relative abundance (ZAC 42, BEL 53, ZXX 54) atomic beam experiments are easier than molecular ones because of the greater beam intensity that is possible with less stringent requirements of deflecting power. In such cases, the $\Delta\nu$'s of both the rare and the normal isotopes are determined and the magnetic moments can be inferred from ratios of the Fermi–Segré formulae of Eqs. (III. 82), (III. 88), etc. Since only ratios are involved most of the uncertainties cancel out in isotopic atoms and the magnetic moment ratio is given by

$$\mu_{I1}/\mu_{I2} = \frac{\Delta\nu_1\, I_1(2I_2+1)}{\Delta\nu_2\, I_2(2I_1+1)}. \qquad \text{(IX. 11a)}$$

Some uncertainties due to the phenomena discussed in the next paragraph remain, but these are very small compared with the uncertainties of the coefficients that must be calculated in the Fermi–Segré formula if the magnetic moment of one of the isotopes is not known. Nuclear moments measured in this way and the corresponding references are

Table IX. 1

Atomic Hyperfine Structure Separations for $J = \tfrac{1}{2}$

This table lists the accurate data on atomic hyperfine structure separations in the ground state of atoms with electronic angular momentum $J = \tfrac{1}{2}$. (He^3, which is in the metastable $^3S_1$ state but for which $I = \tfrac{1}{2}$, is also included.) The atom is in $^2S_{\tfrac{1}{2}}$ except where the state is explicitly given. The number in parenthesis indicates the uncertainty in the last printed numeral of the preceding listed value. The figures below can be converted into cm^{-1} by multiplication with $10^6/c = 3\cdot335636(9) \times 10^{-5}$ cm^{-1} Mc^{-1}. The asterisk (*) indicates a radioactive isotope and a superscript m indicates a metastable isomer. Additional data for H, Na, In, and Au in Appendix G. Results for $J > \tfrac{1}{2}$ are given in Table XI. 3.

Z	Atom	A	$\Delta \nu$ (Mc sec^{-1})	References
1	H	1	1,420·40577(5)	NAF 48, NAG 47, PRO 50, ZZA 28, ZZA 29
	D	2	327·384302(30)	NAF 48, NAG 47, PRO 50, ZZA 28
	T	3*	1,516·702(10)	NEL 49
2	He	3	6,739·71(5)	WEI 53, TEU 54
	($^3S_1$)			
3	Li	6	228·208(5)	KUS 40, KUS 49a
		7	803·512(15)	KUS 40, KUS 49a
5	B$^2P_{\tfrac{1}{2}}$	11	732·4(1)	ZZA 30
11	Na	22*	1,220·64(4)	DAV 48a, DAV 48b, DAV 49a, DAV 49c
		23	1,771·631(2)	KUS 40, KUS 48, MIL 40, KUS 49a, LOG 51
		24*	1,139·35(10)	BEL 53
13	Al	27	1,506·14(5)	LEW 48, LEW 53, KOS 52
17	Cl $^2P_{\tfrac{1}{2}}$	35	2074·383(8)	ZZA 37
		37	1726·700(15)	ZZA 37
19	K	39	461·723(10)	KUS 40, KUS 49a
		40*	1,285·790(7)	ZAC 42, EIS 52, DAV 49a, DAV 49c
		41	254·018(6)	KUS 40, OCH 50
		42*	1,258·9(1)	BEL 53
29	Cu	64*	1,278(20)	HAM 53, BRO 54
31	Ga	69	2,677·9875(10)	REN 40, KUS 48, BEC 48, KOS 52, ZZA 57
	($^2P_{\tfrac{1}{2}}$)	71	3,402·6946(13)	REN 40, KUS 48, BEC 48, KOS 52, ZZA 57
37	Rb	81	5,000(125)	NIE 54, BRO 54
		85	3,035·730(5)	KUS 40, MIL 40, KUS 49a, OCH 52
		86*	3,960(20)	BEL 51, BEL 53
		87	6,834·681(10)	KUS 40, MIL 40, KUS 49a, OCH 52
47	Ag	107	1,712·56(4)	WES 53
		109	1,976·94(4)	WES 53
		111*	2,180(100)	HAM 53, BRO 54
49	In	113	11,387(4)	HAM 39, HAR 42, MAN 50
	($^2P_{\tfrac{1}{2}}$)	115	11,409·50(10)	HAM 39, HAR 42, TAU 49, KUS 48, MAN 50
55	Cs	131	13,200(110)	BEL 53
		133	9,192·63183(1)	MIL 40, TAU 49, KUS 49a, SHE 52, ZZA 39
		134*	10,465(12)	JAC 52, BEL 53
		134^{m*}	3684·594(20)	COH 54, BRO 54, ZZA 15, ZZA 16
		135*	9,724(8)	DAV 49, DAV 49a, DAV 49b
		137*	10,115·527(15)	DAV 49, DAV 49a, DAV 49b, DAV 49c, ZZA55
79	Au	197	6,107·1(10)	WES 53
81	Tl	203	21,106·06(49)	BER 50, BER 52
	($^2P_{\tfrac{1}{2}}$)	205	21,311·48(19)	BER 50, BER 52

given in the Table VI. I. An alternative means of obtaining nuclear magnetic moments from atomic resonance experiments is described in Section IX. 7.

The signs of the nuclear magnetic moments can be obtained as in Chapter IV with the oscillatory magnetic field providing the desired transitions. Alternatively, Hardy and Millman (HAR 42) have determined the sign of nuclear magnetic moments by identifying the F values of the focusing zero moment states. In Section IV. 5 it was shown that the focusing zero-moment states were those for which there are minima in Fig. III. 4. By showing that for one of these focusing states a transition to $m = -(I+J)$ could be induced, Hardy and Millman showed that the configuration in Fig. III. 4 corresponding to a positive magnetic moment applied.

Another means for determining the signs of nuclear moments has been used by Zacharias (ZAC 42) and others (DAV 48, BEL 53, ZXX 54). From Eq. (III. 120) and for the (F, m) transition

$$(I+\tfrac{1}{2}, -I-\tfrac{1}{2}) \leftrightarrow (I+\tfrac{1}{2}, -I+\tfrac{1}{2}),$$

the value of the hyperfine structure $\Delta \nu$ can be calculated from the transition frequency ν by means of

$$\Delta \nu = \frac{(\nu - g_I \mu_0 H_0/h)(g_J \mu_0 H_0/h - \nu)}{\nu - g_J \mu_0 H_0/(2I+1)h - 2Ig_I \mu_0 H_0/(2I+1)h}. \quad \text{(IX. 11b)}$$

The resonance may then be observed at a number of different values of the external field H_0 and the sign of g_I that leads to a consistent set of $\Delta \nu$'s is assumed to be correct. The signs of the nuclear magnetic moments determined in this way are included in Table VI. I along with the corresponding references.

The case of the hyperfine structure of hydrogen-like atoms is particularly important, since for these the hyperfine structure interaction is amenable to detailed theoretical calculation. Consequently the experiments with hydrogen atoms are considered separately in Section IX. 6.

IX. 5. Atomic magnetic moments; the anomalous electron moment

IX. 5.1. *Atomic magnetic moments.* Essentially the same procedures as discussed in the previous sections of this chapter can be applied to the measurement of atomic magnetic moments provided a field-sensitive rather than a field-insensitive transition is selected. Thus by Eq. (III. 120) the transition frequency between two states varies in a known fashion with x if $\Delta \nu$ has once been determined as in the preceding section. However, x in turn, by Eq. (III. 121), depends on $g_J \mu_0 H_0$ except for the small and easily made correction term $g_I \mu_0 H_0$. Therefore, if H_0 has been calibrated as by the measurement of a known moment in the field, g_J can be determined.

In the early days of the molecular-beam magnetic-resonance method there was little interest in the measurement of g_J since it was presumed for the $^2S_{\frac{1}{2}}$ atoms used in most of the experiments that the atomic magnetic moment simply had the theoretical value of exactly one Bohr magneton. However, as discussed in Sections IX. 2 and IX. 6, the experiments of Nafe, Nelson, and Rabi (NAF 48, KOE 51) and Nagle, Julian, and Zacharias (NAG 47) on the hyperfine structure separation in atomic hydrogen first suggested to Breit (BRE 47a) that the spin magnetic moment of the electron might be slightly different from exactly one Bohr magneton.

As a result of this suggestion Kusch and Foley (KUS 48) measured the ratios of various atoms in different atomic states to determine by a direct experiment the existence of any anomalous moment. They assumed that the electron spin g_S and its orbital g_L might depart from their simple theoretical values in accordance with

$$g_S = 2(1+\delta_S)$$
$$g_L = 1+\delta_L. \qquad \text{(IX. 12)}$$

From this assumption and that of Russell–Saunders coupling with atomic Paschen–Back effects neglected, one can easily show (KUS 48) by procedures analogous to those in the derivation of Eq. (III. 91) that the ratio of the g_J values of two atomic states in the same or different atoms is

$$g_{J_1}/g_{J_2} = (g_L \alpha_{L_1}+g_S \alpha_{S_1})/(g_L \alpha_{L_2}+g_S \alpha_{S_2})$$
$$= [(2\alpha_{S_1}+\alpha_{L_1})/(2\alpha_{S_2}+\alpha_{L_2})] +$$
$$+ 2[(\alpha_{S_1}\alpha_{L_1}-\alpha_{L_1}\alpha_{S_2})/(2\alpha_{S_2}+\alpha_{L_2})^2][\delta_S-\delta_L]. \qquad \text{(IX. 13)}$$

In the above for Russell–Saunders coupling

$$\alpha_S = [J(J+1)+S(S+1)-L(L+1)]/2J(J+1) \qquad \text{(IX. 14)}$$

and α_L is the same except for the interchange of S and L. It should be noted that Eq. (IX. 13) depends only on $(\delta_S-\delta_L)$ and not on each separately; consequently both of these cannot be determined separately from the experiments. Therefore δ_L is ordinarily assumed equal to zero and δ_S is obtained from the experiments. This assumption has proved to be consistent with the absolute calibrations of the fields discussed in Section IX. 2.

Kusch and Foley (KUS 48) then measured g_J in the $^2S_{\frac{1}{2}}$ state of Na, in the $^2P_{\frac{1}{2}}$ state of In, and in the $^2P_{\frac{1}{2}}$ and $^2P_{\frac{3}{2}}$ states of Ga. For the last of these the J is $\frac{3}{2}$ instead of $\frac{1}{2}$ so the theoretical expression for the energy levels must be modified from that of Eq. (III. 120) by the use of a Hamiltonian for the atom as in Eq. (III. 122). Calculations based on this Hamiltonian have been given by Becker and Kusch (BEC 48) and by Kusch and Foley (KUS 48); the effects of the nuclear quadrupole interaction are discussed in greater detail in Section IX. 8. The theoretical variation of the energy of the atomic levels of Ga with magnetic field is given in Fig. III. 5. With the aid of this theory the g_J for the $^2P_{\frac{3}{2}}$ of Ga can be inferred as in the above case of $J = \frac{1}{2}$. In this way Kusch and Foley (KUS 48, FOL 48) found experimentally

$$g_J(^2P_{\frac{3}{2}}\text{Ga})/g_J(^2P_{\frac{1}{2}}\text{Ga}) = 2(1\cdot00172\pm0\cdot00006),$$

$$g_J(^2S_{\frac{1}{2}}\text{Na})/g_J(^2P_{\frac{1}{2}}\text{Ga}) = 3(1\cdot00242\pm0\cdot00006),$$

$$g_J(^2S_{\frac{1}{2}}\text{Na})/g_J(^2P_{\frac{1}{2}}\text{In}) = 3(1\cdot00242\pm0\cdot00010). \qquad \text{(IX. 15)}$$

From these and Eq. (IX. 13) with the assumption that $\delta_L = 0$, they found for δ_S in the above three cases $(0\cdot00114\pm0\cdot00004)$, $(0\cdot00121\pm0\cdot00003)$, and $(0\cdot00121\pm0\cdot00005)$ respectively.

There are possibilities that the assumption of perfect Russell–Saunders coupling in Eq. (IX. 13) above and in subsequent calculation of δ_S may be invalid. The effects of perturbations which might modify the coupling slightly as described in Chapter III have been discussed in detail for these cases by Kusch and Foley (KUS 48) and by Phillips (PHI 41); relativistic effects have been discussed by Margenau (MAR 40). These calculations show that the approximation of Eq. (IX. 13) is within the accuracy of the experiment, as is also indicated by the agreement of the values for δ_S. The particularly close agreement between $^2P_{\frac{1}{2}}$ measurements on Ga and In would suggest that these states especially are unaffected by perturbations. The net conclusion of Kusch and Foley (KUS 48) was that $g_S = 2(1\cdot00119\pm0\cdot00005)$ in agreement with more accurate values obtained later.

Kusch and Taub (KUS 49a) have measured the ratios of the g_J values of a number of alkali atoms in the manner given above. They find experimentally that the g_J's for Li[6], Li[7], Na[23], and K[39] are in agreement to within 1 part in 40,000. On the other hand, for Cs[133] and for either Rb[85] or Rb[87] they find $g_J(\text{Cs})/g_J(\text{Na}) = 1\cdot000134\pm0\cdot000007$ and $g_J(\text{Rb})/g_J(\text{Na}) = 1\cdot00005\pm0\cdot00001$. The fact that the g_J's of Li, Na, and K are identical indicates that, within the precision of their experiment, the g_J of these atoms is equal to the spin gyromagnetic ratio, g_S.

For if perturbations were to affect the value of g_J, they would presumably affect different atoms by different amounts. The larger g_J values of Rb and Cs presumably arise from some such perturbation (KUS 48, KUS 49a).

In a similar fashion, Mann and Kusch (MAN 50a) with In in $^2P_{\frac{3}{2}}$ and $^2P_{\frac{1}{2}}$ states have found $g_J(^2P_{\frac{3}{2}})/g_J(^2P_{\frac{1}{2}}) = 2(1{\cdot}00200\pm0{\cdot}00006)$. From this, from the assumption that g_L is exactly unity, and from the assumption of Russell–Saunders coupling which leads to Eq. (IX. 13), they inferred that $g_S = 2(1{\cdot}00133\pm0{\cdot}000014)$. This was largely a confirmation of the earlier experiments of Kusch and Foley; the slight discrepancy was presumably due to the effects of perturbations on the coupling in the $^2P_{\frac{1}{2}}$ state.

Nelson and Nafe (NEL 49) have shown for the $^2S_{\frac{1}{2}}$ ground states of atomic hydrogen and atomic deuterium that

$$g_J(\text{H})/g_J(\text{D}) = 0{\cdot}999991\pm0{\cdot}000010,$$

whence the g value of the electron when bound to the proton differs from that when bound to the deuteron by less than 0·001 per cent. Likewise Franken and Koenig (FRA 52, POH 51) have shown that $g_J(\text{K}^{39})/g_p = 658{\cdot}2274\pm0{\cdot}0023$, where g_p is the proton g value measured in a spherical sample of mineral oil. When this is combined with the value of $g_J(\text{H}^1)/g_p$ obtained by Koenig, Prodell, and Kusch (KOE 51) in Eq. (IX. 17) below, one obtains $g_J(\text{K}^{39})/g_J(\text{H}^1) = 1{\cdot}000016\pm0{\cdot}000004$. Although the difference in the g_J's of K^{39} and H^1 is small it is none the less considerably beyond the experimental error. Franken, Koenig, and Phillips (FRA 52, PHI 41, PHI 52) successfully attribute the discrepancy to three effects: (a) the relativistic behaviour of the electron, (b) the diamagnetism of the atomic core, and (c) the perturbation of the potassium ground state arising from exchange interactions with the excited states of the electronic core.

The most accurate measurement of an atomic state which should be free of effects of perturbation is that of Koenig, Prodell, and Kusch (KOE 51, POH 51, FRA 52, ZZA 34, ZZA 35). They measured the $g_J(^2S_{\frac{1}{2}}, \text{H})$ of atomic hydrogen in its ground state in a magnetic field that was calibrated by the proton nuclear resonance in water, as discussed in greater detail in Section IX. 6.1. Consequently, with the calibration of the proton magnetic moment in Bohr magnetons by Gardner and Purcell (GAR 49) as discussed in Section IX. 2, the g_J of atomic hydrogen was obtained directly. Since the ground state of hydrogen is pure to considerably better than a part in a million, it is apparent that g_J equals g'_S, the spin value of the bound electron in atomic hydrogen. Breit (BRE 28) and Margenau (MAR 40) have shown that the correction for the relativistic mass change due to binding yields for the g value of the free electron

$$g_S = g'_S \, 3(1+2[1-\alpha^2]^{\frac{1}{2}})^{-1}$$
$$= g'_S(1-17\cdot75\times10^{-6})^{-1}. \qquad \text{(IX. 16)}$$

Koenig, Prodell, and Kusch (KOE 51) found after making corrections that
$$-g_J(^2S_{\frac{1}{2}}, \text{H})/g_p = 658\cdot2171\pm0\cdot0006, \qquad \text{(IX. 17)}$$
$$-g_S/g_p = 658\cdot2288\pm0\cdot0006,$$

where g_p is the proton g value measured in a spherical sample of mineral oil. (See Appendix G.) When this result is combined with that of Franken and Liebes (ZZB 9), as mentioned above, and with Eq. (IX. 16), the following experimental results are obtained (KOE 51, HUG 53):

$$g_J(\text{H},\,^2S_{\frac{1}{2}}) = 2(1\cdot001128\pm0\cdot000012) \qquad \text{(IX. 18)}$$
and $$g_S = 2(1\cdot001165\pm0\cdot000012) \qquad \text{(IX. 19)}$$

provided that g_L, the orbital g for a free electron, is taken as 1. Measurements similar to the above have been made by Franken and Koenig (FRA 52) as listed in Table IX. II.

Hughes, Tucker, Rhoderick, and Weinreich (HUG 53) have measured the magnetic moment of a helium atom in its metastable triplet state $^3S_1$. A schematic diagram of their apparatus is shown in Fig. IX. 6. Atomic hydrogen and atomic helium beams were run successively in the same magnetic field. The $m = \pm 1 \leftrightarrow 0$ transition was studied with He^4 (for which $I = 0$) and the F, $m = 1, 0 \leftrightarrow 1, -1$ transition in atomic hydrogen was studied. The metastable $1s2s\,^3S_1$ state of He was produced in the discharge tube in the fraction of about one metastable atom for every $2\cdot3\times10^4$ atoms in the ground state. The metastable helium atoms were detected by the electron emission which results when they strike a metal surface and become de-excited; the detection efficiency corresponds to about 0·24 electron per atom (HUG 53, COB 44, DOR 42, HAG 53). In this way Hughes, Tucker, Rhoderick, and Weinreich (HUG 53) found experimentally

$$g_J(\text{He},\,^3S_1)/g_J(\text{H},\,^2S_{\frac{1}{2}}) = 1-(11\pm16)\times10^{-6}. \qquad \text{(IX. 20)}$$

From Eq. (IX. 18) then

$$g_J(\text{He},\,^3S_1) = 2(1\cdot001117\pm0\cdot000020). \qquad \text{(IX. 21)}$$

Perl and Hughes (PER 53, PER 53a) have developed the theory for the g_J of He in a $^3S_1$ state. For a theoretical anomalous moment of the free electron of the magnitude discussed in the next section they find theoretically that $g_J(\text{He}, {}^3S_1)$ equals $2(1\cdot 0011044)$ including a non-radiative relativistic bound state correction analogous to the hydrogen correction of Eq. (IX. 16) and including also a mutual radiative correction which arises from the Breit interaction and may be interpreted classically as a diamagnetic correction (HUG 53, PER 53). The above

TABLE IX. II

Atomic Magnetic Moments

In the following table are listed the atomic Landé g factors which have been accurately measured by the molecular-beam magnetic-resonance method. The number in parenthesis indicates the uncertainty in the last printed number of the preceding figure. The magnetic field calibration is based on that of Gardner and Purcell (GAR 49).

Atom	State	$g_J/2$	Reference
Free electron spin		1·001146(12)	KOE 51, ZZA 34, ZZA 35
H[1]	${}^2S_{\frac{1}{2}}$	1·001128(12)	KOE 51, HUG 53, ZZA 34
D[2]	${}^2S_{\frac{1}{2}}$	1·001128(30)	NEL 49, KOE 51
He[4]	${}^3S_1$	1·001117(20)	HUG 53
Li[6]	${}^2S_{\frac{1}{2}}$	1·00114(2)	KUS 49a, FRA 52, BRI 53
Li[7]	${}^2S_{\frac{1}{2}}$	1·00114(2)	KUS 49a
Na[23]	${}^2S_{\frac{1}{2}}$	1·00114(2)	KUS 49a
K[39]	${}^2S_{\frac{1}{2}}$	1·00114(1)	POH 51, FRA 52, BRI 53
Cr[52]	${}^7S_3$	1·00081(5)	BRI 53
Ga[69]	${}^2P_{\frac{1}{2}}$	$(\frac{1}{3})[0\cdot 99872(7)]$	KUS 48, FOL 48, BRI 53
Ga[69]	${}^2P_{\frac{3}{2}}$	$(\frac{2}{3})[1\cdot 00044(7)]$	KUS 48, FOL 48, BRI 53
Ga[71]	${}^2P_{\frac{1}{2}}$	$(\frac{1}{3})[0\cdot 99872(7)]$	KUS 48, FOL 48, BRI 53
Ga[71]	${}^2P_{\frac{3}{2}}$	$(\frac{2}{3})[1\cdot 00044(7)]$	KUS 48, FOL 48, BRI 53
Rb[85]	${}^2S_{\frac{1}{2}}$	1·00119(2)	KUS 49a, BRI 53
Rb[87]	${}^2S_{\frac{1}{2}}$	1·00119(2)	KUS 49a, BRI 53
Ag[107]	${}^2S_{\frac{1}{2}}$	1·00112(10)	WES 53
Ag[109]	${}^2S_{\frac{1}{2}}$	1·00112(10)	WES 53
In[113]	${}^2P_{\frac{1}{2}}$	$(\frac{1}{3})[0\cdot 99872(11)]$	KUS 48, FOL 48, BRI 53
In[113]	${}^2P_{\frac{3}{2}}$	$(\frac{2}{3})[1\cdot 00072(11)]$	MAN 50a
In[115]	${}^2P_{\frac{1}{2}}$	$(\frac{1}{3})[0\cdot 99872(11)]$	KUS 48, FOL 48, BRI 53
In[115]	${}^2P_{\frac{3}{2}}$	$\frac{2}{3}[1\cdot 00072(11)]$	MAN 50a
Cs[133]	${}^2S_{\frac{1}{2}}$	1·00125(3)	KUS 49a, TAU 49, BRI 53
Au[197]	${}^2S_{\frac{1}{2}}$	1·00206(6)	WES 53

theoretical value for g_J agrees with the experimental result of Eq. (IX. 21) to within the experimental error. This experiment therefore also affords valuable support for the existence of the anomalous electron moment as discussed in the next section.

Wessel, Lew, Brix, and Eisinger (WES 53, BRI 53) have measured the atomic magnetic moments of Ag^{107}, Ag^{109}, Au^{197}, and Cr^{52}. They used an electron bombardment detector which they first introduced and which is described in Chapter XIV. A schematic diagram of their apparatus is given in Fig. IX. 7.

The results of the measurements of atomic g_J's that have so far been determined with high precision by the magnetic resonance method are listed in Table IX. II.

IX. 5.2. *The anomalous electron magnetic moment.* The most important result of the precision measurements of atomic magnetic moments as discussed in the preceding section is that the electron spin magnetic moment is not exactly one Bohr magneton. In other words,

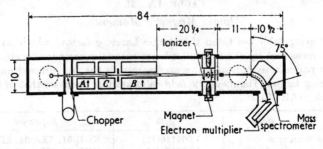

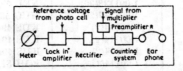

Fig. IX. 7. Atomic-beam apparatus with electron bombardment ionizer. All dimensions are in inches. The amplification system for the electron multiplier signal is shown schematically (WES 53).

g_S is not exactly 2 but instead possesses the experimental value of Eq. (IX. 19)

$$g_S = 2(1 \cdot 001165 \pm 0 \cdot 000012).\tag{IX. 22}$$

Schwinger (SCH 48) has explained this anomalous electron magnetic moment in terms of a relativistic quantum electrodynamics theory. In fact it was the attempt to explain the anomalous electron moment that first led Schwinger to develop his relativistic theory of quantum electrodynamics. According to this theory the observed magnetic moment of the free electron in a vacuum is slightly modified from the Dirac value of exactly one Bohr magneton as a result of the coupling of the electron with the electromagnetic field and with the electrons that occupy the negative energy states of the Dirac theory. Virtual emissions and reabsorption of quanta and of electron-positron pairs are therefore possible; the existence of these modifies the observed electron magnetic moment in much the same way as the energy levels of the atom are affected in the Lamb–Retherford experiment discussed in detail in Chapter XII. The theory of this mechanism is described in greater detail in that chapter. By means of a suitable relativistic second order calculation Schwinger (SCH 48) showed that the magnetic moment of the electron should be $1+\alpha/2\pi$ Bohr magnetons, where α is the fine structure constant. A particularly simple discussion of the physical principles involved is given by Koba (ZZB 1). Others (KAR 50, ZZB 9) extended the calculation to the fourth order with the result that

$$g_S = 2[1+\alpha/2\pi - 0.328\alpha^2/\pi^2 + ...] \quad \text{(IX. 23)}$$

or numerically

$$g_S = 2[1.0011596]. \quad \text{(IX. 24)}$$

The above theoretical result is in excellent agreement with the experimental value in Eq. (IX. 22). As discussed in the next section, a similar theory also agrees very well with experimental measurements of the hydrogen atomic hyperfine structure separation.

Perl and Hughes (PER 53, HUG 53) have extended the theory of the electron magnetic moment to a calculation of the magnetic moment of He⁴ in its metastable ³S₁ state with the result

$$g_J(\text{He}, {}^3S_1) = 2[1+\alpha/2\pi - 0.328\alpha^2/\pi^2 - \langle T\rangle/3mc^2 - \langle e^2/r_{12}\rangle/6mc^2]$$
$$= 2(1.0011044), \quad \text{(IX. 25)}$$

where $\langle T \rangle$ is the expectation value for the kinetic energy of the electrons and $\langle e^2/r_{12}\rangle$ is the expectation value for the electrostatic interaction between the two electrons. The first three terms just give the electron magnetic moment of the free electron, with its anomalous contribution. The fourth term is the non-radiative relativistic bound-state correction and is analogous to the relativistic bound-state correction in atomic hydrogen as in Eq. (IX. 16). The last term is the mutual radiative correction which arises from the Breit interaction between the two electrons and may be interpreted classically as a diamagnetic correction. By a comparison of Eqs. (IX. 25) and (IX. 21) it can be seen that the agreement between theory and experiment is excellent. Teutsch and Hughes (TEU 54) have also calculated the theoretical value for the hyperfine structure separation of He³ in the metastable ³S₁ state and have found agreement with the experimental results of Weinreich and Hughes (WEI 53) to within the accuracy of the calculation.

IX. 6. Hyperfine structure of atomic hydrogen

IX. 6.1. *Atomic hydrogen experiments.* As a result of the simplicity of the hydrogen atom and the consequent feasibility of theoretical interpretations of experimental results with atomic hydrogen, measurements on this atom have been particularly significant. Consequently, the hyperfine structure experiments with atomic hydrogen will be discussed separately in this section. Although some of the atomic hydrogen results have already been described in previous sections, the most detailed discussion of the experiments will be given here.

The first precision hyperfine structure experiments with atomic hydrogen and deuterium were those of Nafe, Nelson, and Rabi (NAF

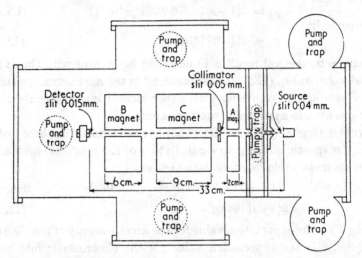

Fig. IX. 8. Schematic diagram of atomic-beam apparatus for atomic hydrogen. Longitudinal dimensions from source to detector are drawn to scale (NAF 48).

48) which were followed shortly by similar experiments by Nagle, Julian, and Zacharias (NAG 47). A schematic diagram of the apparatus used by Nafe, Nelson, and Rabi (NAF 48) is shown in Fig. IX. 8.

For atomic hydrogen Nafe and Nelson studied the σ_1, π_1, and π_2 transitions indicated in Fig. IX. 9a for the $^2S_{\frac{1}{2}}$ ground state. The frequency of these transitions can be seen to vary with magnetic field as shown in Fig. IX. 10a. Likewise for atomic deuterium the possible transitions are as indicated in Figs. IX. 9b and IX. 10b. The atomic hydrogen was produced in a Wood's discharge tube as described in Chapter XIV. Several different transitions were studied for both hydrogen and deuterium and at various magnetic fields. Typical resonances are shown in Fig. IX. 11. All results were consistent with

$$\Delta\nu_H = 1{,}420{\cdot}410 \pm 0{\cdot}006 \text{ Mc}, \quad \Delta\nu_D = 327{\cdot}384 \pm 0{\cdot}003 \text{ Mc},$$

and

$$\Delta\nu_H/\Delta\nu_D = 4{\cdot}33867 \pm 0{\cdot}00004.$$

A similar experiment was performed by Nelson and Nafe (NEL 49)

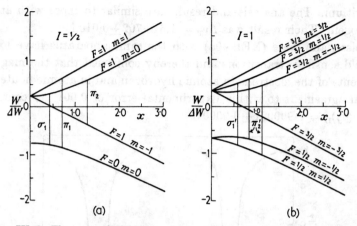

FIG. IX. 9. The magnetic field dependence of the magnetic levels arising from the components of a hyperfine structure doublet (a) for the case in which $J = \frac{1}{2}, I = \frac{1}{2}$ (hydrogen) and (b) for the case in which $J = \frac{1}{2}, I = 1$ (deuterium). Each level is labelled by the quantum numbers F and m that are appropriate in the region of very weak magnetic fields. The σ and π transitions that were studied are indicated. The two $I = 1$ lines marked π'_2 formed an unresolved doublet (NAF 48).

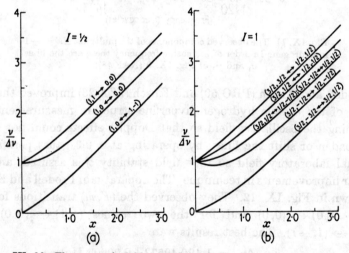

FIG. IX. 10. The magnetic field dependence of the frequencies of transitions between magnetic levels arising from the components of the atomic hyperfine structure doublets for (a) hydrogen and (b) deuterium. The frequencies of transitions between levels which in high field have the same value of m_J have been omitted since these transitions were not observable with the apparatus used in the experiments (NAF 48).

on tritium. The analysis and results are similar to those with atomic hydrogen. Their result was $\Delta\nu_T = 1{,}516{\cdot}702\pm0{\cdot}010$ Mc.

Nelson and Nafe (NEL 49a) with the same apparatus have looked at field sensitive transitions and thereby confirmed that the magnetic moments of the electrons in atomic hydrogen and in atomic deuterium are in agreement to within experimental error of 0·001 per cent., i.e. $(g_J)_H/(g_J)_D = 0{\cdot}999991\pm0{\cdot}000010$.

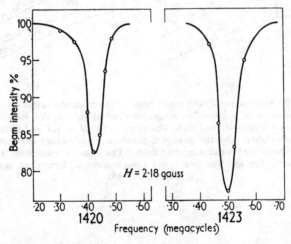

FIG. IX. 11. The observed components of the multiplet $\Delta F = \pm 1$ in hydrogen. In order of increasing frequency these are the lines σ_1 and π_1 of Fig. IX. 9 (NAF 48).

Prodell and Kusch (PRO 50) and Kusch (ZZA 28) improved the precision of the atomic hydrogen hyperfine structure measurements by designing the oscillatory field so that Doppler effects could not occur to broaden or shift the lines, by operating at a 0·3 gauss permanent residual laboratory field so that field stability was assured, and by similar improvements in technique. The apparatus of Prodell and Kusch is shown in Fig. IX. 12. They observed the (F, m) transitions for the σ-line $(1, 0) \leftrightarrow (0, 0)$ and for the two π-lines $(1, 1) \leftrightarrow (0, 0)$ and $(1, 0) \leftrightarrow (1, -1)$. The best results were

$$\Delta\nu_H = 1{,}420{\cdot}40573\pm0{\cdot}00005 \text{ Mc/sec},$$
$$\Delta\nu_D = 327{\cdot}384302\pm0{\cdot}000030 \text{ Mc/sec},$$
$$(\Delta\nu_H/\Delta\nu_D)_{exp} = 4{\cdot}33864947\pm0{\cdot}00000043. \quad (IX. 26)$$

Further results are given in Appendix G.

Koenig, Prodell, and Kusch (KOE 51) and others (ZZA 34, ZZA 35) measured the $g_J(^2S_{\frac{1}{2}}, H)$ of the ground state of atomic hydrogen in a magnetic field that was calibrated by the proton nuclear resonance in water. They studied chiefly the (F, m) transition $(1, 0) \leftrightarrow (1, -1)$ in a field of 1,500 gauss. Their apparatus is shown schematically in Fig. IX. 12. Their results have already been discussed in Section IX. 5.1 and are given in Eqs. (IX. 16, 17, 18, 19) above.

IX. 6.2. *Theoretical interpretation of atomic-hydrogen experiments.* The simplest theoretical interpretation of the atomic hydrogen hyperfine structure separation given in Eq. (IX. 26) could be obtained merely by the use of the Fermi formula of Eqs. (III. 82) and (III. 88) with the

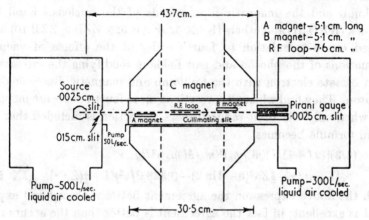

FIG. IX. 12. Schematic diagram of atomic-beam apparatus for the study of atomic hydrogen and deuterium (PRO 50).

reduced mass correction discussed subsequently. When these are combined they give

$$\Delta\nu = (4/3)[(2I+1)/I](m_r/m_0)^3(m_0/M)\mu_I \alpha^2 c Ry_\infty, \quad \text{(IX. 27)}$$

where m_r is the reduced mass and m_0 the rest mass of the electron, M is the proton mass, μ_I is the nuclear moment in nuclear magnetons, $Ry_\infty = 2\pi^2 m_0 e^4/h^3 c$, and α is the fine structure constant $e^2/\hbar c$. Only the Breit reduced mass correction of Section III. 3.2 is used above; the additional one of Low and Salpeter (LOW 51) will be added later. At the time of the original experiments, the best absolute value for the proton magnetic moment that was known for use in this equation was based on the Millman–Kusch absolute calibration (MIL 41) which incorrectly assumed that the atomic magnetic moment of a $^2S_{\frac{1}{2}}$ state was exactly one Bohr magneton. When this value of the proton magnetic moment was used in the above simple theory the predicted result was $\Delta\nu_{\text{Hst}} = 1{,}416{\cdot}97 \pm 0{\cdot}54$ Mc, which disagrees with Eq. (IX. 26) by far more than the experimental error. It was just this disagreement which first suggested to Breit (BRE 47a) and Schwinger (SCH 48) the possibility of an anomalous value for the electron magnetic moment; this suggestion was fully confirmed by the independent experiments discussed in the preceding section. However, the mere use of the correct values for the magnetic moments of the electron and of the proton in the Fermi formula Eqs. (III. 82) and (III. 88) is not adequate for a complete theoretical discussion, though it does provide a good first approximation (SCH 48).

Corrections to the Fermi formula arise from interference between the Coulomb and the magnetic dipole fields of the nucleus. Kroll and Polluck (KRO 51) and others (KAR 52, KAR 52b, ZZB 9, ZZB 10) have carried out a calculation to fourth order of the effects of vacuum fluctuations of the photon and pair fields in modifying the interaction of an S-state electron with the Coulomb and magnetic dipole fields of a proton. They find when all corrections up to fourth order are included and when only the Breit reduced mass correction is included that the Fermi formula becomes

$$\Delta \nu = (4/3)[(2I+1)/I](m_r/m_0)^3(m_0/M)\mu_I \alpha^2 Ry_\infty \times$$
$$\times \{1+\alpha/2\pi-\tfrac{1}{2}Z\alpha^2(5-2\ln 2)-0\cdot 328\alpha^2/\pi^2\}\{1+\tfrac{3}{2}(Z\alpha)^2\}. \quad \text{(IX. 27a)}$$

With the above expression the agreement between theory and experiment is excellent; in fact the agreement is better than the accuracy to which the fine-structure constant α is known. Consequently, the analysis has been inverted and the experimental results have been combined with the above theory to provide an experimental evaluation of the fundamental constant α. Low and Salpeter (LOW 51) and Arnowitt (ZZA 31) have introduced some small additional corrections into the above analysis corresponding to the nuclear recoil as photons are exchanged and to an allowance for the proton not being a point magnetic dipole. With these corrections (ZZB 10) the value of α is given by

$$\alpha^{-1} = 137\cdot 0358 \pm 0\cdot 0016. \quad \text{(IX. 28)}$$

From Eq. (IX. 27), if only the Breit reduced mass correction were applied to give the simplest theoretical value for $(\Delta\nu_H/\Delta\nu_D)_{st}$, one would expect that

$$(\Delta\nu_H/\Delta\nu_D)_{st} = (4/3)(m_H/m_D)^3(\mu_p/\mu_d), \quad \text{(IX. 29)}$$

where m_H and m_D are the reduced masses of the electron in hydrogen and deuterium atoms respectively. With numerical values this then gives (PRO 50, SMA 51)

$$(\Delta\nu_H/\Delta\nu_D)_{st} = 4\cdot 3393876 \pm 0\cdot 0000008. \quad \text{(IX. 30)}$$

This disagrees with the experimental results in Eq. (IX. 26) by far more than the experimental error. This discrepancy can most easily be expressed (LOW 51, PRO 50, ZZA 28) by

$$\Delta \equiv 1 - \left[\frac{(\Delta\nu_D/\Delta\nu_H)_{exp}}{(\Delta\nu_D/\Delta\nu_H)_{st}}\right] = (1\cdot703\pm0\cdot005)\times10^{-4}. \quad \text{(IX. 31)}$$

Low and Salpeter (LOW 51) have accounted for some, though not all, of this discrepancy by introducing an additional reduced mass correction which allows for the nuclear recoil that occurs when virtual photons are interchanged. In this manner Low and Salpeter (LOW 51) can account for a discrepancy of $\Delta = (0\cdot3\pm0\cdot1)\times10^{-4}$.

A number of suggestions (HAL 47, BRE 47b) have been made about the source of the remaining disagreement. One of the most fruitful has been due to A. Bohr (BOH 48, BOH 50, BOH 53) who has pointed out that the electron charge near the nucleus follows the proton in its motion in the deuteron. Hence, in a computation of the magnetic interaction of the neutron with the electrons, the neutron cannot be considered to be at the exact centre of the atom. Low and Salpeter (LOW 51) have made quantitative calculations on Bohr's proposal and find that it leads to $\Delta = (1\cdot98\pm0\cdot10)\times10^{-4}$. The combination of this with the above nuclear recoil correction gives a predicted

$$\Delta = (2\cdot28\pm0\cdot20)\times10^{-4}.$$

Low and Salpeter attribute the discrepancy of $(0\cdot58\pm0\cdot20)\times10^{-4}$ to the structure of the nucleons themselves. Other corrections have been considered by Greifinger (FOL 54a).

In a comparison of the hyperfine structure separations of hydrogen and tritium, one would expect, with merely the Breit reduced mass correction, that from Eq. (IX. 27) in analogy to Eq. (IX. 29) the following should apply (NEL 49):

$$(\Delta\nu_T)_{st} = \Delta\nu_H(\mu_t/\mu_p)(m_T/m_H)^3$$
$$= 1{,}516\cdot709\pm0\cdot015 \text{ Mc}. \quad \text{(IX. 32)}$$

This result is in excellent agreement with the experimental value of Nelson and Nafe (NEL 49) which, as listed in the previous section, was $1{,}516\cdot702\pm0\cdot010$ Mc; the agreement is to one part in 10^5. At first sight it is surprising that this excellent agreement should obtain in the case of tritium, whereas it has just been shown in the preceding paragraphs that for deuterium there was no such agreement until allowances were made for nuclear recoil in the exchange of virtual photons and for the structure of the nucleus. However, by considering the difference in the interaction between a neutron and a proton with parallel spins and a neutron and a proton with antiparallel spins, Fermi and Teller (FER 48) have calculated the fractional shift in the ratio $\Delta\nu_T/\Delta\nu_H$ and find it to be between 0.7×10^{-6} and 2.8×10^{-6}. This effect would reduce the value of the hyperfine structure separation of tritium. From a different point of view and in considering explanations of the triton moment (e.g. exchange moments or large orbital moments), Avery and Sachs (AVE 48) estimate that the tritium anomaly would be roughly 5 per cent. of that in the case of deuterium. Other corrections for tritium and He^3 have been considered by Foley (FOL 54a, ZZA 27) as discussed in Section IX. 10.

IX. 7. Direct nuclear moment measurements with atoms

As discussed in Section IX. 4, the values of nuclear magnetic moments can be inferred from the measured values of the hyperfine structure $\Delta\nu$. Such determinations are particularly effective for the magnetic moment of a rare isotope of an atom for which the normal isotope's $\Delta\nu$ and magnetic moment are both known. In such a case, the magnetic moment can be inferred from Eq. (IX. 11a) with only a small uncertainty.

However, in atomic experiments it is also possible to measure the nuclear magnetic moment from its direct interaction with the external magnetic field. Thus in strong magnetic fields, if the transition $\Delta m_I = \pm 1$, $\Delta m_J = 0$ is induced, a doublet structure is observed as in Eq. (IX. 3), the separation of the doublet being

$$\delta\nu = 2g_I \mu_0 H_0/h. \qquad (IX. 33)$$

The value of H_0 to be used above can be determined from the mean frequency of the doublet and the Breit–Rabi formula Eq. (III. 120). The signs of the magnetic moments can be determined by the methods of Section IX. 4.

In this way Hardy and Millman (HAR 42) have measured the magnetic moments of In^{113} and In^{115}, Becker and Kusch (BEC 48, KUS 50) have measured Ga^{69} and Ga^{71}; Kusch and Taub (KUS 49a) have measured Li^6, Li^7, Na^{23}, and K^{39}; Ochs, Logan, and Kusch (OCH 50) have measured K^{39} and K^{41}, Eisinger and Bederson (EIS 52) have measured K^{40}; Logan and Kusch (LOG 51) have measured Na^{23}, etc. (ZXX 54). These results are included in the tables of Chapter XV.

Although the nuclear magnetic moments of Ga were first determined in this way by Becker and Kusch (BEC 48, KUS 50), they were subsequently measured by Pound (POU 48) and Bitter (BIT 49) by the nuclear resonance absorption method. The results by the two methods, however, did not agree. The discrepancy was 0·7 per cent. Thus, by the above molecular beam method (BEC 48, KUS 50),

$$\mu(\text{Ga}^{71})/\mu(\text{H}^1) = 0{\cdot}9078 \pm 0{\cdot}0015.$$

$$\mu(\text{Ga}^{69})/\mu(\text{H}^1) = 0{\cdot}7146 \pm 0{\cdot}0015. \qquad (\text{IX. 34})$$

On the other hand, by the nuclear resonance-absorption method (POU 48, BIT 49)

$$\mu(\text{Ga}^{71})/\mu(\text{H}^1) = 0{\cdot}9148 \pm 0{\cdot}0004,$$

$$\mu(\text{Ga}^{69})/\mu(\text{H}^1) = 0{\cdot}7203 \pm 0{\cdot}0006. \qquad (\text{IX. 35})$$

For each isotope the discrepancy is about three times the sum of the stated uncertainties. A similar discrepancy exists for In^{113} and In^{115} (HAR 42; LOG 51).

Initially these discrepancies were puzzling since both measurements presumably depended on the interaction of the nuclear magnetic moment with the external magnetic field and did not involve the uncertainties associated with the calculation of μ from $\Delta\nu$. Foley (FOL 50, SCH 54, ZZA 32), however, has discussed the effect of the partial decoupling by the applied magnetic field of the L and S vectors in the $^2P_{\frac{3}{2}}$ state on the nuclear g-value obtained from observational data under the assumption that decoupling does not occur. He concluded that the term $g_I \mu_0 H_0/h$ in Eq. (IX. 3) is modified by a small perturbation term which is, itself, proportional to the applied magnetic field. For the cases of gallium and indium the apparent g-value determined from the hyperfine structure of atoms is thus greater than the g-value obtained in the nuclear resonance experiments. Foley (FOL 50, SCH 54, ZZA 32) has indicated a satisfactory agreement between the observed and calculated values for the discrepancy, especially in view of uncertainties in the theoretical calculations and a rather large experimental error.

A partial confirmation of Foley's theory has been obtained by Logan and Kusch (LOG 51). Foley's effect cannot occur in an atom in a $^2S_{\frac{1}{2}}$ state. Therefore, Logan and Kusch (LOG 51) have measured the nuclear magnetic moment of Na^{23} by the above method and have found excellent agreement with the measurements by nuclear resonance absorption, in agreement with the predictions of Foley's theory (FOL 50). A similar agreement exists for Cs (KUS 49a).

IX. 8. Quadrupole interactions

In almost all of the experiments so far described in this chapter, the atoms have been in atomic states of $J = \frac{1}{2}$. For all atoms in such states nuclear electrical quadrupole moments and other higher nuclear multipole moments cannot be observed for the reasons discussed in Chapter III. However, for larger J's the observation of such multipole moments is not precluded.

From Eqs. (III. 43) and (III. 88a) the interaction Hamiltonian may be written as

$$\mathscr{H} = g_J \mu_0 \mathbf{J} \cdot \mathbf{H}_0 + g_I \mu_0 \mathbf{I} \cdot \mathbf{H}_0 + ha \mathbf{I} \cdot \mathbf{J} + $$
$$+ hb \frac{\frac{3}{2}\mathbf{I} \cdot \mathbf{J}(2\mathbf{I} \cdot \mathbf{J}+1) - I(I+1)J(J+1)}{2I(2I-1)J(2J-1)}, \quad \text{(IX. 36)}$$

where b is given by Eq. (IX. 38) below. Although the above expression is in the form ordinarily used by Zacharias, Lew, and their associates (DAV 49b, LEW 49), a slightly different form corresponding to Eq. (III. 122) was used in the early papers of Hamilton, Kusch, and their associates (HAM 39, BEC 48; MAN 50). The latters' b (here called b_1) is related to the above quantity b by

$$b_1 = 3b/[8I(2I-1)J(2J-1)]; \quad \text{(IX. 37)}$$

sometimes the notation $A_2 = \frac{1}{4}b$ is used instead of b (SCH 54).

With such a Hamiltonian the variations of the energy levels with magnetic field for any choice of the parameters g_I, g_J, a, and b may be calculated as in Chapter III. The variation of the levels of Ga^{69} in the $^2P_{\frac{3}{2}}$ state for the choice of parameters that agrees most closely with experiment is shown in Fig. III. 5. It should be noted, as discussed in detail in Section III. 3.2, that magnetic perturbations from levels of different J can give rise to apparent electric quadrupole effects.

In experiments on nuclear quadrupole moment interactions in atoms, a number of different transition frequencies are ordinarily measured. The parameters g_I, g_J, a, and b are then selected by a process of trial and error to provide the best agreement between theory and experiment. The parameter a is related to the nuclear magnetic moment and the atomic properties by the equations of Chapter III. The parameter b is related to the nuclear quadrupole moment q by the following relation which follows immediately from Eqs. (III. 43) and (III. 41a) and which was first given by Casimir (CAS 36)

$$hb = e^2 q_J Q, \quad \text{(IX. 38)}$$

where q_J is defined in Eqs. (III. 41) and (III. 41a). Q is the nuclear electric quadrupole moment as defined in Chapter III. Means for evaluating q_J will be discussed in detail in Chapter XI.

Becker and Kusch (BEC 48) and Daly, Holloway, and Schwartz

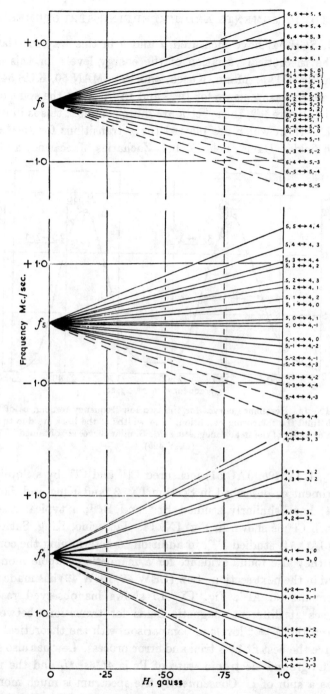

Fig. IX. 13. The theoretical field dependence in the Zeeman region of the frequencies of the π lines ($\Delta F = \pm 1$, $\Delta m_F = \pm 1$) in the spectrum of the $^2P_{\frac{3}{2}}$ state of In^{115} (MAN 50).

(DAL 54, SCH 54) have measured a and b in this way for Ga^{69} and Ga^{71}. The theoretical variation of the energy levels for this case is shown in Fig. III. 5. Mann, Kusch, and Eck (MAN 50, KUS 54) have made similar measurements for In^{113} and In^{115}. Since the spins of both In^{113} and In^{115} are 9/2, the spectra are quite complex; this is illustrated in Fig. IX. 13 in which are plotted the π transitions for In^{115} at low magnetic fields. Davis, Feld, Zabel, Zacharias, Jaccarino, and King

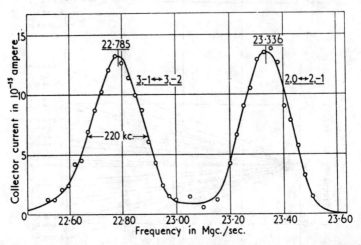

Fig. IX. 14. Resonance curves for the two low-frequency resonances of Cl^{35} which fulfil the refocusing condition. The widths of the lines are due to the inhomogeneity of the steady magnetic field. Similar curves are obtained for Cl^{37} (DAV 49b).

(DAV 48, DAV 49b, JAC 51) measured Cl^{35} and Cl^{37} by a flop-in type of experiment as described in Section IX. 3, and King and Jaccarino (KIN 54) have similarly studied Br^{79} and Br^{81}; a typical resonance curve with Cl^{35} is shown in Fig. IX. 14. Jaccarino, King, Satten, and Stroke (JAC 54) studied I^{127}; in addition to determining the constants a and b they also found evidence for a magnetic octupole moment as discussed in the next section. Lew (LEW 48, LEW 49) has made similar measurements on Al^{27}; Fig. IX. 15 shows his observed resonance frequencies at different magnetic fields for transitions between the $F = 3$ and the $F = 2$ levels in comparison with the theoretical curves which give the best fit by a trial and error process. Lew has also studied Pr^{141}. The ground electronic state of Pr is $4f^36s^2\ {}^4I_{\frac{9}{2}}$ and the nucleus Pr^{141} has a spin of $\frac{5}{2}$. Consequently the spectrum is much more complex, as is illustrated in Fig. IX. 16 in which only a few of the many possible lines are completed.

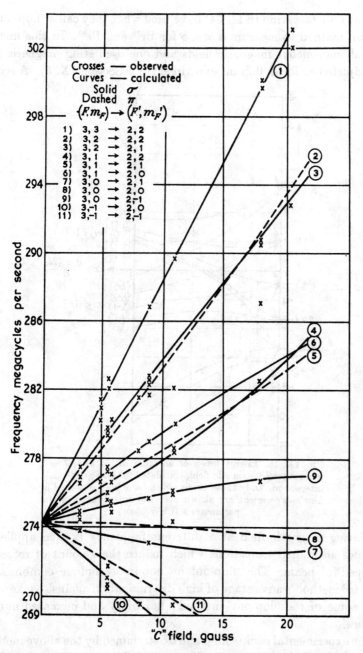

FIG. IX. 15. Observed transitions between $F = 3$ levels and $F = 2$ levels of the $^2P_{\frac{3}{2}}$ state of Al plotted as a function of the external magnetic field (LEW 49).

King and Jaccarino (KIN 54) have used what they call a 'flop-out on flop-in' method to measure a and b for Br^{79} and Br^{81}. In this method the inhomogeneous magnetic fields and one oscillating magnetic field are adjusted as for a 'flop-in' experiment (see Section IX. 3). A second

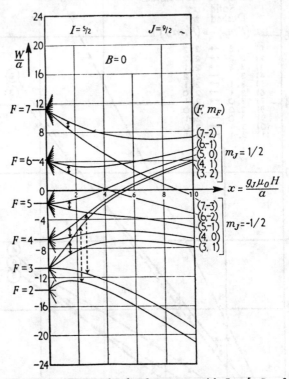

FIG. IX. 16. Energy levels of an atom with $I = \frac{5}{2}$, $J = \frac{9}{2}$ in an external magnetic field, calculated for zero quadrupole interaction. Only those levels which have been observed in Lew's experiment are shown over the entire range of the parameter x (LEW 53a).

oscillating magnetic field at a different frequency is then applied to produce additional transitions which reduce the amount of refocused 'flopped-in' beam. The 'flop-out on flop-in' technique combines the 'flop-in' method's advantage of high discrimination against background with some of the 'flop-out' method's flexibility and increased number of lines.

The experimental results for a and b determined by the above methods are given in Table XI. III. In Chapter XI there is a discussion of the means for determining the nuclear quadrupole moments Q from the values of b. Tables of Q determined in this way are given in Chapter VI.

IX. 9. Magnetic octupole interactions

Jaccarino, King, Satten, Stroke, and Zacharias (JAC 54) have recently detected a nuclear magnetic octupole interaction with iodine in an atomic-beam magnetic-resonance experiment. An octupole moment interaction in the same atom was earlier reported by Tolansky (TOL 39) in optical hyperfine structure experiments. However, Casimir and Karreman (CAS 42) showed that an octupole interaction anywhere near as large as that reported by Tolansky was very unlikely; the experiment of Jaccarino and associates indeed confirmed Casimir and Karreman's theoretical analysis since their observed octupole interaction was much smaller than that reported by Tolansky. Nevertheless, they did find a small octupole moment interaction.

The theory of nuclear magnetic octupole interactions is included implicitly in Chapter III above as can be seen by the combination of Eqs. (III. 15), (III. 57), and (III. 61). A detailed and explicit discussion of the theory of nuclear magnetic octupole interactions has been given by Casimir and Karreman (CAS 42), by Jaccarino, et al. (JAC 54), and by Schwartz (SCH 54). They show that the octupole interaction can be written in an F, m_F representation as

$$W_{M3} = hc\,\frac{5}{4}\,\frac{C^3+4C^2+\frac{4}{5}C[-3I(I+1)J(J+1)+ \\ +I(I+1)+J(J+1)+3]-4I(I+1)J(J+1)}{I(I-1)(2I-1)J(J-1)(2J-1)}$$

(IX. 39)

where C is defined by Eq. (III. 55) and where c is the nuclear magnetic octupole constant. The quantity c can be expressed as

$$hc = \Omega \int_{r_e} r_e^{-4} P_{3e}(\nabla \cdot \mathbf{m}_e)_{JJ}\,d\tau_e = \Omega \int \frac{5z_e^3 - 3z_e r_e^2}{2r_e^7}(\nabla \cdot \mathbf{m}_e)_{JJ}\,d\tau_e,$$

(IX. 40)

where P_{3e} is the Legendre polynomial of degree 3 in the angular coordinates of the electron charge density and $(\nabla \cdot \mathbf{m}_e)_{JJ}$ is the divergence of the electron magnetization in the state $m_J = J$. The quantity Ω is defined by

$$\Omega = \int_{\tau_n} r_n^3 P_3(\nabla \cdot \mathbf{m}_n)_{II}\,d\tau_n = \int_{\tau_n} \tfrac{1}{2}(5z_n^3 - 3z_n r_n^2)(\nabla \cdot \mathbf{m}_n)_{II}\,d\tau_n$$

(IX. 41)

and is called the octupole moment. The symbol $A_3 = c$ is sometimes used instead of c in the octupole expression (SCH 54).

Jaccarino, Zacharias, and their associates (JAC 54) have therefore analysed their experimental results on atomic I^{127} with a Hamiltonian of the form of Eq. (IX. 36) plus Eq. (IX. 39). In this way they obtained values for a, b, and c. However, as discussed in detail by Schwartz (SCH 54), second order perturbations from nuclear dipole and quadrupole interactions also contribute to c. Schwartz has calculated these second order perturbations both from the $^2P_{\frac{1}{2}}$ state and from configuration interactions. When these corrections are made in c so that the remaining value of c is presumably due exclusively to a nuclear octupole interaction, the results (SCH 54) are

$a = 827 \cdot 265 \pm 0 \cdot 003$ Mc/sec.

$b = 1{,}146 \cdot 356 \pm 0 \cdot 010$ Mc/sec.

$c = 0 \cdot 00287 \pm 0 \cdot 00037 - 0 \cdot 00053$ Mc/sec $= 0 \cdot 00234 \pm 0 \cdot 00037$ Mc/sec.

(IX. 42)

The first term of the first expression for c is the value uncorrected for perturbations and the last term in the first expression is the theoretical correction. These values are also listed in Table XI. III. The value of the octupole moment Ω corresponding to the above value of c is (SCH 54)

$$\Omega_{127} = (0 \cdot 17 \pm 0 \cdot 03) \times 10^{-24} \, \mu_N \text{ cm}^2. \qquad (IX. 43)$$

Kusch and Eck (KUS 54) have found evidence for a nuclear magnetic octupole interaction with In^{115} while Daly and Holloway (DAL 54) have done likewise with Ga^{69} and Ga^{71}. Their corrected values for a, b, and c are tabulated in Table XI. III. The corresponding values of the nuclear magnetic octupole moments are

$$\Omega_{115} = (0 \cdot 31 \pm 0 \cdot 01) \times 10^{-24} \, \mu_N \text{ cm}^2$$

$$\Omega_{69} = (0 \cdot 107 \pm 0 \cdot 004) \times 10^{-24} \, \mu_N \text{ cm}^2$$

$$\Omega_{71} = (0 \cdot 146 \pm 0 \cdot 004) \times 10^{-24} \, \mu_N \text{ cm}^2.$$

The magnitudes of the experimental magnetic octupole interactions in Eq. (IX. 42) and Table XI. III are qualitatively very reasonable as can be seen from the following discussion. The order of magnitude of the magnetic dipole interaction is $W_{M1} \sim \mu_0 \mu_N \langle 1/r^3 \rangle$. On the other hand the order of magnitude of the magnetic octupole interaction is $W_{M3} \sim \mu_0 \mu_N R^2 \langle 1/r^5 \rangle$, where R is the nuclear radius. Hence, one would expect
$$W_{M3}/W_{M1} \sim R^2 \langle 1/r^5 \rangle / \langle 1/r^3 \rangle \sim R^2/a^2,$$

where a is the effective radius for the electron which provides the interaction. At first sight one might take for a the Bohr radius a_0, but on further consideration it is apparent that the octupole interaction occurs chiefly when the electron is close to the nucleus in which region its orbit is only slightly affected by the shielding of the other electrons;

in other words a should be taken to have its approximately unshielded value $a = a_0/Z$, so
$$W_{M3}/W_{M1} \sim Z^2 R^2/a_0^2 \sim 10^{-5}$$
for large Z.

Schwartz (SCH 54) has generalized the calculations of the Schmidt lines derived in Section VI. 10.6 and has obtained corresponding expressions for any nuclear multipole moment. In particular he has obtained theoretical values for nuclear octupole moments as functions of nuclear spin I on the assumption that the octupole moment is due to a single odd nucleon. Just as in the case of the Schmidt lines two predictions are made for each value of I depending on whether $I = L+\tfrac{1}{2}$ or $I = L-\tfrac{1}{2}$. Schwartz (SCH 54) finds that the experimental nuclear octupole moments fall between these two theoretical curves in much the same way that the nuclear magnetic dipole moments fall between the Schmidt lines.

IX. 10. Anomalous hyperfine structure and magnetic-moment ratios for isotopes

In Chapter III it was assumed that the entire hyperfine structure separation in atoms of $J = \tfrac{1}{2}$ was proportional to the nuclear magnetic moment as in Eq. (III. 82). In such a case if the ratio of the magnetic moments of two isotopes of the same atom were known, the ratio of the hyperfine structure separations should be directly calculable from Eq. (IX. 11a), or from

$$\left(\frac{\Delta\nu_1}{\Delta\nu_2}\right)_{\text{calc}} = \frac{\mu_{I1} I_2 (2I_1+1)}{\mu_{I2} I_1 (2I_2+1)}. \qquad \text{(IX. 44)}$$

The above equation is based on the assumption that the nuclei are sufficiently heavy for the electron reduced mass correction to be negligible. If this is not so, the right-hand side should be multiplied by the reduced mass correction factor. For s electrons as in Li it is presumed in analogy to Eq. (IX. 29) that this correction factor is $(m_{r1}/m_{r2})^3$, where m_{r1} is the reduced mass of the $2s$ electron in Li atom 1, for example.

However, Nafe, Nelson, and Rabi (NAF 48) and Prodell and Kusch (PRO 50, SMA 51) have shown that there is a discrepancy between the experimental and theoretical ratios of the $\Delta\nu$'s of H and D of 0·017 per cent. as already discussed in detail in Section IX. 6. Likewise, Bitter (BIT 49a) first pointed out in the case of Rb^{85} and Rb^{87} that his measurements of the ratios of the μ's in a magnetic resonance absorption experiment were inconsistent with the ratios of the $\Delta\nu$'s for the same isotopes as measured earlier by Millman and Kusch (MIL 40), the discrepancy being $\frac{1}{3}$ per cent. A convenient and conventional means of expressing the discrepancy between two isotopes of mass numbers α and β is as

$$^\alpha\Delta^\beta = \frac{(\Delta\nu^\alpha/\Delta\nu^\beta)_{\text{obs}} - (\Delta\nu^\alpha/\Delta\nu^\beta)_{\text{calc}}}{(\Delta\nu^\alpha/\Delta\nu^\beta)_{\text{calc}}}. \quad\text{(IX. 45)}$$

With the later and more accurate measurements of Yasaitis and Smaller (OCH 52) on the magnetic moment ratio, and of Ochs and Kusch (OCH 52) on the hyperfine structure $\Delta\nu$'s, the discrepancy for rubidium is

$$^{87}\Delta^{85} = 0 \cdot 003501 \pm 0 \cdot 000006. \quad\text{(IX. 46)}$$

Kusch and Mann (KUS 49c) have studied the Li^7 and Li^6 ratios and

TABLE IX. III

Magnetic Hyperfine Structure Anomaly

The quantity $^\alpha\Delta^\beta$ listed below is defined by

$$^\alpha\Delta^\beta = [(\Delta\nu^\alpha/\Delta\nu^\beta)_{\text{obs}} - (\Delta\nu^\alpha/\Delta\nu^\beta)_{\text{calc}}]/(\Delta\nu^\alpha/\Delta\nu^\beta)_{\text{calc}}$$

See also Appendix G.

Atom	α	β	$^\alpha\Delta^\beta \times 10^4$	Reference
H	1	2	$1 \cdot 703 \pm 0 \cdot 005$	NAF 48, PRO 50, ZZA 28
H	1	3	$0 \cdot 04 \pm 0 \cdot 14$	NEL 49a
Li	7	6	$1 \cdot 2 \pm 0 \cdot 3$	KUS 49c
K	39	41	$-22 \cdot 6 \pm 1 \cdot 0$	OCH 50
K	39	40	$46 \cdot 6 \pm 1 \cdot 9$	EIS 52
Rb	87	85	$35 \cdot 01 \pm 0 \cdot 06$	BIT 49a, MIL 40, OCH 52
Tl	203	205	$1 \cdot 2 \pm 0 \cdot 8$	BER 50, BER 52

have also found a small anomaly in the ratios. When they include the reduced mass correction described under Eq. (IX. 44) in $(\Delta\nu)_{\text{calc}}$ they find for lithium $^7\Delta^6 = 0 \cdot 00012 \pm 0 \cdot 00003$. Berman, Kusch, and Mann (BER 50, BER 52) with Tl^{205} and Tl^{203} in a $^2P_{\frac{1}{2}}$ state have found $^{203}\Delta^{205} = 0 \cdot 00012 \pm 0 \cdot 00008$. For potassium, Ochs, Logan, and Kusch (OCH 50) have measured $^{39}\Delta^{41}$ and Eisinger, Bederson, and Feld (EIS 52) have determined $^{39}\Delta^{40}$. These results are included in Table IX. III. At one time (DAV 49b) it was thought that there was a similar discrepancy between the ratios of the a's for Cl^{35} and Cl^{37} in $^2P_{\frac{3}{2}}$ states and the ratios of the nuclear magnetic moments. However, this discrepancy disappeared in a more careful study by Jaccarino and King (JAC 51).

Casimir (CAS 36), Kopfermann (KOP 40), and Bitter (BIT 49a) have pointed out that if the finite size of the nucleus is taken into account,

$\Delta \nu$ will depend on the spatial distribution of the nuclear magnetism. Bohr and Weisskopf (BOH 50, BOH 51, BOH 53, SCH 54) have shown how the measurement of $^{\alpha}\Delta^{\beta}$ leads to information about the structure of the nuclei involved. Their theory accounts for the hyperfine structure anomalies between Rb^{85} and Rb^{87} and between K^{39} and K^{41}. Their theories have been extended to K^{40} by Eisinger, Bederson, and Feld (EIS 52). The theories of Bohr (BOH 48, BOH 50) and Low and Salpeter (LOW 51) for proton-deuteron discrepancy in atomic hydrogen have already been discussed in detail in Section IX. 6.

It is assumed in the Bohr–Weisskopf theory that the hyperfine structure anomaly arises from the finite size of the nuclei and from the difference for the two isotopes involved between the distributions of the nucleon spins and currents which give rise to the nuclear moments. The magnetic hyperfine splitting, resulting from the interaction of the nuclear magnetic moment with an s-electron, is given by the Fermi–Segré formula of Eqs. (III. 82) and (III. 88). For a nucleus of finite size, the factor $\mu_I |\psi_{n0}(0)|^2$ must be averaged over the nuclear volume, and the resulting $\Delta \nu$ depends on the nuclear size, on the nuclear charge, and on the distribution of the magnetic moment in the nucleus. If $\Delta \nu_{\mathrm{pt}}$ is the $\Delta \nu$ that would occur for the same magnetic moment in a point nucleus, the quantity ϵ may be defined by

$$\Delta \nu_{\mathrm{obs}}/\Delta \nu_{\mathrm{pt}} = \langle \mu_I(r)|\psi(r)|^2 \rangle / \mu_I |\psi(0)|^2 = 1+\epsilon. \quad \text{(IX. 47)}$$

The quantity ϵ is related to the previously defined quantity $^{\alpha}\Delta^{\beta}$ by

$$^{\alpha}\Delta^{\beta} \cong \epsilon^{\alpha} - \epsilon^{\beta}. \quad \text{(IX. 48)}$$

In particular, a nuclear moment caused by the intrinsic moments of one or more nucleons will cause a greater anomaly than a moment caused by orbital motion, since the moment arising from an orbital current is more strongly concentrated at the centre of the nucleus. Bohr and Weisskopf (BOH 50, BOH 51) have shown approximately that

$$\epsilon = -\langle \kappa_S \rangle \alpha_S - \langle \kappa_L \rangle \alpha_L, \quad \text{(IX. 49)}$$

where α_S and α_L represent the fractions of the nuclear moment due to spin moment and orbital moment respectively. By procedures analogous to those in Eq. (III. 91) one can easily show that

$$\alpha_S = \frac{g_S}{g_I}\frac{g_I - g_L}{g_S - g_L}, \qquad \alpha_L = 1 - \alpha_S = \frac{g_L}{g_I}\frac{g_S - g_I}{g_S - g_L}. \quad \text{(IX. 50)}$$

By a direct averaging process in accordance with the above description,

Bohr and Weisskopf (BOH 50, BOH 51) have shown that the quantities $\langle \kappa_S \rangle$ and $\langle \kappa_L \rangle$ which measure the decrease in hyperfine interaction of the spin and orbital moments respectively as a consequence of the deviation of the nuclear magnetization from a point dipole may be expressed by

$$\langle \kappa_S \rangle = (1+0.38\zeta)b(Z, R_0)\langle R^2/R_0^2 \rangle,$$
$$\langle \kappa_L \rangle = 0.62 b(Z, R_0)\langle R^2/R_0^2 \rangle, \qquad (IX.\ 51)$$

where $b(Z, R_0)$ is a parameter dependent only on Z and R_0 and tabulated by Bohr and Weisskopf (Table I of BOH 50) for $S_{\frac{1}{2}}$ and $P_{\frac{1}{2}}$ atomic states, ζ is a constant defined and evaluated by Bohr (BOH 51), R_0 is the radius of the nucleus, often taken as equal to $1.5 A^{\frac{1}{3}} \times 10^{-13}$ cm, and $\langle R^2 \rangle$ is the mean square radius of the nucleon or nucleons producing the magnetization (for a uniform distribution over a sphere $\langle R^2/R_0^2 \rangle = \frac{3}{5}$ whereas for a $d_{\frac{3}{2}}$ proton $\langle R^2/R_0^2 \rangle \cong 0.66$, etc.).

Eisinger, Bederson, and Feld (EIS 52) have applied this theory to account for the Δ's of the three K isotopes and the two Rb isotopes that have been measured. Various models of the nucleus including a strict single particle model, a model of less extreme independent particle character, and an asymmetric model due to Bohr (BOH 51, EIS 52) in which an odd nucleon in the nucleus is assumed to be coupled to a rotating core which deviates from spherical symmetry (BOH 51, EIS 52). All of these models were found to agree fairly well with the experimental values for Δ, though the agreement of the Bohr asymmetric model was the closest.

Foley (FOL 54a, ZZA 27) has considered theoretically various additional corrections to the hyperfine structure interaction, especially for light nuclei such as He^3. These include allowances for two protons attracting the electrons in He^3, for failure of adiabatic approximation, for exchange currents, for effects of two electrons, etc.

In atomic states of $L > 0$ the electron wave function vanishes at the position of the nucleus, so the finite nuclear size and the distribution of spin and moment in the nucleus have negligible effects on the hyperfine structure. However, owing to relativistic effects, the wave function of an electron in a $^2P_{\frac{1}{2}}$ state has, for heavy nuclei, an appreciable $^2S_{\frac{1}{2}}$ component and can therefore give rise to a hyperfine structure anomaly (the above quantity b in a $^2P_{\frac{1}{2}}$ state is of the order of $Z^2\alpha^2$ of that for a $^2S_{\frac{1}{2}}$ state). For an electron of total angular momentum higher than $\frac{1}{2}\hbar$ the influence on the hyperfine structure of the finite size of the nucleus is negligible. For this reason it was puzzling when a hyperfine structure anomaly was originally reported in the $^2P_{\frac{3}{2}}$ state of Cl, and gratifying when the suspected hyperfine structure anomaly was later shown to be non-existent by Jaccarino (JAC 51, ZZA 37) as mentioned above.

With Tl^{203} and Tl^{205}, the Bohr–Weisskopf (BOH 50, BOH 51) theory predicts the low value of $^{203}\Delta^{205} = 1\cdot4 \times 10^{-5}$. The smallness of the predicted effect results from the great similarity of the magnetic moments and of the theoretical magnetic distribution in the two isotopic nuclei and from the fact that the effect in general should be less for $P_{\frac{1}{2}}$ states than for $S_{\frac{1}{2}}$ states. The above predicted value is considerably less than the experimental value for $^{203}\Delta^{205}$ of $(12\pm8)\times 10^{-5}$. Berman (BER 52) has pointed out that an effect first suggested by Crawford and Schawlow (CRA 49) can account for the discrepancy by providing a much larger contribution to Δ than the Bohr–Weisskopf contribution which is so small in this case. Rosenthal and Breit (ROS 32) calculated that in the deduction of nuclear magnetic moments from hyperfine structure splittings there should be a correction factor due to the departure of the electronic wave functions from those for a pure Coulomb field at positions inside the nucleus where there is no longer a pure Coulomb field from a point nuclear source. Crawford and Schawlow (CRA 49) have pointed out that for two isotopes of the same atom a differential correction exists. Crawford and Schawlow have shown that this differential correction should lead to values for $^{203}\Delta^{205}$ of approximately $9\cdot3\times 10^{-5}$ and $10\cdot4\times 10^{-5}$ with nuclei of a uniform electrical charge and with a spherical shell charge respectively. When the above Bohr–Weisskopf and Crawford–Schawlow effects are combined the theoretical value for $^{203}\Delta^{205}$ becomes $(11\pm1)\times 10^{-5}$ in excellent agreement with the experimental value of $(12\pm8)\times 10^{-5}$.

IX. 11. Frequency standards and atomic clocks

The field insensitive hyperfine structure transitions in an atomic-beam magnetic-resonance experiment can be used as frequency and time standards. For this purpose they possess the advantages over piezo-electric crystal resonators of being clearly constant in time, of temperature independence, of obvious reproducibility, and of large Q where the Q of the resonator is given by $\nu_0/\Delta\nu$, ν_0 being the resonance frequency and $\Delta\nu$ the full width of the resonance at half intensity. The atomic-beam standard also possesses the advantages over a microwave absorption experiment, such as the NH_3 clock, of being free of Doppler broadening and of broadening due to collisions of molecules either with each other or with the walls of the absorption cell; Dicke

(DIC 54) has, however, succeeded in overcoming Doppler broadening in some microwave spectroscopy experiments by a 'collision narrowing' process. Typical feasible Q's for the three methods are 10^6 for special piezoelectric crystal resonators, 3×10^5 for an NH_3 microwave absorption clock, 10^6 for an O_2 microwave absorption clock, 3×10^7 for a Cs atomic-beam clock at a ν_0 of 9,192·632 Mc, and 5×10^7 for a Tl atomic-beam clock at a ν_0 of 21,368 Mc (LYO 52).

Sherwood, Lyons, McCracken, and Kusch (SHE 52, LYO 52), Zacharias

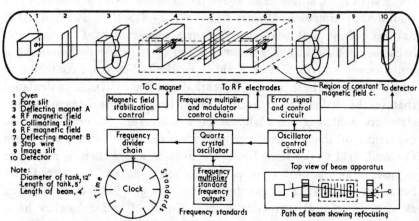

FIG. IX. 17. Schematic diagram of proposed atomic-beam clock (LYO 52).

and Yates (ZZA 38), and Essen and Parry (ZZA 39) have completed many of the preliminary experiments required for the development of such a clock. They have used Ramsey's separated oscillatory field method (RAM 50c, RAM 51d) because of its greater precision and its greater freedom from Doppler broadening. They have used the (F, m) transition $(4, 0) \leftrightarrow (3, 0)$ of Cs. For this transition the resonance frequency is

$$\nu = \Delta\nu + (g_J - g_I)^2 \mu_0^2 H_0^2 / 2h^2 \Delta\nu$$
$$= (9,192\cdot 63183 + 0\cdot 000427 H^2) \times 10^6 \text{ sec}^{-1}. \qquad \text{(IX. 52)}$$

If a field of 0·1 gauss is used and held constant only to 10 per cent., the frequency will be constant to one part in 10^{10}.

A schematic view of the atomic-beam clock of Sherwood, Lyons, McCracken, and Kusch (SHE 52) is shown in Fig. IX. 17. With a 50 cm separation of the separated oscillatory fields in Ramsey's method Q's of 3×10^7 were achieved with Cs. The precision of an atomic clock can of course be much greater than that directly indicated by the Q since the position of a line can be measured to a small fraction of its width, with the fraction depending in part on the signal-to-noise ratio in the apparatus. The results with Cs indicated a possible accuracy of one part in 10^{10} (LYO 52, ZZA 38).

The recent proposals of Zacharias (ZAC 53) discussed in Section V. 5 and in Chapter XIV for the use of very slow molecules to achieve much greater precision in atomic-beam experiments should be well adapted to provide an even more accurate frequency and time standard. With this method, an accuracy of approximately one part in 10^{13} should ultimately be attainable (ZAC 53). It is of interest to note that this accuracy would be sufficient to detect the relativistic gravitational red shift on the surface of the earth. As shown in any of the standard works on relativity theory (EDD 30, BER 42), the gravitational frequency shift $\delta\nu$ at a distance R from the centre a spherical body of mass M of radius less than R is

$$\delta\nu = (G/c^2)(M/R)\nu,$$

where G is the gravitational constant $6{\cdot}67 \times 10^{-8}$ dyne cm² gm⁻², c is the velocity of light, and ν is the frequency observed. On the surface of the earth $\delta\nu$ is $10^{-9}\nu$. The variation of the shift with a 6-mile difference in altitude on the earth is $1{\cdot}5 \times 10^{-12}\nu$ which should be just detectable with an accuracy of one part in 10^{13}.

A molecular beam clock with direct detection of molecular oscillations has recently been employed by Townes (ZZA 36) as described on page 309.

IX. 12. Atomic-beam resonance method for excited states

Most of the experiments described above are on the ground states of the atoms concerned since the lifetime of an excited state is ordinarily too short to persist as the atom traverses the length of the beam. In some cases, however, where metastable atomic states exist whose lifetimes are greater than the time for the atom to traverse the apparatus, the resonance spectrum of the metastable states have been studied. Such experiments are illustrated by the cases of H, He, Ga, Cl, and In discussed above and in Tables IX. I and XI. III.

Rabi (RAB 52) has proposed experiments on atomic excited states which are not metastable but for which the excited states are continuously replenished by absorption of radiation at the appropriate frequency for excitation, which is quite different from the frequency of the resonance transitions being studied. In these experiments there is a necessary loss in precision which arises from the fact that excited states have lifetimes of the order of 10^{-8} sec, which brings with it an inevitable line width of approximately 10 megacycles. It is difficult therefore to measure hyperfine-structure intervals to a higher accuracy than a fraction of a megacycle.

In essence the method is similar to other molecular-beam magnetic-resonance experiments. However, the C field region in addition to being illuminated by the radiofrequency field is also illuminated with optical radiation of the principal series of the atoms under investigation. This resonance radiation brings a substantial fraction of the atoms into an upper level in which they have a lifetime of about 10^{-8} sec. When they return to the lower state by radiation there is a preference for the original hyperfine structure state from which they started, as can be shown by a simple calculation (BIT 49b). The application of a radiofrequency field of sufficient amplitude and of appropriate resonance frequency causes transitions among the upper hyperfine structure or magnetic levels. Since the atoms are essentially tagged by the deflexion which they have suffered in the first deflecting field these hyperfine transitions in the upper state cause a change in the refocused beam intensity by the second deflecting field even though the atoms are then in the lower state.

To reduce the background the deflecting fields are arranged to produce deflexions that are equal and of the same sign at the detector. In this way only very few atoms reach the detector. However, when the beam is illuminated with the optical radiation, some of the atoms which have absorbed this radiation come down to the ground state with moments pointing in a direction opposite to the original and are therefore focused on to the detector by the second deflecting field. When the radiofrequency field is resonant in the excited state, there is a further change in the refocused beam intensity. This molecular-beam method is related to the optical methods of Brossel, Kastler, Bitter, and others (BIT 49b) which use the depolarization of resonance radiation or the so-called 'double resonance' method (BIT 49b, ZZA 2, ZZA 9, ZZA 3).

With excitation by the Na D lines from a sodium vapour lamp, Perl and Rabi (BRO 54, ZZA 23) have studied the $^2P_{\frac{3}{2}}$ state of Na^{23} successfully. The observed radiofrequency resonances were of course quite broad owing to the short lifetime of the state that was studied. The results were in general agreement with those of Sagalyn (SAG 54) on Na by the optical 'double resonance' method. Senitzky, Perl, and Rabi (ZZA 24) have recently studied the $5P_{\frac{3}{2}}$ atomic excited state of Rb^{85} in a similar fashion in order to measure the Rb^{85} nuclear quadrupole moment, as listed in Table VI. I.

X
ELECTRIC DEFLEXION AND RESONANCE EXPERIMENTS

X. 1. Introduction

MOLECULES consist of positive and negative particles and are symmetrical to widely varying degrees. The centre of charge of the positive nuclei may or may nor coincide with the centre of charge of all the negative electrons, i.e. the molecule may either have a permanent electric dipole moment or the dipole moment may vanish. A molecule like O_2 for which the electric dipole moment vanishes by symmetry is called non-polar and one like HCl which possesses a non-vanishing dipole moment is called a polar molecule. The dipole moment μ_ϵ is of the order of the electronic charge times a distance of 1 Å or of the order of $4\cdot 8 \times 10^{-18}$ esu. Consequently electric dipole units are measured conventionally in Debye units where 1 Debye unit $= 10^{-18}$ esu. Even when a molecule possesses no permanent electric dipole moment, its electronic structure is not completely rigid and it possesses a polarizability, so a dipole moment $\alpha_\epsilon E$ may be induced proportional to the strength of the electric field.

The designation of μ_ϵ as a permanent electric dipole moment is subject to an important qualification; it is the dipole moment that the molecule possesses relative to a coordinate system which rotates with the molecule. In the absence of an electric field, the average electric dipole moment in the laboratory coordinate system is zero. Physically, this can be seen in the above example of HCl from the fact that the electric dipole moment is along the internuclear axis, but this axis is necessarily perpendicular to the molecular angular momentum J about which the molecule rotates and hence averages its electric dipole moment to zero. Formally, this can be seen from the general theorem of Chapter III that a non-degenerate system of a definite angular momentum J can have no odd electrical multipole moment and hence in particular no electric dipole moment. An apparent exception to this theorem can occur for polyatomic molecules: a symmetric top molecule, for example, in a state of rotation such that the resultant angular momentum J is along the molecular axis of symmetry, could clearly have a permanent average value of electric dipole moment. However, even this case is consistent with the general theorem since the system is degenerate, as the state with an equal and opposite rate of rotation about the internuclear axis clearly has the same energy. Indeed if the electric field is so weak that its interaction energy is small compared with the Λ type doubling energy that separates the above two almost degenerate rotational states, the average electric moment will be zero even in this case (VAN 32, p. 154).

Although the electric moment of a polar diatomic molecule averages to zero in the absence of an electric field it does not do so in the presence of such a field. Consider for example a rotating diatomic molecule with its angular momentum perpendicular to the electric field. The field tends to twist the dipole and give it a faster rotation when the dipole is oriented in the direction of the field and a slower orientation when it is pointed oppositely to the field. The dipole is consequently oriented away from the field more often than with it, so that, on the average, the dipole is directed oppositely to its static tendency. If such a classical motion is analysed it can easily be shown that the fractional difference between the time the dipole points in the two directions is proportional to $\mu_\epsilon E/\tfrac{1}{2}A\omega^2$, where A is the moment of inertia of the molecule and ω its rotational angular velocity; this result is indeed the physically reasonable one. Consequently the change in energy due to the field is proportional to $\mu_\epsilon E$ times the above fraction, or

$$W_E \propto (\mu_\epsilon E)^2/\tfrac{1}{2}A\omega^2. \quad (X.\ 1)$$

The change of energy in this case is an increase. On the other hand if the molecule rotates with its angular momentum either parallel or antiparallel to the electric field, the rotating dipole is twisted slightly in the direction of the field. The energy is thereby decreased; the decrease can also be shown to be proportional to Eq. (X. 1). In fact it can be shown quantum mechanically, that for $J \neq 0$ the energy change averaged over random orientations of a rotating molecule gives no net change in energy since the various positive and negative changes just cancel. Of course, for a gas in an electric field the collisions lead to a thermal relaxation process so that the different orientation states are not equally occupied but, in accordance with the Boltzmann distribution, the lower states are more densely populated.

The electric dipole moment, which the above discussion shows that most molecules possess in an electric field, can be used as the basis of molecular-beam experiments that are the electric analogue of the previous magnetic experiments. Thus beam molecules can be deflected by inhomogeneous electric fields and electric resonance experiments can be performed if oscillatory electric fields induce electric dipole transitions between different states of the molecule in a uniform electric field. Much of the analysis of the various possible experiments can be carried out in strict analogy to the corresponding magnetic experiment. However, the relevant energy states are those for a molecule in an electric field. The nature of these states will therefore be summarized in the next section.

X. 2. Molecular interactions in an electric field

X. 2.1. *Molecular Hamiltonian.* The theory of molecules in electric fields, i.e. of the molecular Stark effect, has been discussed by a number of authors (TOW 54, GOR 53, HER 50a, HER 45). The discussions related to microwave spectroscopy (TOW 54, GOR 53) are particularly relevant to the present problem. Molecular beam experiments are possible not only with diatomic but also with linear, symmetric top, and asymmetric molecules. However, the high-precision molecular-beam experiments that have been performed so far have all been on diatomic $^1\Sigma$ molecules. For this reason the subsequent discussion will be limited to such molecules. For a theoretical discussion of polyatomic molecules in general and for a more detailed discussion of even the diatomic case, the reader is referred to the books on microwave spectroscopy by Townes (TOW 54), by Gordy, Smith, and Trambarulo (GOR 53), and by Herzberg (HER 50a, HER 45).

The Hamiltonian for a vibrating and rotating diatomic molecule in an electric field can be written as

$$\mathcal{H} = \mathcal{H}_{VR} + \mathcal{H}_\epsilon + \mathcal{H}_{1J} + \mathcal{H}_{2J} + \mathcal{H}_{Q1} + \mathcal{H}_{Q2} + \mathcal{H}_S. \quad (\text{X. 2})$$

$\mathcal{H}_{VR}$ is the component of the Hamiltonian due to the vibration and rotation of the molecule. It may be written approximately as

$$\mathcal{H}_{VR} = \frac{p_1^2}{2M_1} + \frac{p_2^2}{2M_2} + U(R), \quad (\text{X. 3})$$

where p_1 is the momentum operator for the first nucleus of mass M_1 and $U(R)$ is the potential between the two nuclei. For rigid rotators Eq. (X. 3) reduces, except for a constant term, to

$$(\mathcal{H}_{VR})_{\text{rigid}} = (\hbar^2/2A)\mathbf{J}^2, \quad (\text{X. 4})$$

where $\mathbf{J}$ is the rotational angular momentum in units of $\hbar$ and where A is the moment of inertia of the molecule. Therefore,

$$A = \mu' R^2, \quad (\text{X. 5})$$

R being the internuclear spacing and μ' being the reduced mass of the molecule

$$\mu' = M_1 M_2/(M_1+M_2). \tag{X. 6}$$

$\mathscr{H}_\epsilon$ is the component of the Hamiltonian due to the interaction of the electric dipole moment with an external electric field E and is given by

$$\mathscr{H}_\epsilon = -\mathbf{\mu}_\epsilon \cdot \mathbf{E}. \tag{X. 7}$$

$\mathscr{H}_{1J}$ and $\mathscr{H}_{2J}$ are the Hamiltonian components for the spin-rotational interaction which by Eqs. (VIII. 17) and (VIII. 18a) can be written as

$$\mathscr{H}_{1J}+\mathscr{H}_{2J} = c'_1 \mathbf{I}_1 \cdot \mathbf{J} + c'_2 \mathbf{I}_2 \cdot \mathbf{J}. \tag{X. 8}$$

$\mathscr{H}_{Q1}$ is the component of the Hamiltonian for the quadrupole interaction of the first nucleus and $\mathscr{H}_{Q2}$ is that for the second. From Eqs. (VIII. 25) and (VIII. 26) the first of these, for example, is given as long as J is a good quantum number by

$$\mathscr{H}_{Q1} = -(eqQ_1) \frac{3(\mathbf{I}_1 \cdot \mathbf{J})^2 + \tfrac{3}{2}\mathbf{I}_1 \cdot \mathbf{J} - \mathbf{I}_1^2 \mathbf{J}^2}{2I_1(2I_1-1)(2J-1)(2J+3)}. \tag{X. 9}$$

For very large values of $\mu_\epsilon E$, so that there is a considerable mixing of J states, the quadrupole component of the Hamiltonian must be taken in the more complicated form of Eqs. (III. 26) and (III. 57). $\mathscr{H}_S$ is the spin-spin magnetic interaction component of the Hamiltonian which by Eqs. (VIII. 10) and (VIII. 11) can be written as

$$\mathscr{H}_S = \frac{2\mu_1\mu_2}{I_1 I_2} \langle 1/R^3 \rangle \frac{3(\mathbf{I}_1 \cdot \mathbf{J})(\mathbf{I}_2 \cdot \mathbf{J}) - \mathbf{I}_1 \cdot \mathbf{I}_2 \mathbf{J}^2}{(2J-1)(2J+3)}. \tag{X. 10}$$

Once the parameters and the potential $U(R)$ in the above equation have been given, the energy levels for any particular electric field can be obtained by the normal process of diagonalizing the Hamiltonian matrix as discussed in Chapter III. This process, however, in general is complicated when all terms are included and it is usually convenient to consider the effects of the various terms separately.

X. 2.2. *Energy of vibrating rotator.* The energy levels of a vibrating and rotating molecule when only the term $\mathscr{H}_{VR}$ is included have been discussed in a number of books (HER 50a, TOW 54, GOR 53). The potential $U(R)$ may be selected in the form of the Morse potential (MOR 29)

$$U(R) = D[1-e^{-a(R-R_e)}]^2. \tag{X. 11}$$

For such a potential, one can show (HER 50a, TOW 54, GOR 53, JEN

53) that the energy $W_{J,v}$ of rotational state J and vibrational state v may be written approximately as

$$W_{J,v}/hc = \omega_e(v+\tfrac{1}{2}) - \omega_e x_e(v+\tfrac{1}{2})^2 + \omega_e y_e(v+\tfrac{1}{2})^3 +$$
$$+ B_v J(J+1) - D_v J^2(J+1)^2 + ..., \qquad (X.\ 12)$$

where
$$B_v = B_e - \alpha_e(v+\tfrac{1}{2}) + \gamma_e(v+\tfrac{1}{2})^2 + ... \qquad (X.\ 13)$$
and
$$D_v = D_e + \beta_e(v+\tfrac{1}{2}) + ... \qquad (X.\ 14)$$

The first term in Eq. (X. 12) corresponds to simple harmonic vibration, the second and third terms provide the anharmonic correction, the fourth term gives the rotational kinetic energy of a vibrating rotator and the fifth term corresponds to centrifugal stretching.

Ordinarily the energy is merely taken in the above form and the coefficients are determined empirically. On the other hand, for a Morse potential, the coefficients can be related to the parameters of Eqs. (X. 3) and (X. 11) by (HER 50a, TOW 54)

$$B_e = h/8\pi^2 c A_e$$
$$A_e = [M_1 M_2/(M_1+M_2)]R_e^2 = \mu' R_e^2$$
$$\omega_e = \frac{a}{2\pi c}\sqrt{(2D/\mu')}$$
$$x_e = h\omega_e c/4D$$
$$D_e = 4B_e^3/\omega_e^2 = h^3/128\pi^6\mu'^3\omega_e^2 c^3 R_e^6$$
$$\alpha_e = \frac{6\sqrt{(\omega_e x_e B_e^3)}}{\omega_e} - \frac{6B_e^2}{\omega_e} = \frac{3h^2\omega_e}{16\pi^2\mu' R_e^2 D}\left(\frac{1}{aR_e} - \frac{1}{a^2 R_e^2}\right).$$
$$(X.\ 15)$$

From the above it can be seen that, inversely,

$$aR_e = \sqrt{(\omega_e x_e/B_e)} = 1 + \alpha_e \omega_e/6B_e^2$$
$$R_e = (h/8\pi^2 c\mu' B_e)^{\frac{1}{2}}$$
$$D = 2\pi^2 c^2 \omega_e^2 \mu'/a^2 = h\omega_e c/4x_e. \qquad (X.\ 16)$$

An alternative development is due to Dunham (DUN 32, HER 50a, TOW 54). In place of Eq. (X. 11), Dunham assumed a power series for the potential in the form

$$U(R) = hca_0\xi^2(1+a_1\xi+a_2\xi^2+...), \qquad (X.\ 17)$$

where
$$\xi = (R-R_e)/R_e. \qquad (X.\ 18)$$

Dunham then showed that for this potential the energy could be expressed by
$$W_{J,v}/hc = \sum_{l,j} Y_{lj}(v+\tfrac{1}{2})^l J^j(J+1)^j, \qquad (X.\ 19)$$

where l and j are positive integral summation indices. Dunham (DUN

32, TOW 54) has given an extensive table showing the relation between the first fifteen Y_{ij}'s and the coefficients a_0, a_1, etc., of Eq. (X. 17). If B_e/ω_e is small, as it usually is, the Y_{ij}'s of Eq. (X. 19) can be related to the coefficients of Eq. (X. 12) by the following approximate relations:

$$Y_{10} \sim \omega_e \qquad Y_{20} \sim -\omega_e x_e \qquad Y_{30} \sim \omega_e y_e$$
$$Y_{01} \sim B_e \qquad Y_{11} \sim -\alpha_e \qquad Y_{21} \sim \gamma_e$$
$$Y_{02} \sim -D_e \qquad Y_{12} \sim -\beta_e. \qquad (\text{X. 20})$$

In experiments on $^1\Sigma$ diatomic molecules, the results are expressed conventionally in an evaluation of the coefficients of either Eq. (X. 12) or Eq. (X. 19).

X. 2.3. *Stark effect.* The addition of an electric field may now be discussed. Initially consideration will be limited to the case of a rigid rotator for which all terms in Eq. (X. 2) beyond the first two are assumed to vanish. Then from Eqs. (X. 4) and (X. 7)

$$\mathcal{H} = (\hbar^2/2A)\mathbf{J}^2 - \boldsymbol{\mu}_\epsilon \cdot \mathbf{E}. \qquad (\text{X. 21})$$

As long as the electric interaction energy is small compared with the separation of the energy levels, the energies can be calculated by perturbation theory. Eqs. (III. 125) may then be used to yield the energy provided $\mathcal{H}'$ is replaced by $\mathcal{H}_\epsilon$. For diatomic $^1\Sigma$ molecules the first-order perturbation vanishes for the reasons discussed physically in Section X. 1. Hence the first non-vanishing term is the second-order one and it varies as the square of the electric field. The relevant matrix elements can be calculated from the rotator wave functions (DEB 29, CON 35, VAN 32) and the second-order perturbation becomes (VAN 32, TOW 54, SCH 40)

$$W_n^{(2)} = \frac{\mu_\epsilon^2 E^2}{(\hbar^2/2A)} \frac{J(J+1) - 3m_J^2}{2J(J+1)(2J-1)(2J+3)} \quad (J \neq 0),$$

$$W_n^{(2)} = \frac{\mu_\epsilon^2 E^2}{(\hbar^2/2A)} (-\tfrac{1}{6}) \quad (J = 0). \qquad (\text{X. 22})$$

It can be noted that this expression is of the quadratic form anticipated in Eq. (X. 1).

Higher-order perturbation terms for the rigid rotator have been calculated by Brouwer (BRO 30) and summarized by Hughes (HUG 47, HUG 49). W. E. Lamb has derived an implicit expression for W in the form of a continued fraction that is applicable even at very strong electric fields. This expression as given by Hughes (HUG 47, TOW 54) is

$$\frac{W}{(\hbar^2/2A)} = m_J(m_J+1) -$$

$$\frac{\mu_\epsilon^2 E^2/(\hbar^2/2A)^2 A'_{m_J,m_J}}{(m_J+1)(m_J+2) - \frac{W}{(\hbar^2/2A)} - \frac{\{\mu_\epsilon^2 E^2/(\hbar^2/2I)^2\} A'_{m_J+1,m_J}}{(m_J+2)(m_J+3) - \{W/(\hbar^2/2A)\} - \ldots}}$$

(X. 23)

where
$$A'_{x,y} = \frac{(x+1)^2 - y^2}{(2x+1)(2x+3)} \qquad \text{(X. 24)}$$

This equation, like the others discussed here, neglects the distortion polarizability of the molecule which is quite small. It also, of course, neglects the quadrupole and other similar terms that are omitted from Eq. (X. 21). Although J does not appear explicitly in Eq. (X. 23), different solutions of this implicit equation are obtainable and one of these corresponds to the desired J. It should be noted that for $E = 0$ one of the fractions in the continued fraction is indeterminate; for example, with $m_J = 1$ and $J = 2$ the first fraction, i.e. the second term on the right side of Eq. (X. 23) is indeterminate for $E = 0$. For small E it is best to rewrite the continued fraction as an expansion about the numerator of the fraction that becomes indeterminate for $E = 0$. The theoretical variation of the energy levels of a rigid rotating polar molecule as a function of electric field has been calculated by Hughes (HUG 47) to be as shown in Fig. X. 1.

X. 2.4. *Hyperfine structure interactions.* The calculations are more complicated when other perturbations are included as in Eq. (X. 2). When the other perturbations are small compared with the electrical one, Eq. (X. 22), Eq. (X. 23), or the relation of Brouwer (BRO 30) and of Lamb and Hughes (HUG 47, HUG 49) can still be applied to calculate approximately the unperturbed centre of gravity of the split energy levels (HUG 49) while the displacements of the split levels from the centre of gravity may be calculated in an m_I, m_J representation as in Chapter III provided $\mu_\epsilon^2 E^2/\hbar^2 \ll \hbar^2/2A$ so there is little mixing of states with different J. In the opposite extreme when the quadrupole coupling is large compared with the electric perturbation, the quadrupole displacement of the energy level can be determined in the low-field limit and then relations such as Eq. (X. 22) can be used to aid in the calculation of the electrical displacement of energy levels in accordance with the mixture of states involved therein (HUG 49).

Calculations in the intermediate case when the Stark and hyperfine perturbations are comparable are much more complicated. However, the procedures for perturbation calculations and for the solution of secular equations given in Chapter III are applicable. The relevant matrix elements are given in the literature (DEB 29, CON 35, VAN 32), in Chapter III, and in Appendix C. For example, if only the spin-rotational term $\mathcal{H}_{1J}$ is added to Eq. (X. 21), Swartz and Trischka (SWA

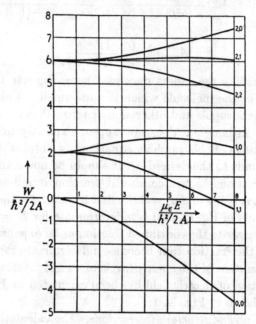

FIG. X. 1. Energy levels of a rotating polar linear molecule with negligible hyperfine structure in an electric field (HUG 47).

52) have shown the energy variation to be as in Fig. X. 2 for $J = 2$. The transition frequencies in a resonance experiment are given in the same figure.

When both quadrupole and spin rotation interactions are present simultaneously the calculations are complicated by the fact that there are four interacting vectors I_1, I_2, J, and E to consider. It is ordinarily convenient to consider various cases each of which can be treated approximately with greater ease than can the intermediate cases. Ordinarily if there is a quadrupole moment at all

$$\hbar^2/2A \gg eqQ \gg c' \tag{X. 25}$$

and the contributions of $\mathcal{H}_S$ are so small as to be negligible. In such circumstances it is conventional to designate fields that are so strong that $\mu_\epsilon^2 E^2/\hbar^2$ is comparable to or greater than $\hbar^2/2A$ as 'very strong fields', those with $|eqQ| \ll \mu_\epsilon^2 E^2/\hbar^2 \ll \hbar^2/2A$ as 'strong fields', those with $|c'| \ll \mu_\epsilon^2 E^2/\hbar^2 \ll |eqQ|$ as 'weak fields', and those with $\mu_\epsilon^2 E^2/\hbar^2 \ll |c'|$ as 'very weak fields'. In all except the very strong field case there is

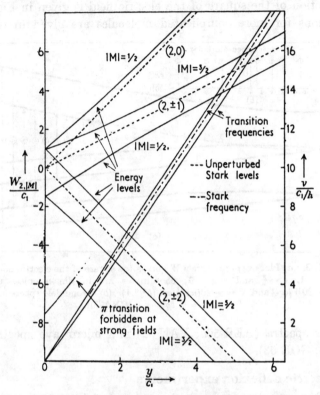

FIG. X. 2. Theoretical curves showing the variation of energy levels W and transition frequencies with electric field, assuming the $c_1 \mathbf{I}_1 \cdot \mathbf{J}$ interaction to be the only perturbing nuclear-molecular interaction. $J = 2$, $I = \tfrac{1}{2}$, $c_1 > 0$. $y \equiv (1/84)\mu^2 E^2/(\hbar^2/2A)$. Since the abscissa is proportional to E^2, the unperturbed Stark levels are linear (SWA 52).

a negligible mixing of J states and the expressions such as Eqs. (III. 43) and (X. 9) which assume J is a good quantum number are valid. In the very strong field case, Eqs. (III. 26) and (III. 57) or their equivalent would have to be used. If the quadrupole coupling is due exclusively to nucleus number one as in RbF (HUG 50), the calculation is simplified in the weak and very weak field cases as a result of there being a well-defined $\mathbf{F}_1 = \mathbf{I}_1 + \mathbf{J}$; consequently a representation based in part on the quantum number $\mathbf{F}_1$ is convenient.

Theoretical calculations of the above nature in the presence of several mutual interactions among I_1, I_2, J, and E have been made by Fano (FAN 48), Nierenberg, Rabi, and Slotnick (NIE 48), Trischka (TRI 49, LUC 53), Grabner and Hughes (GRA 50, HUG 50, HUG 50a), and others (CAR 52, LUC 53, SWA 52). A typical plot of the energy levels as a function of the square of the electric field is given in Fig. X. 3. Calculations for more complicated molecules are given in texts on

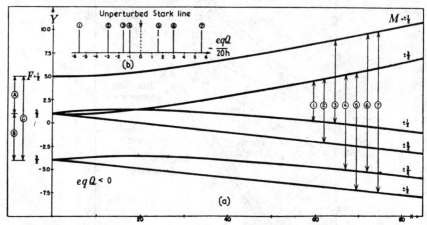

Fig. X. 3. (a) Plot of energy levels W versus the square of the electric field E^2 for $J = 1$, $I = \frac{3}{2}$ and $eqQ < 0$; coordinates are in dimensionless units $Y = -20W/eqQ$ and $X = -20(\mu E)^2/eqQ(h^2/2A)$. (b) Strong field spectrum for $J = 1, I = \frac{3}{2}, eqQ < 0$ (GRA 50).

molecular spectra (HER 50a, VAN 32) and microwave spectroscopy (TOW 54, GOR 53).

X. 3. Electric deflexion experiments

Since the energy of a molecule in an electric field varies as a function of the field, it can be considered by analogy with Eq. (IV. 4) to possess an effective electric dipole moment $\mu_{\epsilon\text{eff}}$ given by

$$\mu_{\epsilon\text{eff}} = -\frac{\partial W}{\partial E} \quad (X. 26)$$

where W is the energy of the state as determined in the preceding section. The force on such a molecule in an inhomogeneous electric field is given as in Eq. (IV. 3) by

$$\mathbf{F} = \mu_{\epsilon\text{eff}} \nabla E = \mu_{\epsilon\text{eff}} \frac{\partial E}{\partial z} \quad (X. 27)$$

if the z axis is taken in the direction of the gradient. The acceleration of the molecule is then

$$a = F/m = (\mu_{\text{eff}}/m)(\partial E/\partial z). \quad (X. 28)$$

With this value for a, the deflexions of the molecule under different circumstances may be calculated from Eqs. (IV. 9) to (IV. 22). The deflected beam shape is then given by such relations as Eq. (IV. 43) with s_α determined from Eqs. (IV. 20) and (X. 28).

Typical numerical values of the electric deflexions can easily be calculated from the above formulae; for example, with $\mu_\epsilon = 10^{-18}$ esu,

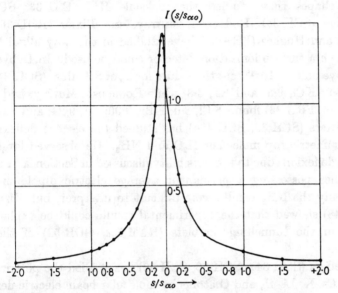

FIG. X. 4. Calculated intensity distribution in an electrically deflected beam for a linear rotator at high temperature. The curve is normalized so $\int_{-\infty}^{\infty} I\,d(s/s_{\alpha 0}) = 1$ (FRA 31).

large J, $E = 500$ esu, $(\hbar^2/2A)J(J+1) \approx kT$, $T = 300°$ K, $\partial E/\partial z = 10^4$ esu cm^{-1}, and $l = $ length of deflecting region $= 10$ cm there results $\mu_{\epsilon\text{eff}} = 3 \times 10^{-3}\mu_\epsilon = 3 \times 10^{-21}$ esu, and $s = $ deflexion $= \tfrac{1}{2}at^2 = 7\cdot5 \times 10^{-2}$ mm.

A typical theoretical deflexion pattern for polar molecules at high temperatures, as calculated by Feierabend (FRA 31) for a Rabi type (RAB 29, RAB 29a) field of Chapter XIV, is shown in Fig. X. 4. At low temperatures the deflexion pattern is obtained by a separate calculation for each rotational state which may subsequently be summed over all rotational states (FRA 31, FRA 37).

Wrede (WRE 27a) early in molecular beam studies, observed a beam deflexion of KI when the beam was shot parallel to a charged wire as shown in Fig. X. 5. Estermann and Wohlwill (EST 28, EST 29, WOH 33) obtained more quantitative data with their apparatus, which possessed an improved inhomogeneous electric field.

Estermann and Fraser (EST 33) have pointed out that measurements of electrical dipole moments can be obtained merely by observing the diminution of beam intensity at the undeflected position as a function of electric field. In this fashion and also by observations of deflected beam shapes they studied the molecule HCl (EST 33, SCH 40). Scheffers (SCH 34), Rodebush, Murray, and Bixler (ROD 36), and Fraser and Hughes (FRA 36) have studied in this way alkali halides for which a surface ionization detector could be used. In Debye units (1 Debye unit = 10^{-18} esu) the values for μ_ϵ of Scheffers (SCH 34) were NaI, 4·9; KCl, 6·3; and KI, 6·8, while Rodebush, Murray, and Bixler (ROD 36, FRA 36) found KCl, 8·0; KBr, 9·06; KI, 9·24; and CsI, 10·2.

Scheffers (SCH 39, SCH 40a) has studied the electric deflexions of the symmetric top molecules H_2O and NH_3. He observed large symmetric deflexions due to the possibility discussed in Section X. 1 of such molecules possessing a permanent average electric dipole moment. Originally the NH_3 results were difficult to interpret, but Mizushima (MIZ 48) showed that the experimental results could be explained in terms of the tunnelling doublets (RAM 53e, GOR 53) of the NH_3 molecule.

Scheffers (SCH 34a, SCH 36, SCH 39) has studied the polarizability of the Cs, K, Li, H, and O atoms by molecular-beam electric-deflexion techniques. In this case the symmetry of the atom is such that it possesses no permanent electric dipole moment. However, as discussed in Section X. 1 the electric polarizability α_ϵ may be measured.

Paul (ZZA 40) has produced focusing electric fields for polar molecules. Townes (ZZA 36) has made electric focusing molecular amplifiers and oscillators as described on page 309.

X. 4. Electric resonance experiments

X. 4.1. *Introduction.* An electric resonance method analogous to the magnetic resonance method can be applied to polar molecules. A typical apparatus for such an experiment is shown schematically in Fig. X. 6. The beam is deflected by an inhomogeneous electric field A and refocused with an inhomogeneous electric field B. A stop wire is introduced in the B field region to eliminate those molecules which have a small $\mu_{\epsilon\text{eff}}$. Since the lengths of the deflecting fields are usually different, a state is ordinarily refocused for different values of electric field E in deflecting field A and in deflecting field B. However, from Eq. (X. 26)

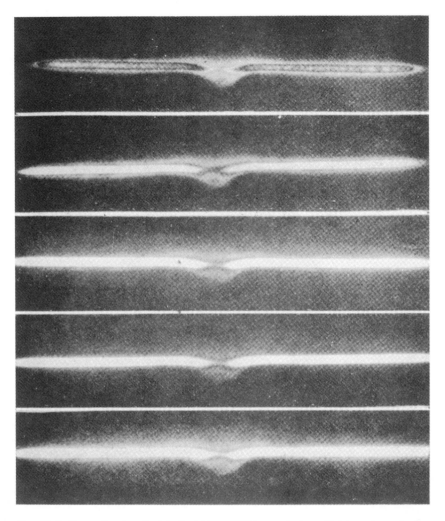

FIG. X. 5. Trace showing electric deviation of KI beam shot parallel to a charged wire. The trace is shown at different stages of dry development (see Chap. XIV). The central portions of the traces show the influence of the electric field around the wire (WRE 27a).

and from the equations similar to Eq. (X. 23), the value of $\mu_{\epsilon\text{eff}}$ depends on E and ordinarily does so differently for different states. Therefore, the refocusing condition ordinarily applies only to a single state and no other state will be refocused.

The above selection of a particular state for refocusing is illustrated in Hughes's experiment on CsF (HUG 47). The energy levels of the

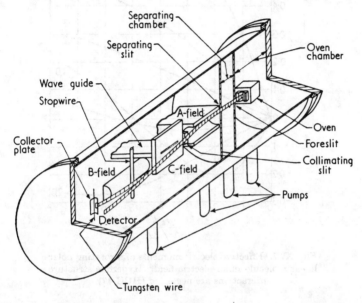

FIG. X. 6. Schematic diagram of molecular-beam electrical-resonance apparatus for the study KCl (LEE 52).

molecule as a function of E have been given in Fig. X. 1. From Eq. (X. 26) the values of $\mu_{\epsilon\text{eff}}$ can be calculated from these curves by differentiation to yield the results shown in Fig. X. 7. From this figure it is apparent that at low fields molecules in the (J, m) state $(1, 0)$ have a negative moment while at high fields their moment is positive. If then $\mathbf{E}$ and $d\mathbf{E}/dx$ point in the same direction in both the A and B fields and if values of $\mu_\epsilon E/(\hbar^2/2A)$ are greater and less than 5 (at which $\mu_{\epsilon\text{eff}} = 0$) in the A and B fields respectively, $(1, 0)$ molecules [and some $(2, \pm 1)$ molecules] will be refocused but all others will hit either the wire stop or some other part of the apparatus. This is indicated schematically by the sigmoid paths in Fig. X. 8 where the signs of $\mu_{\epsilon\text{eff}}$ in the A and B fields are listed.

Between the deflecting and refocusing fields a homogeneous electric field is usually applied and an oscillatory electric field is provided in

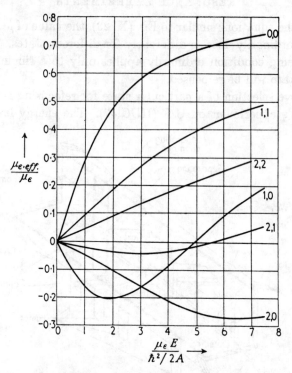

FIG. X. 7. Effective electric moments of a rotating polar linear molecule in an electric field. Hyperfine structure interactions are neglected (HUG 47).

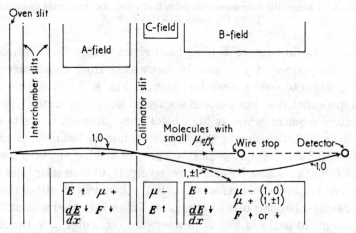

FIG. X. 8. Schematic diagram of electric-resonance apparatus showing paths of molecules. The transverse scale of the drawing is much larger than the longitudinal scale (HUG 47).

the same region. The detected beam intensity then shows a decrease when the frequency of the oscillating field is a Bohr frequency as in Eq. (I. 1) for an allowed electric dipole transition. The selection rules for allowed electric dipole transitions in the absence of hyperfine structure are

$$\Delta J = 0, \pm 1$$
$$\Delta m_J = 0, \pm 1. \quad (X.\ 29)$$

The transitions with $\Delta m_J = 0$ occur when the static and oscillatory fields are parallel and the $\Delta m_J = \pm 1$ when they are mutually perpendicular. On the other hand, if the hyperfine structure is large compared with the Stark effect splitting, the most suitable quantum numbers for the specification of the molecular state are I, J, F, and m_F as in Chapter III. The selection rules for electric dipole transitions are then

$$\Delta J = 0, \pm 1$$
$$\Delta F = 0, \pm 1$$
$$\Delta I = 0$$
$$\Delta m_F = 0, \pm 1. \quad (X.\ 30)$$

In most of the early experiments the transitions with $\Delta J = \pm 1$ were not observed because there were no suitable high-frequency generators (HUG 47, etc.); however, in later experiments after the development of suitable microwave oscillators, transitions with $\Delta J = \pm 1$ were observed (LEE 52).

X. 4.2. *Electric resonance experiments for molecules with negligible hyperfine structure interactions.* The first experiment by the molecular-beam electric-resonance method was that of H. K. Hughes (HUG 47) with CsF. For this molecule the effects of hyperfine structure interactions proved to be negligible with his relatively poor resolution. Hughes's apparatus is shown schematically in Figs. X. 6 and X. 8 with the horizontal scale in Fig. X. 8 being indicated by the fact that the length of the A field was 10 cm while the B field length was 18 cm.

For CsF, Hughes studied the (J, m_J) transitions $(1, 0) \to (1, \pm 1)$ and $(2, \pm 1) \to (2, \pm 2)$. A typical resonance is shown in Fig. X. 9. The dependence of the resonance on electric field is shown in Fig. X. 10 where the frequency of the $(1, 0) \to (1, \pm 1)$ resonance is plotted as a function of E on log-log coordinates. By an analysis of his results, Hughes was able to infer values of B_e, the equilibrium internuclear distance R_e, and μ_e. These values are included in Table X. 1.

Trischka (TRI 49) with the same CsF molecule but with an apparatus

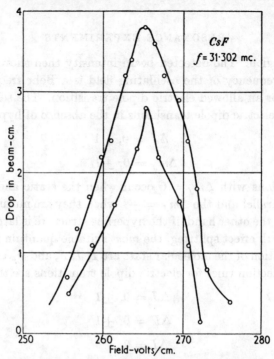

FIG. X. 9. Decrease in CsF beam intensity as a function of the homogeneous electric field intensity at a fixed frequency. The upper and lower curves are for relatively large and small amplitudes of the oscillating electric field. The width is due chiefly to field inhomogeneities (HUG 47).

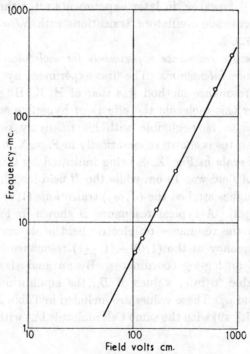

FIG. X. 10. Plot of frequency against field for the $(1, 0) \to (1, \pm 1)$ transition of CsF (HUG 47).

of higher resolution showed that hyperfine structure interactions could be observed. These experiments will be discussed in the next section.

X. 4.3. Electric resonance experiments with hyperfine structure.

Trischka (TRI 48) improved Hughes's apparatus described in the preceding section by making its homogeneous electric C field much more uniform in the region of the oscillatory field. Trischka's means for accomplishing this are described in Chapter XIV. With this increased

TABLE X. I

Molecular Data From Molecular-beam Electric-resonance Experiments

The tabulated quantities are defined in Section X. 2.1. Electric-resonance molecular data are also included in Tables VIII. I, XI. I, and XI. II. Mc indicates megacycles per second.

Molecule	B_e (Mc)	R_e (10^{-8} cm)	μ_e (Debyes)	ω_e (Mc)	Y_{01} (Mc)	$-Y_{11}$ $\approx \alpha_e$ (Mc)	Y_{21} $\approx \gamma_e$ (Mc)	References
$Cs^{133}F^{19}$	5,550	2·34	7·88	8,100	..	55·5	..	TRI 48, TRI 49 HUG47,HUG49
$Rb^{85}F^{19}$	..	..	..	10,200	..	..	..	HUG 50
$K^{39}F^{19}$	6,080	2·55	7·33	11,700	..	..	..	GRA 50
$TlCl^{35}$	2,620·1	2·541	4·444	..	..	13·1	..	CAR 52
$CsCl^{35}$	2,180	2·88	10·46	..	..	16	..	LUC 53
$K^{39}Cl^{35}$	3,856·399	2·6666	10·4	..	3,856·370	23·680	0·050	LEE 52
$K^{39}Cl^{37}$	3,746·611	2·6666	10·4	..	3,746·583	22·676	0·047	LEE 52
$K^{41}Cl^{35}$	3,767·421	2·6666	10·4	..	3,767·394	22·865	0·048	LEE 52
$K^{39}Br^{79}$	2,434·953	2·8207	10·5	..	2,434·947	12·136	0·023	LEE 52
$K^{39}Br^{81}$	2,415·081	2·8207	10·5	..	2,415·075	11·987	0·022	LEE 52
$Rb^{85}Cl^{35}$	2,627·414	2·78670	..	..	2,627·394	13·601	0·021	TRI 54
$Rb^{87}Cl^{35}$	..	..	..	..	2,609·779	13·464	0·021	TRI 54

Li^6F^{19} ($\mu_\epsilon^2 A = 747·2 \times 10^{-76}$ c.g.s. for $v = 0$ and 1·043 times this for $v = 1$ [SWA 52])
$K^{39}Cl^{35}$ ($\mu_\epsilon[v = 0] = 10·48$; $\mu_\epsilon[v = 2] = 10·69$ [LEE 52])
$K^{39}Br^{79}$ ($\mu_\epsilon[v = 0] = 10·41$; $\mu_\epsilon[v = 2] = 9·93$ [LEE 52]), Li^7F^{19} (see ZZA 22).

uniformity the resolution of the apparatus was much improved. Trischka also studied the resonances at different values of frequency and field.

At high electric fields Trischka (TRI 48) found for CsF—the same molecule as that used by Hughes (HUG 47)—that what had appeared to be a single line in Hughes's experiment was resolved into five separate lines as shown in Fig. X. 11. This structure was attributed to the different vibrational states as indicated in the figure. The displacements of the levels for molecules with different amplitudes of vibration arises from the differences in the average values of the moment of inertia A in Eqs. (X. 22), (X. 23), etc.

At lower electric fields the resolution of the apparatus was greater

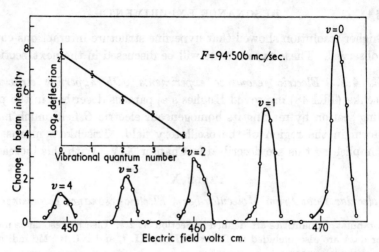

Fig. X. 11. Electric resonance spectrum of CsF in strong electric fields. The different lines are produced by molecules in five different vibrational states. The fine structure resulting from nuclear-molecular interactions is not resolved. The graph in the upper left-hand corner is a semi-log plot showing the variation of intensity with vibrational quantum number v (TRI 48).

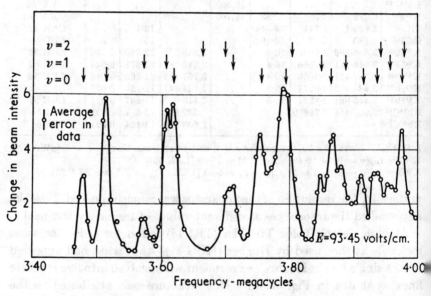

Fig. X. 12. Strong electric field spectrum of CsF showing the fine structure produced by the nuclear-molecular interactions for molecules in several vibrational states. Arrows at the top of the figure indicate theoretical line positions. v is the vibrational quantum number (TRI 48).

and a hyperfine structure of the resonance could be observed as shown in Fig. X. 12. The structure could be attributed to nuclear electric quadrupole and spin-rotational interaction as in Eq. (X. 2). At even much lower fields satisfying the 'weak field' conditions of Section X. 2 the resonance spectrum was as shown in Fig. X. 13. Trischka (TRI 48)

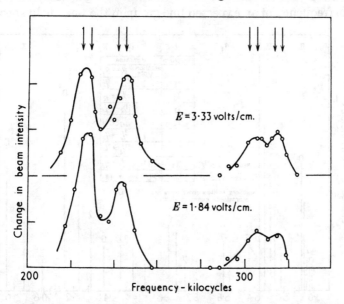

FIG. X. 13. Weak field spectra of CsF for two different values of the field. Arrows show the theoretical positions of lines at zero field (TRI 48).

was able to analyse his observations with the Hamiltonian of Eq. (X. 2) and he thereby determined the hyperfine structure interaction parameters eqQ and c. These values are included in the tables of Chapters VIII and XI.

Hughes and Grabner (HUG 50, HUG 50a, GRA 51) have studied the hyperfine structure of the molecular beam electric resonance spectra of $Rb^{85}F$ and $Rb^{87}F$ in detail. The observed spectral lines are shown in Fig. X. 14. They were able to analyse their experiment with the Hamiltonian of Eq. (X. 2). They obtained values for the Rb nuclear electric quadrupole interaction parameter eqQ and for the fluorine spin-rotational interaction constant c. The quadrupole interaction could be determined separately in each of the first four vibrational states and showed a 4 per cent. variation between different states. The results for these parameters are included in the tables of Chapters VIII and XI. As discussed in Chapter VIII the same molecule has been measured with the molecular-beam *magnetic* resonance method by Bolef and Zeiger (BOL 52). The results for the quadrupole interaction parameters as determined by the two methods were in agreement.

In the course of their electric resonance experiments on RbF, Hughes and Grabner (HUG 50) found an unpredicted line group at exactly one-half the frequency of an expected line group in the weak field spectrum.

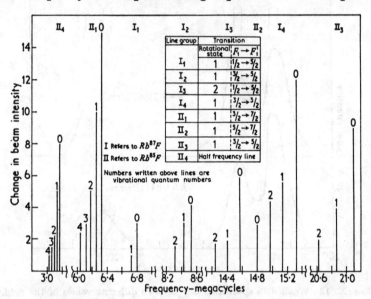

FIG. X. 14. Spectrum observed for $Rb^{85}F$ and $Rb^{87}F$ under low electric field conditions ($\mu^2 E^2/\hbar^2 \ll |eq_1 Q|$). Change in beam intensity is expressed in cm of galvanometer deflexion. All the lines were not taken under comparable conditions so line intensities cannot be compared indiscriminately. Fine structure of the lines is not indicated (HUG 50).

They attributed this line group to a two-quantum transition (HUG 50, HUG 50a) in which two quanta are absorbed in the transition. They showed (HUG 50a) that such two-quanta transitions should be expected theoretically for sufficiently intense oscillatory fields (HUG 50a), i.e. for oscillatory electric fields of amplitude comparable to the static electric field. In a subsequent report Grabner and Hughes (GRA 51) confirmed that their extra line group indeed corresponded to a two-quantum transition by the use of two different oscillatory frequencies ν_1 and ν_2 in which case an extra line group was found when

$$\nu_1 + \nu_2 = (E_n - E_m)/h. \quad (X.\,31)$$

These transitions occur between two levels for which a single quantum transition is also allowed. They are different therefore from the multiple transitions of Sections V. 3.1 and V. 4.3 which are between two states for which a single quantum transition cannot ordinarily occur. The intensities of the multiple quantum transitions of Sections V. 3.1 and V. 4.3 depend greatly on the location of an intermediate state whereas no additional state is involved here.

Grabner and Hughes (GRA 50) have studied $K^{39}F$ in a fashion similar to their studies in the preceding paragraph with RbF. The chief variation of the experiment was that in some experiments they refocused a specific molecular state by the method described above while in others they refocused only those molecules which had undergone a nonadiabatic transition between the A and B fields from the (J, m_J) state $(1, \pm 1)$ to $(1, 0)$ (GRA 50). Their results are included in the tables of Chapters VIII, X, and XI.

LiF and CsCl have been studied by Swartz, Luce, and Trischka (SWA 52, LUC 53, GOL 53) and RbCl by Trischka and Braunstein (TRI 54); their results are included in the tables listed above. One puzzling result of the LiF experiment is that the value for the spin-rotational interaction constant c is almost double that inferred for the same molecule from the molecular-beam magnetic-resonance experiments of Nierenberg and Ramsey (NIE 47). Swartz and Trischka have pointed out that a similar discrepancy exists for CsF. This discrepancy is not yet fully understood but possible explanations include the dependence of c on the rotational quantum number J and the various uncertainties discussed on page 211 and in Appendix G.

X. 4.4. *Electric resonance experiments with change of J.* In all of the molecular-beam electric-resonance methods so far described the transitions were such that $\Delta J = 0$, where J is the rotational quantum number of the molecule. However, transitions with $\Delta J = \pm 1$ can also be observed. Thus Carlson, Lee, Fabricand, and Rabi (CAR 52, LEE 52) have studied in detail the electric resonance spectra of thallium monochloride, potassium chloride, and potassium bromide, utilizing transitions with $\Delta J = \pm 1$ as well as transitions with $\Delta J = 0$. A schematic view of their apparatus is shown in Fig. X. 15 and a typical set of resonances is given in Fig. X. 16. The results are included in the tables of Chapters VIII, X, and XI. One of the most surprising results from this experiment is that the Cl quadrupole interaction in KCl increases by more than 200 per cent. between the first and second molecular vibrational states whereas the quadrupole interaction of K in the same molecule changes only 1 per cent. between the same two vibrational states. From Table X. I it is apparent that more molecular information

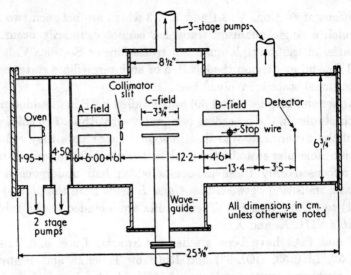

FIG. X. 15. Dimensions of apparatus for electric-resonance studies of KCl (LEE 52).

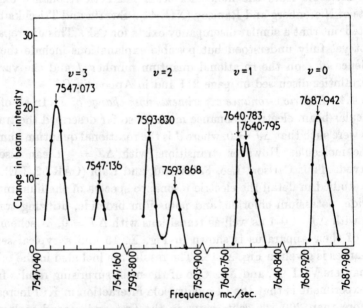

FIG. X. 16. Variation of the hyperfine structure due to chlorine quadrupole interaction in KCl with vibrational quantum number v. The transitions are $J = 0 \to J = 1$ with the upper state $F_1 = \frac{3}{2}$ (LEE 52).

is obtainable from experiments in which transitions with $\Delta J = \pm 1$ as well as with $\Delta J = 0$ are observed.

Trischka and Braunstein (TRI 54) have studied RbCl utilizing both $\Delta J = 0$ and $\Delta J = \pm 1$ transitions. Their results are included in the tables of Chapters VIII, X, and XI.

Molecular-beam electric-resonance experiments with $\Delta J = 0$, ± 1 should be of great value in future molecular studies.

X. 5. Molecular amplifier

Gordon, Zeiger, and Townes (ZZA 36) have recently used molecular-beam electric deflexion techniques to make a molecular microwave amplifier and oscillator. The device, as used on the ammonia inversion spectrum, depends on the emission of energy inside a high Q cavity by a beam of ammonia molecules that are mostly in the upper states of the inversion levels. A beam of molecules passes through a system of quadrupolar cylindrical electrostatic focusing electrodes similar to the magnetic six-pole systems of Section XIV. 4.7. Of the inversion levels, the upper states experience a radial inward (focusing) force, while the lower states experience an outward force. The molecules arriving at the cavity are then virtually all in the upper states. If transitions are induced in the cavity, power is given up by the molecules to the cavity since stimulated emission exceeds absorption when the high-energy states are occupied more than the low. If the power emitted from the beam is enough to maintain the field strength in the cavity at a sufficiently high level to induce transitions in the following beam, then self-sustained oscillations will result. Such oscillations have been produced with an estimated power level of 10^{-8} watts. The frequency stability of the oscillation promises to compare well with other 'atomic clocks'. Under conditions such that oscillations are not maintained, the device acts like an amplifier of microwave power near a molecular resonance. Such an amplifier may have a noise figure very near unity. The device is sometimes called a 'maser', which is an acronym for 'microwave amplification by stimulated emission of radiation'.

XI
NUCLEAR ELECTRIC QUADRUPOLE MOMENTS

XI. 1. Introduction

IN several of the previous chapters nuclear electric quadrupole interaction measurements have been discussed in connexion with the principal topics of the chapters concerned. In the present chapter, the results of measurements of nuclear quadrupole interactions by the various molecular-beam methods will be consolidated. Since the methods of measurement have already been described, they will not be repeated, although an indication will be given as to whether the experiment is by the magnetic or the electric-resonance method.

The measured quadrupole interaction is the product of the desired quadrupole moment Q with an unknown gradient of electric field, ordinarily indicated as in Eq. (IX. 38) in the case of atoms and usually called q in the case of diatomic molecules as in Eq. (III. 50). Therefore, means for the estimation of the electric gradients are discussed in this chapter. These gradient values when combined with the measured quadrupole interactions provide values for the nuclear quadrupole moments. The theoretical interpretations of these quadrupole moments are discussed in the final section of the chapter.

XI. 2. Nuclear electric quadrupole interactions

XI. 2.1. *Nuclear quadrupole interactions in molecules.* The results of measurements of nuclear electric quadrupole interactions in molecules by the various molecular-beam magnetic and electric resonance experiments are listed in Tables XI. I and XI. II. The symbol M following a reference indicates experiments by the molecular-beam magnetic-resonance method whereas E indicates electric resonance experiments.

In Table XI. I are listed the various nuclear quadrupole interaction constants; where possible the rotational and vibrational states corresponding to the measurements are designated. In Table XI. II are listed the empirical coefficients of the quadrupole interaction expansion of Eq. (VIII. 37) in the cases where such analyses have been made.

XI. 2.2. *Nuclear quadrupole interactions in atoms.* In Table XI. III are listed the empirical values of the magnetic dipole, electric quadrupole, and magnetic octupole interaction parameters a, b, and c of

Table XI. I
Nuclear Electric Quadrupole Interactions

Nuclear quadrupole interactions measured in molecular-beam experiments are listed in this table. Where possible, the molecular rotational and vibrational state for the measurement is indicated. The asterisk (*) under v and J indicates average measurements not attributable to a single state. The dagger (†) indicates $e\bar{q}Q/h$ of the expansion of Eq. (VIII. 37); the remaining terms in this expansion are given in Table XI. II. Mc indicates megacycles per second. M under method indicates magnetic resonance experiments while E indicates electric experiments.

Molecule	Nucleus	v	J	eqQ/h (Mc)	Method	References
HD	D^2	0	1	$+0.227 \pm 0.004$	M	KEL 40
D_2	D^2	0	1	$+0.224992 \pm 0.000100$	M	KEL 40, KOL 52, HAR 53
LiCl	Li^6	*	*	$+0.0044$	M	KUS 53, KUS 49
Li_2	Li^7	*	*	$+0.060$	M	KUS 49, KUS 49b, LOG 52
LiF	Li^7	*	*	$+0.408$	M	KUS 49, KUS 49b, LOG 52
LiF	Li^7	0, 1	1	$+0.412$	E	ZZA 22
LiCl	Li^7	*	*	$+0.192$	M	KUS 49, KUS 49b, LOG 52
LiI	Li^7	*	*	$+0.172$	M	KUS 49, KUS 49b, LOG 52
LiBr	Li^7	*	*	$+0.184$	M	KUS 49, KUS 49b, LOG 52
Na_2	Na^{23}	*	*	-0.423	M	KUS 39a, LOG 52
NaF	Na^{23}	*	*	-8.12	M	LOG 52
NaF	Na^{23}	†	†	-8.40 ± 0.02	M	ZEI 52, LOG 52
NaCl	Na^{23}	*	*	-5.40	M	NIE 47, LOG 52
NaCl	Na^{23}	†	†	$\pm 5.608 \pm 0.02$	M	ZEI 52
NaBr	Na^{23}	*	*	-4.68	M	NIE 47, LOG 52
NaI	Na^{23}	*	*	-3.88	M	NIE 47, LOG 52
KCl	Cl^{35}	*	*	-0.420	M	LOG 52
TlCl	Cl^{35}	†	†	-15.788 ± 0.02	M	ZEI 52, LOG 52
TlCl	Cl^{35}	*	*	-15.795 ± 0.040	E	CAR 52
KCl	Cl^{37}	*	*	-0.366	M	LOG 52
TlCl	Cl^{37}	†	†	$\pm 12.425 \pm 0.02$	M	ZEI 52
TlCl	Cl^{37}	*	*	-12.446 ± 0.003	E	CAR 52
KF	K^{39}	†	†	-7.938 ± 0.040	E	GRA 50
K_2	K^{39}	*	*	-0.158	M	KUS 39a, LOG 52
RbF	Rb^{85}	†	†	-70.31 ± 0.10	E	HUG 50
Rb_2	Rb^{85}	*	*	-1.10	M	LOG 52
RbF	Rb^{87}	†	†	-34.00 ± 0.06	E	HUG 50
RbF	Rb^{87}	†	†	-33.96 ± 0.02	M	BOL 52
RbCl	Rb^{87}	†	†	-25.485 ± 0.006	M, E	BOL 52, TRI 54
Rb_2	Rb^{87}	*	*	-0.580	M	LOG 52
CsF	Cs^{133}	†	†	$+1.240 \pm 0.008$	E	TRI 48
Cs_2	Cs^{133}	*	*	$+0.23$	M	KUS 39a, LOG 52
$K^{39}Cl^{35}$	K^{39}	0	1	-5.656 ± 0.006	E	LEE 52
$K^{39}Cl^{35}$	K^{39}	1	1	-5.622 ± 0.006	E	LEE 52
$K^{39}Cl^{35}$	K^{39}	2	1	-5.571 ± 0.008	E	LEE 52
$K^{39}Cl^{35}$	K^{39}	3	1	-5.511 ± 0.008	E	LEE 52
$K^{39}Cl^{37}$	K^{39}	0	1	-5.660 ± 0.006	E	LEE 52
$K^{39}Cl^{37}$	K^{39}	1	1	-5.628 ± 0.010	E	LEE 52

Molecule	Nucleus	v	J	eqQ/h (Mc)	Method	References
K^{41}Cl^{35}	K^{41}	0	1	−6·899±0·006	E	LEE 52
K^{41}Cl^{35}	K^{41}	1	1	−6·840±0·010	E	LEE 52
K^{39}Cl^{35}	Cl^{35}	0	1	0·040	E	LEE 52
K^{39}Cl^{35}	Cl^{35}	1	1	±0·075±0·010	E	LEE 52
K^{39}Cl^{35}	Cl^{35}	2	1	±0·237±0·010	E	LEE 52
K^{39}Cl^{35}	Cl^{35}	3	1	±0·393±0·010	E	LEE 52
K^{39}Br^{79}	K^{39}	0	1	−5·003±0·003	E	LEE 52
K^{39}Br^{79}	K^{39}	1	1	−4·984±0·003	E	LEE 52
K^{39}Br^{79}	K^{39}	2	1	−4·915±0·003	E	LEE 52
K^{39}Br^{79}	Br^{79}	0	1	+10·244±0·006	E	LEE 52
K^{39}Br^{79}	Br^{79}	1	1	+11·224±0·006	E	LEE 52
K^{39}Br^{79}	Br^{79}	2	1	+12·204±0·006	E	LEE 52
K^{39}Br^{81}	K^{39}	0	1	−5·002±0·003	E	LEE 52
K^{39}Br^{81}	Br^{81}	0	1	8·555±0·006	+	LEE 52
Rb^{85}Cl^{35}	Rb^{85}	0	1	−52·675±0·005	E	TRI 54
Rb^{85}Cl^{35}	Rb^{85}	1	1	−52·306±0·030	E	TRI 54
Rb^{85}Cl^{35}	Rb^{85}	2	1	−51·903±0·040	E	TRI 54
Rb^{85}Cl^{35}	Cl^{35}	0	1	+0·774±0·009	E	TRI 54
Rb^{85}Cl^{35}	Cl^{35}	1	1	+0·612±0·013	E	TRI 54
Rb^{85}Cl^{35}	Cl^{35}	2	1	+0·470±0·010	E	TRI 54

TABLE XI. II

Coefficients of Nuclear Quadrupole Interaction Expression

In this table are listed the experimental coefficients of Eq. (VIII. 37) for those molecules which have been analysed in this fashion. These coefficients are in cycles per second and are defined by Eq. (VIII. 37) which is as follows:

$$eqQ/h = e\bar{q}Q/h + eq^{(J)}QJ(J+1)/h + eq^{(v)}Qv/h.$$

The symbol ($\pm$) indicates sign unknown; ($+$) indicates the same sign as $\bar{q}$; ($-$) indicates the opposite sign to $\bar{q}$. Plus or minus signs without parenthesis are used when the absolute sign is known. The letter M under method indicates magnetic resonance experiments and E electric resonance experiments.

Molecule	Nucleus	$e\bar{q}Q/h$ (c/s)	$eq^{(J)}Q/h$ (c/s)	$eq^{(v)}Q/h$ (c/s)	Method	References
NaF	Na^{23}	−(8·40±0·02)×10^6	+32	+80×10^3	M	ZEI 52, LOG 52
NaCl	Na^{23}	(±)(5·608±0·02)×10^6	(−)24	(−)88×10^3	M	ZEI 52, NIE 47
TlCl	Cl^{35}	−(15·788±0·020)×10^6	−18·4	−96×10^3	E, M	ZEI 52, CAR 52
TlCl	Cl^{37}	−(12·425±0·020)×10^6	−16·4	−76×10^3	E, M	ZEI 52, CAR 52
KF	K^{39}	−(7·938±0·040)×10^6	∼0	+100×10^3	E	GRA 50
RbF	Rb^{85}	−(70·31±0·10)×10^6	∼0	+770×10^3	E	HUG 50
RbF	Rb^{87}	−(33·96±0·02)×10^6	+26±4	+(410±40)×10^3	M, E	BOL 52, HUG 5
RbCl	Rb^{87}	−(25·485±0·006)×10^6	+12±2	+(200±40)×10^3	M, E	BOL 52, TRI 54
CsF	Cs^{133}	+(1·240±0·008)×10^6	∼0	<1%	E	TRI 48

TABLE XI. III

Nuclear Multipole Interactions in Atoms

The nuclear magnetic dipole, electric quadrupole, and magnetic octupole moment interactions that have been measured by molecular-beam methods for atoms with $J > \frac{1}{2}$ are listed below. Hyperfine structure interactions in atoms with $J = \frac{1}{2}$ are listed in Table IX. I. The magnetic dipole interaction constant a, the electric quadrupole constant b, and the magnetic octupole constant c are defined in Eqs. (IX. 36) and (IX. 39). The listed values of c are corrected for second-order perturbations (SCH 54, JAC 54) so that the listed value for c is presumably due exclusively to nuclear magnetic octupole interaction. The correction is the third term under c and the uncorrected value is the first term (SCH 54). All atoms in lowest $^2P_{\frac{3}{2}}$ state except Pr^{141} in $^4I_{\frac{9}{2}}$. For Na^{23} and Rb^{85} see ZZA 23 and ZZA 24.

Atom	a (Mc)	b (Mc)	c (Mc)	References
B^{11}	+73·347(6)	2·695(16)	..	ZZA 13
Al^{27}	94·25±0·04	18·76±0·25	..	LEW 48, LEW 49
Cl^{35}	205·050±0·005	54·873±0·005	..	DAV 48, DAV 49b, JAC 51
Cl^{37}	170·681±0·010	43·255±0·010	..	DAV 48, DAV 49b, JAC 51
Ga^{69}	190·794270 ±0·000055	62·522490± ±0·000100	0·000050±0·000003 +0·000034	REN 40, BEC 48, SCH 54, DAL 54
Ga^{71}	242·433955 ±0·000055	39·399030± 0·000100	0·000086±0·000003 +0·000029	REN 40, BEC 48, SCH 54, DAL 54
Br^{79}	884·810±0·003	−384·878±0·008	< 0·0001	KIN 54
Br^{81}	953·770±0·003	−321·516±0·008	< 0·0001	KIN 54
In^{113}	241·624±0·024	443·102±0·043	..	HAM 39, HAR 42, MAN 50
In^{115}	242·16485 ±0·00006	449·5524±0·0006	(0·000011±0·000032 +0·00109)	HAM 39, HAR 42, MAN 50, KUS 54, SCH 54
I^{127}	827·265±0·003	1146·356±0·010	0·00287±0·00037 −0·00053	JAC 54
Pr^{141}	926·03±0·10	−13·9±1·0	..	LEW 53a

Eqs. (IX. 36) and (IX. 39) that have been measured by the atomic-beam magnetic-resonance method.

XI. 3. Electric-field gradients

The above measured quadrupole interactions are products of the desired quadrupole moments Q with unknown gradients of electric field ordinarily called q in the case of diatomic molecules as in Eq. (III. 50) and indicated by eq_J as in Eq. (IX. 38) in the case of atoms. Therefore, means for the estimation of electric gradients are discussed in this section. Of course it should be noted as in Eq. (III. 53) that the term electric field gradients is an abbreviation for the gradient of the electric field arising from charges external to a small sphere surrounding the nucleus.

XI. 3.1. *Electric-field gradients in atoms.* For many atoms, q_J can be attributed dominantly to a single electron, in which case Eq. (III. 41a) applies. If, furthermore, the angular and radial wave-functions are separable for that electron, Eq. (III. 41a) may be written as

$$q_J = -\langle r_e^{-3}\rangle R\langle 3\cos^2\theta_e - 1\rangle_{JJ}, \quad (\text{XI. 1})$$

where R provides a small relativistic correction evaluated by Casimir (CAS 36, DAV 49b, KOP 40). The value of $\langle 3\cos^2\theta_e - 1\rangle_{JJ}$ may be directly calculated theoretically from the wave-function of the assumed electronic state (KOP 40, p. 67). In particular, for $^2S_{\frac{1}{2}}$, $^2P_{\frac{1}{2}}$, and $^2P_{\frac{3}{2}}$ states it is easy to show that

$$\langle 3\cos^2\theta_e - 1\rangle_{JJ} = 0 \quad (^2S_{\frac{1}{2}}, {}^2P_{\frac{1}{2}})$$

$$\langle 3\cos^2\theta_e - 1\rangle_{JJ} = -\tfrac{2}{5} \quad (^2P_{\frac{3}{2}}). \quad (\text{XI. 2})$$

Except for the relativistic correction R, all that remains to be evaluated is $\langle r_e^{-3}\rangle$.

The quantity $\langle r_e^{-3}\rangle$ may be evaluated empirically from the experimental value δ of the optical fine (not hyperfine) structure separation δ since by the theory of optical fine structure (KOP 40, pp. 20 ff., but note his omission of a factor $[2l+1]$ on pp. 67 ff.)

$$\langle r_e^{-3}\rangle = \frac{\delta}{\mu_0^2(2l+1)Z_i\lambda}, \quad (\text{XI. 3})$$

where Z_i is the inside effective charge of Section III. 3.2 and λ provides a small relativistic correction (BRE 30, BRE 33, RAC 31, KOP 49). From the above it is apparent that for a single electron in a $^2P_{\frac{3}{2}}$ state

$$q_J = \frac{2}{15}\frac{\delta}{\mu_0^2 Z_i}\frac{R}{\lambda}. \quad (\text{XI. 3}a)$$

An even better means for evaluating $\langle r_e^{-3}\rangle$ is from the measured magnetic hyperfine structure parameter a of Section III. 3.2 and IX. 8. This method was first introduced by Davis, Feld, Zabel, and Zacharias (DAV 49b, JAC 51) in their studies of atomic Cl. From Eq. (III. 83)

$$\langle r_e^{-3}\rangle = \frac{ha}{\mu_0(\mu_I/I)\mathscr{F}}\frac{J(J+1)}{2L(L+1)}, \quad (\text{XI. 3}b)$$

where $\mathscr{F}$ provides a small relativistic correction that has been calculated by Casimir (CAS 36, DAV 49b). The desired q_J is then obtainable directly from Eqs. (XI. 1, 2, 3b). This method is particularly effective since it does not depend on Z_i, which is estimated only approximately.

In all of the above methods allowance should be made for the mixing of higher configurations with the ground configuration assumed in the calculations of the above type (FER 33, KOS 52). Koster (KOS 52) has discussed the necessary corrections in detail. For Ga, whose ground configuration is $(4s)^2 4p$, Koster has shown that configuration interaction mixes in enough of the $4s4p5s$ configuration to provide an important correction in the above calculation. Koster has tabulated these corrections for various atomic states. The correction is ordinarily smaller if $\langle r_e^{-3} \rangle$ is determined from the magnetic hyperfine structure of a $^2P_{\frac{1}{2}}$ state instead of a $^2P_{\frac{3}{2}}$ state.

Sternheimer (STE 50, STE 54, FOL 54) has pointed out that in calculations of q_J for most atoms by any of the above methods, the contributions of the inner electron shells are omitted because filled shells which make up the atomic core are presumed to be spherically symmetric. Sternheimer, however, showed that the inner atomic shells are distorted by the nuclear quadrupole moment and that this distorted core in turn interacts with the valence electrons in a way similar to the interaction of the nuclear quadrupole moment. Thus the atomic core depending on the sign of the effect, produces either a shielding or antishielding effect on the nuclear quadrupole interaction. This effect is discussed in greater detail in Section XI. 3.3.

XI. 3.2. *Electric-field gradients in molecules.* In this important case of D_2, Nordsieck (NOR 40), Newell (NEW 50), and others (ZZB 11) have calculated q directly from the molecular wave function. Since both of the molecular electrons are explicitly included in the calculation there is no uncertainty due to the shielding phenomena of the preceding and following sections. Extension of this method to other molecules, however, is numerically quite difficult.

Townes (TOW 47, TOW 49, TOW 49a) has suggested for heavy atoms covalently bonded in *molecules* with p-orbitals that $\partial^2 V^e / \partial z_0^2$ can be calculated from *atomic* fine structure separations similar to the calculation of q_J in atoms. If the electronic state of the atom were a pure p state, the calculation would be quite simple and reliable. However, two uncertainties arise to give difficulty. One uncertainty is that the bond may be ionic instead of covalent (PAU 41). In such a case the ion probably has a closed electron shell and the quadrupole interaction is greatly reduced, because the $\partial^2 V^e / \partial z_0^2$ from the other ion is much less than that of a p electron in the atom itself. Thus NaCl and similar molecules are almost wholly ionically bonded, whereas other molecules are partially ionic and partially covalent. The percentage of ionic bond can often be estimated from the electronegativity difference of the atoms being bonded (PAU 41, TOW 49a). The other uncertainty is that the bond may not be a pure one with the electron in a pure p

state. There may be hybridization of the bond with some s as well as p state, in which case the magnitude of $\partial^2 V^e/\partial z_0^2$ is decreased. On the other hand, if the hybridization is with a d state $\partial^2 V^e/\partial z_0^2$ is increased. There is no simple rule to determine the amount of hybridization. In fact, from a molecular point of view, one of the chief applications of quadrupole moment interaction measurements is the determination of the degree of hybridization by measurements of the quadrupole interactions of the same atom in different molecules (TOW 54).

Except for the case of molecular D_2 and HD, the effects of the atomic cores discussed in the following section modify the quadrupole interactions in molecules as well as atoms.

In molecules, allowance must be made for the effects upon q of molecular vibration and centrifugal stretching. These effects have been discussed in detail by Nordsieck (NOR 40), Newell (NEW 50), Ramsey (KOL 51, RAM 52f), and Zeiger and Bolef (ZEI 52). Their results are summarized above in Section, VIII. 7, VIII. 5, and X. 4.

XI. 3.3. *Effect of the atomic core on the nuclear quadrupole interaction.* R. M. Sternheimer (STE 50, STE 54) and Foley, Sternheimer, and Tycko (FOL 54), at the original suggestion of Rabi, have discussed the effects of the inner filled electron shells in modifying the nuclear electric quadrupole interaction. Superficially it might appear that this inner atomic core should not affect the quadrupole interaction due to its spherical symmetry. However, this spherical symmetry is distorted by the nuclear quadrupole moment and then interacts with the field gradients from the valence electrons or remainder of the molecule.

The effect of the atomic core on the nuclear quadrupole coupling can be represented as the interaction of the valence electron or charges of the remainder of the molecule with the quadrupole moment induced in the core by the nuclear quadrupole moment Q. In his original calculations, Sternheimer (STE 50) used a Fermi–Thomas model and found a shielding for Q of about 10 per cent., i.e. the quadrupole coupling was decreased in absolute magnitude by this amount. This effect corresponds to an angular rearrangement of the electronic charge. For a positive Q the electrons tend to concentrate along the axis of Q where the potential energy is a minimum. This angular rearrangement is due to an excitation of s electrons into higher d states, and of p electrons into f states.

In subsequent calculations Sternheimer (STE 50, STE 54) and others (FOL 54) showed in addition that the perturbation by the nuclear Q gives rise to an excitation of p electrons into higher p states and of d

electrons into higher d states. This effect corresponds to a radial redistribution of the charge of the electron core. If $Q > 0$, then along the axis of Q the electrons tend to lie closer to the nucleus than in the unperturbed state, while at right angles to the axis of Q the electrons are on the average farther away than without the perturbation. The radial redistribution thus tends to reinforce the effect of the nuclear Q if the valence electron is sufficiently far outside the core, and for this reason is referred to as antishielding. This effect depends sensitively on the principal and azimuthal quantum numbers of the external shells of the core; it cannot be obtained from the Fermi–Thomas model but the Hartree–Fock wave-functions are suitable and have usually been used. The existence of exchange between the valence electron and the core electrons gives rise to an excitation of the core electrons to higher states so there are exchange terms due to this and the induced quadrupole moment of the core (STE 50, STE 54).

The sign and magnitude of the nuclear quadrupole shielding depends on the mean distance r from the charge producing the field gradient to the nucleus which is also the centre of the shell. This can be seen by considering the quantity $\gamma(r)$ defined by the assumption that in the presence of the core the quadrupole interaction quantity q for a charge e at position $\mathbf{r}$ from the nucleus is given by

$$q = (2e/r^3)[1-\gamma(r)]. \tag{XI. 4}$$

$\gamma(r)$ then vanishes if there is no atomic core shielding and a negative γ corresponds to antishielding. Foley, Sternheimer, and Tycko (FOL 54) have calculated the variation of $\gamma(r)$ as a function of r for several typical ionic cores with the results shown in Fig. XI. 1.

From this figure it is apparent that for large r, $\gamma(r)$ is large and negative corresponding to a large antishielding effect. For smaller distances $\gamma(r)$ becomes much smaller and may be slightly positive.

For polar molecules the charge of the other atom is relatively distant and $\gamma(\infty)$ can be taken as a measure of the quadrupole antishielding. For Cl⁻, Cu⁺, Rb⁺, and Cs⁺ ions $-\gamma(\infty)$ is $+46\cdot5$, $+8\cdot7$, $+50\cdot2$, and $+86\cdot8$. The existence of the negative sign shows that there is antishielding and the dominant effect is the radial distribution one discussed above. These results mean, for example, that owing to the antishielding effect of the core the nuclear quadrupole interaction in a Cs polar molecule is about $87\cdot8$ times bigger than it would have been with no core antishielding.

In atoms the mean distance r of the valence electron is much less, so for lighter atoms in their ground states the shielding effect from angular rearrangements often dominate. Since in atoms the valence electron is at different positions r it is the expectation value $\langle \gamma(r_e)/r_e^3 \rangle$ that determines the average quadrupole shielding of the core. Consequently, the average shielding is measured by the quantity R_S which

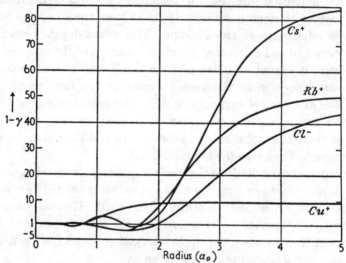

FIG. XI. 1. $[1-\gamma(r)]$ for Cl^-, Cu^+, Rb^+, and Cs^+. The function
$(1-\gamma)eQ(3\cos^2\theta - 1)/2r^3$
gives the total quadrupole potential (FOL 54).

is the negative of the ratio of the induced moment interaction energy to the nuclear Q interaction energy. R_S is then given by

$$R_S = \langle \gamma(r_e)/r_e^3 \rangle / \langle 1/r_e^3 \rangle. \tag{XI. 5}$$

Theoretical values of R_S for different atoms have been calculated and tabulated by Sternheimer (STE 50, STE 54) for a number of atoms in different states. A typical result is that R_S for the $2p$ ground state of boron is 0·14, corresponding to a shielding instead of an antishielding effect.

When the value of $\langle 1/r_e^3 \rangle$ is evaluated from the magnetic hyperfine structure as in Section XI. 3.1, allowance for the core shielding of the magnetic hyperfine structure interaction must be included as well. Sternheimer (STE 50, STE 54) has calculated the procedures for making this correction. One portion of the core effect—part of the exchange effect—approximately cancels between the magnetic and quadrupole corrections in this case. However, the other portions remain and require both the magnetic and electric corrections described by Sternheimer.

XI. 4. Nuclear electric quadrupole moments

From the quadrupole interactions tabulated in Section XI. 2 and from calculations of the electric field gradients by the methods of the preceding section, a number of nuclear electric quadrupole moments, Q, have been evaluated. The results of such measurements are included in Table VI. 1. However, many of the quadrupole interaction measurements in Table XI. 1 do not contribute usefully to the knowledge of Q, because we lack a reliable q. Also included in Table VI. 1 are values of Q obtained from such non-molecular beam experiments as microwave spectroscopy. However, almost all the values of Q suffer from considerable uncertainty since q or eq_J are unreliable; this uncertainty often arises principally from the electric-shielding effects as discussed in the preceding section.

The quadrupole moment of the deuteron, whose $Q = (0.00282 \pm 0.00002) \times 10^{-24}$ cm^2 (KOL 52, HAR 53, ZZB 11), is probably the most important nuclear quadrupole moment for the reasons discussed in the next section. Also, fortunately, this quadrupole moment suffers least as a result of uncertainties from the effects of inner closed shells in shielding the quadrupole interaction.

The theoretical significance of various nuclear electric quadrupole moments is discussed in the next section.

A small but interesting anomaly has been observed in isotopic ratio measurements with nuclear quadrupole interactions. Dehmelt and Krüger (DEH 50, DAV 49b) found that the ratio of the nuclear quadrupole moments of Cl^{35} and Cl^{37} varied in different molecular compounds. Most of the proposals to account for the original anomalies proved unsatisfactory. A proposal by Gunther-Mohr, Geschwind, and Townes (GES 51, GUN 51) that the anomaly was due to the polarizability of the nucleus appeared feasible. However, recent experiments (JAC 51, LIV 51, WAN 52) indicate that the anomaly is smaller than was thought originally but still definite (about 0.02 per cent.). With the diminished size of the anomaly, it may be due to molecular distortion instead of nuclear polarizability.

XI. 5. Theoretical interpretation of nuclear quadrupole moments

XI. 5.1. *Theory of deuteron quadrupole moment.* Since the deuteron is the simplest nucleus with more than a single nucleon its study has been particularly revealing as to the nature of nuclear forces. In particular, prior to the discovery of the electrical quadrupole moment of the deuteron by Kellogg, Rabi, Ramsey, and Zacharias (KEL 39, KEL

40, KOL 52), it was presumed that nuclear forces were central forces, i.e. that the nuclear potential depended only upon the magnitude of the nucleon spacing r as in $V(r)$. For such forces the ground state of the deuteron could easily be shown to be a $^3S_1$ state (RAM 53g). Since such a state is spherically symmetric, there should be no deuteron quadrupole moment.

Therefore, following the experimental discovery of the deuteron quadrupole moment, Schwinger (RAR 41) introduced a tensor force which coupled the spin and orbital motion together as a means for producing a spherically asymmetric state as required by the existence of the deuteron quadrupole moment. Schwinger assumed the interaction potential to be proportional to the tensor force operator

$$S_{12} = \frac{3(\sigma_1 \cdot \mathbf{r})(\sigma_2 \cdot \mathbf{r})}{r^2} - \sigma_1 \cdot \sigma_2, \tag{XI. 6}$$

where σ_1 and σ_2 are the Pauli spin operators for the proton and neutron, respectively. It is of interest to note that the angular dependence of Eq. (XI. 6) is just the same as that for the classical magnetic interaction of two magnetic dipoles as in Eq. (III. 95). The assumption of this form of tensor interaction can be justified on quite general grounds, since it has been proved by Wigner (WIG 41, RAR 41) that, if the interaction between two particles of spin $\frac{1}{2}$ is assumed to be velocity independent and invariant to the choice of the observer's coordinate system, the most general interaction can be reduced to the form

$$V = -\left[J(r) + L(r)\frac{\sigma_1 \cdot \sigma_2 - 1}{4} + K(r)S_{12} \right]. \tag{XI. 7}$$

In quantum theory the constants of motion whose eigenvalues may be used to specify the eigenstates of the energy must all correspond to operators which commute with the Hamiltonian and with each other (CON 35). When the Hamiltonian contains a potential of the form of Eq. (XI. 7) which couples the spin and orbital angular momenta together, the operators (CON 35) for $\mathbf{L}^2$ and $\mathbf{L}_z$ no longer commute with the Hamiltonian. Consequently the eigenstates of the energy can no longer correspond to definite values of the orbital quantum numbers L and m_L but correspond instead to a mixture of several values, for example of $L = 0$ and $L = 2$ or of S and D orbital states. In a similar way if $\mathbf{S} = \frac{1}{2}(\sigma_1 + \sigma_2)$, $(\mathbf{S})_z$ will not in general commute with the Hamiltonian so m_S will not be a good quantum number either.

However, there are other quantities which are constants of motion even when a tensor force is present. Corresponding to the conservation of the total angular momentum $\mathbf{I} = \mathbf{L}+\mathbf{S}$, the quantum numbers I and m_I may be used along with the energy E. In addition there are two other valid constants of motion. In the case of only two particles, if P_σ is an operator corresponding to the interchange of σ_1 and σ_2, then, as can be seen by inspection, V is not altered by the application of the operator P_σ, so P_σ commutes with the Hamiltonian, whence its eigenvalues are constants of motion. If ψ is a simultaneous eigenstate of both the Hamiltonian and P_σ, then $P_\sigma \psi = k\psi$, where k is a constant, and $P_\sigma^2 \psi = k^2 \psi$. However, P_σ^2 corresponds to interchanging the two spins twice, which is equivalent to doing nothing, so $k^2 = 1$ and $k = \pm 1$. Consequently, even with tensor forces, there is still a sharp split into symmetrical and antisymmetrical spin functions. However, just as with two electrons (PAU 35), symmetrical spin functions correspond to $S = 1$ and antisymmetrical to $S = 0$, therefore in the case of two particles each having a spin $\tfrac{1}{2}$, S can be used as a good quantum number even with a tensor force. The remaining constant of motion can be the eigenvalues of the parity operator P_r, which replaces $\mathbf{r}$ by $-\mathbf{r}$. Since this interchange does not affect Eq. (XI. 7), P_r commutes with the Hamiltonian and the eigenvalues of P_r are suitable constants. By exactly the same argument as with P_σ above, the only possible eigenvalues p of P_r are ± 1; states with $+1$ are said to have even parity, and states with -1 odd parity. Therefore a set of constants to specify the different states is E, I, S, p, and m_I. If there is no degeneracy (other than spatial) of nuclear states, all these quantities (except m_I) are uniquely determined by a specification of the energy of the state.

Since the effects of the tensor force are small, it is reasonable to expect that the lowest energy state should be one which is as nearly similar as possible to the $^3S_1$ state that exists as the ground state in the absence of tensor forces. In particular the state with tensor forces should have the same value of I, S, and p. Since $I = 1$ and $p = 1$ in a $^3S_1$ state, the ground state of the deuteron in the presence of a relatively weak tensor force can be taken as specified by these same values. A single value of L will not be associated with the state; however, this state can be represented as a superposition of eigenstates to the central force problem since these form a complete set. The contributing eigenstates are, however, severely limited in number by the fact that they must have the same value of I, S, and parity. A value of $I = 1$ when $S = 1$ can be achieved only with $L = 0$, 1, and 2. However, $L = 1$ has odd parity, so the ground state of the deuteron for which $I = 1$,

$S = 1$, and the parity is even must be a superposition of a $^3S_1$ state and a $^3D_1$ state. Therefore, we may write

$$\psi(E, I = 1, S = 1, p = +1, m_I) = a_1\psi(^3S_1) + a_2\psi(^3D_1), \quad \text{(XI. 8)}$$

where each of the above three ψ's is assumed to be normalized individually to unity.

Since $S = 1$ for the ground state of the deuteron,

$$\sigma_1 \cdot \sigma_2 = 2S^2 - \tfrac{1}{2}\sigma_1^2 - \tfrac{1}{2}\sigma_2^2 = 1. \quad \text{(XI. 9)}$$

Then Eq. (XI. 7) for the ground state of the deuteron becomes

$$V = -[J(r) + K(r)S_{12}]. \quad \text{(XI. 10)}$$

Schwinger (RAR 41, SCH 46, FES 51), Rarita (RAR 41), Feshbach (FES 51, PEA 51), and others (PEA 51, ZXX 54) have made extensive calculations on the deuteron ground state. The $^3S_1$ portion of the wave function is easily represented as a product of the S orbital wave function by the spin wave function with S equal 1 and m_S equal to the desired m_I. On the other hand, the $^3D_1$ portion is more complicated in that with $L = 2$, $S = 1$, and $I = 1$ a particular value of m_I can be achieved in three different ways depending on the relative orientation of **L** and **S**. Therefore, the $^3D_1$ portion is itself a superposition of three states with different spin and orbital relative orientation. This would make the writing of the wave function complicated though the procedure is straightforward with the use of the Clebsch–Gordon or vector addition coefficients (CON 35, SIM 54). Schwinger (RAR 41, SCH 46, FES 51) and others, however, have shown that the wave function can be written much more easily with the aid of the operator S_{12}. This operator acting on one product of spin and orbital wave functions indeed converts it into a linear sum of different product combinations. It can be shown (RAR 41, SCH 46, FES 51) that the linear sums are of exactly the correct form for the desired ground-state wave functions. In this way Schwinger and others have shown that the wave function for ground state of the deuteron can be written as

$$\psi(E, 1, 1, +1, m_I) = \frac{1}{\sqrt{(4\pi)}}\left[\frac{u(r)}{r} + 2^{-\tfrac{1}{2}}S_{12}\frac{w(r)}{r}\right]\chi_1(1, m_I), \quad \text{(XI. 11)}$$

where $\chi(1, m_I)$ is the spin wave function of the two nucleons for states with $S = 1$ and $m_S = m_I$, and where $u(r)$ and $w(r)$ are the radial wave functions of the S and D states, u and w being normalized by

$$\int_0^\infty (u^2 + w^2)\, dr = 1. \quad \text{(XI. 12)}$$

The fraction P_D of D state mixed in with the wave function by Eqs. (XI. 8) or (XI. 12) can then be expressed in either of the following forms:

$$P_D = |a_2|^2/[|a_1|^2+|a_2|^2] = \int_0^\infty w^2\, dr. \qquad (XI.\ 13)$$

With the wave function of Eq. (XI. 11) the quadrupole moment of the deuteron may be calculated directly from Eq. (III. 37) as

$$Q = \frac{1}{e}\int_\tau \rho_{nII}(3z_n^2 - r_n^2)\, d\tau_n$$

$$= \tfrac{1}{4}\sum_s \int (3z^2-r^2)|\psi(E,1,1,+1,m_I)|^2\, d\tau, \qquad (XI.\ 13a)$$

where $\mathbf{r}_n$ is the nuclear coordinate relative to the centre of mass of the deuteron, while $\mathbf{r} = 2\mathbf{r}_n$ is the coordinate of the proton relative to the neutron and $\sum_s$ indicates a summation over spin coordinates. From Eqs. (XI. 11) and (XI. 13a) it can be shown (RAR 41, SCH 46, FES 51) that

$$Q = \frac{\sqrt{2}}{10}\int_0^\infty r^2(uw - 2^{\frac{1}{2}}w^2)\, dr. \qquad (XI.\ 14)$$

With any particular form of potential—for example a Yukawa potential—assumed for $J(r)$ and $K(r)$, the wave equation can be formed and u and w can be determined by the requirement that Eq. (XI. 11) satisfy the equation. These values can in turn be used to determine the theoretical value of Q in Eq. (XI. 14). Since in practice the forms and magnitudes of $J(r)$ and $K(r)$ are not known in advance, the procedure is usually reversed and the requirement that the Q of Eq. (XI. 14) should equal the known experimental result for Q is taken as a restriction to limit the range of values considered for $J(r)$ and $K(r)$. Such calculations have been made by Schwinger, Feshbach, and Pease (FES 51, PEA 51) and others (ZXX 54), often with the use of high-speed computers; forms of $J(r)$ and $K(r)$ which yield the experimental (KOL 52, HAR 53) value for Q of $2\cdot 738\times 10^{-27}$ cm^2 and which satisfy other experimental data are listed in their publications. The most favoured satisfactory potentials and wave functions correspond to values for P_D of approximately 0·04, i.e. the deuteron ground state is approximately 4 per cent. a D state. As discussed in Chapter VI this result agrees with the results of magnetic moment measurements.

XI. 5.2. *Theories of quadrupole moments of complex nuclei.* The variation of the nuclear quadrupole moments with number of nucleons in the nucleus has been studied by Kligman (KLI 48); Gordy (GOR 49); Townes, Foley, and Low (TOW 49); and others (ZXX 54). Townes, Foley, and Low have plotted the quadrupole moment divided by the square of the theoretical nuclear radius as a function of the number

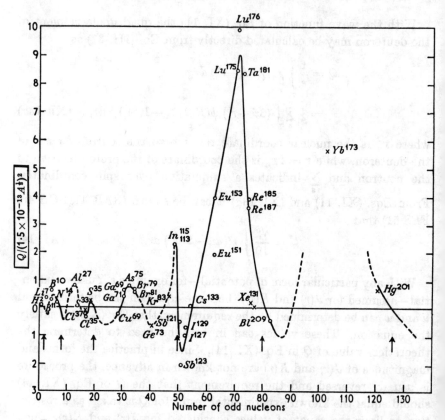

FIG. XI. 2. The plotted points are quadrupole moments divided by the square of the nuclear radius $(1\cdot5\times 10^{-13}A^{\frac{1}{3}})^2$. Moments of odd proton nuclei and of odd proton-odd neutron nuclei (except Li^6 and Cl^{36}) are plotted as circles against number of protons. Moments of odd neutron nuclei are plotted as crosses against number of neutrons. The arrows indicate closing of major nucleon shells. The solid curve represents regions where quadrupole moment behaviour seems established. The dashed curve represents more doubtful regions (TOW 49).

of odd nucleons in the nucleus as in Fig. XI. 2. This figure indeed shows marked periodicities. The arrows on the curve indicate closing of odd nucleon shells at the 'magic numbers' of like nucleons—2, 8, 20, 28, 50, 82, or 126.

The signs and magnitudes of many nuclear quadrupole moments can be accounted for satisfactorily on a simple nuclear shell model (TOW

49, RAI 50, BOH 51, BOH 53). However, other nuclear quadrupole moments do not fit such a simple theory. Thus the sign of the Li^7 quadrupole moment seems to be positive, which is opposite to that expected theoretically (KUS 49b, LOG 52). However, this discrepancy may be due to the theoretical difficulty in the estimation of the sign of q (STE 53). Moszkowski and Townes (MOS 54) point out that experimentally there seems to be a natural tendency for the formation of positive nuclear quadrupole moments. Their examination of the signs of nuclear quadrupole moments and of the nuclear states which produce them showed that positive quadrupole moment states tended to lower energy and positive quadrupole moments dominated beyond normal expectations for nuclei several particles from closed shells. Such a tendency for positive quadrupole moments is theoretically reasonable from the electrostatic and other forces of core deformation (BOH 51, BOH 53) and from tensor forces (MOS 54).

Some nuclear quadrupole moments are experimentally much too large to fit a simple nuclear shell theory; this is particularly true of such nuclei as In^{113}, In^{115}, Sb^{123}, Lu^{176}, etc. Rainwater (RAI 50), Bohr (BOH 51, BOH 53), and others (ZXX 54) have attempted to account for the large magnitudes of many nuclear quadrupole moments by considering shell models in which the nucleons moved in fields which were not spherically symmetrical.

A particularly effective approach to the theory of nuclear quadrupole moments has been made by Bohr and Mottelson (BOH 53) who attempt a unified description of nuclear structure which takes into account both the individual particle aspects of the nuclear shell model as well as the collective features associated with oscillations of the nucleus as a whole. The most important of the collective types of motion at low nuclear excitation energies are oscillations in the nuclear shape, which resemble surface oscillations. The collective motion is associated with variations of the average nuclear field, and is therefore strongly coupled to the particle motion.

The particle-surface coupling implies an interweaving of the two types of motion, which depends on the particle configuration as well as on the deformability of the surface. In the immediate vicinity of major closed shells, the high stability of the spherical nuclear shape makes the coupling relatively ineffective. In such a weak coupling situation, the nucleus can be described in terms of approximately free surface oscillations and the motion of individual nucleons in a spherical potential.

With the addition of particles, the coupling becomes more effective, and the nucleus acquires a deformed equilibrium shape. For sufficiently large deformations, a simple limiting coupling scheme is realized, which bears many analogies with that of linear molecules. In the strong coupling situation, the nucleus performs small vibrations about an axially symmetric equilibrium shape. The particles moving in the deformed field are decoupled from each other and precess rapidly about the nuclear axis, following adiabatically the slow rotations of the nuclear shape.

For nuclei with major closed shell configurations, or with a single extra particle, the expected weak coupling situation is confirmed by the small quadrupole moments and the approximate validity of simple shell models.

For configurations with a few particles, the empirical data give evidence for a major effect of the particle-surface coupling, and in regions further removed from closed shells, a rather fully developed strong coupling situation is found. This is indicated, for example, by the large nuclear quadrupole moments that exist far from closed shells (BOH 53).

XII

ATOMIC FINE STRUCTURE

XII. 1. Introduction

In the development of quantum mechanics, studies of atomic hydrogen have always played an important role since only a single electron and proton are principally involved so that the problem is favourable to accurate calculations which can be compared with experimental measurements.

Some of the most informative and most recent studies on hydrogen have been the atomic-beam experiments of Lamb, Retherford, and others (LAM 50a, LAM 51, LAM 50) on the atomic fine structure of hydrogen in its first atomic excited state. As a preliminary to this experiment a brief introductory review will be given of the state of knowledge concerning atomic hydrogen prior to the Lamb-Retherford experiment. In this introductory review the small reduced mass corrections will not be included and the nucleus will be treated as a fixed point charge Ze exerting an inverse square law Coulomb force upon the electron of mass m and charge $-e$.

According to the Bohr theory of 1913, the electron could be bound to the nucleus only in certain discrete stationary states of energy

$$W_n = -Z^2 hc Ry/n^2, \qquad \text{(XII. 1)}$$

where Ry is the Rydberg constant $Ry = e^4 m/4\pi \hbar^3 c$ and n is the principal quantum number possessing only the values 1, 2, 3,.... According to the Bohr theory, no fine structure separation was expected for the energy states of the same n but different degrees of ellipticity. Such a fine structure for atomic hydrogen had been discovered in 1887 by Michelson and Morley (LAM 51).

In 1916 Sommerfeld found that the relativistic variation of the electron mass with velocity could account for a dependence of the energy on the eccentricity of the orbit. For $n = 2$ and small atomic number, there were two possible modes of motion, one circular and one elliptic, the energy separation of the levels being

$$\Delta W_2 = (1/16)\alpha^2 Z^4 hc Ry, \qquad \text{(XII. 2)}$$

where α is the fine structure constant $e^2/\hbar c \sim 1/137$. For $Z = 1$, ΔW_2 is 0·365 cm^{-1}.

When wave mechanics was first applied to atomic hydrogen the same energy levels of Eq. (XII. 1) were predicted as for the simple Bohr theory. When the relativistic mass variation of the electron was included in wave mechanics a splitting of otherwise degenerate levels was obtained, but the separation was 8/3 times the Sommerfeld value of Eq. (XII. 2). However, when the effects of electron spin and magnetic moment postulated by Uhlenbeck, Goudsmit, and Thomas were included the fine structure separation turned out to have the Sommerfeld value of Eq. (XII. 2). Even so, there was then the difference that the extra degree of freedom associated with spin increased the number

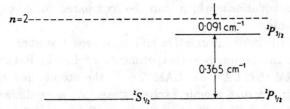

FIG. XII. 1. Fine structure of $n = 2$ levels of hydrogen according to the Dirac theory. The dotted line indicates the position of the $n = 2$ level according to the Bohr theory (LAM 51).

of quantum states; nevertheless the predicted number of spectral fine structure lines was not altered since degeneracies were also predicted. Dirac's theory of atomic hydrogen predicted the same level spacings and the same degeneracies but in a more rigorous fashion than was the case of the wave mechanical theory to which the electron spin was an *ad hoc* addition. The energy levels predicted by the Dirac theory for $n = 2$ are shown in Fig. XII. 1, which illustrates the theoretical degeneracy of the $2\,^2S_{\frac{1}{2}}$ and $2\,^2P_{\frac{1}{2}}$ states while the $2\,^2P_{\frac{3}{2}}$ state is 0.365 cm^{-1} higher. In general, the predicted energy levels according to the Dirac (DIR 47) theory should be given by the following expansion

$$W_{nj} = -Z^2 hc\,Ry/n^2 - (\alpha^2 Z^4 hc\,Ry/n^3)[(j+\tfrac{1}{2})^{-1} - (3/4n)] + \ldots. \quad \text{(XII. 3)}$$

Extensive spectroscopic studies have been made of the H_α lines (the hydrogen lines emitted in the transition $n = 3$ to $n = 2$) (LAM 51). The experimental results of Houston and of Williams (LAM 51) disagreed with the Dirac theory predictions, but the disagreement was not clearly defined because of the difficulty of the experiment and the small size of the discrepancy in comparison with the spectral line widths. Pasternack (PAS 38) pointed out that the results could be interpreted if the $2\,^2S_{\frac{1}{2}}$ level were raised about 0.03 cm^{-1} relative to the $2\,^2P_{\frac{1}{2}}$ level. However, Drinkwater, Richardson, and Williams (LAM 50a) later concluded that there was no clear evidence for departures from Dirac's theory; the predictions of the Dirac theory were therefore generally believed at the time of the Lamb–Retherford experiment. Attempts were made between 1932 and 1935 to detect in a Wood's hydrogen discharge tube the absorption of 2·74 cm microwave radiation, corresponding to the transition between the $2\,^2S_{\frac{1}{2}}$ and $2\,^2P_{\frac{3}{2}}$ states (LAM 50a). These attempts, however, were unsuccessful. Subsequent to the precision Lamb–Retherford experiment described in the next section, much less accurate but independent confirmations of the Lamb–Retherford results have been obtained in the optical spectroscopic experiments of Mack (MAC 47), Kopfermann (KOP 49a), Murakawa (MUR 49), Kuhn (SER 50), Series (SER 50), and their associates.

XII. 2. Hydrogen atomic-beam method

In the case of atomic hydrogen, the $2P$ states decay to the $1S$ state with the emission of a photon of wavelength 1,216 Å in $1\cdot6\times10^{-9}$ sec; the atom could move only $1\cdot3\times10^{-3}$ cm in that time, assuming a speed of 8×10^5 cm sec^{-1}. On the other hand the possibility exists that the $2\,^2S_{\frac{1}{2}}$ state would be sufficiently metastable so that a beam of particles in this state could be used in an atomic beam experiment. If a transition were then induced by radiofrequency or otherwise to a $2P$ state, a decay to $1\,^2S_{\frac{1}{2}}$ would take place so quickly that the number of excited atoms in the beam would be reduced. If one could then find a detector which responded selectively to the excited hydrogen atoms, one would have the possibility of measuring the energy difference between the metastable $2\,^2S_{\frac{1}{2}}$ state and the various $2P$ states.

To carry out this programme, Lamb and Retherford (LAM 50, LAM 50a, LAM 51) had to solve the following problems: (1) production of a beam of atoms in the $2\,^2S_{\frac{1}{2}}$ state, (2) stimulation of the desired microwave transition, and (3) detection of the $2\,^2S_{\frac{1}{2}}$ atoms by a means which discriminates between these and the much more abundant ground state atoms. They accomplished this with the apparatus shown in Fig. XII. 2.

Molecular hydrogen was admitted to a tungsten oven a that was heated to a temperature of 2,500° K to dissociate the hydrogen molecules into atoms. A larger scale drawing of the hydrogen dissociator is shown in Fig. XII. 3. For a pressure of 10^{-3} atmospheres inside the oven, the dissociation is 64 per cent. complete. The atoms which emerge from the oven are all in the atomic ground state and hence not yet suitable for the radiofrequency resonance.

The hydrogen atoms that emerged from the oven were excited to the

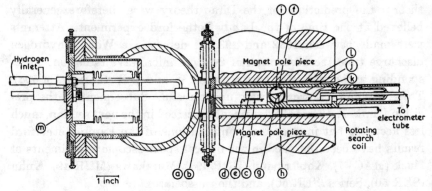

FIG. XII. 2. Cross-section of the second hydrogen fine structure apparatus showing (a) tungsten oven of hydrogen dissociator, (b) movable slits, (c) electron bombarder cathode, (d) grid, (e) anode, (f) transmission line, (g) slots for passage of metastable atoms through interaction space, (h) plate attached to centre conductor of radiofrequency transmission line, (i) d.c. quenching electrode, (j) target for metastable atoms, (k) collector for electrons ejected from target, (l) pole face of magnet, and (m) window for observation of tungsten oven temperature (LAM 50a).

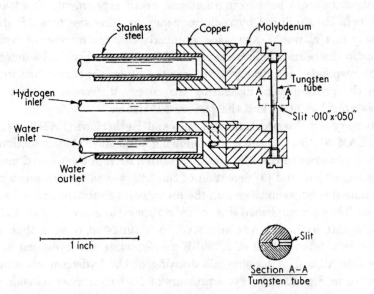

FIG. XII. 3. Detail of end of hydrogen dissociator (LAM 50a).

$n = 2$ state by electron bombardment in the region of e in Fig. XII. 2. The electrons were at 10·8 volts for which the cross-section for $2\,^2S_{\frac{1}{2}}$ excitation was approximately 10^{-17} cm². It was estimated that approximately one atom in forty million would be excited to the $2\,^2S_{\frac{1}{2}}$ state.

The beam next passed into the radiofrequency region in which the transitions were induced. This consisted essentially of an enlarged coaxial line with apertures for the admission of the beam and with a molybdenum sheet welded to the central conductor to increase the effective length of the radiofrequency region and hence to diminish the line width. An auxiliary electrode i was placed in the same region to permit quenching of the metastable atoms when desired by the application of the electric field that arises when 125 volts is applied to the electrode i. The quenching mechanism is that the Stark effect from the electric field mixes the $2\,^2S_{\frac{1}{2}}$ and the $2\,^2P_{\frac{1}{2}}$ states whereby the atom can decay to the ground $1\,^2S_{\frac{1}{2}}$ state via the P state. The change in the detected beam when this quenching was induced was very convenient as a measure of the number of $2\,^2S_{\frac{1}{2}}$ atoms in the beam.

The detector consisted of a tungsten target j and electron collector k. When a metastable hydrogen atom comes close to the target surface it can undergo an inelastic collision of the second kind in which the atom returns to the ground state and an electron is liberated from the system to take up the excess energy (LAM 50a) as discussed in greater detail in Section XIV. 3.1. The electron current to the collector k was then measured with an FP54 electrometer which used an input resistance of 7×10^{10} ohms and possessed an overall current sensitivity of about $1\cdot9 \times 10^{-16}$ amp mm^{-1} with fluctuations of about 1 mm. The measured electron current diminished when the microwaves induced transitions resonantly from the $2\,^2S_{\frac{1}{2}}$ state to a $2P$ state, since an atom in the latter state quickly decayed to the ground $1\,^2S_{\frac{1}{2}}$ state.

As shown in Fig. XII. 2, the excitation, transition, and detection regions were all in a uniform magnetic field whose value could be varied from near zero to 4,000 gauss. In this way the Zeeman effect of the resonance lines could be observed.

Numerous experimental details and necessary precautions are discussed in detail in the original papers of Lamb, Retherford, Triebwasser, and Dayhoff (LAM 50a).

XII. 3. Hydrogen and deuterium results

Since the experiments have ordinarily been performed in a magnetic field, the Zeeman effect on the atomic levels needs to be considered as a preliminary to the discussion of experimental results. According to the Dirac theory, the Zeeman splitting of the fine structure of the $n = 2$ states for hydrogen should be as shown in Fig. XII. 4 in analogy to the earlier derived Fig. III. 4. However, since the most significant

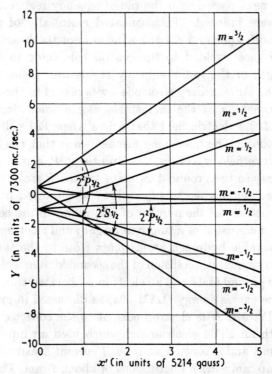

FIG. XII. 4. Zeeman splitting of the fine structure states of $n = 2$ of hydrogen according to the Dirac theory. $x' = 3\mu_0 H/2(W_{P_{\frac{3}{2}}} - W_{P_{\frac{1}{2}}})_0 \approx H(\text{in gauss})/5214$ and Y is the energy (LAM 50a).

result of the Lamb–Retherford experiment is the demonstration that the $2\,^2S_{\frac{1}{2}}$ and $2\,^2P_{\frac{1}{2}}$ states are not degenerate, a similar theoretical plot is given in Fig. XII. 5 on the assumption that the $2\,^2S_{\frac{1}{2}}$ state is raised 1,000 Mc. It should be noted that the scale for Fig. XII. 5 is somewhat larger than for Fig. XII. 4. The level designations α, β, a, b,... are used in subsequent figures to identify the observed transitions. The above figures are applicable only when nuclear moment effects are neglected. When hyperfine structure is included each line becomes multiple with characteristic structure at its low field end. For atomic hydrogen in a $2\,^2S_{\frac{1}{2}}$ state this characteristic structure has been derived in Chapter III and is given in Fig. XII. 6. This structure provides an

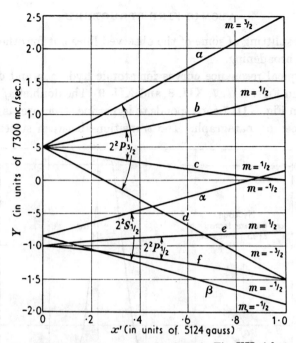

Fig. XII. 5. Zeeman energy levels as in Fig. XII. 4 but with $2\,^2S_{\frac{1}{2}}$ pattern raised by 1,000 Mc/sec. x' and Y are defined as in Fig. XII. 4 (LAM 50a).

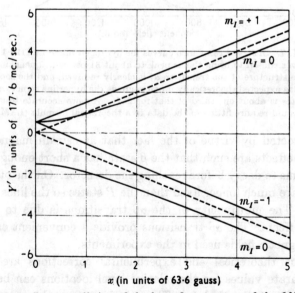

Fig. XII. 6. Zeeman splitting of the hyperfine structure of the $2\,^2S_{\frac{1}{2}}$ level. The zero field separation is taken to be one-eighth that measured for the ground state by Nafe and Nelson. The dotted lines show the energy levels obtained if hyperfine structure is ignored. x is as defined in Eq. (III. 120) and Y' is the energy (LAM 50a).

observable splitting of some of the observed lines but for others merely provides a broadening.

Some typical resonance curves for atomic hydrogen and deuterium are shown in Figs. XII. 7, XII. 8, and XII. 9. The doublet αf of atomic hydrogen in Fig. XII. 7 corresponds to hyperfine structure, as discussed in the preceding paragraph. The transitions between states α and β

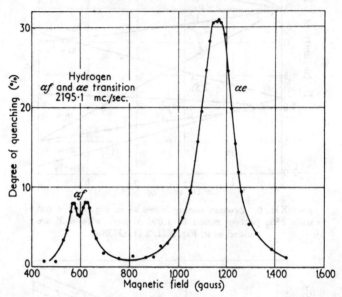

FIG. XII. 7. Observed resonance curves for hydrogen taken at 2195·1 Mc/sec with power chosen to bring the αe peak to about 31 per cent. quenching. The hyperfine structure of the transition αf is clearly resolved, and the separation is about as expected theoretically if one assumes the hyperfine structure of the $2^2P_{\frac{1}{2}}$ states is about one-third of that for $2^2S_{\frac{1}{2}}$. A more accurate test of this ratio would require fitting of the data to a theoretical formula (LAM 50a).

can be detected by virtue of the fact that at certain magnetic fields quenching effects are such that the β state has a short enough lifetime for the atoms to decay before reaching the detector. On the other hand, the states are much longer lived than the P states so the lines are much narrower. The multiplicity of the $\alpha\beta$ transitions is due to hyperfine structure effects. The $\alpha\beta$ transitions provide a convenient calibration for the magnetic fields used in the experiments.

Numerous theoretical and experimental corrections are required before accurate values for the relative level locations can be inferred from the observational data. These include quenching asymmetry, corrections for line asymmetry, Stark effects, radiative and non-radiative corrections to line shape, hyperfine structure effects, and damping

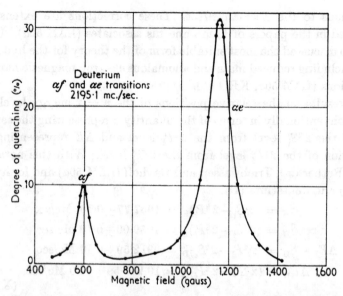

Fig. XII. 8. Observed resonance curves for deuterium, taken at 2195·1 Mc/sec with power chosen to bring the αe peak to about 31 per cent. Each peak is the composite of three unresolved peaks separated by about 9 gauss (compared to about 59 gauss for hydrogen) (LAM 50a).

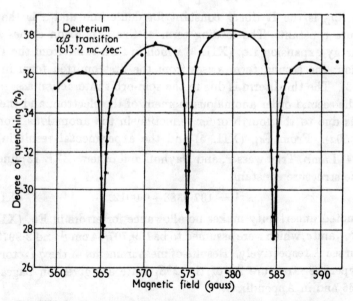

Fig. XII. 9. Sharp resonance curves for transitions αβ in deuterium at a frequency of 1613·2 Mc/sec, for which the broad resonance αe overlaps the sharp resonance αβ (LAM 50a).

corrections to the Zeeman effect. These corrections are extensively discussed in the papers of Lamb and his associates (LAM 50a). Lamb has also discussed the most suitable form of the theory for the hydrogen atom including reduced mass and anomalous electron magnetic moment corrections (LAM 50a, KRO 49).

The results on the relative positions of the $n = 2$ energy levels are given conventionally in terms of the quantity S representing the separation of the $2\,^2S_{\frac{1}{2}}$ level from the $2\,^2P_{\frac{1}{2}}$ level and ΔE representing the separation of the $2\,^2P_{\frac{3}{2}}$ level from the $2\,^2P_{\frac{1}{2}}$ level. With this notation, Lamb, Retherford, Triebwasser, and Dayhoff (LAM 50a) find for hydrogen (H) and deuterium (D):

$$S_H = (2\,^2S_{\frac{1}{2}} - 2\,^2P_{\frac{1}{2}})_H = 1057 \cdot 77 \pm 0 \cdot 10 \text{ Mc/sec}$$
$$S_D = (2\,^2S_{\frac{1}{2}} - 2\,^2P_{\frac{1}{2}})_D = 1059 \cdot 00 \pm 0 \cdot 10 \text{ Mc/sec}$$
$$\Delta E_D - S_D = (2\,^2P_{\frac{3}{2}} - 2\,^2S_{\frac{1}{2}})_D = 9912 \cdot 59 \pm 0 \cdot 10 \text{ Mc/sec}$$
$$\Delta E_D = (2\,^2P_{\frac{3}{2}} - 2\,^2P_{\frac{1}{2}})_D = 10{,}971 \cdot 59 \pm 0 \cdot 20 \text{ Mc/sec.}$$

(XII. 4)

As discussed in greater detail by Lamb (LAM 50a), the theoretical expression for ΔE in deuterium is

$$\Delta E_D/h = \frac{cRy_D}{16} \alpha^2 \left[1 + \tfrac{5}{8}\alpha^2 + \frac{\alpha}{\pi} - \frac{5 \cdot 946}{\pi^2}\alpha^2 \right], \qquad \text{(XII. 5)}$$

where Ry_D is the Rydberg constant for deuterium and α is the fine structure constant. The leading term of this expression comes from the energy expansion Eq. (XII. 3) which in turn came from the Dirac equation; the second term arises from the first omitted term in Eq. (XII. 3). The third term is due to the spin-orbit interaction associated with the second order anomalous moment of the electron, and the last term is due to the fourth order reduction in the anomalous moment (LAM 50a). From Eq. (XII. 5) and the experimental result in Eq. (XII. 4) Lamb, Triebwasser, and Dayhoff and others (ZZB 13) find for the fine structure constant

$$\alpha^{-1} = 137 \cdot 0388 \pm 0 \cdot 0012. \qquad \text{(XII. 6)}$$

The quoted uncertainty makes no allowance for errors in Eq. (XII. 5) or in Ry_D and c, which were assumed to be 109,707·44 cm^{-1} and $2 \cdot 99790 \times 10^{10}$ cm sec^{-1}, respectively. Results of measurements of the g-factor and the hyperfine structure $\Delta \nu$ of the $2\,^2S_{\frac{1}{2}}$ state are given on pages 260 and 266 and in Appendix G.

XII. 4. Fine structure of singly ionized helium

Lamb and Skinner (LAM 50, LAM 51) have studied the fine structure of singly ionized helium (He$^+$ or He II) by a microwave method. Since this experiment is not of a molecular beam type, it will not be described in detail but some description will be given because of the close relation of this experiment to the atomic hydrogen fine structure experiment described previously, which was by an atomic-beam method.

The relevant energy levels for He and He⁺ are shown in Fig. XII. 10. The resonance transition that was chiefly studied was

$$2\,{}^2S_{\frac{1}{2}}(m = +\tfrac{1}{2}) \to 2\,{}^2P_{\frac{1}{2}}(m = -\tfrac{1}{2}),$$

which occurs at approximately 14,000 Mc/sec.

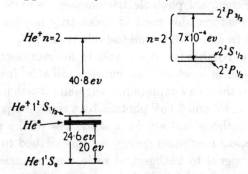

FIG. XII. 10. Energy levels of He and He⁺ (LAM 50).

In contrast to the case of hydrogen it is difficult to form a beam of metastable helium ions. Even if one could do so, the detection scheme used for hydrogen would not be satisfactory. It is known that ions may eject electrons from metal surfaces, so that there would be a large background detector signal due to the helium ions in the ground state which could not be separated from the much smaller number of metastable ions.

For this reason Lamb and Skinner (LAM 50) adopted another method. Helium gas was bombarded by electrons of several hundred volts energy. The various spontaneous transitions which occur in the bombardment region are accompanied by the emission of ultraviolet radiation and may be detected photoelectrically. As can be seen from Fig. XII. 10, this radiation consists of (a) photons of about 20–24·6 volts emitted in transitions from excited states of un-ionized helium to its ground state, (b) radiation of energy 41–54·4 eV emitted in transitions from excited states to the ground state of the singly ionized He⁺, and (c) radiation emitted in other transitions. Of primary interest is the radiation of 41 eV energy emitted in transitions $n = 2 \to n = 1$ of the ion. This may arise either from states $2\,{}^2P_{\frac{1}{2}}$ or $2\,{}^2P_{\frac{3}{2}}$. The $2\,{}^2S_{\frac{1}{2}}$ state, being metastable with a life of 2×10^{-3} sec, would not ordinarily contribute to the 41 eV radiation since the atoms in this state would usually diffuse to the walls or to another part of the chamber where a perturbing electric field would cause them to decay. If quenching fields are present or if a microwave field induces a transition from $2S$ to $2P$, an additional amount of 41 eV radiation is produced. The experiment consists of a detection of the increase in this radiation when such transitions are induced by radio waves of a frequency corresponding to the energy difference $2\,^2S_{\frac{1}{2}} \to 2\,^2P_{\frac{1}{2}}$.

Since the efficiency of photoelectric emission is quite high both for 20 and 41 eV photons, this method necessarily implies a large background signal. In addition, the metastable atoms $2\,^1S_0$ of un-ionized He produced in the bombardment are able to liberate electrons from the photosensitive surface which are indistinguishable from the photoelectrons due to the 20 eV radiation. Originally it was hoped to distinguish between 41 eV and 20 eV photons by a retarding potential placed between photocathode and anode, but too few of the photoelectrons had the theoretical maximum energy for this method to succeed.

Instead, the signal to background ratio was increased by inserting a film of collodion a few hundred Ångströms thick between the bombardment and detector regions. This completely absorbed the metastable atoms and, furthermore, absorbed the softer photons preferentially. Even with the help of the collodion film, however, there was a background eight times as large as the signal due to the 41 eV radiation from the $2\,^2P$ states of the ion. With this background, nevertheless, significant resonance curves could be obtained.

A cross-sectional diagram of the apparatus used by Lamb and Skinner (LAM 50) is shown in Fig. XII. 11. A and B are pole pieces of a magnet, C is a waveguide, and F is a collodion film. G is a hot filament maintained at a negative potential, so that electrons of several hundred volts energy stream across the waveguide into the opposite pole piece and are collected on a metal plate at H. The photons are detected with a photoelectric detector, which consists of a copper plate D and an electron collector E. The current from D to E is measured with an electrometer tube and a sensitive galvanometer.

As in the experiment with hydrogen the measurements are carried out in the presence of a magnetic field so that Zeeman splittings are observed. A typical resonance curve for the αf transition with

$$2\,^2S_{\frac{1}{2}}(m = \tfrac{1}{2}) \to 2\,^2P_{\frac{1}{2}}(m = -\tfrac{1}{2})$$

is shown in Fig. XII. 12.

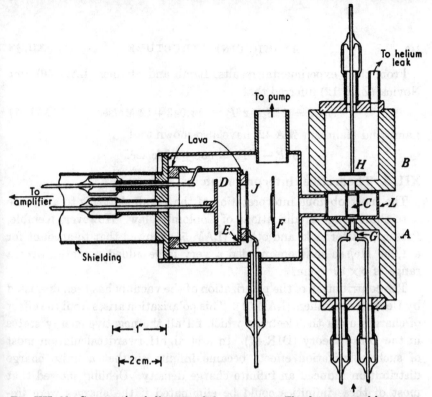

FIG. XII. 11. Cross-sectional diagram of the apparatus. The data reported here were taken with radiofrequency radiation of about 1·6 cm wavelength transmitted through the K band waveguide C. This guide could be slipped out of the apparatus, however, and long wavelengths could be transmitted through the x-band guide L. J, a disk with a circular hole, defined the solid angle subtended by the detector plate at the excitation region. J was maintained at a positive potential with respect to D and E so that electrons ejected from J would not reach D (LAM 50).

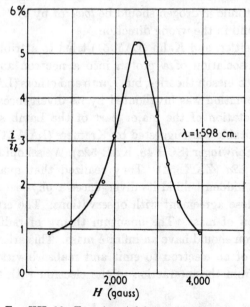

FIG. XII. 12. Experimental fine structure resonance curve for singly ionized helium (LAM 50).

From their experimental results, Lamb and Skinner (LAM 50) and Novick (ZZB 12) inferred that

$$S_{\text{He}^+} = 2\,^2S_{\frac{1}{2}} - 2\,^2P_{\frac{1}{2}} = 14{,}043 \pm 12 \text{ Mc/sec}. \qquad (\text{XII. 7})$$

Lamb and Maiman (ZZA 42) have also shown that

$$3\,^3P_1 - 3\,^3P_2 = 658 \pm 1 \text{ Mc/sec}.$$

XII. 5. Theoretical interpretation

The most obvious interpretation of the Lamb shift S is to attribute it to short range modifications of Coulomb's law. However, Kemble, Present, Pasternack, and others (LAM 51) showed that to account for a 1,000 Mc/sec shift one needed a very large added potential over a range of 5×10^{-12} cm.

The contribution of the polarization of the vacuum has been discussed by Uehling and others (LAM 51). This polarization arises from the effect of charges upon the electrons which fill all the negative energy states in the Dirac theory (DIR 47). In a straightforward calculation most of such polarization effects became infinite so that a finite charge distribution induced an infinite charge density. Uehling showed that most of these infinities could be eliminated if the charges were 're-normalized', i.e. if the inducing charge and the polarization charge were combined together and set equal to the experimental charge. In this way the charge divergences were attributed to the renormalization of charge, and the remaining finite polarization charge density could be taken as physically observable. In this way Uehling showed that the $2\,^2S_{\frac{1}{2}}$ level of atomic hydrogen should be *lowered* by 27 Mc/sec, which is far too small and in the wrong direction.

Fröhlich, Heitler, and Kahn (LAM 51) tried to attribute the shift to the partial dissociation of a proton into a neutron and a meson in accordance with meson theories, but Lamb and others (LAM 51) showed that their calculation was invalidated by its divergences.

The interpretation of the major part of the Lamb shift that now seems most suitable was suggested by Kramers (LAM 51), Bethe (BET 50, BAR 53), Schwinger (SCH 48, KAR 52a), Weisskopf (WEI 49), and others (KAR 52a, ZXX 54). They realized that quantum electrodynamics had hidden behind its divergences a physical content which was in very close agreement with observations. The crucial idea was renormalization of mass. The quantum theory of radiation predicts that the electron should have an infinite mass. This arises in part from the tendency of an electron to emit and reabsorb virtual quanta or equivalently from the interaction of the electron with the zero point oscillations of the electromagnetic field. Some explanation should indeed be found for the fact that the observed mass is finite, but Bethe and others recognized that this was temporarily outside the scope of the theory and proposed to ignore the electromagnetic mass completely. For an electron bound in a hydrogen atom, an infinite energy occurs, but this is merely a manifestation of the infinite electromagnetic mass which should be eliminated in some future theory. If the mass terms are properly subtracted, a finite remainder is obtained which would be zero for a free electron. In the case of a bound electron the force field which produces the binding modifies the effect of the electromagnetic field and a finite displacement of energy levels results.

The calculation of all the corrections that enter into the theory of the Lamb shift has been the subject of many papers by Bethe (BET 50), Schwinger (SCH 48, KAR 52a), Weisskopf (WEI 49), Feynman (KAR 52a), Lamb and Kroll (KRO 49), Salpeter (SAL 53), and others (ZXX 54). Their principal results will be reviewed at the end of this section, but a detailed discussion of these calculations would be quite inappropriate in the present volume. However, Weisskopf and Welton (WEI 49) have suggested a simple means by which the principal part of the Lamb shift can be derived and an adaptation of this discussion is given below.

In quantum electrodynamics (HEI 53, SCH 48) the vector potential is ordinarily expanded in terms of the normal standing wave modes inside a large box of volume Ω in which the radiation is assumed to be confined. In such an expansion it is easily shown (HEI 53, SCH 48) that the coefficients of the expansion satisfy the simple harmonic oscillator differential equation. Consequently quantization is achieved by quantizing the amplitudes as simple harmonic oscillators. One result of this is the existence of zero point oscillations in the electromagnetic fields.

The fluctuating field strength can contribute to the self energy of an electron in the hydrogen atom in two ways. One is that it can interact with it directly regardless of the presence of the proton's Coulomb field. This interaction turns out theoretically to be infinite in the present state of the theory. However, it is also of no interest since it only contributes to the electron's rest mass; this divergence is therefore ignored in the process of mass renormalization. The second effect of the fluctuating electric fields is to cause the electron to vibrate in such a way that its effective charge is smeared out. Consequently the average electrostatic interaction of the electron with a non-uniform electric field is modified. It is this change of the electron's potential energy in a hydrogen atom that gives rise to most of the Lamb shift.

Consider a stationary state n of the electron in an atom where the Coulomb potential is
$$V(\mathbf{r}) = Ze/r. \qquad (\text{XII. 8})$$
The average potential energy of the electron in this field is
$$W_n = -e \int \overline{V} |\psi_n(\mathbf{r})|^2 \, d\tau, \qquad (\text{XII. 9})$$
where $\overline{V}$ is the average potential of the electron when at average position $\mathbf{r}$. When the effects of zero point vibration are then included the instantaneous position of the electron is at $\mathbf{r}+\boldsymbol{\delta}$ where $\boldsymbol{\delta}$ is the displacement due to zero point vibration. Then
$$\overline{V} = \overline{V(\mathbf{r}+\boldsymbol{\delta})} = V(\mathbf{r}) + \nabla V \cdot \boldsymbol{\delta} + \tfrac{1}{2} \sum_{ij} \left(\frac{\partial^2 V}{\partial r_i \partial r_j} \right) \overline{\delta_i \delta_j}, \qquad (\text{XII. 10})$$
where δ_1 is the x component of $\boldsymbol{\delta}$, etc. Then as $\overline{\delta_i} = 0$, $\overline{\delta_i \delta_j} = 0$ for $i \neq j$, and $\overline{\delta_i^2} = \overline{\delta^2}/3$,
$$\overline{V} = V(\mathbf{r}) + \tfrac{1}{2} \nabla^2 V \overline{\delta^2}/3. \qquad (\text{XII. 11})$$
Consequently, the additional electrostatic potential energy due to the vibration of the electron is
$$\delta W_n = -\frac{e}{6} \int \nabla^2 V \overline{\delta^2} |\psi_n(\mathbf{r})|^2 \, d\tau. \qquad (\text{XII. 12})$$
However, if ρ_0 is the nuclear charge density and if the nucleus is approximated to a point charge
$$\nabla^2 V = -4\pi\rho_0 = -4\pi Ze\, \delta(\mathbf{r}). \qquad (\text{XII. 13})$$
Then
$$\delta W_n = \frac{2e^2}{3} \pi Z \int \delta(\mathbf{r}) \overline{\delta^2} |\psi_n(r)|^2 \, d\tau$$
$$= \frac{2e^2}{3} \pi Z \overline{\delta^2} |\psi_n(0)|^2. \qquad (\text{XII. 14})$$

If a free particle whose position is designated by $\boldsymbol{\delta}$ is subject to an oscillatory field $E_0 \cos \omega t$
$$m\ddot{\boldsymbol{\delta}}_\omega = -eE_0 \cos \omega t,$$
so
$$\delta_\omega = \frac{eE_0}{m\omega^2} \cos \omega t$$
and
$$\overline{\delta_\omega^2} = \frac{e^2 E_0^2}{m^2 \omega^4} \overline{\cos^2 \omega t} = \frac{e^2 E_0^2}{2m^2 \omega^4}. \qquad (\text{XII. 15})$$

The amplitude E_0 of the zero point oscillation of the electromagnetic mode can be obtained from the fact that the energy of its oscillation inside the volume Ω, in which it is assumed to be confined, is equal to the normal zero point energy $\frac{1}{2}\hbar\omega$ of a simple harmonic oscillator. Then

$$\tfrac{1}{2}\hbar\omega = \frac{1}{8\pi}\int (\mathbf{E}^2+\mathbf{H}^2)\,d\tau = \frac{E_0^2}{8\pi}\Omega,$$

so
$$E_0^2 = \frac{4\pi}{\Omega}\hbar\omega$$

and
$$\overline{\delta_\omega^2} = \frac{2\pi}{\Omega}\frac{\hbar e^2}{m^2\omega^3}. \tag{XII. 16}$$

In a volume Ω, with the inclusion of the two possible polarizations the number of modes of oscillation between ω and $\omega+d\omega$ is (PAU 35)

$$\Omega Z(\omega)\,d\omega = 4\pi\frac{\nu^2}{c^3}d\nu\,.\,2\,.\Omega = \frac{\omega^2\,d\omega}{\pi^2 c^3}\Omega. \tag{XII. 17}$$

For independent modes of oscillation

$$(\boldsymbol{\delta}_{\omega'}+\boldsymbol{\delta}_{\omega''})^2 = \delta_{\omega'}^2+\delta_{\omega''}^2.$$

Therefore,
$$\overline{\delta^2} = \Omega\int \overline{\delta_\omega^2}\,Z(\omega)\,d\omega = \frac{2}{\pi}\frac{\hbar e^2}{m^2 c^3}\int_{\omega_1}^{\omega_2}\frac{d\omega}{\omega}$$
$$= \frac{2}{\pi}\alpha(\hbar/mc)^2\ln(\omega_2/\omega_1), \tag{XII. 18}$$

where α is the fine structure constant $e^2/\hbar c$.

However, the wave function of the hydrogen atom is such that for S states (PAU 35, CON 35)

$$|\psi_n(0)|^2 = Z^3/\pi a_0^3 n^3, \tag{XII. 19}$$

where a_0 is the Bohr radius $\hbar^2/me^2$.

From Eqs. (XII. 14), (XII. 18), and (XII. 19) then

$$\delta W_n = \frac{2e^2}{3}\pi Z\,.\,\frac{2}{\pi}\alpha\left(\frac{\hbar}{mc}\right)^2\ln(\omega_2/\omega_1)\,.\,\frac{Z^3}{\pi a_0^3 n^3}. \tag{XII. 20}$$

The constants in the above expression can be recombined and expressed in terms of α and the Rydberg constant $Ry = (me^4/2\hbar^2)/hc = (e^2/2a_0)/hc$. With the above approximations the shift of the S level in frequency units is then

$$\delta W_n/h = \left(\frac{\alpha^3 Ry\,c}{3\pi}\right)\frac{8Z^4}{n^3}\ln\frac{\omega_2}{\omega_1}. \tag{XII. 21}$$

The determination of the limits ω_2 and ω_1 of the integral in Eq. (XII. 18) is quite complicated and has been discussed extensively by Bethe and others (BET 50, SAL 53). The upper limit arises from the fact that for $\hbar\omega > 2mc^2$ pair creation becomes possible, so as this frequency is approached one might expect the properties of a vacuum pertaining to virtual electron-positron pair creation to affect the results. A complete calculation (BET 50, SCH 48, KAR 52a, SAL 53, BAR 53, ZXX 54) indeed shows that this is the case and that the effect of virtual pair creation or of polarization of the vacuum is to induce electron vibrations that are out of phase with the vibrations caused by the electromagnetic field in such a way that the integral converges at a frequency of the order of mc^2. For this reason, ω_2 may be assumed to be given by

$$\omega_2 = mc^2/\hbar\kappa, \qquad (XII.\ 22)$$

where κ is a constant of the order of unity. If now the new parameter k_1 is introduced in accordance with $k_1 = \hbar\omega_1\kappa$, the shift of the S level may be written as

$$\delta W_n/h = \left(\frac{\alpha^3 Ry\, c}{3\pi}\right)\frac{8Z^4}{n^3}\ln\frac{mc^2}{k_1}. \qquad (XII.\ 23)$$

The magnitude of k_1 can be estimated since the lower integration limit ω_1 arises from the fact that the electron can no longer be regarded as free when ω is in the neighbourhood of atomic orbital angular frequencies, $2\pi Ry\, c$. Furthermore, since the electron contributes most importantly to the Lamb shift when it is nearest to the nucleus, the effective lower limit ω_1 should be considerably larger than $2\pi Ry\, c$ and in fact Bethe (BET 50) has shown that for this theory $\kappa\omega_1$ should be $17\cdot 8(2\pi Ry\, c)$ whence k_1 is $17\cdot 8 Ry\, hc$.

Relativistic treatments of the Lamb shift have been carried out by Kroll and Lamb (KRO 49, LAM 51), Schwinger (SCH 48), Weisskopf (WEI 49), Feynman (BAR 53), and others (SAL 53, ZXX 54). The result to lowest order in both α and $Z\alpha$ is that the Lamb shift $S^{(1)}_\infty$ between the $n\,^2S_{\frac{1}{2}}$ and $n\,^2P_{\frac{1}{2}}$ states in a fixed Coulomb potential is

$$S^{(1)}_\infty = \left(\frac{\alpha^3 Ry\, c}{3\pi}\right)\frac{8Z^4}{n^3}\left\{\left[\ln\frac{mc^2}{k_0(2,0)} - \ln 2 + \tfrac{5}{6}\right] - \tfrac{1}{5} - \left[\ln\frac{Ry\, hc}{k_0(2,1)} - \tfrac{1}{8}\right]\right\},$$
$$(XII.\ 24)$$

where the k_0's are average excitation energies evaluated by Bethe, Brown, and Stehn (BET 50) to have the following values for hydrogen in its $n = 2$ state:

$$k_0(2, 0) = (16\cdot 646 \pm 0\cdot 007) Ry\, hc,$$
$$k_0(2, 1) = (0\cdot 9704 \pm 0\cdot 0002) Ry\, hc. \qquad (XII.\ 25)$$

The first term, in the first square brackets, of Eq. (XII. 24) comes from the shift of the $^2S_{\frac{1}{2}}$ level, excluding effects of vacuum polarization. It is similar in form to the expression derived classically in Eq. (XII. 23). The term involving $(-\tfrac{1}{5})$ is the contribution of vacuum polarization to the $^2S_{\frac{1}{2}}$ level. The last term, in square brackets, comes from the shift of the $^2P_{\frac{1}{2}}$ level. For $Z = 1$ and $n = 2$, $S_\infty^{(1)}$ acquires (SAL 53) the numerical value 1052·14 Mc/sec. This value of $S_\infty^{(1)}$ includes a contribution of $-27\cdot13$ Mc/sec from the vacuum polarization and of $+67\cdot82$ Mc/sec from the anomalous magnetic moment of the electron.

The next term in the expansion in powers of $Z\alpha$ (Coulomb potential acting twice, etc.) has been calculated by Baranger (BAR 53), by Karplus and Klein (KAR 52a), and by others to be $+7\cdot14$ Mc/sec for $Z = 1$ and $n = 2$ (SAL 53, ZZA41). Of the same order in α is the term corresponding to two virtual photons. This consists of three parts: (a) the fourth-order anomalous magnetic moment of the electron contributes $-0\cdot94$ Mc/sec; (b) the fourth-order vacuum polarization contributes $-0\cdot24$ Mc/sec; and (c) the remaining fourth-order radiative correction terms of $+0\cdot24$ Mc/sec. The sum of the terms of this order in α or $Z\alpha$ is $(6\cdot20\pm0\cdot10)$ Mc/sec (SAL 53). All of these terms are fourth-order terms whereas those in Eq. (XII. 24) are second order. When these are combined with the value of $S_\infty^{(1)}$, the best theoretical value for S_∞ with the omission of still higher orders is

$$S_\infty = 1058\cdot34 \pm 0\cdot12 \text{ Mc/sec}. \quad \text{(XII. 26)}$$

For this to be applied to either hydrogen or deuterium, corrections of $-1\cdot175$ Mc/sec must be made for the effect of the nuclear mass on the level shift in hydrogen, of $-0\cdot588$ Mc/sec for the same effect in deuterium, of $+0\cdot733$ Mc/sec for the effect of the deuteron radius, and of $+0\cdot025$ Mc/sec for the effect of electron-nucleon interaction in hydrogen as discussed by Salpeter (SAL 53). When these are all combined the theoretical values for the Lamb shift in hydrogen and deuterium are (ZZB 9)

$$S_H = (1057\cdot99 \pm 0\cdot13) \text{ Mc},$$
$$S_D = (1059\cdot23 \pm 0\cdot13) \text{ Mc}. \quad \text{(XII. 27)}$$

A comparison between Eqs. (XII. 4) and (XII. 27) shows that the experimental values are at present about 0·2 Mc lower than the theoretical values. This is excellent agreement and confirms not only the major interpretation of the Lamb shift but also establishes experimentally such small effects as the vacuum polarization term of -27 Mc. The 0·6 Mc discrepancy has not as yet been resolved. It might be due to deviation from the Coulomb field at short distances, but it seems unlikely that the deviation would produce so large an effect and that it would be the same in hydrogen and deuterium. The discrepancy may be due to relativistic corrections of the next order; such corrections might reasonably be of the required magnitude (BAR 53).

XIII

MOLECULAR-BEAM DESIGN PRINCIPLES

'It's all a matter of intensity.' S. Millman

XIII. 1. Introduction

THERE is no simple general procedure by which a molecular-beam experiment can be designed optimally. Each proposed experiment usually has its own special problems which prevent the rigid application of general design principles. Thus in some cases only small amounts of source material may be available while in others there may be unlimited amounts of source material but the isotopic dilution may be great. In some cases a small magnetic moment may need to be measured in a molecule which also possesses a large moment. In some cases a surface ionization detector can be used, in some a Pirani gauge is required, while in others the detection may be by nuclear radioactivity. In some experiments the requirements for narrow resonances and high precision dominate while in others the dominant design criteria may be determined by the need for adequate beam intensity for any observations to be made at all. In each of these cases the design criteria are quite different.

On the other hand, there are many design relations that apply in one way or another to most experiments. Therefore, the present chapter will be devoted to a discussion of these design principles. To prevent excessive generality, the discussion will primarily be applied to molecular-beam magnetic-resonance experiments. However, most of the design criteria can also be adapted to quite different experiments, such as electric-resonance experiments or molecular-beam scattering experiments.

Most of the fundamental design principles for molecular-beam experiments were laid down by Stern and Knauer in their early fundamental papers (STE 26, KNA 26, KNA 29). What has been added to the subject since then has in some cases never been published and in many cases has appeared merely incidentally in various published articles on molecular-beam experiments. The chief collected discussions of molecular-beam design principles are contained in Fraser's two books (FRA 31, FRA 37) and in review articles by Estermann (EST 46), Kellogg and Millman (KEL 46), and Kusch (KUS 50).

XIII. 2. Formulae useful in design of molecular-beam experiments

Most of the molecular-beam intensity and deflexion formulae have been derived in earlier chapters of this book. However, those which are particularly relevant to molecular-beam design problems will be collected together here. The definitions of the quantities which enter into the equations may be found with the original equations, to which cross-references are given below.

The beam intensity I, or number of molecules which strike the detector per second, from Eqs. (II. 15) and (II. 16) is given by

$$I = \frac{1}{4\pi} \frac{A_d}{l_0^2} n_s \bar{v} A_s \qquad (XIII. 1)$$

$$= 1 \cdot 118 \times 10^{22} \frac{p' A_s A_d}{l_0^2 \sqrt{(MT)}} \text{ molecules sec}^{-1},$$

where M is the molecular weight and p' is the source pressure in mm of Hg. The total rate at which molecules emerge from the source is given by Eq. (II. 7)

$$Q = (1/\kappa)\tfrac{1}{4} n \bar{v} A_s, \qquad (XIII. 2)$$

where $1/\kappa$ is unity for very short apertures and is given by Eqs. (II. 8) to (II. 14) for apertures of various shapes and dimensions.

The source pressures that may be used are, however, limited to being below those given by Eqs. (II. 3) to (II. 5) or by

$$w_s = \lambda_{Ms} = \frac{1}{n\sigma\sqrt{2}} = 7 \cdot 321 \times 10^{-20} \frac{T}{p'\sigma} \text{ cm}. \qquad (XIII. 3)$$

For long source channels an even lower source pressure is required since the mean free path needs to be more nearly comparable to the length of the channel so as to avoid excessive scattering in the channel.

If the pressure in the system is sufficient to produce attenuation by scattering, the resultant beam intensity I' as in Sections II. 2.1 and II. 4.1 is given approximately by

$$I' = I \exp(-\sum_k l_k/\lambda_k), \qquad (XIII. 4)$$

where λ_k is the average mean free path in section k of the apparatus. For a discussion of the effects of such attenuation on the velocity distribution the reader is referred to Sections II. 2.1 and II. 4.1. There is some inconclusive evidence (BAR 52, BRA 52a) that the beam intensity achieved experimentally with molecular hydrogen may be less by a factor of up to 50 in some experiments than that estimated from Eqs. (XIII. 1) and (XIII. 4) with the λ_k's estimated from the ionization gauge pressures of the apparatus.

As an indication of the beam intensities required it may be mentioned that a current of one electron per second equals $1{\cdot}602\times10^{-19}$ amp. With a good FP 54 amplifier in conjunction with a surface ionization detector, current changes as low as about 10^{-15} amp can be conveniently observed. With electron multipliers and digital counters the sensitivity can be increased as discussed in Chapter XIV; on the other hand with a Pirani detector the sensitivity is much less.

If the beam is defined by a source of slit width w_s and a collimator of width w_c and if it is measured with a very narrow detector, the beam shape is in the form of a trapezoid whose narrow width is $2p$ and whose base width is $2d$, where from Eq. (II. 17)

$$p = \tfrac{1}{2}|w_c+(w_c-w_s)a|,$$
$$d = \tfrac{1}{2}[w_c+(w_c+w_s)a],$$
$$a = l_{cd}/l_{sc} = (l_{sd}/l_{sc})-1 = r-1. \qquad (XIII.\ 5)$$

If $w_s = w_c = w$ and if $a = 1$,

$$p = \tfrac{1}{2}w, \qquad d = \tfrac{3}{2}w. \qquad (XIII.\ 6)$$

Analytical expressions for the beam intensity at different parts of the trapezoid are given in Eqs. (II. 21) to (II. 23).

The velocity distribution in the beam by Eq. (II. 26) is given by

$$I(v) = \frac{2I_0}{\alpha^4}v^3\exp(-v^2/\alpha^2), \qquad (XIII.\ 7)$$

where α is the most probable velocity in a volume of gas and is given by Eq. (II. 27)
$$\alpha = \sqrt{(2kT/m)}. \qquad (XIII.\ 8)$$

The deflexion s_α of a molecule of velocity α at the detector position of a magnetic resonance type of molecular beam apparatus is given by Eq. (IV. 22),

$$s_\alpha = \frac{\mu_{\text{eff}\,A}}{4kT}\left(\frac{\partial H_0}{\partial z}\right)_A l_A(l_A+2l_1)(l_6/l_5) + \frac{\mu_{\text{eff}\,B}}{4kT}\left(\frac{\partial H_0}{\partial z}\right)_B l_B(l_B+2l_4).$$
$$(XIII.\ 9)$$

The corresponding formula for electric deflexions is obtained from Eqs. (IV. 20) and (X. 28). Deflexions at other positions and other velocities are given by Eqs. (IV. 9) to (IV. 16). If a molecular transition is induced between the A and B fields or if these fields are of different strengths, the values of $\mu_{\text{eff}}(\partial H_0/\partial z)$ will, of course, be different in the two deflecting fields.

The value that can be achieved for $\partial H_0/\partial z$ depends of course, on the magnet design as is discussed in the next chapter. However, as shown

there, the maximum value of $\partial H_0/\partial z$ is ordinarily limited by the height h of the beam since for too large a value of $\partial H_0/\partial z$ the variation over the height of the beam is excessive. As shown in Chapter XIV a good approximate working rule is that the maximum value of $\partial H_0/\partial z$ is given by
$$\partial H_0/\partial z = k_1/h \approx 1\cdot 6 H_0/h. \qquad \text{(XIII. 10)}$$
The beam shape observable with a very narrow detector in the presence of a deflecting field that produces a deflexion s_α for the molecule of most probable velocity α is given in Eq. (IV. 43) as

$$I(s) = I_{00} \sum_i w_i' \frac{1}{(d-p)} [(s+d)e^{-|s\alpha|/(s+d)} + |s-d|e^{-|s\alpha|/|s-d|} - $$
$$-(s+p)e^{-|s\alpha|/(s+p)} - |s-p|e^{-|s\alpha|/|s-p|}]. \qquad \text{(XIII. 11)}$$

It should be noted as in Chapter IV that this relation applies only when the states are symmetrically populated along the z-axis; otherwise Eq. (IV. 29) must be used. The effective weight factor w_i' of a state of energy W_i is given by

$$w_i' = e^{-W_i/kT} \Big/ \sum_i e^{-W_i/kT} \quad \text{for } s_\alpha > 0, \qquad \text{(XIII. 11}a\text{)}$$

$$w_i' = \tfrac{1}{2} e^{-W_i/kT} \Big/ \sum_i e^{-W_i/kT} \quad \text{for } s_\alpha = 0, \qquad \text{(XIII. 11}b\text{)}$$

$$w_i' = 0 \qquad\qquad\qquad\qquad \text{for } s_\alpha < 0. \qquad \text{(XIII. 11}c\text{)}$$

With atoms, if only a single electronic level of angular momentum J is appreciably excited and if the nuclear spin is I, Eq. (XIII. 11a) reduces to $1/[(2J+1)(2I+1)]$ while Eq. (XIII. 11b) is just half as large.

If $s = 0$, Eq. (XIII. 11) reduces as with Eq. (IV. 44) to

$$I(0) = 2I_{00} \sum_i w_i' \frac{1}{d-p} [de^{-|s\alpha|/d} - pe^{-|s\alpha|/p}]. \qquad \text{(XIII. 12)}$$

When the conditions of Eq. (XIII. 6) apply, $I(0)$ equals $\tfrac{1}{2}I_{00}$ when $s_\alpha = 1\cdot 6w$ as in Eq. (IV. 45) and equals $0\cdot 1 I_{00}$ for

$$s_\alpha = 4\cdot 05w. \qquad \text{(XIII. 13)}$$

If allowance for the finite width of the detector is desired, Eq. (IV. 46a) can be used instead of Eq. (XIII. 12). If the contributions of only certain specific states are desired, the sums in the above equations should be carried out only over those states.

In Rabi's magnetic resonance method the optimum value for the perturbation of Eq. (V. 2) which induces the transition is by Eq. (V. 23)

$$b = 0\cdot 600\pi\alpha/l. \qquad \text{(XIII. 14)}$$

In the case of a simple transition of a nucleus with magnetic moment

μ_I and spin $\tfrac{1}{2}$, the optimal value for the amplitude H'_1 of the oscillatory magnetic field and of H_1 for a rotating field are by Eqs. (VI. 6), (VI. 7), and (XIII. 14) as

$$H'_1 = 2H_1 = 2\cdot 400\pi(\alpha/l\omega_0)H_0 = 1\cdot 200(\alpha/l\nu_0)H_0, \quad \text{(XIII. 15)}$$

where ν_0 is the nuclear resonance frequency in a field H_0. For such optimal perturbations the full width of the resonance at half maximum is by Eq. (V. 25)

$$\Delta\nu = 0\cdot 568 b = 0\cdot 893\nu_0 H'_1/H_0 = 1\cdot 072\alpha/l. \quad \text{(XIII. 16)}$$

In Ramsey's method with separated oscillatory fields the optimal value for the perturbation by Eq. (V. 42) is

$$b = 0\cdot 300\pi\alpha/l, \quad \text{(XIII. 17)}$$

where l is the length of a single transition field. Alternatively, therefore,

$$H'_1 = 2H_1 = 0\cdot 600(\alpha/l\nu_0)H_0. \quad \text{(XIII. 18)}$$

The full resonance width at half maximum in this method is given by Eq. (V. 42a)

$$\Delta\nu = 0\cdot 65\alpha/L. \quad \text{(XIII. 19)}$$

Optimal transition probabilities for nuclei with higher spins can be inferred from the above with the aid of Eqs. (VI. 38) ff.

Since vacuum problems play a dominant role in most molecular-beam experiments, anyone working in the field should study one of the good books on vacuum techniques (STR 38, GUT 49) and no attempt will be made to provide a substitute for such study in the present book. However, a few particularly useful relations of vacuum technology will be reproduced here.

If S_0 is the pumping speed of a vacuum pump in litres per second at the low-pressure side of the pump, and if A_p is the area of the throat of the pump in cm^2 (STR 38),

$$S_0 = fA_p\sqrt{(13\cdot 2T/M)} \text{ litres sec}^{-1}, \quad \text{(XIII. 20)}$$

where M is the molecular weight, T the absolute temperature, and f the efficiency factor of the pump, which is unity for a perfect vacuum at the throat of the pump, is often around $0\cdot 5$ for a good oil diffusion pump, and varies between $0\cdot 1$ and $0\cdot 3$ for mercury diffusion pumps. For air at room temperature (STR 38)

$$S_0 = 11\cdot 7 fA_p \text{ litres sec}^{-1}. \quad \text{(XIII. 21)}$$

If the pump is at the end of a tube of length l_t in cm and diameter d_t, the effective pumping speed S at the other end of the tube is given by

$$\frac{1}{S} = \frac{1}{S_0} + W, \quad \text{(XIII. 22)}$$

where W is the resistance of the tube which at low pressures is given approximately by (STR 38)

$$W = 0{\cdot}0159\sqrt{\left(\frac{273M}{T}\right)}\left(\frac{l_t}{d_t^3}+\frac{4}{3d_t^2}\right) \text{ sec litre}^{-1}. \quad \text{(XIII. 23)}$$

A very rough working rule is that the pumping speed for a metal kinetic vacuum system should be at least equal to the volume of the apparatus per second plus the speed required to handle all known gas flows at the desired pressure. Of course, if the system contains known gassy surfaces or has a large amount of exposed surface inside, the pumping speed should be increased accordingly.

XIII. 3. Design considerations

XIII. 3.1. *Optimum collimator position.*

There is no single collimator and deflecting-magnet arrangement that is equally optimal for all molecular-beam experiments. Thus in the electric experiments of Chapter XI a different strength of electric field, and consequently a different length of field, is desired for special reasons in the focusing and refocusing portions of the apparatus. However, some general principles applicable to a great many cases can be seen by considering some special problems.

Optimal arrangements can most easily be discussed in the case of very weak deflecting power, s_α, and it is in just such cases that optimal conditions are most needed. For $p > \tfrac{1}{2}w_d$, where w_d is the detector width, and for such a small s_α that $s_\alpha \ll p-\tfrac{1}{2}w_d$, Eq. (XIII. 11) on expansion of the exponentials becomes

$$I(s) = I_{00}\sum_i w_i'\left[1-\frac{s_\alpha^2}{4(d-p)}\left\{\frac{1}{p-s}+\frac{1}{p+s}-\frac{1}{d-s}-\frac{1}{d+s}\right\}\right].$$

Therefore, as in Eq. (IV. 46a), (XIII. 24)

$$I = 2\int_0^{\tfrac{1}{2}w_d} I(s)\,ds = 2I_{00}\sum_i w_i'w_d\left[1-\frac{s_\alpha^2}{2w_d(d-p)}\left\{\ln\frac{(p+\tfrac{1}{2}w_d)(d-\tfrac{1}{2}w_d)}{(p-\tfrac{1}{2}w_d)(d+\tfrac{1}{2}w_d)}\right\}\right].$$

(XIII. 25)

This equation could also have been obtained as an approximation to Eq. (IV. 46a) in the case of small s_α. Provided $w_c > w_s/(1+1/a)$, Eq. (XIII. 5) shows that $d-p = aw_s$.

Now consider the special case of $w_s = w_d = w$, while $w_c = (1+\beta)w$ where $\beta > 0$. Then Eq. (XIII. 25) becomes

$$I = 2I_{00}\sum w_i'w_d\left[1-\frac{s_\alpha^2}{2aw^2}\left\{\ln\frac{(2+\beta+a\beta)(2a+\beta+a\beta)}{(1+a)(2+2a+\beta+a\beta)}+\ln\frac{1}{\beta}\right\}\right].$$

(XIII. 26)

For low values of β

$$I = 2I_{00} \sum w'_i w_d \left[1 - \frac{s_\alpha^2}{2aw^2}\left\{\ln\frac{2a}{(1+a)^2} + \ln\frac{1}{\beta}\right\}\right]. \quad \text{(XIII. 27)}$$

The fractional reduction in beam R by the deflexion field is then given by

$$R = \frac{s_\alpha^2}{2aw^2}\left\{\ln\frac{2a}{(1+a)^2} + \ln\frac{1}{\beta}\right\}. \quad \text{(XIII. 27a)}$$

For very small β the last term dominates and the dependence of R on s_α and a can be taken as

$$R \sim s_\alpha^2/aw^2. \quad \text{(XIII. 28)}$$

The quantity R should be maximized to provide the best sensitivity for a deflexion apparatus.

With these criteria, the optimal adjustments for a specific type of apparatus may be discussed. As a particularly simple example, consider the case of the deflecting field extending the entire length of the system so that in Fig. (I. 3) $l_A = l_5$, $l_B = l_6$, $l_1 = l_4 = l_c = 0$. Then by Eq. (XIII. 9)

$$s_\alpha = K[l_A^2(l_B/l_A) + l_B^2] = Kl_0 l_B = Kl_0^2 \frac{a}{a+1}, \quad \text{(XIII. 29)}$$

where K is the constant coefficient of Eq. (XIII. 9). Then, from either Eq. (XIII. 27) or the approximate (XIII. 28), R is a maximum when $a/(1+a)^2$ is a maximum, or when $a = 1$. Therefore, in this case the most suitable location for the collimator is midway between the source and detector as pointed out by Manley (MAN 36). This result is also reasonable in view of the reversibility of the molecular paths, in which case if the problem is otherwise symmetrical an extremum would be expected for the collimator at the mid-point.

XIII. 3.2. *Beam widths, heights, and lengths.* There is no single optimum beam width, height, or length for all experiments. Furthermore, the criteria for the beam width are quite dependent on the nature of the detector and its noise, and upon the question of whether the deflecting magnets are the dominant source of the beam length or only minor contributors thereto. These considerations will be illustrated by a few examples. To simplify the discussions, a symmetric beam-deflexion arrangement with the collimator at the mid-point and with equal length A and B fields will be assumed in all cases. Other less symmetric problems can be estimated in a similar fashion with the aid of the formulae of Section XIII. 2.

From Eq. (IV. 46a), a given percentage of reduction in beam intensity is achieved for a given ratio of s_α/w. Therefore, in the present

comparison it will be assumed that
$$s_\alpha = k_2 w. \quad (XIII.\ 30)$$
Likewise, from Eq. (XIII. 3),
$$p' = k_3/w \quad (XIII.\ 31)$$
where $k_3 = 7\cdot 3 \times 10^{-20} T/\sigma$ mm of Hg cm. From these and from Eqs. (XIII. 9), (XIII. 10), and (XIII. 1)

$$\begin{cases} k_2 w = \dfrac{\mu_{\text{eff}}}{4kT} k_1 \dfrac{1}{h} l_A(l_A + 2l_1) \\ I = k_4(k_3/w)(hw)^2/(2l_A + 2l_1 + 2l_{c'})^2, \end{cases} \quad (XIII.\ 32)$$

where $k_4 = 1\cdot 118 \times 10^{22}/\sqrt{(MT)}$ molecules sec^{-1} cm^{-2} mm of Hg^{-1}. From these two equations
$$I = k_5 \frac{(1 + 2l_1/l_A)}{(2 + 2l_{c'}/l_A + 2l_1/l_A)^2} h. \quad (XIII.\ 33)$$
When $l_1 + l_{c'} \ll l_A$, the above reduces to
$$I = \tfrac{1}{4} k_5 h. \quad (XIII.\ 34)$$
In this case, then, the beam increases in proportion to the beam height. Of course, if the designed height is increased, according to the first Eq. (XIII. 32), either the beam width must be decreased or the beam length must be increased.

A quite different limit of Eq. (XIII. 33) is achieved if $l_1 \ll l_A \ll l_{c'}$. Then
$$I = \tfrac{1}{4} k_5 (l_A/l_c)^2 h. \quad (XIII.\ 35)$$
This result indicates a rapid increase of intensity with l_A and h. Of course, if these are increased until l_A becomes comparable with $l_{c'}$, the approximations on which Eq. (XIII. 35) are based cease to be valid and Eq. (XIII. 33) must be used.

A mere maximization of the above intensities is not the only important design criterion. In particular, the limitation to most measurements ordinarily depends on the signal to noise ratio. Therefore, the source of the detector noise and its dependence on w and h must be considered in determination of the optimal apparatus design. Thus, if the chief noise with a surface ionization detector using a circular wire arose from a random emission of particles of an amount proportional to the detector area, the noise would increase proportionately with $w_d h$. On the other hand, if the limiting noise arose from the input resistor to the electrometer circuit, the noise would be independent of h and w_d. With a Pirani detector and a constant κ factor, on the other hand, the rate at which the gas leaks from the source is proportional to wh. Consequently, the equilibrium pressure in a Pirani detector is proportional to Eq. (XIII. 33) divided by wh. However, the noise pressure may be decreased with a larger wh. Detector design problems will be discussed in greater detail in the next chapter.

Various practical considerations often determine the final design. Thus the cost and inconvenience of an apparatus ordinarily increases with its length and these factors often provide the length limitation. The specific dimensions of an apparatus are often determined by such practical considerations as the lengths of lathe beds, shapes of commercially available milling cutters, room dimensions, etc.

If multiple slit sources are used, as in some of Zacharias's (ZAC 53) experiments discussed in Section V. 5, the design limitation of Eq. (XIII. 31) is modified. Consequently, the subsequent equations which are dependent on this are then not applicable. Design problems in experiments of this type are discussed in Section V. 5.

Once the beam widths, lengths, heights, etc., have been selected, the anticipated sensitivity and related characteristics of a proposed experiment can be calculated with the aid of the equations of Section XIII. 2. Since there is always the danger that an excessive amount of the beam may be cut off by the magnet or other objects at positions of large deflexion, the loss to be expected in this fashion may be calculated at the most critical points by the use of Eqs. (XIII. 9) and (XIII. 11) as discussed in Section IV. 1.3. Likewise, if a beam stop is introduced to eliminate the undeflected beam in a flop-in experiment, its effectiveness can be estimated in the same manner.

XIII. 4. Design illustration

As an illustration of the application of the design procedures, a more detailed discussion will be given of the molecular-beam apparatus that has been used in the high resolution studies of molecular hydrogen by Kolsky, Phipps, Ramsey, and Silsbee (KOL 50*b*, KOL 52).

Certain apparatus parameters in this experiment were selected somewhat arbitrarily. Thus the resonance width was selected to ensure adequate precision in the anticipated measurements of molecular diamagnetic shifts, of the isotopic mass dependence of the spin-rotational interaction, and of the nuclear quadrupole interaction as discussed in Chapter VIII. A full width at half value of 300 c/s with D_2 appeared satisfactory. This precision at high magnetic fields required the use of Ramsey's separated oscillatory field method. Because of the inconvenience of a liquid hydrogen cooled source, it was decided to operate the source at liquid nitrogen temperature or 77·4° K. From Eq. (XIII. 8), α for D_2 is then $5\cdot66\times10^4$ cm sec^{-1}. From Eq. (XIII. 19) then L

should be at least 121 cm long. A C magnet length of 150 cm was therefore selected.

A symmetrical apparatus with all slit widths equal was chosen for lack of any very good reason for doing otherwise. l_1, l_2, l_3, and l_4 were made as small as geometrically convenient with the result that $l_1 = l_4 = 14 \cdot 3$ cm and $l_2 = l_3 = 6 \cdot 04$ cm. The pumping speed was chosen to be such that the effect of Eq. (XIII. 4) should have been negligible, although some effect with small-angle scattering apparently persisted because the collimated beam intensity proved to be only about half as great as the uncollimated intensity; in the following calculation the attenuation by scattering will be omitted.

A Stern–Pirani detector with a single slit channel and a single slit source were used, so these were made narrow for the reasons discussed in Sections XIII. 3.2 and XIV. 3.5. On the basis of past experience and of convenience in construction and alignment this width was selected to be 0·0015 cm. The deflecting power was determined from the requirement that $I(0) = 0 \cdot 1 I_{00}$ for the $\Delta m_I = \pm 1$ transition of the deuteron. By Eq. (XIII. 13) this required $s_\alpha = 4 \cdot 05 w = 0 \cdot 0061$ cm for $\Delta \mu_\text{eff} = 4 \cdot 33 \times 10^{-24}$ erg gauss^{-1}.

Permendur pole tips that normally saturate at 23,000 gauss were used on the A and B magnets, so an assumption of $H_0 = 14{,}000$ gauss in Eq. (XIII. 10) appeared to be quite safe. On the basis of viscosity measurements, the kinetic theory collision cross-section of molecular D_2 was assumed to be $\sigma = 4 \cdot 5 \times 10^{-16}$ cm^2. Finally it was decided on the basis of experience with previous Stern–Pirani detectors and with anticipated improvements including a large κ-factor (see Section XIV. 3.5) that a total beam flux density at the detector of $I/w_d h = 5 \cdot 12 \times 10^{13}$ molecules cm^{-2} sec^{-1} should be adequate to make the small peaks due to the rotational state $J = 1$ usable.

With the above evaluations, the relevant remaining relations from Eqs. (XIII. 1), (XIII. 3), (XIII. 9), and (XIII. 10) are

$$I/wh = 1 \cdot 118 \times 10^{22} \frac{p'wh}{l_0^2 \sqrt{(MT)}},$$

$$w = 7 \cdot 32 \times 10^{-20} \frac{T}{p'\sigma},$$

$$s_\alpha = \frac{\Delta \mu_\text{eff}}{4kT} \frac{\partial H_0}{\partial z} l_A(l_A + 2l_1),$$

$$\frac{\partial H_0}{\partial z} = 1 \cdot 6 H_0 / h,$$

$$l_0 = 2l_1 + 2l_2 + 2l_4 + l_c, \qquad \text{(XIII. 36)}$$

TABLE XIII. I

Characteristics of a High-resolution Molecular Hydrogen-beam Apparatus

The following characteristics have been determined by Kolsky (KOL 50b) to apply to the apparatus used by Kolsky, Phipps, Ramsey, and Silsbee (KOL 52, KOL 50b) in high-resolution studies of molecular hydrogen. These are listed in detail because many of them are typical of other molecular-beam experiments as well. However, many of the parameters vary greatly from one experiment to another. Illustrations of this apparatus are given in Figs. I. 2, I. 3, I. 4, XIV. 18, XIV. 19, XIV. 20, XIV. 29, and XIV. 30.

Symbol	Meaning	Value
A_s	Area of source slit.	$1 \cdot 2 \times 10^{-3}$ cm².
A_d	Area of detector slit.	$1 \cdot 2 \times 10^{-3}$ cm².
h	Beam height.	$0 \cdot 8$ cm.
w	Beam width.	$0 \cdot 0015$ cm.
p'	Source pressure.	$3 \cdot 5$ mm Hg.
T	Temperature of source.	$77 \cdot 4°$ K.
Q	Rate of gas flow from source.	1×10^{19} molecules sec⁻¹.
I/wh	Beam flux density of detector.	$4 \cdot 4 \times 10^{13}$ molecules sec⁻¹ cm⁻².
I	Beam to detector.	$5 \cdot 3 \times 10^{10}$ molecules sec⁻¹.
κ	Kappa factor of detector slit.	302.
κ_s	Kappa factor of source slit.	10.
$\kappa_d I/wh$	—	$1 \cdot 3 \times 10^{16}$ molecules sec⁻¹ cm⁻².
s_α	Lateral deflexion of D_2 molecule of velocity α for $\Delta\mu_{\text{eff}}$ of $4 \cdot 33 \times 10^{-24}$ erg gauss⁻¹.	$0 \cdot 0068$ cm.
n	Molecular density in source.	$4 \cdot 42 \times 10^{17}$ molecules cm⁻³.
$\alpha =$	$\sqrt{(2kT/M)} = $ most probable molecule velocity in source	
	For H_2	$8 \cdot 0 \times 10^4$ cm sec⁻¹.
	For D_2	$5 \cdot 66 \times 10^4$ cm sec⁻¹.
$\bar{c}_b =$	$\sqrt{(9\pi kT/8m)} = $ average velocity of molecule in beam	
	For H_2	$10 \cdot 6 \times 10^4$ cm sec⁻¹.
	For D_2	$7 \cdot 5 \times 10^4$ cm sec⁻¹.
σ	Geometrical cross-section for H_2 molecule.	$4 \cdot 52 \times 10^{-16}$ cm².
	Average distance between molecules in beam at detector.	$6 \cdot 2 \times 10^{-3}$ cm.
$l_A = l_B$	Lengths of A and B magnets.	$38 \cdot 2$ cm.
l_C	Length of C magnet.	150 cm.
l_0	Total length of beam.	269 cm.

Symbol	Meaning	Value
l_1, l_4	Distance source to end of A magnet.	14·3 cm.
l_2, l_3	Distance between A and C magnets.	6·03 cm.
l_5, l_6	Distance source to collimator and collimator to detector.	134·5 cm.
a	Radius of the 'bead' of A and B magnets.	0·476 cm ($\tfrac{3}{16}$ in.).
a'	Radius of the groove of A and B magnets.	0·556 cm ($\tfrac{7}{32}$ in.).
$(H_0)_A$, $(H_0)_B$	Magnetic field strength for A and B.	About 14,000 gauss maximum.
H_0	Magnetic field strength for C.	About 10,000 gauss maximum.
$\left\{\dfrac{\partial H_0}{\partial z}\right\}_{A,B}$	Gradient in A and B.	About 26,000 gauss cm^{-1}.
	Thickness of iron magnet yokes.	5·0 cm (2 in.).
d	Gap width of C magnet.	0·728 cm.
N	Number of turns for each magnet.	16.
I_A, I_B	Currents in A and B magnets.	75 to 500 amp.
I_C	Current in C magnet.	10 to 500 amp.
NI	Ampere turns.	About 8,000 amp turns maximum.
	Total width of magnets.	17·8 cm.
	Total height of magnets.	17·8 cm.
	Centre hole in magnet yoke.	7·62 cm × 7·62 cm (3 × 3 in.).
	Weight of C magnet iron.	662 lb.
	Weight of C magnet copper windings.	145 lb.
	Weight of A or B magnet iron.	178 lb.
	Weight of A or B magnet copper windings.	46 lb.
	Length of one turn of winding.	
	A or B	99 cm.
	C	320 cm.
	Cross-sectional area of copper windings.	
	1 turn	1·39 cm^2 ($\tfrac{3}{4}$ in. × $\tfrac{1}{4}$ in.)
	16 turns	22·25 cm^2.
	Cross-sectional area of cooling water per turn.	0·312 cm^2.
	Resistance of C winding.	0·0060 ohm.
	Resistance of A and B winding.	0·0021 ohm.
	Resistance water-cooled stainless steel resistors.	0 to 0·25 ohm.
	Resistance of constantan fixed resistors.	0·1, 0·2, or 0·3 ohm.
	Resistance of constantan shunts for monitoring C current.	0·00462, 0·00117, and 0·000475 ohm.
	Resistance of all other leads in C circuit.	0·00609 ohm.
	Voltage across A or B magnet for 400 amps.	0·88 volts.

Symbol	Meaning	Value
	Voltage across C magnet for 400 amps.	2·76 volts.
	Heat generated in A or B.	About 330 watts.
	Heat generated in C.	About 1,100 watts.
	Flow rate of cooling water.	45 cm^3 sec^{-1}.
	Auxiliary field equalizing coils 1 in. back from magnet gap.	41 turns of no. 18. wire of 0·368 ohms resistance each coil normally carrying about 0·2 amp.
	Kind of 2 volt magnet storage batteries of which 10 are used.	U.S. Navy submarine Exide storage batteries of approximately 10,000 ampere hours capacity per battery.
	Battery discharge rate.	Normal 400 amps for 10 hrs. Peak 3,400 amps for $\frac{1}{2}$ hr.
	Normal specific gravity at 80° F.	1·210.
	Normal continuous charging rate from Lincoln 600 'Shield Arc' welder used to charge batteries.	500 amps at 20 volt.
	Monitoring for C magnet.	Leeds and Northrop potentiometer model 7551 and galvanometer model 2430 of sensitivity 0·0005 microamp/millimeter.
	Rated pumping speed of Hypervac 20 mechanical pump.	3·3 litres sec^{-1} with ultimate vacuum of 10^{-4} mm Hg.
	Source chamber National Research Corp. E-10 type 100 oil-diffusion pump using Narcoil.	1,420 litres sec^{-1} with ultimate vacuum of 10^{-5} mm Hg.
	Separating chamber Distillation Products VMP-260 oil-diffusion pump using Octoil S.	250 litres sec^{-1} with ultimate vacuum of 4×10^{-7} mm Hg.
	Main chamber Distillation Products MCF-700 oil diffusion pump using Octoil S.	700 litres sec^{-1} with ultimate vacuum of 3×10^{-7} mm of Hg.
	Typical operating vacua:	
	Foreline measured with National Research type 501 thermocouple gauge.	2×10^{-2} mm Hg without beam and 7×10^{-2} with.
	Source chamber with Distillation Products type VG-1A ion gauge.	3×10^{-6} mm Hg without beam and $2·3 \times 10^{-5}$ with.
	Separating chamber.	$1·4 \times 10^{-6}$ without and $5·7 \times 10^{-6}$ with.
	Main chamber (with liquid nitrogen trap).	$3·5 \times 10^{-7}$ without and $9·6 \times 10^{-7}$ with.

Symbol	Meaning	Value
	Volume of system with magnets removed.	270 litres.
	Volume of rest of system with magnets installed.	137 litres.
	Lengths of the different 12-in. diameter chambers:	
	Source chamber	11 in.
	Separating chamber	1-⅞ in.
	Main chamber	106-⅝ in.
	Distance source slit to first separating chamber slit.	1·59 cm.
	Distance source slit to second separating chamber slit.	9·52 cm.
	Widths and heights of separating chamber slits.	0·1 cm × 1·25 cm.
	Measured pumping speed at source at 10^{-5} mm Hg.	910 litres sec^{-1}.
	Measured pumping speed of main chamber at 5×10^{-7} mm Hg.	265 litres sec^{-1}.
	Dimensions of each Pirani detector block (see Fig. XIV. 18).	8·8 cm high, 7·5 cm long, and 1·85 cm thick.
	Dimensions of each cavity in block.	7·9 cm high, 0·476 cm in beam direction, and 0·064 cm across beam.
	Beam channel.	2·5 cm long, 0·8 cm high, and 0·0015 cm wide.
	Dimensions each platinum Pirani strip.	7·6 cm long, 0·00010 cm thick, 0·05 cm wide, and 0·76 cm² total surface area.
	Resistance each strip.	11·6 ohms at 195° K and 18·9 ohms at 317° K.
	Effective temperature of Pirani strips.	317° K.
	Temperature of Pirani blocks.	195° K (dry ice).
	Total current through bridge.	24 milliamps.
	Presumed accommodation coefficient for Pt.	¼.
	Usual galvanometer deflexion for full beam.	31 cm uncollimated and 16 cm collimated or 4·2 microvolt across galvanometer.
	Sensitivity of detector galvanometer, Leeds and Northrop HS with 9 ohm CDRX, 15·6 ohm internal resistance, and 7·5 sec period.	7·32 cm microvolt^{-1}.
	Observed full width at half maximum for radiofrequency resonance with D_2.	285 c/s.

where values for I/wh, w, T, σ, s_α, $\Delta\mu_{eff}$, H_0, l_1, l_2, and l_c are given in the preceding paragraphs; for D_2, $M = 4$. Consequently, Eqs. (XIII. 36) provide five relations among the five unknowns p', h, l_0, $\partial H_0/\partial z$, and l_A. When the numerical values of the preceding paragraphs are inserted, the relations reduce to

$$5\cdot 4 \times 10^{-5} = hp'/l_0^2$$
$$0\cdot 12 = 1/p'$$
$$6\cdot 1 \times 10^7 = \left(\frac{\partial H_0}{\partial z}\right) l_A (l_A + 28\cdot 6)$$
$$\partial H_0/\partial z = 2\cdot 24 \times 10^4/h$$
$$l_0 = 2l_A + 190\cdot 7. \qquad \text{(XIII. 37)}$$

A solution of the simultaneous Eqs. (XIII. 37) is

$$l_A = 19\cdot 6 \text{ cm}$$
$$l_0 = 229\cdot 9 \text{ cm}$$
$$h = 0\cdot 345 \text{ cm}$$
$$p' = 8 \text{ mm of Hg}$$
$$\frac{\partial H_0}{\partial z} = 6\cdot 5 \times 10^4 \text{ gauss cm}^{-1}. \qquad \text{(XIII. 38)}$$

Actually h was selected arbitrarily to be 8 mm since the increased beam height gave an increased beam intensity by the first relation in Eqs. (XIII. 36) and by the fact that l_A contributes only slightly to l_0. The remaining quantities were then determined from this value of h and the last four of Eqs. (XIII. 37).

As an illustration of the characteristics of a typical molecular beam apparatus, in Table XIII. 1 are listed a number of properties of the final apparatus discussed above (KOL 50b). This equipment is the one illustrated in Figs. I. 2, I. 3, I. 4, XIV. 18, XIV. 19, and XIV. 30 that was used in the high-resolution studies of molecular hydrogen discussed in Chapter VIII. It should be noted that in different equipments for different experiments many of these numbers will be greatly changed. For example, if a surface ionization detector combined with a mass spectrometer and electron multiplier as in Fig. I. 6 can be used, very much weaker beams can be detected successfully; the ultimate limitation being essentially counting statistics. Successful experiments have been done in such cases with as few as five or ten atoms in appropriate states striking the detector per second. Likewise, with condensable vapours much lower source chamber pumping speeds can be used than with non-condensable gases like H_2.

XIV

MOLECULAR-BEAM TECHNIQUES

XIV. 1. Introduction

EACH molecular-beam experiment ordinarily incorporates a number of different techniques, but the combination of techniques chosen for one experiment is often quite different from that appropriate to another. Also, the most valuable techniques are frequently not explicitly molecular-beam ones but they may be common to the general fields of electronics, high vacua, etc. For these reasons a clear, complete, and ordered presentation of molecular-beam techniques is not possible. What will be presented in this chapter will, therefore, be a miscellaneous array of techniques that are especially applicable to molecular beams. A reader who plans to undertake active experimental molecular-beam research should supplement this book with reading of original molecular-beam research papers (STE 26, STE 27, EST 46, ZZA 58, ZXX 54), and with study in related fields, notably high vacua (GUT 49), mechanical design (STR 38), and electronics (TER 37, ELM 49). He would also be well advised to visit and preferably work in a laboratory already engaged in molecular-beam research.

In the remainder of this chapter, the various techniques will be subdivided into sections in accordance with the major molecular-beam component to which the techniques most contribute: sources, detectors, deflecting magnets, uniform fields, oscillatory fields, vacuum system, miscellaneous devices, and alignment.

XIV. 2. Sources

XIV. 2.1. *Sources for non-condensable gases.* For non-condensable gases like H_2 the source problem is so simple as to be essentially trivial. A tube carrying gas to a small volume with an adjustable slit opening for the beam to emerge is all that is required. Ordinarily the source is attached to a liquid air trap so that it may be cooled. Because of the thermal contractions on cooling, the source should be made of a single material (stainless steel is often used) and should not be soft soldered. The trap walls should be kept thin to diminish liquid air evaporation by thermal conduction along the walls. A typical non-condensable gas source is shown at the bottom of the liquid air trap in Fig. XIV. 1.

Often the source slits are made of appreciable thickness so that the

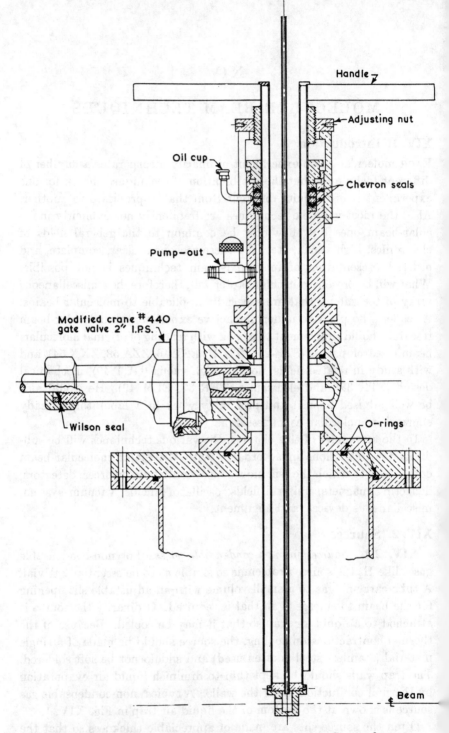

Fig. XIV. 1. Interchangeable molecular-beam source. Source for non-condensable gases is shown in position, but this may be replaced by a similar tube with a heated oven at the bottom. The distance from the handle to the beam line is 19 inches (RAM 54).

aperture is sufficiently long in the direction of the beam to diminish the total amount of gas flow for a given amount of beam intensity in accordance with Eqs. (II. 7) and (II. 11) (FRA 37, NOL 49). Typical dimensions for an H_2 source aperture are 0·015 mm wide, 3 mm high and 0·25 mm thick (KEL 39a).

Zacharias and Jaccarino (ZAC 53) have successfully used many small tubes, such as a stack of hypodermic needles, in parallel in the source to allow for a greater source pressure with wider beams and to provide greater directivity to the emerging molecules by an increase in the flow resistance for those molecules which are incorrectly directed. This technique is discussed in greater detail in Sections V. 5 and XIV. 2.2. Kantrowitz and Grey (KAN 51) have also proposed the achievement of initial directivity by the use of a high velocity jet as the source. Kistiakowsky and Slichter (KIS 51) have applied the proposal of Kantrowitz and Grey to an NH_3 beam and have obtained an intensity gain of a factor of 20 but have experienced considerable difficulty from the large volume of gas flow (ZZA 43).

With H_2 an increase in precision could presumably be obtained if a liquid H_2 reservoir were used to cool the source because of the lower velocities of the emerging gas at the lower temperature. However, so far no such source has been used. An alternative low-temperature source which might be more convenient in laboratories with liquid helium but no liquid hydrogen might be one to use a thermostatically controlled helium gas flow from a liquid helium reservoir as has been done in nuclear-scattering experiments by Stahl, Swenson, and Ramsey (STA 54).

In some cases it may be desirable to change from a non-condensable gas source to a heated-oven source without loss of the system vacuum. A mechanism which permits such changes has been designed by Ramsey and Wanagel (RAM 54) and is illustrated in Fig. XIV. 1.

One complication exists with non-condensable gases that is more severe than with substances whose vapour pressure is low at room temperature. Ordinarily much greater pumping speed in the source chamber is required in the non-condensable gas case since all the gas which emerges from the source must be pumped away and no effective addition to the pumping speed is attained by condensation of the emerging molecules on the walls of the source chamber. In the hydrogen experiments of Kolsky, Phipps, Ramsey, and Silsbee, for example, a source-chamber diffusion pump of a rated speed of 1,420 litres per second was used. Because of the effects of source-chamber pressure there is an optimal distance between the source slit and the first fore slit to the separating chamber. If this distance is too great excessive amounts of scattering will occur in the source chamber and if the distance is too small the separating chamber pressure will be too great. From Eq. (XIII. 4) the separation should be such that, if the subscript 1 indicates the source chamber and 2 the separating chamber, the quantity

$$(l_1/\lambda_1) + (l_2/\lambda_2)$$

should be a minimum. The criteria become more complicated when effects of molecular reflection on the separating slits and of appreciable slit thickness are included as discussed by Fraser (FRA 31) and Esterman (EST 46).

XIV. 2.2. *Heated ovens.* Many different designs of heated ovens have been used in molecular-beam experiments depending on the material to be vaporized, the length of the run desired, etc.

The choice of material from which the oven is constructed is usually determined by the criteria that it must not melt at the desired temperature, that it must not react with the material to be studied, that it will not alloy with the material, and that excessive creep to the slit jaws must be avoided. For most molecular-beam substances iron, nickel, or monel metal have been found to form suitable ovens with iron being cheaper and more easily machined while nickel may be somewhat more permanent. A gold-plated iron oven with gold-plated slits often provides a more stable beam than a plain iron one. Melted hydroxides react vigorously with iron, but Millman and Kusch (MIL 41, TAU 49) have found that silver ovens are suitable with hydroxides. Indium and gallium alloy with iron, but indium beams have been produced effectively in a pure molybdenum oven at 1,000° C (MIL 38, HAM 39). A satisfactory beam of gallium cannot be produced in this way, but Renzetti (REN 40) has shown that graphite ovens are suitable. Alternatively a graphite insert into a molybdenum oven, with channel, wells, and slit jaws of graphite has been found suitable (KUS 50a) for gallium. A graphite insert has also been found to improve the stability of indium beams, although molybdenum slit jaws have been used in such cases since they are then less easily broken. An Al_2O_3 crucible inside a graphite oven with the crucible and oven separated by a tantalum foil to prevent the reduction of the crucible has been found by Lew (LEW 49) to be especially suitable for atomic aluminium. For praseodymium Lew (LEW 53a) has used a ThO_2 crucible inside a molybdenum oven, since this diminishes the chances of the slits being clogged with molten metal.

The typical design of oven ordinarily used at Columbia by Kusch and others is shown in Fig. XIV. 2 (KUS 50). It should be noted that the heating elements are closer to the slit jaws than to the well so that the slit is at a higher temperature than the rest of the oven. This diminishes clogging of the slits. The cut under the slit also diminishes cooling in the slit region. The hole above the well is ordinarily sealed

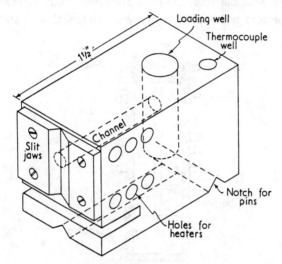

FIG. XIV. 2. Schematic view of a typical molecular-beam oven (KUS 50).

with a fitted conical plug that is inserted by pushing it hard in a vice and that is removed either mechanically or by being bored out. The slit jaws are usually held in place by four screws made of the same material as the oven and slit jaws. The slit jaws and the mating oven surface are lapped to prevent leakage (improved design in ZZA 45). The channel to the slit is of a diameter slightly greater than the desired beam height. The oven is usually mounted on three tapered tungsten pins which fit into the V-shaped notches shown in Fig. XIV. 2. The thermocouple for measuring the oven temperature is inserted into the hole indicated. A chromel alumel thermocouple is ordinarily used. Sometimes it is insulated from the oven with ceramic tubing and sometimes it is not. In the latter case a small hole may be used and the thermocouple may be peened into position so that it cannot fall out accidentally. The heater inserted into each hole of the oven ordinarily consists of 50 cm of 0·010 in. wolfram (tungsten) wire coiled about a mandril of 0·075 in. diameter and placed within a Stupakoff ceramic tube of about 3·5 mm outer diameter. A convenient means of winding the coil is by the use of a hand-drill with a drill of suitable size serving as the mandril. Quartz tubing has also been used as an insulator but it is more expensive and more fragile. The number of heaters used has varied from two to twenty depending on the desired temperature. Ordinarily two adjacent coils are wound from a single piece of tungsten wire. Connexions between the coils are often made mechanically with nickel connectors possessing set screws. Alternatively, and often with

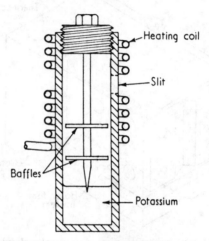

Fig. XIV. 3. An oven used for potassium (EIS 52).

greater reliability, the interconnexions are made by spot welding the tungsten wires to intermediate nickel wires (tungsten does not weld to tungsten as effectively as to nickel). Alternating current, controlled with a variac and stabilized with a saturable reactor in the primary, is ordinarily used for heating. A one-to-one insulating transformer is often used to diminish the seriousness of a short of the heater wires to ground. Tantalum heater wires are superior to wolfram (ZZA 44).

A quite different oven design has been used by Eisinger and Bederson (EIS 52) as shown in Fig. XIV. 3. The oven was approximately $2\frac{1}{2}$ inches high and $\frac{3}{4}$ inch in diameter. After being freshly loaded with 2 grammes of K metal, the oven gave large steady beams for approximately 60 hours. The oven was supplied with a pair of baffles to counteract any tendency of the charge to spatter the slit jaws (to 'spritz') and thereby clog them. The oven was heated by a tungsten heating coil near the top of the oven. The 0·006-inch wide slit was milled with a very thin milling cutter.

A somewhat similar design by Lew (LEW 49) for atomic Al is shown in Fig. XIV. 4. An aluminium oxide crucible is placed in a thin-walled graphite jacket. Tantalum foil is wrapped around the crucible to prevent it from being in contact with the graphite. The aluminium to be evaporated is placed in the crucible in the form of wire. The heating of the oven is accomplished by passing alternating current directly through the graphite jacket. With a water-cooled radiation shield surrounding the oven, 800 watts of power is necessary to raise the temperature in the vicinity of the slit to the 1,670° K operating temperature. Above 1,700° K the aluminium oxide begins to decompose. A charge of 0·17 gramme lasts about six hours. In some cases ovens have been heated by electron bombardment (BER 52).

For his studies of praseodymium, Lew (LEW 53a) has used the oven arrangement shown in Fig. XIV. 5. Two ovens were mounted on a rotatable platform since he wished to study two different kinds of beams interchangeably. The cell used for the vaporization of Pr was essentially a thin-walled molybdenum cylinder with a slit cut longitudinally in the wall. The unit was machined from a ⅜-inch solid molybdenum rod to the shape shown in the figure. The thin-walled section is ¼ inch inside diameter and $\tfrac{5}{16}$ inch outside diameter and the slit is 0·004 inch by ⅛ inch. The unit is clamped by its lower stem between two blocks of copper which are in thermal and electrical contact with the bottom of the water-cooled radiation shield surrounding the unit. The top end of the cell, i.e. the stem of the cap that closes the cylindrical chamber, protrudes through the radiation shield and makes contact with a water-cooled copper block which is electrically insulated from the rest of the oven assembly. Electrical power is applied between this electrode and the radiation shield by way of the tubes which lead cooling water to these parts. The praseodymium metal to be vaporized, instead of being placed in direct contact with the molybdenum, is contained in a ThO_2

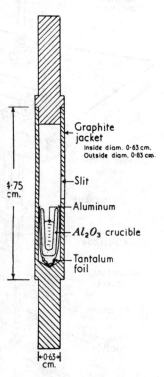

FIG. XIV. 4. An oven for generating a beam of aluminium atoms (LEW 49).

crucible. This greatly reduces the chances of the oven slit being clogged by the molten metal. A current of about 240 amps at 10 volts is required to bring the oven to the operating temperature of 2,000° K (1,570° C brightness temperature at a wavelength of $0·65\mu$). The oven

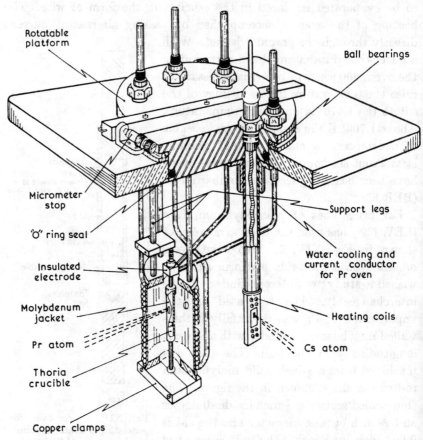

FIG. XIV. 5. Pr and Cs ovens mounted on a rotatable platform for simple and accurate interchange of positions (LEW 53a).

for the Cs beam consists of an iron cell of square cross-section $\frac{1}{4}$ in. × $\frac{1}{4}$ in. × $\frac{7}{8}$ in. with a longitudinal slit in one face $0·004$ in. × $\frac{1}{8}$ in. This fits into a rectangular hole in a block of iron in which are embedded some molybdenum heating coils. A charge of Na metal and CsCl in the cell yields a very steady beam of Cs atoms when the oven is raised to a temperature of about 200° C.

A different oven has been used by Davis (DAV 48b) in his study of Na^{22}. Since the Na^{22} was produced by bombardment in a cyclotron, a maximum of economy of source material was required in this experiment. Therefore a long narrow canal was used to diminish the number of molecules emerging in the wrong direction, in accordance with Eqs. (II. 7) and (II. 11). The oven was constructed of monel metal to diminish absorption of the sodium by the metal of the oven and a very small oven cavity of less than 25 cubic millimetres volume was used.

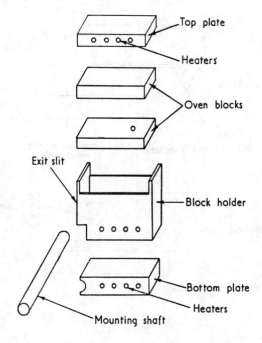

Source

FIG. XIV. 6. Oven for Na[22]. Details of the oven blocks are shown in Fig. XIV. 7 (DAV 48b).

The oven assembly is shown in Fig. XIV. 6 while details of the two oven blocks are shown in Fig. XIV. 7. As illustrated, the oven assembly consists of simply constructed oven blocks and an oven block holder. Heaters of molybdenum wire coated with aluminium oxide for insulation were mounted in the holes of the oven block to permit heating. The oven block holder was mounted on a horizontal shaft so that the beam could be directed in a vertical plane to strike the detector, as the beam was quite directional having a width at half intensity of $1\frac{1}{2}$ degrees which was of the order of the angle of the detector height subtended at the source. The oven blocks were two monel metal blocks with joining surfaces lapped flat to within 0·00001 inch to prevent leakage. On the bottom of the upper block was milled a slit 0·010 in. × 0·010 in. × 1·125 in. as shown in Fig. XIV. 7 while a cavity of 25 mm³ capacity was drilled in the lower block. The oven pressure had to be kept below 3×10^{-3} mm of Hg to prevent the mean free path from becoming less than the length of the canal. Sodium azide (NaN_3) was

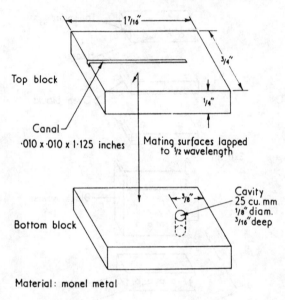

Fig. XIV. 7. Details of oven blocks for Na^{22} oven shown in Fig. XIV. 6 (DAV 48).

found to be convenient for handling the Na since it was stable to air and water at normal temperatures and since it decomposed to yield free sodium at 320° C.

Means for providing even greater economy of source material have been suggested by Zacharias and Jaccarino (ZAC 53, ZZA 38) who have successfully used stacks of hypodermic needles or of channel crimped foil in the source to provide high directivity. Such sources are also discussed in Section V. 5.

With materials which require high oven temperatures, it may be necessary to use cooled radiation shields to prevent excessive heating of the rest of the apparatus. A typical means for providing such cooling is illustrated incidentally in Fig. XIV. 5. Similar shields may also be used for the collection of precious source materials as shown in one of the details of Fig. I. 6. The bulkhead between the oven and separating chambers is frequently water cooled.

Many molecular-beam source materials like K, Rb, and Cs react with water and air with explosive violence, so care must be exercised in handling the materials and in opening used ovens that may not be completely empty. Some materials like Na can be loaded by cleaning them under a liquid like petroleum ether and then quickly inserting them in the source before the petroleum ether has evaporated. Such a procedure, however, is not adequate for Rb and Cs which are usually distilled in vacuum into a small glass capsule which is broken when the oven plug of Fig. XIV. 2 is inserted into an oven flushed with helium, nitrogen, or other inert atmosphere. An alternative procedure has been used by Millman and Kusch (MIL 41), Lew (LEW 53a), and others. For example, a mixture of Na metal and CsCl is inserted in the oven and when this is heated to 200° C a steady beam of Cs atoms emerges (LEW 53a). Calcium metal and CsCl may also be used (MIL 41) to produce a Cs beam. A Na beam may be produced from a mixture of NaCl and fresh Ca chips.

When most alkali beams like those of Na, K, etc., are produced, molecular beams of Na_2, etc., are ordinarily present to amounts of about $\frac{1}{2}$ per cent. At higher oven temperatures this fraction usually diminishes. Also with salts such as NaCl and NaF, some polymerization of the beam occurs and dimers, such as $(NaCl)_2$, are emitted (OCH 53, KUS 53a, RAM 48).

In Table XIV. I are given a list of many of the substances from which satisfactory beams have been formed, along with the required oven temperature for a suitable beam (KUS 50). In cases where the oven temperature is ordinarily measured with a chromel-alumel thermocouple, the temperature is given in terms of the e.m.f. of such a thermocouple with the cold junction at room temperature. An e.m.f. of 1 millivolt corresponds to a temperature difference of about 25° C; exact calibrations are given in the handbooks (HOD 54). These temperatures are only approximate and vary with the needs of the experiment. Ovens should only slowly be brought to their final temperature.

As an alternative to an oven, the surface of a heated filament can be used as the source for a molecular-beam experiment with a substance that can be vaporized only at high temperatures (STE 20a, FRA 31, STA 49). However, owing to the limited amount of material that can be stored on such a source, filaments are only rarely used now.

Provision for lateral translation of ovens and other sources must be provided. In the case of ovens similar to that in Fig. XIV. 2, this is often accomplished by mounting the three pins which support the oven on an inverted sliding dovetail designed so that the differential expansion with heating will loosen rather than tighten it.

TABLE XIV. I

Oven Temperatures

This table lists the oven temperatures that are suitable for the formation of molecular beams with the listed substances (KUS 50). Since the oven temperatures are ordinarily measured in millivolts of e.m.f. with a chromel-alumel thermocouple whose cold junction is at room temperature, the oven temperature is listed in such units. One millivolt corresponds to a temperature difference of approximately 25° C; exact calibrations are given in the handbooks (HOD 54). In a few cases temperatures in degrees Kelvin are listed directly in which case the units are given explicitly. Temperatures in parentheses are brightness temperatures as determined with an optical pyrometer at 0.65μ.

Substance	Temperature (millivolts)	Substance	Temperature (millivolts)
Li	26–29	CsCl	25
Na	14–16	LiI	22·5
K	10–11	RbI	26
Rb	9	Li, Na, and K meta	35–38
Cs	7–8	and tetraborates	
In	36–45	$NaFBeF_2$	33
Ga	40–45	$KFBeF_2$	33
LiF	37	$NaClAlCl_3$	13
NaF	35	$KClAlCl_3$	13
KCl	35	NaOH	25–28
KF	32	KOH	25–28
RbF	28	$InCl_3$	10
CsF	27	Al	1,670° K
LiCl	28	Pr	2,000° K (1,840° K)

XIV. 2.3. *Sources for dissociated atoms.* Often atoms such as H, D, Cl, etc., need to be studied despite the fact that these atoms normally come combined in molecules such as H_2, D_2, Cl_2, etc. Therefore some means for dissociating the molecules into atoms is required.

One means for dissociating H_2 molecules was discussed in detail in Chapter XII. This method involves the use of a heated tungsten oven; from such a source much of the hydrogen emerges in atomic form as discussed in Chapter XII. Although this method was used in the Lamb–Retherford experiment, it is not the one that has been most frequently used in atomic-beam experiments.

In most atomic-beam experiments the atoms have been produced in a Wood's discharge tube (WOO 20). The chief difficulty with such a tube is in obtaining a large yield of atoms. Any foreign matter or contamination such as pieces of metal, dust, wax, sharp or broken edges of glass, etc., greatly reduces the yield of atomic hydrogen. Even the electrodes must be kept a long way away from the region in which the atomic hydrogen is desired.

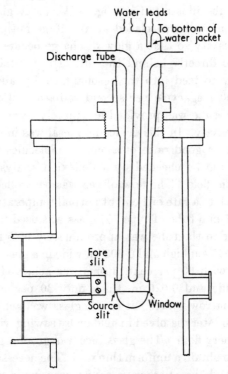

FIG. XIV. 8. Section of atomic hydrogen source chamber to show construction of Wood's discharge tube and of eccentric brass adjusting collar (KEL 36).

One discharge tube for the formation of beams of atomic hydrogen is shown in Fig. XIV. 8. A discharge tube about 4 metres long was used to keep the electrodes at a distance. The hydrogen reservoir at atmospheric pressure is saturated with water vapour and is connected to the discharge tube through a controlled leak near one of the electrodes. The central portion of the discharge tube is at the beam source position and is water cooled. An aperture in the discharge tube at the source position is closed by two microscope cover glasses waxed to the tube and defining a slit about 0·001 inch wide. The pressure in the discharge tube is about 1 mm of Hg with approximately

10,000 volts and 50 milliamps being applied to the tube. An atomic discharge is characterized by a red colour while a blue discharge characterizes molecular hydrogen. About 80 per cent. of atoms have been obtained from such a discharge-tube source.

As an alternative source for atomic hydrogen and the halogens, Zacharias and others (NAG 47, DAV 49b) have developed the electrodeless discharge tube illustrated in Fig. XIV. 9. A glass tube with a narrow slit at one end was placed at a voltage antinode in a microwave resonant cavity so that a glow discharge occurs in the gas for a centimetre or so directly behind the slit. To maintain the discharge it was necessary to feed into the resonant cavity about 50 watts of continuous 3,000 megacycle per second radiation. Cooling the outer wall of the glass tube was provided adequately by an air blast inside the microwave cavity. In this device no metal was in contact with the atomic gases and a good ratio of atoms to molecules was obtained in the beam, as could be checked by a deflexion analysis in the homogeneous magnetic field. The flow of gas was controlled by a Bourdon type variable leak to a rate of 1 cm^3 at normal temperature and pressure per minute. A 7-mm tube of type 707 glass was used for the discharge. The gas pressure in the tube was approximately 0·25 mm of Hg. The source slit was 0·45 cm high and 0·003 in. wide in a glass of wall 0·010 in. thick. A beam with 60 per cent. atoms was achieved, though with a slit 0·001 in. wide and 0·020 in. thick only 30 per cent. atoms were obtained. The narrow source slit in the glass was achieved by sealing in a covar ribbon later dissolved in acid or by sawing with a copper wire charged with emery flour. The glass face was ground gently internally and externally to obtain a uniform thin wall. The necessary seal between the glass tube and the microwave cavity was made with Apiezon W wax. One-inch coaxial arc sources have been found to be even better.

XIV. 3. Detectors

XIV. 3.1. *Miscellaneous detectors.* The fundamental problem of molecular beams is the quantitative detection of the molecules in the beams. For this purpose many different kinds of detectors have been used. However, of these, four are the chief ones currently used. These are (a) surface-ionization detector, (b) electron-bombardment ionizer, (c) Pirani gauge detector, and (d) radioactivity detection. Detection by either of the first two methods may be supplemented with a mass spectrometer or an electron multiplier tube. Separate subsections will be devoted to each of these four methods. The present subsection will

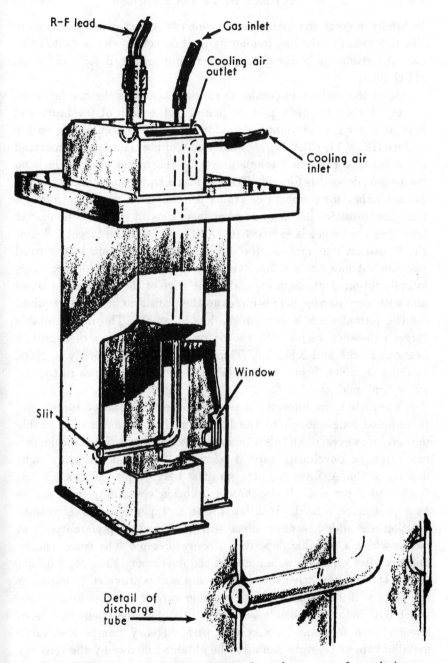

Fig. XIV. 9. Cut-away sketch resonant cavity and gas system for producing atomic Cl from Cl_2 (DAV 49b).

be briefly devoted to other methods which have occasionally been used. The discussion of the less frequently used detectors will be quite brief, and the reader is referred to other references for details (FRA 31, HUG 53).

One of the earliest molecular-beam detectors was the condensation target. A glass or metal plate is placed in the path of the beam and kept at a low enough temperature for a deposit to accumulate on the surface (DUN 11, FRA 31). For each kind of beam and target material there exists a transition temperature for the target above which no permanent deposit is formed. There is also a transition intensity (FRA 31) such that for numbers of atoms per unit area per unit time less than the transition intensity no permanent deposit is formed no matter how long the target is exposed (FRA 31). Even if the target is below the transition temperature, the molecules do not remain in one fixed position but move at random over the surface to a slight degree. This lateral motion (Rutscheffekt) makes the edge of the deposit less sharp and with very narrow beams increases the transition intensity required for the formation of a permanent deposit at all. The most suitable target substance varies with the nature of the beam as discussed in Sections II. 5.4 and XIV. 3.6. Thus sodium adheres strongly to glass, mercury to silver, bismuth to nickel, and in a lesser degree to copper, silver, and gold, etc.

If adequate beam intensity is available, the condensation targets can be exposed long enough to the beam for the formation of a visible deposit. However, when this is impossible, an otherwise invisible deposit can often be developed until it is visible. Thus an invisibly light deposit of Cu, Ag, Au, Ni, etc., on glass may be placed in a not too fresh 1 to 2 per cent. hydroquinone solution containing gum arabic as a protective colloid. If a few drops of 1 per cent. silver nitrate solution are added, atomic silver will be deposited preferentially at places where a metallic deposit is already present. The trace usually appears after only a few minutes in the developer (FRA 31, EST 23, GER 24). An alternative means of development is provided by exposing the plate with invisible traces to metallic vapour at a low pressure, in which case the vapour condenses preferentially on the nuclei that have already been formed. In a system with mercury pumps a suitable metallic vapour pressure can often be obtained merely by the removal of liquid air from the mercury traps (FRA 31). This method of development is usually preferable to the wet method described above (FRA 31, EST 23, KNA 26). Typical traces obtained in this way are shown in

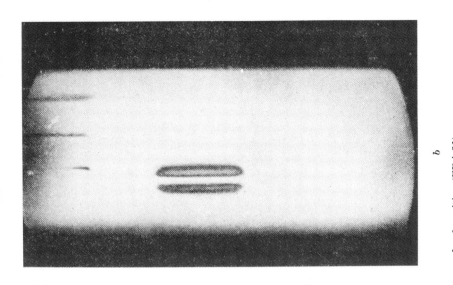

Fig. XIV. 10. Typical molecular beam traces by deposition (FRA 31).

Fig. XIV. 10. Alternatively the last method can be used to sensitize the target. If mercury vapour is allowed access to the cooled target before its exposure to the beam, a visible trace appears in a much shorter time. The molecules of a molecular-beam deposit are not distributed uniformly over the area of the deposit but are concentrated in isolated condensation nuclei, each consisting of many thousands of molecules. This characteristic makes the condensation target a much more sensitive detector; in favourable cases a deposit which on the average is only 2 molecules thick is visible as a result of the aggregation. The wet method of development may increase the sensitivity by a factor of ten and the dry metallic vapour method may under favourable conditions increase the sensitivity by a factor of a thousand. It is difficult to make quantitative measurements of beam intensity by the deposit method because of the many factors upon which the trace visibility depends. The best quantitative success has been achieved by the use of the time for first appearance of the trace as a measure of the reciprocal of the beam intensity (FRA 31, EST 23, KNA 26) or by comparisons of two portions of a trace of equal intensity (RAB 34).

Molecular-beam measurements have also been made with chemical targets which depend on the impinging molecules reacting chemically with the target. One of the most important examples of this method is the use of a MoO_3 target to detect atomic hydrogen; the atomic hydrogen gives a blue MoO_2 trace against the pale yellow MoO_3 background. Johnson (JOH 29) has shown that if over-exposure is avoided the darkening as measured with a densitometer is proportional to the number of atoms striking the target, to within 25 per cent. Quantitative measurements have also been made by measuring the length of time for the first appearance of the trace and by taking the beam intensity as inversely proportional to this time and by comparisons of portions of a trace of equal intensity (RAB 34). Atomic oxygen may also be detected chemically by using a litharge target. A brown PbO_2 trace is then formed against the pale-yellow PbO background. Kerschbaum claims that a Schumann plate makes a sensitive atomic hydrogen detector as a result of a chemical reaction (FRA 31). Suggestions have been made for detecting beams by the poisoning of electron emission or similar surface phenomena by the beam molecules, but no such detector has as yet been realized in practice (ZXZ 54, MAR 50, MAY 35). Conductivity detectors have been made with semi-conductors like CuO_2 (ZZA 49).

Sensitive manometers to measure the pressure inside a small receiving vessel have been used effectively for beam detection. One of these, the Stern–Pirani detector, is used at the present time for the detection of non-condensable gases like hydrogen. For this reason, the Pirani detector will be the subject of the entire Section XIV. 3.5 and the fundamental principles common to all manometers will be discussed in that section.

The ionization gauge is a form of manometer detector which has been used much less extensively than the Pirani gauge. Kingdon (KIN 23), Johnson (JOH 28a, JOH 29), Knauer and Stern (KNA 29), and others (COP 35, ELL 31, EST 33b, HUN 36, JEW 34, BRA 51), however, have used the ionization gauge to detect molecular beams. In principle the ionization gauge can be made quite sensitive for molecules with a low ionization potential. However, in practice ionization detectors have proved to be much less stable than Pirani detectors and more subject to zero drift; consequently, they have only been used occasionally.

Thermopiles and bolometers may be used for the detection of molecular beams. The kinetic energy of the beam molecules is not sufficient to be detected, but if atomic hydrogen or oxygen combines chemically at the surface with the release of about 10^5 calories per mole the effect on the thermopile or bolometer can be detected. Wohlwill (WOH 33) has detected beams in this fashion utilizing merely the heat of condensation.

Molecular beams can be detected with a radiometer or a sensitive balance which measures the momentum flux of the beam (FRA 31, COP 37, PAU 48). Paul and Wessel (PAU 48) have in fact measured the mean velocity of a silver beam by measuring the pressure on a sensitive balance and also by weighing the mass of silver deposited.

Electron ejection detectors have been used in several important molecular-beam experiments with atoms in metastable states. Such a detector as applied to metastable atomic hydrogen by Lamb and Retherford (LAM 50a) has already been described in Chapter XII. Hughes, Tucker, Rhoderick, and Weinreich (HUG 53) have used the same principle in the detection of He atoms in the metastable $1s2s^3S_1$ state. When a metastable He atom strikes the surface of a cold W wire, it may make an inelastic collision of the second kind, in which it may be considered that an electron from the metal goes into the K shell of the atom, and the L shell electron is ejected and carries off the excess energy (COB 44). The yield of this process has been measured to be 0·24 electrons per atom (DOR 42, HAG 53). Detection of metastable He was then accomplished by the insertion of a movable unheated W wire surrounded by a nickel electron collector plate attached to the grid of an FP 54 electrometer tube as in Section XIV. 3.3 with a 10^{11} ohm resistance to ground. A schematic view of the apparatus is shown in Fig. IX. 6; a supplementary Pirani detector is used for alignment with the unexcited beam. The overall sensitivity of the electrometer tube and its associated circuit was 10^{-16} amps per mm.

XIV. 3.2. *Surface ionization detectors.* One of the most effective and widely used detectors is the surface ionization detector first developed for molecular beams by Taylor (TAY 29a, TAY 30) on the basis of the observation by Langmuir and Kingdon (LAN 25) that every Cs atom which strikes a heated wolfram wire comes off as a positive ion. Once the atom is ionized it can be measured in any of several different ways, including current measurement with a sensitive electrometer, current measurement with the aid of an electron multiplier tube, and electronic counting with an electron multiplier tube. The ion may also before it is counted be passed through a mass spectrometer for purposes of identification and for reduction of background noise from other atoms. Details of the various measuring techniques will be given in the next section while the process of ionization will be discussed in this section.

If I is the ionization potential of the incident atom and ϕ is the work function of the metal the ratio s^+/s^0 of positive ions to neutral atoms that emerge from the detector is given in equilibrium by

$$s^+/s^0 = e^{-e(I-\phi)/kT}. \tag{XIV. 1}$$

Therefore, if ϕ exceeds I by 0·5 volts, all of the atoms will emerge from the heated wire as ions (KNA 49, FRA 31, KOD 39, STA 49).

Since the work function of pure wolfram (tungsten) is 4·5 volts, a pure wolfram wire can detect Cs ($I = 3·87$ V), Rb ($I = 4·16$ V), and K ($I = 4·3$ V) (HOD 54). It is important that the tungsten be thorium free since thoriated tungsten has a work function of only 2·6 volts so that none of the above atoms could be ionized. Ordinarily a tungsten wire of 0·001 inches diameter has been used which operates at 1,260° K with a 43 to 50 mA current, at 1,800° K with 100 mA, and 2,600° K with 200 mA. For Cs, Rb, and K operation at 50 mA is suitable. The filament may be cleaned by flashing it at 150 mA. The detector wire is usually surrounded by a copper shield cooled to liquid nitrogen temperature. Extensive data for wolfram have been tabulated (ZZA 52).

In addition to the detection of the above atoms, a molecule which contains one of the atoms can also be detected in this way (ROD 32).

Presumably this detection results from the partial dissociation of the molecule on the hot wire and the surface ionization of the resulting detectable atom. Although the equilibrium concentration of neutral atoms may be quite low, as indicated by the non-dissociation in the oven, the ionized atoms may be relatively abundant since the ionized atoms are not nearly as effective as neutral atoms in suppressing the molecular dissociation.

Some elements of higher I can be detected with oxidized W whose work function is about 6 volts. Na ($I = 5\cdot12$ V), Li ($I = 5\cdot36$ V), In ($I = 5\cdot76$ V), and Ga ($I = 5\cdot97$ V) (HOD 54) have been detected on oxidized wolfram; compounds containing one of these atoms can be detected similarly. The procedure for oxidizing the filament is to flash it clean at a current of 150 mA. Then it is operated for a while at 85 mA or 1,500° K in a slight oxygen pressure which may be produced either by a stream of oxygen flowing directly to the filament or by the admission of about $\frac{1}{2}$ a c.c. of oxygen at atmospheric pressure to the detector chamber of the apparatus. In some cases during actual observations a stream of oxygen is directed continuously on to the filament by passing air through a capillary tube from a reservoir at a pressure (approximately 5×10^{-3} mm of Hg) such that the effect on the system pressure is less than 10^{-7} mm of Hg; such a continuous coating does not always seem necessary, however. Even in the absence of a beam an oxygenated filament provides some ion current. Oxidized filaments for the detection of indium are ordinarily operated slightly above 50 mA.

With the aid of a mass spectrometer and an electron multiplier, Lew (LEW 49) has detected an Al beam ($I = 5\cdot96$ V) by surface ionization on a W detector wire at 1,780° K, at which temperature the surface was probably free of oxygen and had a work function of only 4·5 volts so that the ionization efficiency by Eq. (XIV. 1) should be small. The oxidized W efficiency was even lower, probably as a result of the formation of Al_2O_3 when the aluminium atoms strike the oxidized wire. With W at 1,780° K the ionization efficiency was estimated to be 10^{-6}. Hay (HAY 41) has detected Ba ($I = 5\cdot19$ V) with a clean filament operated at a high temperature corresponding to 130 mA or 2,000° K. Lew (LEW 53a) with considerable difficulty has detected Pr ($I = 5\cdot76$ V) with a Mo ribbon ($\phi = 4\cdot27$ V) at 2,070° K for which the ionization efficiency by Eq. (XIV. 1) should be 2×10^{-4}. The Mo ribbon had been cleaned before use by prolonged heating in vacuum 12 hours at 2,000° K) to remove impurity ions which would otherwise be emitted by the detector.

Davis, Feld, Zabel, and Zacharias (DAV 49b) have detected Cl with a modified form of surface ionization detector. In this case, Cl has an electron affinity so that it can form Cl⁻ ions. What is then required is a high electron density at the filament to enhance the probability of electron capture. On the other hand the filament must not be too hot or the electron and Cl⁻ backgrounds become excessive. The accelerating potentials must, of course, be the reverse of that for positive ion detection. The negative ion current cannot be measured directly as easily as positive ion current because of confusion with the electron current which is also emitted from the filament. However, these two negative currents may easily be separated with a mass spectrometer. About one Cl atom in 10^4 is ionized to Cl⁻ in this method with a clean tungsten wire, though the detection efficiency varies markedly with the W wire that is used and is sometimes a hundred times worse. With thoriated tungsten at 1,200° K to 1,500° K 1 per cent. of the atoms may be ionized but more background of Cl⁻ atoms comes out of the filament. Atomic I, Br, and F have been detected with thoriated W in this fashion. Cl_2, I_2, and Br_2 have been detected in this way but the ions produced have the charge to mass ratio appropriate to Cl⁻. Trischka, Marple, and White (TRI 52) have detected up to 60 per cent. of Cl⁻ atoms from CsCl, 20 per cent. Br⁻ from CsBr, and 4 per cent. I⁻ from CsI. No trace of molecular ions was discovered. These results suggest that there is dissociation of the caesium halides on the thoriated W surface (TRI 52, MAS 50, HEN 37).

XIV. 3.3. *Electrometers, mass spectrometers, and electron multiplier tubes.* Once charged ions are formed, as in the preceding section, the ion current may be measured in many different ways. With positive ions it is frequently adequate to measure the ion current directly with a sensitive electrometer circuit or D.C. amplifier. An FP 54 electrometer tube with an input resistance R_0 of 10^{10} ohms is often used in a Du Bridge and Brown (DUB 33, STR 38, EST 46) balanced circuit along with a sensitive galvanometer as in Fig. XIV. 11. A galvanometer of sensitivity 5×10^{-10} amps per mm, 6 sec period, 10,000-ohm CDRX, and 500-ohm coil resistance like the Leeds and Northrup Type R 2500b is often used.

The procedure for balancing the circuit is as follows. With the galvanometer shunted to one-tenth or one-hundredth of its full sensitivity and R_1 adjusted so that the galvanometer reads zero when I_f is near its rated value, R_5 is slowly varied until a deflexion extremum with R_5 is found. If the galvanometer goes off scale before the extremum is reached it may be brought back with an adjustment of R_1. The extremum adjustment is finally made with the galvanometer at full sensitivity. The advantage of the circuit is that effects of fluctuations of the battery voltage or of filament emission balance out to first order. The sensitivity of such a circuit is often about 3×10^{-14} amp/mm but with special care, a very high input resistor, and with a 10-second time constant of the amplifying system a usable sensitivity of about 3×10^{-17}

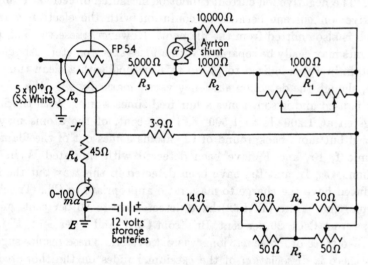

Fig. XIV. 11. Typical Du Bridge–Brown electrometer circuit used in molecular-beam detection (EST 46).

amp/mm can be obtained (LEW 49). The Ferranti DBM 8A double electrometer valve has been used instead of an FP 54 to diminish zero drift. Alternatively a vibrating reed electrometer (APP 54) may be used.

Davis, Nagle, and Zacharias (DAV 48b, DAV 49a) introduced the use of a mass spectrometer following the surface ionizer. Not only does this identify the atom concerned but also it improves the signal to noise ratio by eliminating the contributions from other isotopes in the original beam and from contaminants in the surface-ionization detector or in the vacuum of the system. A schematic view of their apparatus which incorporates such a mass spectrometer is shown in Fig. XIV. 12. An artist's perspective view of the same apparatus is shown in Fig. I. 6.

The ion source they used for the mass spectrometer is shown in Fig. XIV. 13. It consists of a flat ribbon of W wire 0·015 inches wide mounted flush with the surface of a plane electrode. The curvature and spacing of the accelerating electrode is such as to cause ions from

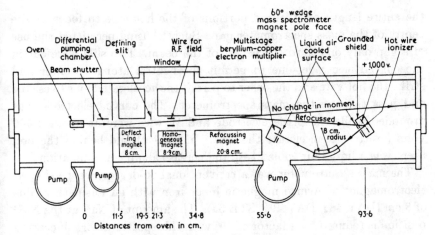

Fig. XIV. 12. Schematic diagram of Na^{22} atomic-beam apparatus showing mass spectrometer and electron multiplier. A perspective sketch of this apparatus is shown in Fig. I. 6 (DAV 48b).

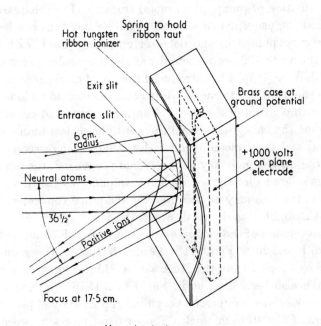

Fig. XIV. 13. Hot wire ionizer used with a mass spectrometer in the detection of Na^{22} (DAV 48b).

the entire length of the used portions of the hot wire to focus at the centre of the mass spectrometer gap. Such focusing permitted the use of a gap only one-third the height of the beam. Two slits were cut in the curved electrode, one to permit the neutral atoms to enter and strike the hot wire and the other to permit the accelerated ions to escape and pass through the mass spectrometer. The curved electrode was grounded while the plane electrode and W wire were at about 1,000 volts positive to accelerate the ions. In front of and behind the box there was a liquid-air cooled trap to decrease alkali contamination.

The mass spectrometer was a conventional 60-degree, wedge-shaped electromagnet of Armco magnetic ingot iron with an ion path radius of 8 cm (DAV 48b, DAV 49a, STE 34). The amount of Na^{23} at the Na^{22} position is reduced by a factor of 10^4 with this spectrometer. Jaccarino and King (JAC 54) have pointed out the importance of shielding all insulators from the ion beam from its time of ionization until its detection; otherwise, electrical charges that accumulate on the insulators may deflect the ion beam erratically. Care must even be taken to avoid the accumulation of pump oil on metal surfaces. For this reason they recommend the use either of a metal shielding box that can be heated or of mercury pumps in the detector chamber. Paul (ZZA 48) has replaced the magnetic spectrometer by an electric quadrupole mass filter.

Lew (LEW 49, DAV 48b) introduced the use of an electron multiplier following the mass spectrometer. With this the individual ions may be counted with conventional nuclear counting circuits and counting-rate meters while the background from the multiplier is less than one count per second and the time constant is less than a microsecond. When the counting rate is high, say above 10,000 per second, the ion current may be measured directly with a D.C. amplifier as discussed above; in this case the ions may be deflected on to a probe connected directly to the D.C. amplifier as shown in Fig. XIV. 14.

The construction of the multiplier used by Lew (LEW 49) and others (DAV 48b) is shown in Figs. XIV. 14 and XIV. 15. The geometry of the plates was copied from the RCA 931–A. Linear electron multiplier arrays have also been used as in Fig. IX. 4 (EIS 52). The material of the electrodes was commercial beryllium copper alloy (4 per cent. Be, 96 per cent. Cu) 0·012 cm thick. The two end pieces between which the dynodes were mounted were of Lavite, though mica templates have also been used successfully (EIS 52).

Just before the dynodes were mounted in position they were scrubbed with a fine emery cloth and washed in acetone or alcohol. The assembly was then placed in a vacuum and heated for about fifteen minutes at 900–1,100° K, as recommended by J. S. Allen (LEW 49); the vacuum should be about 10^{-5} mm of Hg and the heating may be by induction or radiation. Alternatively, heating in a hydrogen furnace may be substituted (EIS 52). If these temperatures were greatly exceeded it was found that the Be–Cu deposited metal on the lavite with resulting

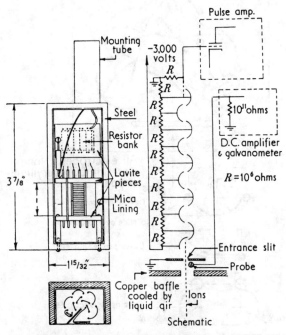

Fig. XIV. 14. Electron multiplier and circuit (LEW 49).

electrical leakage. All matter with a higher vapour pressure than Be or Cu such as solder should be excluded during the heat treatment. After firing, the multiplier was immediately mounted in its housing which consisted of a rectangular iron box lined with mica; subsequent to heat treatment the multiplier should be exposed to air as little as possible since the multiplier plates deteriorate on exposure.

The housing held, in addition to the multiplier, a bank of eleven resistors mounted on a piece of Lavite. Ten of the resistors constituted a voltage divider whereby proper operating voltages for the multiplier were obtained and the eleventh was the output load resistor as in Fig. XIV. 14. The resistors were ½ watt 10 megohm commercial ones with composition bodies and paint removed. The dynodes were connected to the resistors by 0·06 cm nickel wire inside glass tubing. A total of

3,000 volts were applied to the voltage divider which thereby provided 300 volts between successive stages of the multiplier. For positive ion detection the first dynode was at −3,000 volts with respect to the box or ground so ions entering the front slit of the box from the

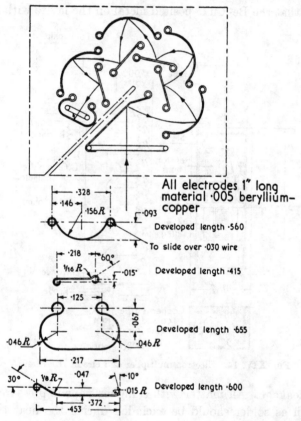

FIG. XIV. 15. Geometry of electron multiplier (LEW 49, DAV 48b).

mass spectrometer acquired an additional 3,000 electron volts of energy before striking the multiplier.

The electron multiplier was found to be 15 per cent. efficient in the detection of Na and K ions of 3,600 electron volts total energy; Eisinger and Bederson (EIS 52) subsequently produced multipliers that were almost 100 per cent. efficient with a background count of less than one per second even after 450 hours of operation. When the multiplier was near the part of an oil diffusion pump the background counting rate was 30 a second, which was ten or twenty times greater than when it was several feet away from a well-baffled pump; when a copper baffle cooled with liquid air was placed in front of the multiplier the background count dropped to less than one a second. With the combination of the mass spectrometer and electrometer, beam intensities as low as five or ten atoms per second could be measured.

In a later experiment Wessel and Lew (WES 53), instead of counting the pulses from the multiplier tube, modulated the beam mechanically and measured the output modulation from the multiplier with an audiofrequency amplifier. For most favourable signal-to-noise ratio a narrow band lock-in amplifier or phase sensitive detector was used (DIC 46, SCH 51, WES 53, EPP 53). Modulation of the radiofrequency transition signal instead of mechanical modulation of the beam can be used and is particularly effective when only a small fraction of the beam is affected by the transition. The modulation procedure in combination with a lock-in amplifier is particularly helpful when there is a large background, as with the universal ionizer discussed below in Section XIV. 3.4, since the unmodulated background is not detected. Wessel and Lew (WES 53) in this fashion have obtained an improvement in signal-to-noise ratio by a factor of a hundred.

XIV. 3.4. *Electron bombardment ionizer.* The great virtue of the surface ionization detector is that the detected atom is converted to an ion, in which case any of the observational methods discussed in the preceding section can be used. The great disadvantage of the surface ionization detector is that its application is limited to relatively few substances, as discussed above.

Wessel and Lew (WES 53, HEI 43, ION 48, ZZA 50) have achieved many of the advantages of a surface-ionization detector with an electron-bombardment ionizer that is applicable to many other atoms. With this, ionization of the atoms in the beam is achieved by means of electrons oscillating back and forth transversely across the beam. The electrons are guided in their oscillations by a magnetic field of a few hundred gauss directed parallel to the oscillations. An exploded view of the device is shown in Fig. XIV. 16, while the relation of the ionizer to the rest of the apparatus is shown in Fig. IX. 7. The ionizer assembly is at the end of a tubular liquid air trap.

Ionization of the atoms takes place in the chamber a through which the atomic beam passes. This chamber is a rectangular box which is open at both ends and which has a vertical slit in each of the broad faces. W filaments, measuring $0 \cdot 005$ cm $\times 0 \cdot 038$ cm in cross-section, are mounted opposite these slits. One of these filaments b is shown stretched between these two stainless-steel clamps c. The lower of these two clamps can slide up and down in a groove. A coil spring d attached to the lower clamp keeps the W filament taut. All these parts are mounted on a fired lava piece l.

Electrons emitted by the heated W filaments are accelerated into the chamber a at an energy of 30 to 100 electron volts. The total current

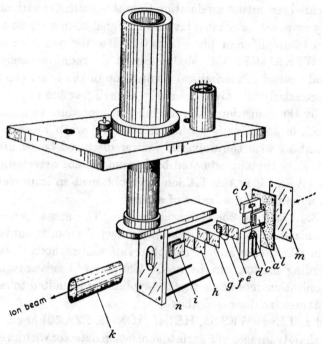

Fig. XIV. 16. Electron bombardment ionizer (WES 53).

is usually between 3 and 15 milliamps. The ions that are formed are pulled out of the ionization region by an electrode e and further accelerated by electrodes g and h. The potential of e relative to chamber a is 0 to 20 volts. The potential of chamber a relative to h, which is necessarily at ground potential, depends on the mass of the ion and the strength of the magnetic field in the mass spectrometer magnet. For silver isotopes it is usually around $+1,300$ volts. The electrode g is at some intermediate potential adjusted experimentally for maximum ion intensity. The exit slit in h as well as the entrance slit in m measures $0 \cdot 038$ cm $\times 1 \cdot 27$ cm. The electrostatic deflecting plates i permit the ion beam to be directed more accurately at the mass spectrometer magnet.

The copper tube k is merely a shield between the ions and the stray fields of the lead-in wires.

Located between the electrodes e and g and at the same potential as e is a tungsten grid f which can be heated electrically. Its purpose is to serve as a surface ionization detector for the alkalis. Although it intercepts only 8 per cent. of the beam, its efficiency of ionization for alkali atoms which do strike it is so high that the net efficiency for alkalis is much higher than that of the electron bombardment ionizer. The efficiency of the electron bombardment ionizer with K^{39} has been estimated from a comparison of the relative effectiveness of the W grid and the bombardment ionizer in the production of ions; from this it is estimated that the bombardment ionizer ionizes one out of every 3,000 incident K^{39} atoms. Fricke (ZZA 50) ionizes more with improved design.

Following the ionization, the ions may be accelerated through a mass spectrometer to an electron multiplier tube as discussed in the preceding section. The use of a mass spectrometer is essential with the electron bombardment ionizer to separate out the many undesired ions that are also produced. The electron multiplier may be fed into a conventional counting system for aural monitoring or into a rectifier and narrow band 10 c/s or so amplifier of the lock-in type, as described in the preceding section. In most cases, despite the selectivity of the mass spectrometer, the background is so large that the 10 c/s system is necessary to improve the signal-to-noise ratio. The background, being unmodulated, does not get through the amplification system. The improvement in signal-to-noise ratio may be as high as a hundredfold.

Some instabilities have been found with the above ionizer and these may be due to exposed insulators acquiring electrostatic charges which deflect the ion beam. For these reasons, a design of ionizer in which the insulators were electrostatically shielded from a direct view of the beam and in which oil deposits were prevented would probably be better, as discussed in the last section.

The atoms which have so far been detected with an electron bombardment ionizer include Ag, Au, K, etc.

The beam chopper used by Wessel and Lew (WES 53) for beam modulation consisted of a vertical tube with four vertical slots cut in it equally spaced around the circumference. It is driven by an external motor through a shaft which enters the vacuum chamber through a Garlock 'Klozure' seal.

XIV. 3.5. *Stern–Pirani detector.* Such gases as molecular and atomic hydrogen, helium, etc., cannot be ionized with a surface-ionization detector. However, since these gases are non-condensable they may be detected with a sensitive pressure manometer. Ordinarily the manometer is of the hot wire or Pirani type (PIR 06, HAL 11) and is placed inside a small volume chamber with a channel to define and admit the beam. The channel is ordinarily long and narrow to increase the outward flow resistance of the gas so a higher equilibrium pressure can be achieved.

The equilibrium pressure that should be attained in a manometer type of detector may easily be calculated. By Eq. (XIII. 1) the rate at which gas enters the detector is

$$I = \frac{1}{4\pi} \frac{A_d}{L^2} n_s \bar{v}_s A_s. \tag{XIV. 2}$$

On the other hand, from Eq. (XIII. 2) applied to the detector, the rate of emergence from the detector is

$$Q_d = (1/\kappa)\tfrac{1}{4} n_d \bar{v}_d A_d. \tag{XIV. 3}$$

At equilibrium these may be equated. The resulting equation with the aid of Eqs. (II. 27), (II. 28), and (II. 1a) reduces to

$$p_\infty = \kappa \frac{p_s A_s}{\pi l_0^2} \sqrt{\left(\frac{T_d}{T_s}\right)}. \tag{XIV. 4}$$

Values of κ, ordinarily called the kappa factor of the detector, may be obtained theoretically from Eqs. (II. 8)–(II. 14).

Although the equilibrium pressure rises proportionally with the kappa factor, there is a disadvantage from too large a kappa factor. The time constant τ of the gauge—the time for the gauge to empty to $1/e$ of its initial value with a similar relation applying on filling—increases with κ. In fact, if V is the volume of the manometer detector, from Eqs. (XIV. 3), (II. 27), and (II. 28)

$$\tau = \frac{V n_d}{Q_d} = \kappa V \frac{4}{\bar{v}_d A_d} = \frac{\kappa V}{A_d} \sqrt{\left(\frac{2\pi m}{k T_d}\right)}. \tag{XIV. 5}$$

Consequently if κ is greatly increased V must be decreased correspondingly to prevent a large increase in the filling time constant τ of the gauge. In addition to the time constant for the filling of the gauge there is a time constant of the associated circuits and a thermal time constant of the heated wire of the Pirani. There is no need for the filling time constant to be appreciably less than the other time constants.

An approximate theory of a Pirani gauge sensitivity can be developed as follows if one assumes that the only cooling of the wire arises from gaseous conduction or from thermal radiation. Let ϵ be the emissivity of the Pirani ribbon, σ the radiation constant, T_k the Pirani ribbon temperature, T_d to the temperature of the detector walls, s' the surface area of the Pirani ribbon per unit length, α' the thermal accommodation coefficient of Eq. (II. 55), C_v the specific heat of the gas per mole at constant volume, and N_0 the number of molecules per mole. Then, from Eq. (II. 2) the energy dissipated by gas conduction per unit length of ribbon is $\tfrac{1}{4}n_d\bar{v}_d\alpha'(C_v/N_0)(T_p-T_d)s'$ while that from radiation is $\epsilon\sigma(T_p^4-T_d^4)s'$. Let e be the electrical energy per unit length supplied to the wire, regarded as constant over the small pressure range that occurs. Then for temperature equilibrium

$$e = \epsilon\sigma(T_p^4-T_d^4)s'+\alpha'b(T_p-T_d)s'p, \qquad (\text{XIV. 6})$$

where
$$b = \frac{C_v}{N_0\sqrt{(2\pi mkT_d)}} \qquad (\text{XIV. 7})$$

from Eqs. (II. 27), (II. 28), and (II. 1a). By differentiation one therefore obtains

$$\frac{dT_p}{dp} = -\frac{\alpha'b(T_p-T_d)}{4\epsilon\sigma T_p^3+\alpha'bp} \qquad (\text{XIV. 8})$$

or in the limit of small p and $T_p \gg T_d$

$$\frac{dT_p}{dp} \approx -\frac{\alpha'b}{4\epsilon\sigma}\frac{1}{T_p^2} = -\frac{\alpha'}{4\epsilon\sigma T_p^2}\frac{C_v}{N_0\sqrt{(2\pi mkT_d)}}. \qquad (\text{XIV. 9})$$

From this result, from the temperature coefficient of the Pirani ribbon, and from the sensitivity of the galvanometer circuit one can calculate directly the beam sensitivity of the Pirani detector. More elaborate theoretical discussions including such effects as metallic thermal conductivity along the ribbon have been given by Julian (JUL 47), Kolsky (KOL 50b), and Silsbee (SIL 50, KNU 10).

A horizontal cross-section of one typical form of Pirani detector as used by Kellogg, Rabi, Ramsey, and Zacharias is shown in Fig. XIV. 17. The nickel ribbons make up four arms of a Wheatstone bridge such that the two detector ribbons are in two electrically opposite arms while the same is true of the balancing ribbons. The two detectors are symmetrically constructed, as shown in the illustration, so that one can be used to detect the beam while the other balances out vacuum fluctuations as in Fig. IX. 6. Each set of gauge chambers is formed by inlaying in solder two 1·6 mm inner diameter copper tubes in a rectangular trough cut in a brass block. A transverse hole is drilled from the outside to the inside to interconnect the tubes as shown. The faces of the block are then surface ground and lapped. The slit jaws are carefully lapped brass blocks. The slit is formed by forcing these slit blocks against a 0·015-mm aluminium foil spacer to form a channel 0·015 mm wide, 2 mm high, and 4·5 mm long. The gauge ribbons are made of nickel ribbon 9 cm long, 0·25 mm wide, and 0·004 mm thick. The ribbons are kept taut by spiral springs of 6 mil nickel wire formed on the point of a needle. Nickel leads are brought out through glass tubes and the ends are sealed with a drop of clear glyptal lacquer. Finally the gauge is baked at a temperature of 150° C for three days.

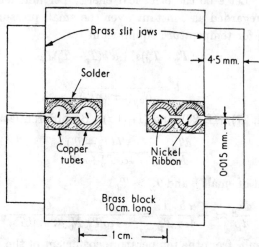

Fig. XIV. 17. Horizontal cross-section of a Pirani detector. The block and tubes are 10 cm long (KEL 39a).

Zacharias and Julian (JUL 47) have introduced a new method of Pirani gauge construction which permits smaller internal volumes and consequently greater kappa factors and sensitivity. In this construction, the Pirani ribbons are mounted in small channels in a flat lapped block. These channels are then closed with another lapped block and a gold foil spacer which defines the beam channel. A Pirani detector of this design used by Kolsky, Phipps, Ramsey, and Silsbee (KOL 50b) is shown photographically in Fig. XIV. 18. The dimensions of the cavities, ribbons, etc., for this detector are included in Table XIII. I along with its operating characteristics. The kappa factor of this gauge is 300. A photograph of the same detector mounted on a movable trap attached to a porthole cover is shown in Fig. XIV. 19. Pirani gauges of the same general type have been made from glass blocks instead of brass ones. A typical Wheatstone bridge circuit used with a Pirani detector is shown in Fig. XIV. 20. It is important that all leads be well shielded to avoid electromagnetic pick-up either from

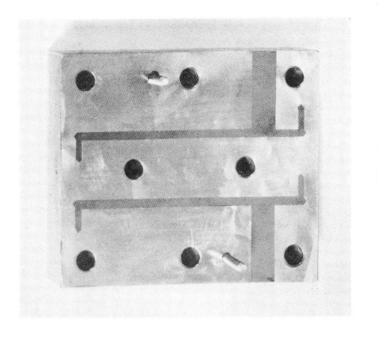

FIG. XIV. 18. Interior of a Pirani gauge molecular beam detector. Scale in inches. (KOL 50b).

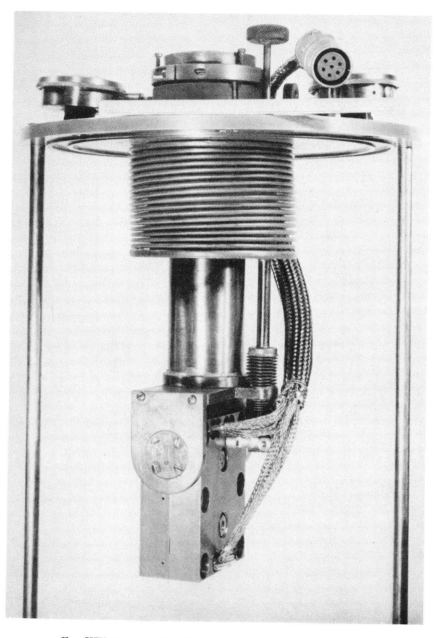

FIG. XIV. 19. Assembled Pirani detector in its mount (KOL 50b).

electrical discharges or from the radiofrequency oscillators of the resonance method.

Pirani detectors suffer from low sensitivity and a considerable amount of unexplained noise, drift, and fluctuations. However, where no other detector has been available they have been used effectively, especially with atomic and molecular hydrogen.

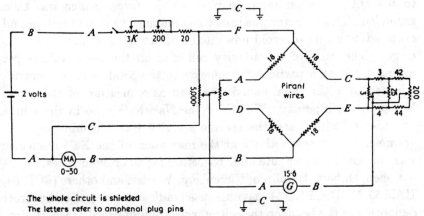

FIG. XIV. 20. Circuit diagram of a typical Pirani gauge detector. The numbers indicate resistances in ohms (KOL 50b).

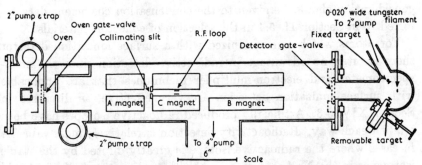

FIG. XIV. 21. A schematic view of atomic-beam magnetic-resonance apparatus with deposition radioactivity detector (BEL 53).

XIV. 3.6. *Radioactivity detection.* Molecular beams of radioactive isotopes may be detected by their radioactivity. The first experiments using radioactivity as the means of detection were by Smith and Bellamy (SMI 51, BEL 53) with Na^{24} and K^{42}. Their apparatus is shown in Fig. XIV. 21. After being collimated by an 0·010-inch slit mounted on the C magnet, the beam fell on a 0·025-inch wide hot oxidized W strip, one end of which was grounded. The positive ions were then attracted to a 1-cm diameter brass target mounted on an insulated support which could be removed from the vacuum system through a gate valve for radioactivity counting. The target, which was 100 volts negative with respect to ground, was connected to the input of a D.C. amplifier and integrator. A second similarly situated, normally grounded collector was mounted in the detector chamber; it could be switched to the D.C. amplifier input in place of the target which was then grounded. This collector was used when making the C field setting and other adjustments to avoid unwanted activity being deposited on the target. The ratio of the activity collected on the removable target during a run at a particular frequency to the total beam current integrated over the same period was used as a measure of the active isotope beam intensity. With Na^{24} the Na^{24}/Na^{23} ratio in the source was about 2×10^{-8} and the specific activity was 200 millicuries per gramme. The target activity at the maximum of the Na^{24} resonance was 150 counts per minute after collecting for 20 or 30 minutes.

Cohen, Gilbert, Hamilton, Nierenberg, Wexler, and others (COH 54, HAM 53, NIE 54, BRO 54) have also used radioactive detection by direct deposition of the beam on the collector without intermediate ionization. As discussed in Section II. 5.4 and XIV. 3.1 the condensation or sticking coefficient varies greatly from one pair of substances to another, so consideration should be given to the condensation coefficient data referred to in Section II. 5.4 in the selection of collector materials.

For atoms which can be ionized with a surface ionization detector there is often no advantage in radioactive detection since the mass-spectrometer and electron-multiplier techniques that are applicable with surface ionization detection are very effective, as discussed in Section XIV. 3.3. Artificially produced radioactive isotopes have been studied each way. Radioactivity detection is relatively more valuable in cases where the radioactive isotope is greatly diluted by the stable isotope as in the Na^{24} case above. Also the radioactive half life should not be too great or insufficient amounts of activity will be produced in reasonable bombardment times. Radioactive detection should be particularly valuable with many radioactive isotopes for which the surface ionization technique is ineffective.

XIV. 4. Deflecting and uniform fields

XIV. 4.1. *Stern–Gerlach and Hamburg deflecting fields.* The earliest atomic-beam magnetic-deflexion experiments were done by Stern and Gerlach (STE 21, GER 24) with iron deflecting magnets whose cross-section was approximately as in Fig. XIV. 22. In order to achieve a large field gradient the beam passed quite close to the knife edge. However, with such a beam position the magnetic-field gradient varied markedly over the height of the beam, which made quantitative measurements difficult.

Subsequent field studies showed that if the beam were located much farther from the knife edge as indicated in Fig. XIV. 22 the field gradient $\partial H/\partial z$ was much more uniform over the beam height; in the beam

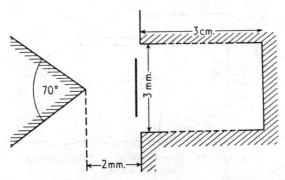

Fig. XIV. 22. Stern–Gerlach deflecting magnet, illustrating Hamburg set-up (FRA 31).

location of the figure the inhomogeneity was constant to within a few per cent. over the height of the beam. The arrangement with the beam far from the knife edge is often known as the Hamburg set-up (FRA 31).

XIV. 4.2. *Rabi deflecting field.* When quantitative-deflexion experiments are made the previous magnetic field has the disadvantage that the field gradient must be measured, which is much more difficult than a measurement of the field itself; gradients are sometimes measured by the deflexion of a quartz reed to the end of which a bismuth test body is attached (FRA 31).

An important improvement for quantitative-deflexion experiments was made in 1929 by Rabi (RAB 29) who pointed out that it was possible to obtain measurable deflexions with atoms by sending the beam at an angle between flat pole pieces as in Figs. XIV. 23 and XIV. 24. In so far as the transition region from the zero field region to the region of field H occurs in a negligibly short distance, the deflexion depends on the value of H but not on the value of the gradient as can be seen from the following argument. Since the force and hence the acceleration is entirely in a direction normal to the boundary,

$$v_0 \cos\theta = v\cos(\theta+\delta). \qquad (XIV.\ 10)$$

Therefore, for small δ and for a fixed magnetic moment whose component along H is μ_1,

$$1-\tan\theta\sin\delta = \frac{v_0}{v} = \sqrt{\left(\frac{T_0}{T_0+\mu_1 H}\right)}, \qquad (XIV.\ 11)$$

where T_0 is the initial kinetic energy of the atom or molecule. Since $\mu_1 H \ll T_0$, this gives approximately

$$s = l\delta = \frac{\mu_1 H l}{2T_0 \tan\theta}. \qquad (XIV.\ 12)$$

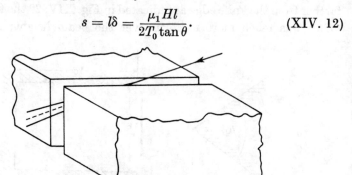

Fig. XIV. 23. Illustration of the Rabi deflecting field (RAB 29a).

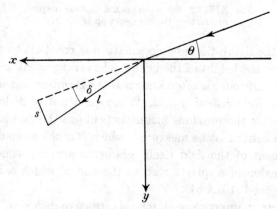

Fig. XIV. 24. Principle of Rabi field (RAB 29a).

The gradient of H does not appear in this as anticipated. Consequently for interpretation of absolute deflexion measurements, the magnetic-field data are much more reliable. On the other hand, for comparative measurements against a standard substance, the Rabi field has no advantage over the deflexion fields of the previous and subsequent sections while it suffers from the disadvantage of less deflecting power for given beam dimensions.

Detailed discussions of the corrections for the finite size of the transition region may be found in Rabi's original paper (RAB 29) and Fraser's book (FRA 31).

XIV. 4.3. *Two-wire deflecting field.* An important improvement in the means for obtaining a magnetic field of known magnitude and gradient with considerable uniformity over the beam height was made by Rabi, Kellogg, and Zacharias (RAB 34) with the introduction of

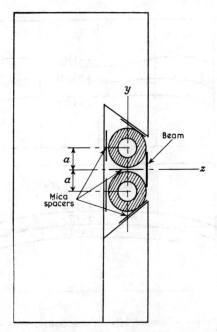

FIG. XIV. 25. Cross-section of two-wire field holder and field wires. The direction of the beam is perpendicular to the plane of the paper (RAB 34, REN 40).

the two-wire field. They used the magnetic field of two long parallel straight wires carrying equal currents in opposite directions. The field and gradient are completely calculable from the geometrical arrangement of the currents as a result of the absence of ferromagnetic materials. Consequently difficult and tedious measurements of these quantities are avoided. Fig. XIV. 25 shows in cross-section the relation of such a pair of wires to the beam and to a set of coordinate axes yz. Graphs of the magnitude of the magnetic field and of the gradient are shown in Fig. XIV. 26. If $2a$ is the distance between the centres of the wires, then it is evident from these curves that for $z = 1 \cdot 2a$ both field and gradient are nearly constant over the region between $y = -0 \cdot 7a$ and

$+0.7a$. The curves of Fig. XIV. 26 may be plotted from the following relations which can be derived in a straightforward though tedious fashion (RAB 34, DAV 53) as in Appendix F, which gives for the field and gradient at any point

$$H = 2I\frac{2a}{r_1 r_2}, \qquad \frac{\partial H}{\partial z} = -2I\frac{2a}{r_1^3 r_2^3}(r_1^2 + r_2^2)z, \qquad \text{(XIV. 13)}$$

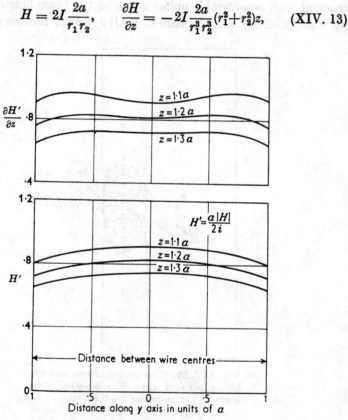

Fig. XIV. 26. Plot of the magnetic field and gradient in the region of the beam with a two-wire field (RAB 34).

where I is one-tenth the current in amperes and r_1 and r_2 are the distances of the point in question from the centres of the two wires. In Appendix F it is also shown that the magnetic equipotentials are circles passing through the centres of the wires as shown for wires of infinitesimal diameter in Fig. XIV. 27. The field lines form sets of circles orthogonal to the equipotentials as also shown in Fig. XIV. 27. With $a = 1.23$ mm and $y = 0$ a current of 100 amps in each wire produces a field of 133 gauss and a gradient of 1,080 gauss per cm at $z = 1.2a$. In some cases currents up to 800 amps have been used in two-wire fields (HAM 39). Such high current densities require the use of hollow water-cooled tubing. The tubes are ordinarily mounted in a duralumin block and insulated with mica spacers, as shown in Fig. XIV. 25. Often submarine storage batteries with characteristics

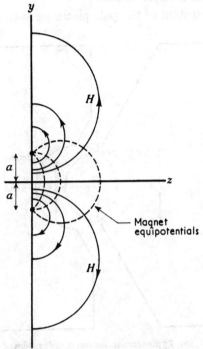

FIG. XIV. 27. Magnetic field lines around two-wire field (full curves) and magnetic equipotentials (dashed curves). Infinitely narrow wires are assumed in the figure but outside the wires the field is independent of the wire diameter (BYE 93).

similar to those in Table XIII. I are used to provide these or greater currents.

Millman (MIL 35) has occasionally used a four-wire system giving a greater ratio of gradient to magnetic field (11 to 1). A detailed description of this system is given in Millman's original paper (MIL 35).

XIV. 4.4. *Two-wire deflecting fields from iron magnets.* Although two-wire fields have many advantages, they have one serious deficiency in that it is very difficult to produce really high fields such as 10,000 gauss without the use of inconveniently high current densities. Millman, Rabi, and Zacharias overcame this difficulty while retaining many (though not all) advantages of a two-wire field. They used an iron magnet such that the gap boundaries were the equipotentials corresponding to the field produced by a two-wire system of Fig. XIV. 27. Frequently the radius of the convex magnet surface is chosen to equal the value of the a of the corresponding two-wire field, in which case the concave surface has a larger radius.

A typical cross-section of the pole pieces for such a magnet is shown

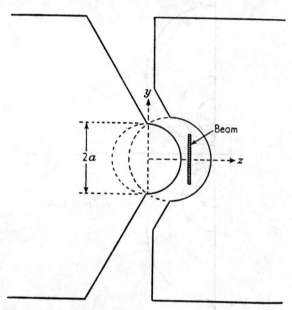

FIG. XIV. 28. Typical cross-section of pole pieces of magnets effectively producing two-wire fields. The direction of the beam is perpendicular to the plane of the paper (KEL 46).

in Fig. XIV. 28. As pointed out in the preceding section and illustrated in Fig. XIV. 27, the magnetic equipotentials are circles which intersect at positions of the centres of the two equivalent wires. This is fortunate in that milling cutters to produce circular cross-sections are readily available.

As long as saturation difficulties are avoided, the field produced by such an electromagnet is equivalent to that of a two-wire field and possesses its advantages of uniformity of gradient and field over the beam height. However, owing to the presence of iron in the magnetic circuit, the field can no longer be calculated accurately from the exciting current. The deflecting magnets in most molecular-beam magnetic-resonance experiments are of this type.

Photographic views of a typical deflecting magnet are shown in Figs.

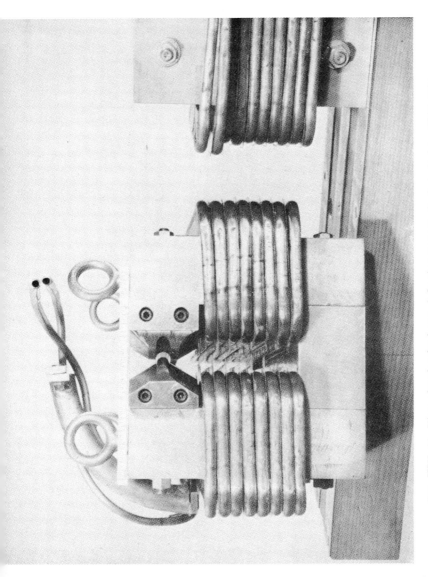

Fig. XIV. 29. View of molecular beam deflecting magnet. Scale in inches. (KOL 50b).

Fig. XIV. 30. The B and C magnets before installation. The current, cooling water, and radiofrequency connexions are shown (KOL 50b).

I. 4 and XIV. 29. For higher fields with less current the coils should be around the portion of the magnet closest to the pole tips. The iron ordinarily used in such magnets is Armco (American Rolling Mill) magnet iron. After machining, the iron is ordinarily annealed in accordance with the Armco procedure B (ARM 45). When especially high fields are desired the pole tips are made of Western Electric permendur which nominally saturates at 23 kilogauss as compared to 16 kilogauss for Armco iron; the pole tips must be specially heat treated after machining. The magnets are often wound with water-cooled copper tubing or bus bar which is insulated with mica sheets or wrapped teflon tape. The magnets are often excited with several hundred amperes of current from large storage batteries with the characteristics given in Table XIII. I; the batteries may be charged by a welding generator as described in the table. Standard magnet design principles may be followed in the design of the magnets (POW 46, BOZ 47, ARM 45, UND 54, IND 54, KOL 50b, PHI 50). Brass spacers are often used above the magnet gap to prevent the pole tips partially drawing together when the field is applied.

The magnet currents are ordinarily controlled by home-made rheostats consisting in part of water-cooled stainless-steel resistors for high currents and in part of constantan resistor strips over which one may slide clips adapted from battery clips by the insertion of brass inserts so that a sliding contact may be obtained. The currents are usually measured with the use of commercial high-current shunts and a Leeds and Northrup Type K potentiometer.

Although the gradients of the A and B fields may be either in the same or in the opposite direction, the directions of the A, B, and C fields are ordinarily in the same direction to diminish the danger of spontaneous transitions or Majorana flops in the regions between magnets. In particular, as shown in Chapter IV, it is important that the direction of a magnetic field as seen by a moving molecule should not change at a rate comparable to or greater than the Larmor frequencies of the magnetic moments concerned. As an aid in achieving this, iron 'ears' as shown in Fig. XIV. 29 are sometimes placed on the ends of the deflecting magnets nearest the C field with a contour to provide a more gradual transition from the strong deflecting field to the much weaker fields between magnets.

XIV. 4.5. *Uniform magnetic fields.* When strong uniform magnetic fields are desired for the C region, they are usually produced by an electromagnet constructed and controlled in a fashion similar to the deflecting electromagnets described in the preceding section. A photographic view of such a magnet is given in Fig. XIV. 30.

When very narrow field dependent resonances are studied the magnetic field drift with slight changes of cooling water temperature, with battery discharge, etc., can become very serious with such a magnet. This difficulty can be alleviated by the use of an electronic control circuit which will ensure constancy in the position of the photoelectrically measured spot of the Type K potentiometer galvanometer. The same circuit can introduce larger corrections when necessary by an associated servomechanism to adjust the rheostats (SIL 50, KOL 52).

Ferromagnetic electromagnets are not very reliable for the production of weak magnetic fields of a few gauss or less because of hysteresis effects which produce field irregularities and uncertainties comparable to the field itself. In such cases the uniform magnetic field may be produced by currents in conducting metal strips or coils without the use of iron; iron electromagnets despite their inherent difficulties at low fields have nevertheless been used in many experiments, especially when the resonance frequency has not been very field sensitive. In all experiments in which an accurately known but weak C field is required, special care must be taken to compensate or to allow for stray magnetic fields from the A and B magnets, from neighbouring magnetized bodies or current carrying wires, and from the earth's magnetization.

One of the important problems with any magnetic field is its calibration. Ordinarily this is done in terms of the resonance frequency of some known magnetic moment as in Chapters VI and IX. Methods for obtaining absolute calibrations of magnetic fields are discussed in Section VI. 5.

Occasionally permanent magnets have been used for the C field because of their great stability and constancy (COR 53, COR 54), despite the obvious disadvantage that they are not easily adjustable. Such permanent magnets are usually made of Armco iron and of Alnico V. The design principles for such magnets are discussed in the literature (BOZ 47, UND 54) and in the pamphlets obtainable from the manufacturers (IND 54, UND 54, ARM 45).

XIV. 4.6. *Electrostatic fields.* Electrostatic fields for electric deflexion and resonance experiments are usually produced with electrostatic electrodes of the same shape as the pole faces used in the analogous magnetic experiments (HUG 47). Fig. XIV. 31, for example, shows a typical electrostatic deflecting field (LEE 52).

Trischka (TRI 48) has shown that a high degree of uniformity for the electrostatic C field is necessary if hyperfine structure effects are to be observed in electric resonance experiments at other than very low electrostatic fields. Swartz and Trischka (SWA 52) have achieved the desired uniformity in some cases with an electrostatic C field formed from two pieces of $\frac{1}{4}$-inch plate glass, coated on adjacent surfaces with an evaporated metal film and separated by three glass spacers. The

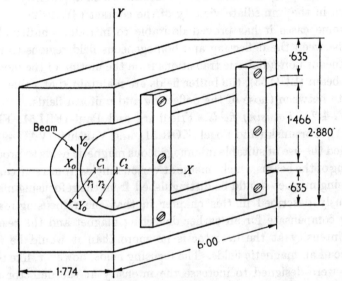

FIG. XIV. 31. Oblique projection of electrostatic deflecting field (LEE 52).

plates were 10 cm long in the beam direction and 6 cm high. These plates, supported by a brass mount, were clamped together over the three spacers by adjustable spring brass clamps. For the alignment of the plates to plane parallelism, light from a Na vapour lamp was transmitted through the field normal to the coated surfaces, and the clamping pressure over each spacer was adjusted while the Haidinger fringes were observed. The physical arrangement and alignment procedure were thus basically the same as for the Fabry–Perot interferometer. The metal films were nearly opaque, giving about 1 per cent. transmission of white light. The segment for the formation of the oscillatory field was formed by making a narrow scratch in one of the films so as to isolate electrically a rectangle 8 cm long in the beam direction by 3 cm high at the top centre of the film. The oscillating voltage was then applied between this rectangle and the rest of the film. Both gold and aluminium films were used with success, though some difficulty was experienced in achieving low resistance at each contact with the film. The contacts were made with silver-plated spring clips and had to be of low resistance compared with the leakage resistance across the spacers if the films were to have the same electrostatic potentials as the clips. Damping resistances in series with the batteries which supplied the voltages to the clips helped preserve good contacts by preventing transient charging and discharging currents which apparently removed the film in the immediate vicinity of the contact (SWA 52).

In some cases it has proved desirable to introduce buffer electric fields between the deflecting and homogeneous field regions to reduce spontaneous non-adiabatic transitions from the motion of the molecules in the beam (LUC 53); the buffer fields are maintained at values intermediate between those of the deflecting and uniform fields.

XIV. 4.7. *Focusing fields.* Friedburg and Paul (FRI 51, FRI 50, ZZA 11), Korsunskii and Fogel (KOR 51), and Vauthier (VAU 49) have proposed the use of suitable inhomogeneous magnetic fields to produce a focusing of the beams which emerge at various angles from a source. Such a focusing magnet should be distinguished from the refocusing magnets previously described in this chapter in that the refocusing magnets merely compensate for an earlier deflecting magnet and the resultant beam intensity at the detector is no more than it would be in the absence of all magnetic fields. The focusing fields, however, like optical lenses, were designed to increase the intensity at the detector above that in the absence of the focusing magnet by bringing together molecules which emerged from the source at different angles. One difficulty with all such magnetic focusing lenses so far proposed is that the focal length is velocity dependent so that the focusing, for a given geometrical configuration, occurs only for molecules in a limited velocity distribution. Despite this important limitation, focusing magnets are very useful in many experiments, including those in which the velocity distribution is already limited for other reasons.

Friedburg and Paul (FRI 51) have used a focusing magnet designed on the following considerations. Consider an inhomogeneous magnet such that the force on a molecule is given by

$$\mathbf{F} = -D\mathbf{r}_1 = -D(x\mathbf{i}+y\mathbf{j}), \qquad (XIV. 14)$$

where $\mathbf{r}_1$ is the projection of the position vector $\mathbf{r}$ on a plane perpendicular to the dominant beam direction which is taken as the z axis. In such a force field the molecule will undergo simple harmonic oscillations with the angular frequency $\omega = \sqrt{(D/m)}$. This oscillatory motion is extended along the z-axis because of the velocity component v_z. All molecules which originate from a source point x_0, y_0, z_0 and diverge with arbitrary velocity components v_x and v_y will converge after half a period at the point $-x_0$, $-y_0$, $z_0 + (\pi/\omega)v_z$. Therefore, for a single velocity v_z, a sharp real image of the source is formed.

For a permanent magnet of spin $\frac{1}{2}$, the potential energy may be taken as one of the two values

$$W = \pm \mu_1 |H|. \qquad (XIV.\ 15)$$

Hence the force is $\quad \mathbf{F} = -\nabla W = \mp \mu_1 \nabla |H|. \qquad (XIV.\ 16)$

It is easily seen by substitution that Eqs. (XIV. 15) and (XIV. 16) are satisfied by

$$|H| = \frac{D}{2\mu_1}(x^2+y^2). \qquad (XIV.\ 17)$$

Furthermore, it is apparent that the following form of $\mathbf{H}$ yields Eq. (XIV. 17) for its absolute value and satisfies $\nabla \cdot \mathbf{H} = 0$ as required by Maxwell's equations in free space:

$$\mathbf{H} = \frac{D}{2\mu_1}(x^2-y^2)\mathbf{i} - \frac{D}{\mu_1}xy\mathbf{j}. \qquad (XIV.\ 18)$$

The field lines and a set of equipotentials for such a field are plotted in Fig. XIV. 32.

The angular width ϕ of the bundle of emergent particles that is refocused may easily be calculated. A molecule with a transverse velocity v_x will go out as far as $\mu_1 |H| = \frac{1}{2}mv_x^2$ before starting to return toward the axis. Therefore, if the typical v_z is taken as the most probable velocity $\sqrt{(2kT/m)}$,

$$\tfrac{1}{2}\phi = \frac{v_x}{v_z} = \sqrt{\left(\frac{\mu_1 H_{\max}}{kT}\right)}. \qquad (XIV.\ 19)$$

For $T = 550°$ K, $H_{\max} = 10^4$ gauss, and μ_1 one Bohr magneton, $\phi = 1/14$ radian. Under these conditions, if the beam is mono-energetic, it is possible at 20 cm from the source to concentrate the same intensity into a point as without focusing would be distributed over a circular disk 1·4 cm in diameter.

The above focusing occurs only for mono-energetic atoms; but in some experiments only such atoms are used. Alternatively, if the beam source is pulse modulated, the magnetic field may be varied as a function of time so that the magnetic field is stronger for the faster atoms which reach the magnet first and weaker for the slower atoms in just such a way that all velocities focus at the same position (FRI 51). Alternatively, the effects of chromatic aberration can be diminished

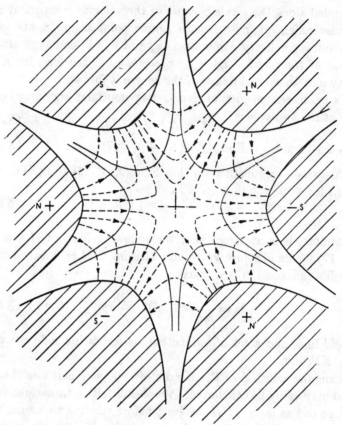

FIG. XIV. 32. Focusing magnetic field. The field and potential lines for the magnetic lens are shown. The field lines are dashed. (FRI 51, HAM 53).

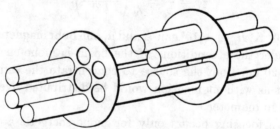

FIG. XIV. 33. Schematic view of current bars for producing desired time variable focusing field (FRI 51).

by using one converging and one diverging lens in series as discussed in greater detail in Section V. 5.1.

In his first experiments Friedburg (FRI 51) produced the desired field by currents in six conductors as in Fig. XIV. 33, the currents in adjacent conductors being oppositely directed. Such a magnetic field could be conveniently varied with time for use with a modulated beam as in the preceding paragraph. Alternatively iron pole tips following the magnetic potential curves of Fig. XIV. 32 may be used as in Hamilton's experiments (HAM 53) discussed in Section V. 5.1.

XIV. 5. Oscillatory fields

XIV. 5.1. *Oscillators.* The oscillators used in molecular-beam resonance experiments are of drastically different designs depending on the desired frequency range, the required power output, the desired frequency stability, etc.

In many of the early molecular-beam resonance experiments the oscillators were simple shunt-fed Hartley oscillators employing an Eimac 250 TL power transmitting tube (TER 37, AME 54, RAB 39, KEL 39a). These proved to be very convenient and covered a wide frequency range. Although such oscillators have continued in use to the present time they have also been supplemented by many other varieties of oscillators, including surplus aircraft transmitters (KOL 52), crystal controlled oscillators, lighthouse-tube circuits, and reflex klystrons (LEE 52) at higher frequencies, etc. Fortunately, in higher frequency experiments, the magnetic moment with which the oscillatory field interacts is ordinarily of Bohr magneton size instead of nuclear magneton size, so much less power is required.

The oscillator design problems for molecular-beam resonance applications are essentially the same as in other fields such as radio transmission, radar, television, etc. For this reason no discussion of oscillator-design principles will be given here. Instead, the reader is referred to any of the many books on radio-frequency and microwave oscillators (TER 37, AME 54, BRA 42, TOW 54, RAD 47).

The results of most molecular-beam resonance experiments depend directly upon a measurement of the oscillator frequency. Ordinarily these frequencies are measured with such commercially available frequency-measuring equipment as the General Radio 620A heterodyne frequency meter good to one part in 10^4 or the General Radio frequency standard 1,100A in conjunction with an interpolation oscillator to an accuracy of one part in 10^8.

XIV. 5.2. *Oscillatory field loops and electrodes.* Many different designs of loops have been used to generate the oscillatory magnetic fields used in molecular-beam magnetic resonance experiments.

In the earliest experiments a simple hairpin loop was employed similar to the one in Fig. VI. 2. As discussed in Chapter VI, such a loop possesses both the advantage and evil of Millman effect.

Millman effect can be reduced with the aid of a loop in the form of two vertical strips joined at the bottom to form a U as in Fig. XIV. 34.

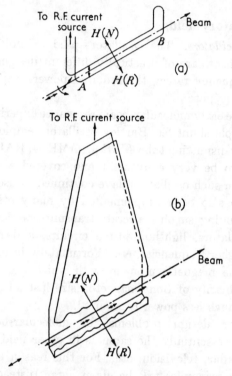

Fig. XIV. 34. Arrangements for superposing an oscillating magnetic field on a fixed magnetic field. The orientations of the fixed and alternating fields are indicated by the arrows. (a) The two-wire arrangement which exhibits marked Millman effect. (b) The parallel-plane arrangement in which the Millman effect is suppressed (TAU 49).

In some cases the ribbons have been as little as $\frac{3}{16}$ inch wide in the direction of the beam as in Fig. I. 6 (DAV 48b), while in other cases strips several inches wide in the direction of the beam have been used as in Figs. IX. 6 and XIV. 34 (HUG 53). For the Ramsey separated oscillatory-field method, two such loops have usually been used.

The loops may either be fed directly from the oscillator or the loops may form part of a tuned resonance circuit that is excited by a transmission line from the oscillator. With the separated oscillatory field, special care must be taken to ensure that the relative phases of the two loops are as desired. The shape of pattern on an oscilloscope trace when the vertical and horizontal traces are fed from the two loops through identical circuits may be used for this purpose. An alternative form of single radiofrequency loop that has been used consists of two copper ribbons forming a horizontal U; typical dimensions for the conductors are $\frac{3}{4}$ inch high, $\frac{1}{16}$ inch apart, and $\frac{1}{32}$ inch thick (EIS 52).

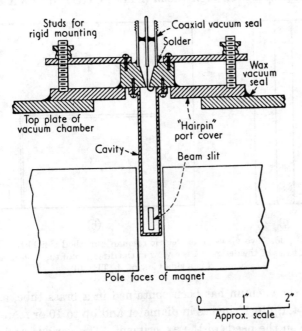

Fig. XIV. 35. Resonant wave guide source of oscillatory magnetic field (WEI 53).

At microwave frequencies, tuned wave guide cavities have often been used to provide the oscillatory magnetic field. A typical wave guide cavity for use in this fashion at 6,850 Mc/sec is shown schematically in Fig. XIV. 35.

In the electric-resonance method, the most important criteria for the oscillatory electric-field electrodes is that they should not distort the constant uniform electric field for the reasons discussed in Section XIV. 4.6. A means for accomplishing this that was introduced by Trischka (TRI 48) is shown in Fig. XIV. 36. The oscillatory potential is applied between sections 3 and 4. An alternative construction utilizing silver-plated glass blocks has been described in Section XIV. 4.6.

XIV. 6. Vacuum system and mechanical design

Most molecular-beam experiments require a vacuum in the main chamber of the apparatus of about 3×10^{-7} mm of Hg as measured by an ionization gauge. This pressure is ordinarily achieved with more or less conventional vacuum techniques and the research worker in the field should be thoroughly familiar with these techniques as described in the various texts on high vacua (STR 38, GUT 49, ZXX 54).

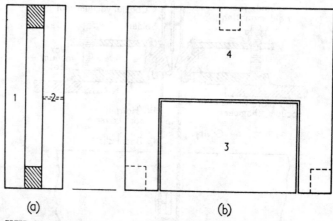

FIG. XIV. 36. Two views of an electric resonance method C field showing the basic features of the design. A is a view in the direction of the molecular beam and B is a side view of A (TRI 48).

Usually the vacuum has been contained in a brass tube, as in Figs. I. 2 and I. 4, up to 12 inches in diameter and up to 10 or more feet long depending on the needs of the experiment. The various ports, flanges, etc., are either soft soldered or hard soldered in place: the latter has the advantage of greater strength and much less likelihood of leak production but the disadvantages of producing great warpage in the brass tube at the high hard-soldering temperatures required and of greater removal difficulty for changes. Stainless steel tubes may be used and have the advantage of being suitable for arc welding; such tubes can be rolled from flat stock which is then welded into a tube. A stainless steel that welds and machines well should be selected. Rectangular vacuum containers made of heavy brass plates bolted and soldered together have been used effectively (LEW 53).

Oil-diffusion pumps have been used in recent molecular-beam experiments but they usually need to be supplemented with liquid nitrogen traps at least in the main chamber. Also additional water cooled, dry ice, or refrigerant-cooled baffles are usually needed between the pumps and the vacuum chamber to diminish accumulation of oil in the latter. In Table XIII. 1 are listed typical pressures, pumping speeds, choices of pumps, etc. These, of course, vary greatly from experiment to experiment; for example, with condensable vapours a much lower source-chamber pumping speed is required. Octoil S, Apiezon diffusion pump oil, and silicone oils have been used with the oil-diffusion pumps. Narcoil has been used in source chambers where higher pressures are tolerable.

Standard leak-hunting procedures (STR 38, GUT 49) are applicable to molecular beams. If a helium leak detector (GUT 49, CON 54, DIS 54) is available, it is very effective for almost all leak hunting. On the other hand, in the absence of a helium leak detector, effective leak hunting can be accomplished by the observation of changes in an ionization gauge reading as the apparatus is painted with water, soap and water, xylene, alcohol, etc., or is covered with a gas like CO_2 or illuminating gas. Many leaks can be repaired temporarily with Apiezon 'Q' putty.

Standard neoprene 'O' rings (GAR 54) or similar gaskets can be used effectively to seal port covers, etc. Typical 'O' ring grooves, for example, are shown in Fig. XIV. 1. Recommended groove dimensions are specified by the manufacturers, but a slight increase in the 'O' ring compression beyond the normal recommendations is often helpful. If no helium leak detector is available, two 'O' rings per port with a pump-out hole between, as in Fig. I. 4, may be helpful since a leak through the inner gasket can immediately be detected by an improvement in the vacuum when the region between the gaskets is partially exhausted. If a helium leak detector is available a single 'O' ring is adequate and is less likely to leak.

Often various water leads, etc., have to be introduced into the apparatus. In such cases, Lew (LEW 53a, ZXZ 54) has found the fittings of Fig. XIV. 37 to be of great use and wide versatility. Alternatively, the commercially available Buggie fittings (BUG 54) can be used effectively.

Often rotary or translatory motions must be provided from the outside to the inside of a vacuum system. Sylphon or metal bellows are effective for this purpose, though they are sometimes more difficult to install than gasket seals; leaks occasionally develop in the soft solder joints to the bellows and sometimes the bellows themselves develop cracks. Typical devices using such bellows are described in high vacuum texts (GUT 49, STR 38). A convenient bellows device for small lateral motions such as moving sources is shown in Fig. XIV. 38. Alternatively,

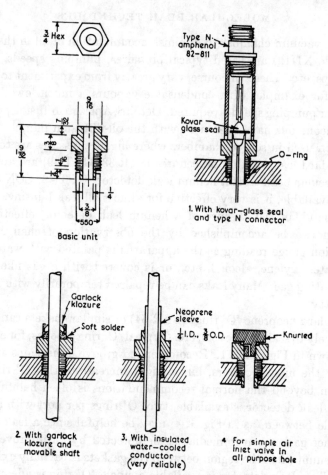

Fig. XIV. 37. Typical vacuum fittings (LEW 53a).

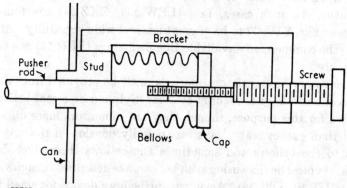

Fig. XIV. 38. Motion transmitted through metal bellows seal. A typical use for such a motion is in the adjustment of an oven position (KUS 50).

chevron seals (Garlock Packing Company) may be used for the same purpose as is illustrated incidentally in Fig. XIV. 1. A still different seal that has been found to be satisfactory is the Garlock 'Klozure' seal (LEW 53a, GAR 54, ZXZ 54), as illustrated in Fig. XIV. 37. A long tapped hole with a tight-fitting screw that is greased with stop-cock grease and covered on the outside with Apiezon 'Q' putty has been found satisfactory for small translatory motions.

Either in this case or with other seals such as sylphons, fine translatory motions can be obtained with differential screw threads as shown in Fig. XIV. 38. If the screw on one half has n_1 threads per inch and on the other end n_2 and if the tapped bracket and cap to which the screw is threaded are arranged to provide opposite directions of translation as in Fig. XIV. 37, the net effective number of threads n per inch is given by $1/n = 1/n_1 - 1/n_2$. Values of 32 and 40 for n_1 and n_2 are often used, in which case n is 160, corresponding to very fine thread.

Different types of valves suitable for use with vacuum systems are either commercially available or may be adapted from commercial valves. A typical adapted valve is included incidentally in Fig. XIV. 1. Sometimes a gate valve is required to close an aperture without the consumption of too much space. In such cases a valve similar in principle to that included in Fig. XIV. 39 can be very effective (LEW 53a, ZXZ 54). Hoke valves, Rockwood ball valves, and modified Crane valves are often useful (ZZA 53, ZZA 54).

In order that adequate vacuum can be achieved in the main chamber, most molecular-beam apparatuses are divided into three separately pumped chambers which open into each other with separating chamber slits which are sufficiently bigger than the beam dimensions to permit easy alignment and at the same time are small enough greatly to restrict the flow of gas between the chambers (0·010 inch separating slit widths are often used with 0·001 inch wide beams). Typical pressures, pumping speeds, etc., in the source, separating chamber, and main chamber are listed in Table XIII. I. In some cases a gate valve is placed between the source and separating chambers as in Fig. XIV. 39 (LEW 53a, ZXZ 54) so that the oven may be removed for a refill without loss of vacuum in the remainder of the apparatus. The use of such a gate valve is an alternative to the device of Fig. XIV. 1 which also permits an oven change without a loss of vacuum.

Glass tubes can be attached to metal apparatus by means of kovar seals which can be soldered directly to the apparatus or to appropriate metal flanges with 'O' ring seals. Kovar seals can also be used to introduce electrically insulated leads into an apparatus as in Fig. XIV. 37. At lower voltages Buggie fittings (BUG 54) or the equivalent shown in Fig. XIV. 37 can be used. Micalex plates can be used as easily machinable and mechanically strong insulating vacuum parts. Sometimes air leaks occur in the micalex plates, but these can be cured by painting them with a solution of polystyrene plastic dissolved in benzene.

The relative positions of detector wires, collimator slits, etc., are sometimes measured with telemicroscopes. However, the use of machinists'

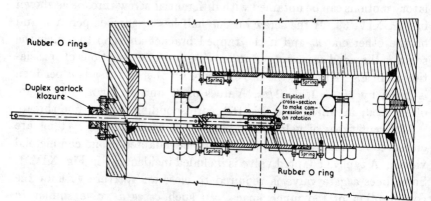

Fig. XIV. 39. Separating chamber bulkheads with gate valve to allow oven changes without loss of vacuum in the remainder of system. The valve is slid into position by sliding the rod which passes through the klozure seal, and the gate valve is sealed by a 90° rotation of the rod which has an elliptical cross-section near the gates (LEW 53a).

dial gauges for this purpose has proved very convenient. The use of micrometer screws to provide motions is also satisfactory provided backlash is avoided by making all final motions in the same direction.

Although some molecular-beam apparatuses have been constructed with the use of 'O' ring seals and very little wax or grease, many have depended heavily on waxed or greased joints. Thus, seals for the lateral translations of collimators, etc., have been made by moving ground and lapped brass plates over similar lapped surfaces which have been greased with stop-cock grease. The waxes and greases that have been most extensively used in molecular-beam work include Apiezon N stop-cock grease which is excellent but very expensive, Cenco stop-cock grease or Apiezon M which are satisfactory for most purposes, Celvacene, Apiezon W40 wax, red sealing wax, and Apiezon Q putty.

Pressures inside the different chambers of a molecular-beam apparatus are ordinarily measured with an ionization gauge such as the Distillation Products VG–1A gauge (DIS 54). These are suitable from

10^{-3} mm of Hg down to 10^{-7} mm. In preliminary leak hunting, thermocouple gauges such as the one manufactured by the National Research Corporation (NAT 54) are used for pressures between 1 mm of Hg and 10^{-3} mm. They are also used to check the pressure in the foreline between the diffusion and mechanical pumps. As a protection of the diffusion-pump oil, the foreline thermocouple gauge may be arranged to turn off the diffusion pumps when the pressure exceeds 0·5 mm of Hg.

XIV. 7. Miscellaneous components

In addition to the above major molecular-beam techniques there are many minor ones that are of value in some experiments though not in others.

Deliberately inserted beam obstacles are frequently of value in determining the signs of various atomic and molecular interactions, since the states concerned in the resonance can then be identified by the direction of their deflexion in one of the deflecting fields (LOG 52). Also at positions of large deflexion, the insertion of selection slits can provide a considerable amount of velocity selection; velocity selection has also been obtained with slotted disks and the other means discussed in Chapter II (FRA 31, p. 73).

Beam flags for mechanically interrupting the beam are often actuated either by an electromagnet or by a direct mechanical linkage as in Section XIV. 6; a permanent magnet moving outside a glass housing containing a ferromagnetic rod inside the housing that is attached to the beam stop has also been used.

XIV. 8. Alignment

XIV. 8.1. *Optical alignment.*
The preliminary alignment necessary to locate the molecular beam is usually done optically. Once the molecular beam has been found most of the final adjustment is done with the beam itself as the alignment element.

The most important instrument for optical alignment is a telescope that can focus the length of the apparatus and that retains the alignment of objects as the focus is changed. The telescope should have an adjustable vertical cross hair. A good engineer's dumpy level such as the Keuffel and Esser P5022 tilting dumpy level with 5092 lateral adjuster has been found to be satisfactory (COH 54). A Gaertner filar micrometer used with such a level greatly increases its flexibility and usefulness. A cathetometer style mount is convenient but not essential (KOL 50b).

Frequently the magnets have been aligned optically by sighting with the telescope along vertical projections that protrude above the magnet, when the system is open to the air and the separating chamber bulkheads are removed. However, Ramsey (RAM 54, KOL 50b) has found it convenient to attach a short plumb-bob of stretched B. and S. No. 40 copper wire below a fiducial mark on each end of the magnet and on similar devices whose locations are critical. Then, with the aid of small windows in each bulkhead and of projection slit jaws on the source, collimator, and detector, the alignment can be checked optically under vacuum and operating conditions.

All slits are made vertical optically with the aid of the telescope cross-hair which can itself be made vertical by a comparison with a plumb-bob of B. and S. No. 40 stretched copper wire with a light weight at one end sitting in a small oil bath to provide damping. Ordinarily, this can be done with sufficient accuracy by optical means that it need not be rechecked with the beam.

The vertical alignment of the beam components is much less critical than the horizontal alignment as a result of the relatively great beam height, so the vertical alignment is ordinarily done only optically. However, it must be done with care since a gross mistake will prevent a good beam.

XIV. 8.2. *Alignment with beam.* After the alignment has been completed optically, the system has been evacuated, and the oven has been raised to an appropriate temperature, the beam hunt can begin. This is done with the collimator removed, the fore slits either removed or more widely opened, and all other removable beam obstacles out of the beam line. If the optical alignment is good, the beam can then usually be found quickly. However, allowance sometimes needs to be made for transverse displacement of the oven due to thermal expansion as it is heated.

Once the beam is found, the remainder of the alignment is done with the beam itself. As an aid to this alignment with the beam, vertical brass knife edges are normally placed at the ends of the deflecting magnets at a known small distance beyond each bead as indicated schematically in Fig. XIV. 40. With these the locations of the magnet positions can be determined and the beam can be adjusted as desired relative to the magnets. Many different geometrical tricks can be used in this alignment, of which a few are illustrated in the following paragraphs.

The exact locations of the magnets relative to the source and detector beam scales and the degree of parallelism of the A and B magnets can be determined as follows. With the collimator removed, the detector can be moved to position D' of Fig. XIV. 40 where the beam is partially cut, say to half value, by edge A_2. One can confirm that it is edge A_2 and not A_1 or some other object that is cutting the beam by the fact that for a given displacement of the source position ΔS the compensating detector displacement ΔD should be given by $\Delta D/\Delta S = \overline{A_2 D'}/\overline{SA_2}$. The partially cut beam can then be pivoted about A_2 until A_1 is partially cut. The transverse position and orientation of the A magnet is then determined in terms of the scales on the source and detector

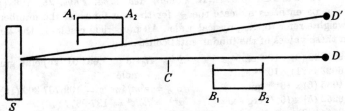

FIG. XIV. 40. Schematic diagram illustrating alignment procedure by use of the molecular beam.

motions. A similar procedure may be applied to the B magnet in which case its degree of parallelism to the A magnet is determined. With the above determinations, the molecular beam may be placed at any desired location relative to the deflecting magnets.

If no good calibration of the source motions is available, a calibration can be obtained in terms of the detector motion by insertion of the collimator C. With the collimator in position and held fixed the transverse displacements of source and detector for maximum beam are related by $\Delta D/\Delta S = \overline{CD}/\overline{SC}$. In fact no accurate scale at all is required on the source motion if all displacements of the source are made with the collimator in position and are followed by the calibrated motion of the detector so that the source displacement can be inferred from the above equation. Likewise, the collimator motions can be calibrated in terms of the detector motions by moving the collimator and detector together so that the beam is retained while the source is held fixed.

In 'flop-in' experiments, such as those described in Section IX. 3, an added alignment precaution is required. The force on an atom in a deflecting field is not completely independent of the transverse displacement in the field. As a result of this and of the different atomic trajectories in the A and B fields, the maximum of the refocused beam is not exactly at the detector position for an undeflected beam. Consequently the best detector position must be determined empirically after the 'flopped-in' beam has been produced.

APPENDIX A

FUNDAMENTAL CONSTANTS

THE following values, with a few exceptions marked by an asterisk (*), are taken from J. W. M. DuMond and E. R. Cohen, *Rev. Mod. Phys.* **25**, 706 (1953). Numbers in parentheses indicate the $\pm$ for the last figure of the number. All atomic weights refer to the physical scale. All numerical results in the text are based on these values of the fundamental constants.

$c = 2{\cdot}997929\ (8) \times 10^{10}$ cm sec^{-1}

$e = 4{\cdot}80288\ (21) \times 10^{-10}$ esu

$m = 9{\cdot}1085\ (6) \times 10^{-28}$ gm

$h = 6{\cdot}6252\ (5) \times 10^{-27}$ erg sec

$\hbar = 1{\cdot}05444\ (9) \times 10^{-27}$ erg sec

$k = 1{\cdot}38042\ (10) \times 10^{-16}$ erg deg^{-1}

$G = 6{\cdot}670\ (5) \times 10^{-8}$ dyne cm^2 gm^{-2} (*)

$N_0 = 6{\cdot}02472\ (36) \times 10^{23}$ gm-mole^{-1}

$R_0 = 8{\cdot}31662\ (38) \times 10^7$ erg mole^{-1} deg^{-1} C

$V_0 = 22{\cdot}4207\ (6) \times 10^3$ cm^3 mole^{-1}

$F' = N_0 e/c = 9{,}652{\cdot}01\ (25)$ emu gm-mole^{-1}

$R_\infty = e^4 m/4\pi \hbar^3 c = 109{,}737{\cdot}309\ (12)$ cm^{-1}

$\alpha^{-1} = \hbar c/e^2 = 137{\cdot}0377\ (16)$

$e/mc = 1{\cdot}75888\ (5) \times 10^7$ emu gm^{-1}

$h/e = 1{\cdot}37943\ (5) \times 10^{-17}$ erg sec esu^{-1}

$a_0 = \hbar^2/me^2 = 0{\cdot}529171\ (6) \times 10^{-8}$ cm

$r_0 = e^2/mc^2 = 2{\cdot}81784\ (10) \times 10^{-13}$ cm

1 atomic mass unit $= 931{\cdot}162\ (24)$ MeV

proton mass $= 938{\cdot}232\ (24)$ MeV

electron mass $= 0{\cdot}510984\ (16)$ MeV

H (atomic wt. of hydrogen) $= 1{\cdot}008142\ (3)$

H$^+$ (atomic wt. of proton) $= \mathrm{H} - N_0 m = 1{\cdot}007593\ (3)$

n (atomic wt. of neutron) $= 1{\cdot}008982\ (3)$ (*)

H$^+/N_0 m$ (proton mass/electron mass) $= 1{,}836{\cdot}13\ (4)$

H$^+/N_0$ (proton mass) $= 1{\cdot}67243\ (10) \times 10^{-24}$ gm

μ_0 (Bohr magneton) $= e\hbar/2mc = 0{\cdot}92732\ (6) \times 10^{-20}$ erg gauss^{-1}

μ_e/μ_0 (electron moment in Bohr magnetons) $= \left[1 + \dfrac{\alpha}{2\pi} - 2{\cdot}973 \dfrac{\alpha^2}{\pi^2}\right]$
$= 1{\cdot}001145356\ (13)$

$\mu_N =$ (nuclear magneton) $= e\hbar N_0/2\mathrm{H}^+ c = 5{\cdot}05038\ (36) \times 10^{-24}$ erg gauss^{-1}

$\mu_N/h = 762{\cdot}295\ (35)$ cycles sec^{-1} gauss^{-1} (nuc. res. in 1 gauss for $g = 1$) (*)

$2\mu_{\mathrm{H}^+}/h = 4{,}257{\cdot}83\ (15)$ cycles sec^{-1} gauss^{-1} (proton res. freq. in 1 gauss) (*)

$\mu_{\mathrm{H}^+} = 2{\cdot}79277\ (6)\ \mu_N$ [or $2{\cdot}792743\ (60)\ \mu_N$(*) as used in text]

1 microampere $= 6{\cdot}24192\ (30) \times 10^{12}$ electrons per second (*)

1 curie $= 3{\cdot}7 \times 10^{10}$ disintegrations per second (*)

Wavelength associated with 1 electron-volt $= 12{,}397{\cdot}8\ (5) \times 10^{-8}$ cm

Temperature associated with 1 electron-volt $= 11{,}605{\cdot}7\ (5)$ ° K

J_{15} (specific heat H$_2$O at 15° C) $= 4{\cdot}1855\ (4)$ abs-joule gm^{-1} deg^{-1} C (*)

$A_0 = 76 \rho_{\mathrm{Hg}0°\mathrm{C}} \delta_{m\mathrm{H}_2\mathrm{O}}\, g_0 = 76 \times 13{\cdot}54542\ (5) \times 0{\cdot}999972\ (2) \times 980{\cdot}665$
$= 1{\cdot}013246\ (4) \times 10^6$ dyne cm^{-2} atm^{-1} (*)

$\pi = 3{\cdot}14159265$ (*) $e = 2{\cdot}71828183$ (*)

1 yd $= 3{,}600/3{,}937$ m (*)   1 lb $= 0{\cdot}4535924277$ kg (*)

$T_0 = 273{\cdot}16\ (1)$ ° K (*)

APPENDIX B

VECTOR AND TENSOR RELATIONS

THE following vector and tensor (dyadic) relations are used in the text. The derivation of these relations is given in many of the standard books on vector and tensor analysis (WEA 44, WIL 31). A dyadic is represented by a symbol such as $\underset{\sim}{F}$.

(B. 1) $\mathbf{a} \cdot \mathbf{b} \times \mathbf{c} = \mathbf{b} \cdot \mathbf{c} \times \mathbf{a} = \mathbf{c} \cdot \mathbf{a} \times \mathbf{b}$

(B. 2) $\mathbf{a} \times (\mathbf{b} \times \mathbf{c}) = (\mathbf{a} \cdot \mathbf{c})\mathbf{b} - (\mathbf{a} \cdot \mathbf{b})\mathbf{c}$

(B. 3) $(\mathbf{a} \times \mathbf{b}) \cdot (\mathbf{c} \times \mathbf{d}) = \mathbf{a} \cdot \mathbf{b} \times (\mathbf{c} \times \mathbf{d})$
$= \mathbf{a} \cdot (\mathbf{b} \cdot \mathbf{d} \mathbf{c} - \mathbf{b} \cdot \mathbf{cd})$
$= (\mathbf{a} \cdot \mathbf{c})(\mathbf{b} \cdot \mathbf{d}) - (\mathbf{a} \cdot \mathbf{d})(\mathbf{b} \cdot \mathbf{c})$

(B. 4) $(\mathbf{a} \times \mathbf{b}) \times (\mathbf{c} \times \mathbf{d}) = (\mathbf{a} \times \mathbf{b} \cdot \mathbf{d})\mathbf{c} - (\mathbf{a} \times \mathbf{b} \cdot \mathbf{c})\mathbf{d}$

(B. 5) $\nabla(\phi + \psi) = \nabla\phi + \nabla\psi$

(B. 6) $\nabla(\phi\psi) = \phi\nabla\psi + \psi\nabla\phi$

(B. 7) $\nabla \cdot (\mathbf{a} + \mathbf{b}) = \nabla \cdot \mathbf{a} + \nabla \cdot \mathbf{b}$

(B. 8) $\nabla \times (\mathbf{a} + \mathbf{b}) = \nabla \times \mathbf{a} + \nabla \times \mathbf{b}$

(B. 9) $\nabla \cdot (\phi\mathbf{a}) = \mathbf{a} \cdot \nabla\phi + \phi\nabla \cdot \mathbf{a}$

(B. 10) $\nabla \times (\phi\mathbf{a}) = \nabla\phi \times \mathbf{a} + \phi\nabla \times \mathbf{a}$

(B. 11) $\nabla(\mathbf{a} \cdot \mathbf{b}) = (\mathbf{a} \cdot \nabla)\mathbf{b} + (\mathbf{b} \cdot \nabla)\mathbf{a} + \mathbf{a} \times (\nabla \times \mathbf{b}) + \mathbf{b} \times (\nabla \times \mathbf{a})$

(B. 12) $\nabla \cdot (\mathbf{a} \times \mathbf{b}) = \mathbf{b} \cdot \nabla \times \mathbf{a} - \mathbf{a} \cdot \nabla \times \mathbf{b}$

(B. 13) $\nabla \times (\mathbf{a} \times \mathbf{b}) = \mathbf{a}\nabla \cdot \mathbf{b} - \mathbf{b}\nabla \cdot \mathbf{a} + (\mathbf{b} \cdot \nabla)\mathbf{a} - (\mathbf{a} \cdot \nabla)\mathbf{b}$

(B. 14) $\nabla \times \nabla \times \mathbf{a} = \nabla\nabla \cdot \mathbf{a} - \nabla \cdot \nabla \mathbf{a} = \nabla\nabla \cdot \mathbf{a} - \nabla^2\mathbf{a}$

(B. 15) $\nabla \times \nabla\phi = 0$

(B. 16) $\nabla \cdot \nabla \times \mathbf{a} = 0$

(B. 17) $\nabla \cdot \mathbf{r} = 3, \quad \nabla \times \mathbf{r} = 0, \quad (\mathbf{a} \cdot \nabla)\mathbf{r} = \mathbf{a}, \quad \nabla r = \frac{\mathbf{r}}{r}, \quad \nabla\frac{1}{r} = -\frac{\mathbf{r}}{r^3}$

(B. 18) $\int_V \nabla\phi \, d\tau = \int_S \phi \, d\mathbf{S}$

(B. 19) $\int_V \nabla \cdot \mathbf{a} \, d\tau = \int_S \mathbf{a} \cdot d\mathbf{S}$

(B. 20) $\int_V \nabla \times \mathbf{a} \, d\tau = \int_S d\mathbf{S} \times \mathbf{a}$

(B. 21) $\int_S d\mathbf{S} \times \nabla\phi = \int_C \phi \, d\mathbf{s}$

(B. 22) $\int_S \nabla \times \mathbf{a} \cdot d\mathbf{S} = \int_C \mathbf{a} \cdot d\mathbf{s}$

(B. 23) $\nabla \times \nabla \mathbf{a} = 0$

(B. 24) $\nabla \cdot \nabla \times \underset{\sim}{F} = 0$

(B. 25) $\nabla \cdot \nabla \mathbf{a} = \nabla^2 \mathbf{a}$

(B. 26) $\nabla \times \nabla \times \underset{\sim}{F} = \nabla\nabla \cdot \underset{\sim}{F} - \nabla^2\underset{\sim}{F}$

(B. 27) $\nabla(\mathbf{a} \cdot \mathbf{b}) = (\nabla\mathbf{a}) \cdot \mathbf{b} + (\nabla\mathbf{b}) \cdot \mathbf{a}$

(B. 28) $\nabla(\mathbf{a} \times \mathbf{b}) = (\nabla\mathbf{a}) \times \mathbf{b} - (\nabla\mathbf{b}) \times \mathbf{a}$

(B. 29) $\nabla(f\mathbf{a}) = (\nabla f)\mathbf{a} + f\nabla\mathbf{a}$

(B. 30) $\nabla \cdot (f\underset{\sim}{F}) = (\nabla f) \cdot \underset{\sim}{F} + f\nabla \cdot \underset{\sim}{F}$

(B. 31) $\nabla \times (f\underset{\sim}{F}) = (\nabla f) \times \underset{\sim}{F} + f\nabla \times \underset{\sim}{F}$

(B. 32) $\nabla \cdot (\mathbf{ab}) = (\nabla \cdot \mathbf{a})\mathbf{b} + \mathbf{a} \cdot \nabla\mathbf{b}$

(B. 33) $\nabla \times (\mathbf{ab}) = (\nabla \times \mathbf{a})\mathbf{b} - \mathbf{a} \times \nabla\mathbf{b}$

(B. 34) $\nabla \times (\mathbf{a} \times \mathbf{b}) = \nabla \cdot (\mathbf{ba} - \mathbf{ab})$

(B. 35) $(\mathbf{ab}):(\mathbf{cd}) \equiv \mathbf{a} \cdot \mathbf{c}\, \mathbf{b} \cdot \mathbf{d}$

(B. 36) $\int_S d\mathbf{S} \times \nabla \mathbf{a} = \int_C d\mathbf{s}\, \mathbf{a}$

(B. 37) $\int_S d\mathbf{S} \cdot \nabla \times \underset{\sim}{\mathbf{F}} = \int_C d\mathbf{s} \cdot \underset{\sim}{\mathbf{F}}$

(B. 38) $\int_V \nabla \mathbf{a}\, d\tau = \int_S d\mathbf{S}\, \mathbf{a}$

(B. 39) $\int_V \nabla \cdot \underset{\sim}{\mathbf{F}}\, d\tau = \int_S d\mathbf{S} \cdot \underset{\sim}{\mathbf{F}}$

(B. 40) $\int_V \nabla \times \underset{\sim}{\mathbf{F}}\, d\tau = \int_S d\mathbf{S} \times \underset{\sim}{\mathbf{F}}$

APPENDIX C

QUADRUPOLE INTERACTION

From Eqs. (III. 31) and (III. 42),

$$\mathcal{H}_{E2} = \frac{1}{6}\frac{e^2 q_J Q}{I(2I-1)J(2J-1)}\sum\left[3\frac{I_i I_j + I_j I_i}{2} - \delta_{ij}\mathbf{I}^2\right]\left[3\frac{J_i J_j + J_j J_i}{2} - \delta_{ij}\mathbf{J}^2\right].$$

(C. 1)

Now consider typical terms in this equation. Thus as **I** and **J** commute with each other

$$\sum_{ij} I_i I_j J_i J_j = \left[\sum_i I_i J_i\right]\left[\sum_j I_j J_j\right] = (\mathbf{I}\cdot\mathbf{J})^2. \tag{C. 2}$$

Likewise,
$$\sum_{ij} I_i I_j \delta_{ij}\mathbf{J}^2 = \left[\sum_i I_i^2\right]\mathbf{J}^2 = \mathbf{I}^2\mathbf{J}^2 \tag{C. 3}$$

and
$$\sum_{ij}\delta_{ij}\mathbf{I}^2\mathbf{J}^2 = 3\mathbf{I}^2\mathbf{J}^2. \tag{C. 4}$$

The only complicated terms are

$$\sum_{ij} I_j I_i J_i J_j = \sum_{ij} I_i I_j J_j J_i. \tag{C. 5}$$

In this, from the commutation rules of Eq. (III. 44),

$$\sum_{ij} I_i I_j J_j J_i = \sum_{ij} I_j I_i J_j J_i + i\sum_{ij} \mathbf{I}_{i\times j} J_j J_i$$

$$= (\mathbf{I}\cdot\mathbf{J})^2 + i\sum_{ij}\mathbf{I}_{i\times j}J_j J_i.$$

But
$$J_j J_i = \frac{J_j J_i + J_j J_i}{2} = \frac{J_j J_i + J_i J_j + i\mathbf{J}_{j\times i}}{2}.$$

So
$$i\sum_{ij}\mathbf{I}_{i\times j}J_j J_i = \tfrac{1}{2}i\sum_{ij}\mathbf{I}_{i\times j}[J_j J_i + J_i J_j] - \tfrac{1}{2}\sum_{ij}\mathbf{I}_{i\times j}\mathbf{J}_{j\times i}.$$

However, the first terms on the right of the above equation are antisymmetric in i and j so their sum over all i and j vanish. Therefore,

$$i\sum_{ij}\mathbf{I}_{i\times j}J_j J_i = -\tfrac{1}{2}\sum_{ij}\mathbf{I}_{i\times j}\mathbf{J}_{j\times i} = \tfrac{1}{2}\sum_{ij}\mathbf{I}_{i\times j}\mathbf{J}_{i\times j}$$

$$= \tfrac{1}{2}\sum_k 2 I_k J_k = \mathbf{I}\cdot\mathbf{J}. \tag{C. 6}$$

Therefore,
$$\sum I_i I_j J_j J_i = (\mathbf{I}\cdot\mathbf{J})^2 + \mathbf{I}\cdot\mathbf{J} \tag{C. 7}$$

and

$$\mathcal{H}_{E2} = \frac{1}{6}\frac{e^2 q_J Q}{I(2I-1)J(2J-1)}\left\{\frac{9}{2}(\mathbf{I}\cdot\mathbf{J})^2 + \frac{9}{2}[(\mathbf{I}\cdot\mathbf{J})^2 + \mathbf{I}\cdot\mathbf{J}] - 3\mathbf{I}^2\mathbf{J}^2 - 3\mathbf{I}^2\mathbf{J}^2 + 3\mathbf{I}^2\mathbf{J}^2\right\}$$

(C. 8)

$$= \frac{e^2 q_J Q}{2I(2I-1)J(2J-1)}[3(\mathbf{I}\cdot\mathbf{J})^2 + \tfrac{3}{2}\mathbf{I}\cdot\mathbf{J} - \mathbf{I}^2\mathbf{J}^2]. \tag{C. 9}$$

In diatomic molecules with either identical or non-identical nuclei the quadrupole interaction can be obtained by applying either (C. 1) or (C. 9) to each nucleus separately. However, in the case of identical nuclei it is usually most convenient to express the result in terms of the resultant angular momentum $\mathbf{I}_R = \mathbf{I}_1 + \mathbf{I}_2$ since I_R is a good quantum number in a rotational state of definite J because of the symmetry requirements on the wave function with identical nuclei. This re-expression is obtained in the following paragraphs.

From Eq. (III. 42) applied to the two identical nuclei with $I_1 = I_2$

$$Q_{ij} = \frac{eQ}{I_1(2I_1-1)}\left[3\frac{I_{1i}I_{1j}+I_{1j}I_{1i}}{2}-\delta_{ij}I_1^2+3\frac{I_{2i}I_{2j}+I_{2j}I_{2i}}{2}-\delta_{ij}I_2^2\right]. \quad \text{(C. 10)}$$

By the general theorem of Section III. 2.3, for calculation of matrix elements diagonal in I_R the symmetric and traceless tensor in the brackets [ ] of the above equation can be written as proportional to a similar tensor based on $\mathbf{I}_R$. Therefore

$$\left[3\frac{I_{1i}I_{1j}+I_{1j}I_{1i}}{2}-\delta_{ij}I_1^2+3\frac{I_{2i}I_{2j}+I_{2j}I_{2i}}{2}-\delta_{ij}I_2^2\right] = E\left[3\frac{I_{Ri}I_{Rj}+I_{Rj}I_{Ri}}{2}-\delta_{ij}I_R^2\right]. \quad \text{(C. 11)}$$

Then E can be evaluated by multiplying both sides on the left by I_{Ri} and on the right by I_{Rj} and summing over i and j. This gives

$$\left[\tfrac{3}{2}\left\{(\mathbf{I}_R\cdot\mathbf{I}_1)^2+\sum_{ij}I_{Ri}I_{1j}I_{1i}I_{Rj}+(\mathbf{I}_R\cdot\mathbf{I}_2)^2+\sum_{ij}I_{Ri}I_{2j}I_{1i}I_{Rj}\right\}-2\mathbf{I}_R^2\,\mathbf{I}_1^2\right]$$
$$= E\left[\tfrac{3}{2}(\mathbf{I}_R^2)+\tfrac{3}{2}\sum_{ij}I_{Ri}I_{Rj}I_{Ri}I_{Rj}-(\mathbf{I}_R^2)^2\right]. \quad \text{(C. 12)}$$

The typical summation term in the above can be evaluated as follows from the commutation rules of Eq. (III. 44):

$$\sum_{ij}I_{Ri}I_{1j}I_{1i}I_{Rj} = \sum_{ij}I_{Ri}I_{1i}I_{1j}I_{Rj}+i\sum_{ij}I_{Ri}I_{1j\times i}I_{Rj}$$
$$= (\mathbf{I}_R\cdot\mathbf{I}_1)^2+i\sum_{ij}I_{1j\times i}I_{Ri}I_{Rj}-\sum_{ij}I_{1i\times(j\times i)}I_{Rj}. \quad \text{(C. 13)}$$

However, $\quad i\times(j\times i) = (i\cdot i)j-(i\cdot j)i = j \quad$ (C. 14)

and in the summation over i this can be achieved twice, once from each non j Cartesian component, whence an extra factor of two appears in the last term of the following equation. The penultimate term in Eq. (C. 13) is evaluated by Eq. (C. 6). Therefore

$$\sum_{ij}I_{Ri}I_{1j}I_{1i}I_{Rj} = (\mathbf{I}_R\cdot\mathbf{I}_1)^2+\mathbf{I}_R\cdot\mathbf{I}_1-2\mathbf{I}_R\cdot\mathbf{I}_1$$
$$= (\mathbf{I}_R\cdot\mathbf{I}_1)^2-\mathbf{I}_R\cdot\mathbf{I}_1. \quad \text{(C. 15)}$$

Eq. (C. 15) and its obvious variants can be used in Eq. (C. 12) with the result that

$$[3(\mathbf{I}_R\cdot\mathbf{I}_1)^2+3(\mathbf{I}_R\cdot\mathbf{I}_2)^2-\tfrac{3}{2}\mathbf{I}_R\cdot\mathbf{I}_1-\tfrac{3}{2}\mathbf{I}_R\cdot\mathbf{I}_2-2\mathbf{I}_R^2\,\mathbf{I}_1^2] = E[3(\mathbf{I}_R^2)^2-\tfrac{3}{2}\mathbf{I}_R^2-(\mathbf{I}_R^2)^2]. \quad \text{(C. 16)}$$

However,

$$\mathbf{I}_R\cdot\mathbf{I}_1 = -\tfrac{1}{2}[(\mathbf{I}_R-\mathbf{I}_1)^2-\mathbf{I}_R^2-\mathbf{I}_1^2] = -\tfrac{1}{2}[\mathbf{I}_2^2-\mathbf{I}_R^2-\mathbf{I}_1^2] = \tfrac{1}{2}\mathbf{I}_R^2 \quad \text{(C. 17)}$$

since $\mathbf{I}_1^2 = \mathbf{I}_2^2$. Therefore (C. 16) becomes

$$[\tfrac{3}{2}(\mathbf{I}_R^2)^2-\tfrac{3}{2}\mathbf{I}_R^2-2\mathbf{I}_R^2\,\mathbf{I}_1^2] = \tfrac{1}{2}E\mathbf{I}_R^2[4I_R(I_R+1)-3] \quad \text{(C. 18)}$$

or $\quad \tfrac{1}{2}\mathbf{I}_R^2[3I_R(I_R+1)-3-4I_1(I_1+1)] = \tfrac{1}{2}E\mathbf{I}_R^2[(2I_R-1)(2I_R+3)] \quad$ (C. 19)

so

$$E = 1-\frac{I_R(I_R+1)+4I_1(I_1+1)}{(2I_R-1)(2I_R+3)}. \quad \text{(C. 20)}$$

With this result then, Eqs. (C. 10) and (C. 11) yield

$$Q_{ij} = \frac{eQ}{I_1(2I_1-1)}\left[1-\frac{I_R(I_R+1)+4I_1(I_1+1)}{(2I_R-1)(2I_R+3)}\right]\left[3\frac{I_{Ri}I_{Rj}+I_{Rj}I_{Ri}}{2}-\delta_{ij}\mathbf{I}_R^2\right]. \quad \text{(C. 21)}$$

This expression is in the desired form and is dependent on the resultant angular momentum operator $\mathbf{I}_R$. Since the factor containing the operator $\mathbf{I}_R$ in Eq. (C. 21) is exactly the same as the factor containing the operator $\mathbf{I}$ in Eq. (C. 1), precisely the same reduction that led to the equation (C. 9) is applicable so that the total quadrupole interaction energy for a homonuclear diatomic molecule may also be written as

$$\mathscr{H}_{E2} = \frac{e^2 q_J Q}{2I_1(2I_1-1)J(2J-1)}\left[1 - \frac{I_R(I_R+1)+4I_1(I_1+1)}{(2I_R-1)(2I_R+3)}\right][3(\mathbf{I}_R\cdot\mathbf{J})^2 + \tfrac{3}{2}\mathbf{I}_R\cdot\mathbf{J} - \mathbf{I}_R^2\mathbf{J}^2].$$

(C. 22)

From Eqs. (C. 9) and (C. 22) it is apparent that it is important to determine the matrix elements of the quantity f defined by

$$f \equiv 3(\mathbf{I}\cdot\mathbf{J})^2 + \tfrac{3}{2}(\mathbf{I}\cdot\mathbf{J}) - \mathbf{I}^2\mathbf{J}^2. \qquad (C.\ 23)$$

As discussed in Section III. 4, these matrix elements may be desired in either an m_I, m_J representation or in an Fm representation, where F and m are the total and magnetic quantum numbers of

$$\mathbf{F} = \mathbf{I} + \mathbf{J}. \qquad (C.\ 24)$$

The $m_I m_J$ representation is suitable to perturbation calculations in strong external magnetic fields while the Fm representation is suitable in the weak field limit.

In the F, m representation, the calculation is trivial. The quantity $(\mathbf{I}\cdot\mathbf{J})$ and hence the operator f commutes with $\mathbf{F}^2$ and $(\mathbf{F})_z$. Therefore the matrix elements of f are diagonal in the Fm representation. As shown in Eq. (III. 56) they are given by

$$(Fm|f|Fm) = \tfrac{3}{2}C(C+1) - I(I+1)J(J+1) \qquad (C.\ 25)$$

where $\quad \tfrac{1}{2}C \equiv \tfrac{1}{2}[F(F+1) - I(I+1) - J(J+1)] = \mathbf{I}\cdot\mathbf{J}. \qquad (C.\ 26)$

In the $m_I m_J$ representation on the other hand the calculation is more difficult. However, as shown by Kellogg, Rabi, Ramsey, and Zacharias (KEL 40) they can easily be derived by writing f in the equivalent form

$$\begin{aligned}f =\ & \tfrac{1}{2}[3J_z^2 - J(J+1)][3I_z^2 - I(I+1)] \\ & + \tfrac{3}{4}[J_z J_+ + J_+ J_z][I_z I_- + I_- I_z] + \\ & + \tfrac{3}{4}[J_z J_- + J_- J_z][I_z I_+ + I_+ I_z] + \\ & + \tfrac{3}{4}[I_+^2 J_-^2 + I_-^2 J_+^2].\end{aligned} \qquad (C.\ 27)$$

The equivalence of Eqs. (C. 23) and (C. 27) can be seen by expansion and suitable use of the commutation rules of Eq. (III. 44). With this the desired matrix elements can be obtained directly with the aid of Eq. (III. 35a). They are

$$(m_J, m_I|f|m_J, m_I) = \tfrac{1}{2}[3m_I^2 - I(I+1)][3m_J^2 - J(J+1)],$$

$(m_J, m_I|f|m_J-1, m_I+1) = \tfrac{3}{4}(2m_J-1)(2m_I+1) \times$
$$\times [(J+m_J)(J-m_J+1)(I-m_I)(I+m_I+1)]^{\tfrac{1}{2}},$$

$(m_J, m_I|f|m_J+1, m_I-1) = \tfrac{3}{4}(2m_J+1)(2m_I-1) \times$
$$\times [(J-m_J)(J+m_J+1)(I+m_I)(I-m_I+1)]^{\tfrac{1}{2}},$$

$(m_J, m_I|f|m_J-2, m_I+2) = \tfrac{3}{4}[(J+m_J)(J+m_J-1)(J-m_J+1)(J-m_J+2) \times$
$$\times (I-m_I)(I-m_I-1)(I+m_I+1)(I+m_I+2)]^{\tfrac{1}{2}},$$

$(m_J, m_I|f|m_J+2, m_I-2) = \tfrac{3}{4}[(J-m_J)(J-m_J-1)(J+m_J+1)(J+m_J+2) \times$
$$\times (I+m_I)(I+m_I-1)(I-m_I+1)(I-m_I+2)]^{\tfrac{1}{2}}.$$

(C. 28)

In calculations of interactions of nuclear magnetic moments with external magnetic fields it is often desirable to calculate the matrix elements of $\mathbf{I}$ in an

Fm representation. These matrix elements have been tabulated by Feld and Lamb (FEL 45). The matrix elements diagonal in F are relatively easy to calculate. For example, this may be done by using the general theorem of Section III. 2.3 as in Eq. (III. 36). Alternatively by the simple vector addition argument as used in Eq. (III. 89) it can be seen that the application of the general theorem will give for matrix elements diagonal in F

$$\mathbf{I} = \mathbf{I} \cdot \frac{\mathbf{F}}{|\mathbf{F}|}\frac{\mathbf{F}}{|\mathbf{F}|} = \frac{I(I+1)-J(J+1)+F(F+1)}{2F(F+1)}\mathbf{F}$$

$$= \frac{I(I+1)-J(J+1)+F(F+1)}{2F(F+1)}[F_z\mathbf{k}+\tfrac{1}{2}F_+(\mathbf{i}-i\mathbf{j})+\tfrac{1}{2}F_-(\mathbf{i}+i\mathbf{j})]. \quad (C.29)$$

The desired matrix elements can then be calculated immediately with the aid of Eq. (III. 35a). They are

$$(F,m|\mathbf{I}|F,m+1) = \frac{I(I+1)-J(J+1)+F(F+1)}{2F(F+1)} \tfrac{1}{2}\{(F-m)(F+m+1)\}^{\frac{1}{2}}(\mathbf{i}+i\mathbf{j}),$$

$$(F,m|\mathbf{I}|F,m) = \frac{I(I+1)-J(J+1)+F(F+1)}{2F(F+1)} m\mathbf{k},$$

$$(F,m|\mathbf{I}|F,m-1) = \frac{I(I+1)-J(J+1)+F(F+1)}{2F(F+1)} \tfrac{1}{2}\{(F+m)(F-m+1)\}^{\frac{1}{2}}(\mathbf{i}-i\mathbf{j}).$$
(C. 30)

In the case of matrix elements non-diagonal in F, the calculation is more difficult. One procedure is with the use of Eq. (III. 57). Alternatively they can be obtained from the formulae of Condon and Shortley (CON 35, pp. 63, 64, 67). The results are

$(F,m|\mathbf{I}|F+1,m+1)$
$$= -\frac{1}{2}\left\{\frac{\begin{matrix}(F+1-I+J)(F+1+I-J)(I+J+2+F)\times\\ \times(I+J-F)(F+m+1)(F+m+2)\end{matrix}}{4(F+1)^2(2F+1)(2F+3)}\right\}^{\frac{1}{2}}(\mathbf{i}+i\mathbf{j}),$$

$(F,m|\mathbf{I}|F+1,m)$
$$= \left\{\frac{(F+1-I+J)(F+1+I-J)(I+J+2+F)(I+J-F)[(F+1)^2-m^2]}{4(F+1)^2(2F+1)(2F+3)}\right\}^{\frac{1}{2}}\mathbf{k},$$

$(F,m|\mathbf{I}|F+1,m-1)$
$$= +\frac{1}{2}\left\{\frac{\begin{matrix}(F+1-I+J)(F+1+I-J)(I+J+2+F)\times\\ (I+J-F)(F-m+1)(F-m+2)\end{matrix}}{4(F+1)^2(2F+1)(2F+3)}\right\}^{\frac{1}{2}}(\mathbf{i}-i\mathbf{j}),$$

$(F,m|\mathbf{I}|F-1,m+1)$
$$= +\frac{1}{2}\left\{\frac{\begin{matrix}(F-I+J)(F+I-J)(I+J+1+F)(I+J+1-F)\times\\ \times(F-m)(F-m-1)\end{matrix}}{4F^2(2F-1)(2F+1)}\right\}^{\frac{1}{2}}(\mathbf{i}+i\mathbf{j}),$$

$(F,m|\mathbf{I}|F-1,m)$
$$= \left\{\frac{(F-I+J)(F+I-J)(I+J+1+F)(I+J+1-F)(F^2-m^2)}{4F^2(2F-1)(2F+1)}\right\}^{\frac{1}{2}}\mathbf{k},$$

$(F,m|\mathbf{I}|F-1,m-1)$
$$= -\frac{1}{2}\left\{\frac{\begin{matrix}(F-I+J)(F+I-J)(I+J+1+F)\times\\ \times(I+J+1-F)(F+m)(F+m-1)\end{matrix}}{4F^2(2F-1)(2F+1)}\right\}^{\frac{1}{2}}(\mathbf{i}-i\mathbf{j}).$$
(C. 31)

APPENDIX D

TABLE OF VELOCITY-AVERAGED FUNCTIONS

WITH the aid of such trigonometric identities as

$$\sin^2 x = \tfrac{1}{2}(1-\cos 2x), \quad \cos^2 x = \tfrac{1}{2}(1+\cos 2x)$$
$$\sin x \cos y = \tfrac{1}{2}[\sin(x+y)+\sin(x-y)]$$
$$\sin x \sin y = \tfrac{1}{2}[\cos(x-y)-\cos(x+y)]$$
$$\cos x \cos y = \tfrac{1}{2}[\cos(x-y)+\cos(x+y)] \tag{D. 1}$$

the velocity averages of the transition probabilities in Chapters V and VI can

TABLE D. I
Values of $I(x)$ and $K(x)$ as Functions of x
$I(x)$ and $K(x)$ are defined in Eqs. (D. 2) and (D. 3).

x	$I(x)$	$K(x)$	x	$I(x)$	$K(x)$
0	0·50000	0	6·4	−0·0493	−0·1535
0·1	0·49751	0·04417	6·6	−0·0307	−0·1515
0·2	0·49019	0·08754	6·8	−0·0132	−0·1476
0·3	0·47832	0·12944	7·0	0·00286	−0·14211
0·4	0·46229	0·16933	7·2	0·01749	−0·13518
0·6	0·41950	0·24139	7·4	0·03061	−0·12709
0·8	0·36543	0·30136	7·6	0·04220	−0·11805
1·0	0·30366	0·34805	7·8	0·05224	−0·10830
1·2	0·23746	0·38122	8·0	0·06078	−0·09808
1·4	0·16972	0·40127	8·5	0·07562	−0·07131
1·6	0·10288	0·40910	9·0	0·08221	−0·04496
1·8	0·03892	0·40592	9·5	0·08191	−0·02082
2·0	−0·02062	0·39314	10·0	0·07626	−0·00010
2·2	−0·0746	0·3722	10·5	0·06684	0·01654
2·4	−0·1221	0·3448	11·0	0·05507	0·02889
2·6	−0·1629	0·3122	11·5	0·04224	0·03707
2·8	−0·1966	0·2759	12·0	0·02937	0·04146
3·0	−0·2233	0·2371	12·5	0·01727	0·04259
3·2	−0·2432	0·1971	13·0	0·00650	0·04109
3·4	−0·2565	0·1569	13·5	−0·00259	0·03758
3·6	−0·2639	0·1173	14·0	−0·00982	0·03268
3·8	−0·2657	0·0792	14·5	−0·01517	0·02696
4·0	−0·2626	0·0430	15·0	−0·01872	0·02089
4·2	−0·2552	0·0094	15·5	−0·02065	0·01487
4·4	−0·2441	−0·0214	16·0	−0·02118	0·00921
4·6	−0·2299	−0·0490	16·5	−0·02057	0·00413
4·8	−0·2132	−0·0734	17·0	−0·01906	−0·00022
5·0	−0·1945	−0·0944	17·5	−0·01691	−0·00377
5·2	−0·1745	−0·1120	18·0	−0·01435	−0·00650
5·4	−0·1536	−0·1263	18·5	−0·01159	−0·00844
5·6	−0·1322	−0·1374	19·0	−0·00879	−0·00965
5·8	−0·1108	−0·1455	19·5	−0·00609	−0·01020
6·0	−0·0896	−0·1507	20·0	−0·00360	−0·01021
6·2	−0·0691	−0·1533			

usually be reduced to a sum of terms involving only the two integral functions

$$I(x) = \int_0^\infty \exp(-y^2) y^3 \cos(x/y)\, dy, \qquad (D.\ 2)$$

$$K(x) = \int_0^\infty \exp(-y^2) y^3 \sin(x/y)\, dy. \qquad (D.\ 3)$$

Tables of these functions have been prepared by Kruse and Ramsey (KRU 51) and are reproduced in Table D. 1.

APPENDIX E

MAJORANA FORMULA

THE derivation in this appendix of the Majorana formula for the transition probability of a nucleus with general spin is based on analysis of Schwinger (SCH 52, ZXZ 54) and Bloch and Rabi (BLO 45). In this derivation advantage is taken of the fact that a resultant angular momentum **I** can be considered as composed of the resultant of $2I$ angular momenta of spin $\frac{1}{2}$ if

$$\mathbf{I} = \sum_{k=1}^{2I} \tfrac{1}{2}\boldsymbol{\sigma}_k, \qquad (E.\ 1)$$

where $\boldsymbol{\sigma}$ are the Pauli spin operators. If α_k is the wave function ϕ_k for the kth angular momentum of spin $\frac{1}{2}$ parallel to the z-axis, if β_k is the wave function in the antiparallel case, and if $\psi(II)$ is the wave function for **I**, parallel to the axis, i.e. for $m = I$, then

$$\psi(II) = \prod_{k=1}^{2I} \alpha_k. \qquad (E.\ 2)$$

However, we will need $\psi(I, m)$ in general. This can be formed from Eq. (E. 2) by the application of the lowering operator $I_- = I_x - iI_y$.

Since by Eq. (III. 35a)

$$(m+1|I_+|m) = (m|I_-|m+1) = \sqrt{\{(I-m)(I+m+1)\}}, \qquad (E.\ 3)$$

$$I_-\psi(I,m) = \sqrt{\{(I-m+1)(I+m)\}}\psi(I,m-1). \qquad (E.\ 4)$$

Hence
$$I_-\psi(II) = \sqrt{\{2I\}}\psi(I, I-1),$$

$$I_-^2\psi(II) = \sqrt{\{1 \cdot 2 \cdot (2I)(2I-1)\}}\psi(I, I-2),$$

$$I_-^k\psi(II) = \sqrt{\left\{\frac{k!(2I)!}{(2I-k)!}\right\}}\psi(I, I-k).$$

Or if $k = I - m$

$$I_-^{I-m}\psi(II) = \sqrt{\left\{\frac{(I-m)!(2I)!}{(I+m)!}\right\}}\psi(I,m)$$

or
$$\psi(I,m) = \sqrt{\left\{\frac{(I+m)!}{(I-m)!(2I)!}\right\}}\, I_-^{I-m}\psi(I, I). \qquad (E.\ 5)$$

However, this can also be written

$$\psi(I,m) = \lim_{\lambda \to 0} \sqrt{\left\{\frac{(I+m)!}{(I-m)!(2I)!}\right\}} \left(\frac{d}{d\lambda}\right)^{I-m} e^{\lambda I_-}\psi(I, I). \qquad (E.\ 6)$$

From Eqs. (E. 2) and (E. 6) we then have the expression for $\psi(Im)$.

Now suppose that at time $t = 0$ the state of the complete system with angular momentum I is $\psi(0)$ where $\psi(0)$ is in general a superposition of the above $\psi(I,m)$'s. Then let a perturbation act for a time t as in Chapter VI and let the perturbation act the same way on all the spin $\frac{1}{2}$'s (as it would if all have the same gyromagnetic ratio). Let the perturbation be such that, if $\phi_k(0)$ is the initial value of the wave function of the kth constituent and $\phi_k(t)$ is the value at time t. Also let U_k be the unitary operator which converts one to the other, i.e.

$$\phi_k(t) = U_k(t)\phi_k(0). \qquad (E.\ 7)$$

Although U_k will have the same matrix for all cases the k subscript will be retained to indicate which function it operates on. Then, with ψ being composed of a suitable combination of ϕ_k's

$$\psi(t) = \prod_{k=1}^{2I} U_k(t)\psi(0). \tag{E. 8}$$

Now, consider the special case of

$$\psi(0) = \psi(I, m). \tag{E. 9}$$

Then
$$\prod_{k=1}^{2I} U_k \psi(I, m) = \psi(t) = \sum_{m'} C_{m,m'}(t)\psi(I, m') \tag{E. 10}$$

and
$$C_{m,m'} = \left(\psi(I, m'), \prod_{k=1}^{2I} U_k \psi(I, m)\right) \tag{E. 11}$$

$$= \lim_{\mu,\lambda \to 0} \frac{1}{(2I)!} \sqrt{\left\{\frac{(I+m)!\,(I+m')!}{(I-m)!\,(I-m')!}\right\}} \left(\frac{d}{d\mu}\right)^{I-m'} \left(\frac{d}{d\lambda}\right)^{I-m} \times$$

$$\times \left(e^{\mu I} \cdot \psi(II), \prod_{k=1}^{2I} U_k e^{\lambda I} \cdot \psi(II)\right). \tag{E. 12}$$

For brevity, the last large parenthesis in Eq. (E. 12) will now be temporarily designated by (A).

From Eq. (E. 1)
$$e^{\mu I_-} = \exp\left(\tfrac{1}{2}\mu \sum_{k=1}^{2I} \sigma_{k-}\right) = \prod_{k=1}^{2I} \exp(\tfrac{1}{2}\mu\sigma_{k-}). \tag{E. 13}$$

Hence
$$(A) = \prod_{k=1}^{2I} (e^{(\mu/2)\sigma_k} \cdot \alpha_k, U_k e^{(\lambda/2)\sigma_k} \cdot \alpha_k)$$

$$= (e^{(\mu/2)\sigma} \cdot \alpha, U e^{(\lambda/2)\sigma} \cdot \alpha)^{2I}. \tag{E. 14}$$

But the meaning of the exponential operators may be defined by their expansion so
$$e^{(\lambda/2)\sigma_-} = 1 + (\lambda/2)\sigma_- + \tfrac{1}{2}(\lambda/2)^2 \sigma_-^2 + \ldots$$

$$= 1 + (\lambda/2)\sigma_-, \tag{E. 15}$$

where the last step arises from the fact that

$$\sigma_-^2 = \sigma_x^2 - \sigma_y^2 - i(\sigma_x \sigma_y + \sigma_y \sigma_x) = 1 - 1 - i(i\sigma_z - i\sigma_z) = 0. \tag{E. 16}$$

Therefore
$$(A) = ([1 + (\tfrac{1}{2}\mu)\sigma_-]\alpha, U[1 + \tfrac{1}{2}\lambda\sigma_-]\alpha)^{2I}$$

$$= ([\alpha + \mu\beta], U[\alpha + \lambda\beta])^{2I} \tag{E. 17}$$

since from Eq. (E. 4) $\sigma_-\alpha = 2s_-\alpha = 2\beta$. Then
$$(A) = (U_{\frac{1}{2},\frac{1}{2}} + \lambda U_{\frac{1}{2},-\frac{1}{2}} + \mu U_{-\frac{1}{2},\frac{1}{2}} + \lambda\mu U_{-\frac{1}{2},-\frac{1}{2}})^{2I}$$

$$= [\lambda(U_{\frac{1}{2},-\frac{1}{2}} + \mu U_{-\frac{1}{2},-\frac{1}{2}}) + (U_{\frac{1}{2},\frac{1}{2}} + \mu U_{-\frac{1}{2},\frac{1}{2}})]^{2I}. \tag{E. 18}$$

Consequently

$$\lim_{\mu,\lambda\to 0} \left(\frac{d}{d\mu}\right)^{I-m'} \left(\frac{d}{d\lambda}\right)^{I-m} (A)$$

$$= \lim_{\mu,\lambda\to 0} \left(\frac{d}{d\mu}\right)^{I-m'} \left(\frac{d}{d\lambda}\right)^{I-r} \left[\sum_{I-m=0}^{2I} \frac{(2I)!}{(I-m)!(I+m)!} \times \right.$$

$$\left. \times (U_{\frac{1}{2},-\frac{1}{2}} + \mu U_{-\frac{1}{2},-\frac{1}{2}})^{I-m} (U_{\frac{1}{2},\frac{1}{2}} + \mu U_{-\frac{1}{2},\frac{1}{2}})^{I+m} \lambda^{I-m} \right]$$

$$= \lim_{\mu\to 0} \left(\frac{d}{d\mu}\right)^{I-m'} \left[\frac{(2I)!}{(I+m)!} (U_{\frac{1}{2},-\frac{1}{2}} + \mu U_{-\frac{1}{2},-\frac{1}{2}})^{I-m} (U_{\frac{1}{2},\frac{1}{2}} + \mu U_{-\frac{1}{2},\frac{1}{2}})^{I+m} \right], \tag{E. 19}$$

APPENDIX E   429

since only the λ^{I-m} term contributes after the differentiation and after the limit is taken. Hence from Eq. (E. 12)

$$C_{m,m'} = \lim_{\mu \to 0} \sqrt{\left\{\frac{(I+m')!}{(I-m)!(I+m)!(I-m')!}\right\}} \left(\frac{d}{d\mu}\right)^{I-m'} \times$$

$$\times [(U_{\frac{1}{2},-\frac{1}{2}}+\mu U_{-\frac{1}{2},-\frac{1}{2}})^{I-m}(U_{\frac{1}{2},\frac{1}{2}}+\mu U_{-\frac{1}{2},\frac{1}{2}})^{I+m}]. \quad \text{(E. 20)}$$

For brevity the square root in the above will be written as $\sqrt{B}$. Then the above can be partially expanded and written as

$$C_{m,m'} = \lim_{\mu \to 0} \sqrt{B} \left(\frac{d}{d\mu}\right)^{I-m'} \left[\sum_{I-m'} \sum_r \frac{(I-m)!}{(I-m-r)!r!} \times \right.$$

$$\times U_{\frac{1}{2},-\frac{1}{2}}^{I-m-r} U_{-\frac{1}{2},-\frac{1}{2}}^r \frac{(I+m)!}{(I+m-I+m'+r)!(I-m'-r)!} \times$$

$$\left. \times U_{\frac{1}{2},\frac{1}{2}}^{m+m'+r} U_{-\frac{1}{2},\frac{1}{2}}^{I-m'-r} \mu^{I-m'} \right]. \quad \text{(E. 21)}$$

Since only the $\mu^{I-m'}$ term contributes after the differentiation and after the limit is taken,

$$C_{m,m'} = \sqrt{B} \sum_r \frac{(I-m)!(I+m)!(I-m')!}{(I-m-r)!r!(m+m'+r)!(I-m'-r)!} \times$$

$$\times U_{\frac{1}{2},\frac{1}{2}}^{m+m'+r} U_{\frac{1}{2},-\frac{1}{2}}^{I-m-r} U_{-\frac{1}{2},\frac{1}{2}}^{I-m'-r} U_{-\frac{1}{2},-\frac{1}{2}}^r$$

$$= \sqrt{\{(I-m)!(I+m)!(I-m')!(I+m')!\}} \times$$

$$\times \sum_r \frac{U_{\frac{1}{2},\frac{1}{2}}^{m+m'+r} U_{\frac{1}{2},-\frac{1}{2}}^{I-m-r} U_{-\frac{1}{2},\frac{1}{2}}^{I-m'-r} U_{-\frac{1}{2},-\frac{1}{2}}^r}{(I-m-r)!(I-m'-r)!(m+m'+r)!r!}. \quad \text{(E. 22)}$$

The $U_{q,p}$, etc., are found in just the same way as the C_q's of Section VI. 2.3 with the Rabi method and Section V. 4.2 with the Ramsey method. In particular, with the Rabi method from Eqs. (E. 7), (VI. 34), and (VI. 36) and with $t_1 = 0$,

$$U_{m',m} = (m'|U|m) = (m'|\exp(i\tfrac{1}{2}\omega t\mathbf{k}\cdot\boldsymbol{\sigma})\exp(i\tfrac{1}{2}at\boldsymbol{\alpha}\cdot\boldsymbol{\sigma})|m)$$

$$= e^{i\omega t m'}(m'|\cos(\tfrac{1}{2}at)+i(\sigma_z\cos\Theta+\sigma_x\sin\Theta)\sin(\tfrac{1}{2}at)|m). \quad \text{(E. 23)}$$

So
$$U_{\frac{1}{2},\frac{1}{2}} = (\cos\tfrac{1}{2}at+i\cos\Theta\sin\tfrac{1}{2}at)\times e^{-i\omega t\frac{1}{2}} \equiv Ae^{i\phi}$$

$$U_{-\frac{1}{2},\frac{1}{2}} = (i\sin\Theta\sin\tfrac{1}{2}at)\times e^{i\omega t\frac{1}{2}} \equiv iBe^{i\psi}$$

$$U_{\frac{1}{2},-\frac{1}{2}} = (i\sin\Theta\sin\tfrac{1}{2}at)\times e^{-i\omega t\frac{1}{2}} \equiv iBe^{-i\psi}$$

$$U_{-\frac{1}{2},-\frac{1}{2}} = (\cos\tfrac{1}{2}at-i\cos\Theta\sin\tfrac{1}{2}at)\times e^{i\omega t\frac{1}{2}} \equiv Ae^{-i\phi}. \quad \text{(E. 24)}$$

With these definitions A and B are real. Since B^2 is less than one we can take $B = \sin\tfrac{1}{2}\alpha$. Then

$$B^2 = \sin^2\Theta\sin^2\tfrac{1}{2}at = P_{\frac{1}{2},-\frac{1}{2}} = \sin^2\tfrac{1}{2}\alpha,$$

$$A^2 = |\cos\tfrac{1}{2}at+i\cos\Theta\sin\tfrac{1}{2}at|^2 = \cos^2\tfrac{1}{2}at+\cos^2\Theta\sin^2\tfrac{1}{2}at$$

$$= 1-(1-\cos^2\Theta)\sin^2\tfrac{1}{2}at = 1-\sin^2\Theta\sin^2\tfrac{1}{2}at = 1-\sin^2\tfrac{1}{2}\alpha = \cos^2\tfrac{1}{2}\alpha.$$

$$\text{(E. 25)}$$

With the above

$$C_{m,m'} = \sqrt{\{(I-m)!(I+m)!(I-m')!(I+m')!\}} \times$$
$$\times \sum_r \frac{e^{i[(m+m')\phi+(m-m')\psi]}\cos^{m+m'+2r}\tfrac{1}{2}\alpha(i\sin\tfrac{1}{2}\alpha)^{2I-m-m'-2r}}{(I-m-r)!(I-m'-r)!(m+m'+r)!r!}$$

$$= \sqrt{\{(I-m)!(I+m)!(I-m')!(I+m')!\}}i^{2I-m-m'}e^{i[(m+m')\phi+(m-m')\psi]} \times$$
$$\times (\sin\tfrac{1}{2}\alpha)^{2I}\sum_r \frac{(-1)^r(\cot\tfrac{1}{2}\alpha)^{m+m'+2r}}{(I-m-r)!(I-m'-r)!(m+m'+r)!r!}. \quad (E.\ 26)$$

Hence

$$P_{m,m'} = |C_{m,m'}|^2 = (I-m)!(I+m)!(I-m')!(I+m')! \times$$
$$\times (\sin\tfrac{1}{2}\alpha)^{4I}\left[\sum_r \frac{(-1)^r(\cot\tfrac{1}{2}\alpha)^{m+m'+2r}}{(I-m-r)!(I-m'-r)!(m+m'+r)!r!}\right]^2. \quad (E.\ 27)$$

From Eq. (E. 21) it is apparent that the $\sum$ is over all values of r for which no argument in a factorial is negative. As noted above

$$\sin^2\tfrac{1}{2}\alpha = P_{\frac{1}{2},-\frac{1}{2}}, \quad (E.\ 28)$$

where $P_{\frac{1}{2},-\frac{1}{2}}$ is the transition probability for a moment with the same gyromagnetic ratio and spin of $\tfrac{1}{2}$; the latter can be evaluated as in Chapter VI.

APPENDIX F

THEORY OF TWO-WIRE FIELD

CONSIDER two wires normal to the yz plane with the wires being at the points $y = \pm a$, $z = 0$ as in Fig. XIV. 25. Then let $\mathbf{r}_1$ be a vector in the yz plane from the point $y = -a$, $z = 0$ to an arbitrary point, let $\mathbf{r}_2$ be a similar vector originating at $y = +a$, $z = 0$, and let $\mathbf{r}$ originate from $y = 0$, $z = 0$. Then if the current I (in abamps) through the wire at $(-a, 0)$ comes out of the paper (in the direction of the positive x axis while the other goes in, the field at any point outside the wires is given (PAG 31) by

$$\mathbf{H} = 2I[\mathbf{i} \times \mathbf{r}_1/r_1^2 + \mathbf{r}_2 \times \mathbf{i}/r_2^2]. \tag{F. 1}$$

Between the two wires and in a plane containing them

$$\mathbf{H} = 2I\left[\frac{1}{r_1} + \frac{1}{r_2}\right]. \tag{F. 2}$$

From Eq. (F. 1) it is apparent that the field is independent of the radii of the wires provided that the calculation is applied only in the region outside the wires. Therefore in the remainder of the theoretical discussion the wires, for simplicity, will be assumed to be of negligibly small radius.

The shapes of the lines of force and magnetic equipotentials can be obtained as follows. Provided the plane containing and between the two wires is used as a cut or impenetrable barrier to make the region singly connected and thereby to avoid multiple values of the potential, the magnetic field may be derived (PAG 31, WEA 44) from a scalar potential U by

$$\mathbf{H} = -\nabla U, \tag{F. 3}$$

where U outside the wires and the above plane satisfies

$$\nabla^2 U = 0. \tag{F. 4}$$

For infinitely long wires and by symmetry U depends only on the coordinates y and z. However, as shown in the elementary theory of complex variables (PAG 31) the real (or imaginary) part of any function of $y + iz$ in a region in which the function is analytic satisfies Eq. (F. 4) in two dimensions. Thus if

$$W(y + iz) = U(y, z) + iV(y, z), \tag{F. 5}$$

where U and V are both real functions, then both U and V satisfy Eq. (F. 4). U and V are called conjugate functions. It may easily be shown (PAG 31) that lines of constant U are orthogonal to lines of constant V so that if the former are considered to be equipotentials the latter represent lines of force, or vice versa.

Now consider the function

$$W = 2iI\ln\left(\frac{y+iz+a}{y+iz-a}\right) = 2iI\ln(r_1 e^{i\theta_1}/r_2 e^{i\theta_2}) = -2I(\theta_1 - \theta_2) + i2I\ln(r_1/r_2) \tag{F. 6}$$

where

$$r_1 = [(y+a)^2 + z^2]^{\frac{1}{2}}, \quad r_2 = [(y-a)^2 + z^2]^{\frac{1}{2}},$$
$$\theta_1 = \tan^{-1}[z/(y+a)], \quad \theta_2 = \tan^{-1}[z/(y-a)]. \tag{F. 7}$$

From Eqs. (F. 5) and (F. 6)

$$U = -2I(\theta_1 - \theta_2); \quad V = 2I\ln(r_1/r_2). \tag{F. 8}$$

Between the two wires and in the vicinity of the plane containing them, $\theta_1 - \theta_2$

is approximately equal to $z/r_1 - (\pi - z/r_2)$. Therefore in that plane from Eqs. (F. 3) and (F. 8), **H** is exactly the same as in Eq. (F. 2). U also vanishes at infinity. Consequently, the U of Eq. (F. 8) satisfies the boundary conditions of the two-wire problem and may be taken as the potential function everywhere. Likewise the lines of constant V are the lines of force for **H** in the two-wire problem.

The shapes of the lines of force may then easily be obtained. From Eq. (F. 8) V is constant if r_1/r_2 equals a constant value, say $1/m$. Then

$$\frac{(y+a)^2+z^2}{(y-a)^2+z^2} = \frac{1}{m^2} \tag{F. 9}$$

and this may easily be reduced to

$$[y+a(m^2+1)/(m^2-1)]^2+z^2 = a^2[2m/(m^2-1)]^2. \tag{F. 9}$$

Therefore the lines of force are circles with centres on the line $z = 0$ which passes through the two wires. The circles of smallest radius ($m = 0$ or ∞) are centred on the two wires. These lines of force are shown in Fig. XIV. 27.

Likewise, from Eq. (F. 8) the magnetic equipotentials correspond to curves for which $\theta_1 - \theta_2$ and hence $\tan(\theta_1 - \theta_2)$ are constant, so

$$\tan(\theta_1-\theta_2) = \frac{\tan\theta_1-\tan\theta_2}{1+\tan\theta_1\tan\theta_2} = \frac{z/(y+a)-z/(y-a)}{1+z^2/(y^2-a^2)} = -\frac{1}{p} \tag{F. 10}$$

and this reduces to

$$y^2+(z-pa)^2 = (1+p^2)a^2. \tag{F. 11}$$

Therefore, the magnetic equipotentials are circles whose centres lie on the line $y = 0$. Since Eq. (F. 11) is always satisfied by the points $y = \pm a$, $z = 0$, all the magnetic equipotentials pass through the centres of the wires as shown in Fig. XIV. 27.

The magnitude of **H** can be calculated from Eq. (F. 1)

$$\mathbf{H}^2 = 4I^2\left\{\frac{1}{r_1^2}+\frac{1}{r_2^2}+\frac{2}{r_1 r_2}[(\mathbf{i}\times\mathbf{r}_1/r_1).(\mathbf{r}_2\times\mathbf{i}/r_2)]\right\}. \tag{F. 11a}$$

By Eq. (B. 3) this becomes

$$\mathbf{H}^2 = 4I^2\left\{\frac{1}{r_1^2}+\frac{1}{r_2^2}-\frac{2\mathbf{r}_1.\mathbf{r}_2}{r_1^2 r_2^2}\right\}$$

$$= 4I^2\left\{\frac{1}{r_1^2}+\frac{1}{r_2^2}+\frac{(\mathbf{r}_1-\mathbf{r}_2)^2-\mathbf{r}_1^2-\mathbf{r}_2^2}{r_1^2 r_2^2}\right\} = 4I^2\frac{4a^2}{r_1^2 r_2^2} \tag{F. 12}$$

whence

$$H = |\mathbf{H}| = 2I\frac{2a}{r_1 r_2} \tag{F. 13}$$

as in Eq. (XIV. 13).

Finally, from Eqs. (F. 13) and (F. 7),

$$\frac{\partial H}{\partial z} = -2I\frac{2a}{r_1^2 r_2^2}\left[r_1\frac{\partial r_2}{\partial z}+r_2\frac{\partial r_1}{\partial z}\right]$$

$$= -2I\frac{2a}{r_1^2 r_2^2}\left[r_1\frac{z}{r_2}+r_2\frac{z}{r_1}\right]$$

$$= -2I\frac{2a}{r_1^3 r_2^3}[r_1^2+r_2^2]z \tag{F. 14}$$

as in Eq. (XIV. 13).

Numerical values of Eqs. (F. 13) and (F. 14) at the point ($y = 0, z = 1\cdot 2a$) can be calculated easily with the result that

$$H = \frac{1\cdot 64 I}{a}, \quad \frac{\partial H}{\partial z} = \frac{1\cdot 612 I}{a^2}, \quad \frac{\partial H}{\partial z}\Big/H = 0\cdot 983/a. \tag{F. 15}$$

APPENDIX G

NOTES ADDED IN PROOF

A NUMBER of changes have been made in page proof to include new scientific developments. Most of these have been incorporated at the appropriate place in the book, but in a few cases there was insufficient space. In such cases the additional matter is listed below and a reference to Appendix G is given at the appropriate place in the book.

Page 168. Recently, Ramsey (ZZA 25) has suggested an entirely new method for molecular beam resonance measurements of rotational magnetic moments. In this proposed method the inhomogeneous magnetic deflecting and refocusing fields would be replaced by scattering gases which could produce collision alignment of the molecule (ZZA 25). The elimination of the deflecting fields should make possible the use of broader beams.

Page 197. Stanford, Stephenson, and Bernstein (ZZA 20) have confirmed that the neutron spin is $\frac{1}{2}$ by observations of the transition probability as a function of the oscillatory magnetic field amplitude.

Page 211. As discussed further in Chapter X, the electric resonance and magnetic resonance values of c_i for heavy molecules do not always agree. The disagreement is probably due to the unreliability of the magnetic measurements with heavy molecules as a result of these measurements being based on statistical averages which may be partially confused by the existence of such phenomena as the molecular polymerization discussed in Section VIII. 11 and vibrational and centrifugal effects (RAM 48, OCH 53, ZZA 26).

Page 255. In addition to the data listed in Table IX. I, the following data have recently been obtained.

Z	Atom	A	$\Delta\nu$ (Mc/sec)	References
1	H($2\,^2S_{\frac{1}{2}}$)	1	177·5566(3)	ZZA 33
11	Na($3\,^2P_{\frac{1}{2}}$)	23	188·90(10)	ZZA 23
17	Cl($^2P_{\frac{1}{2}}$)	35	2,074·383(8)	ZZA 37
		37	1,726·700(15)	ZZA 37
49	In($^2P_{\frac{1}{2}}$)	114*^{m}	9,700	ZZA 15
		116*^{m}	9,000	ZZA 15
79	Au($^2S_{\frac{1}{2}}$)	198*	22,200(350)	ZZA 56
		199*	11,180(130)	ZZA 56

Page 260. More recently with microwave paramagnetic absorption techniques Beringer and Heald (ZZA 34) have found a value of $658·2277\pm0·002$ for $-g_S/g_p$. Likewise White (ZZA 35) has made measurements on the metastable $2S$ state of hydrogen and has found $658·231\pm0·008$ for $-g_S/g_p$.

Page 266. Wittke and Dicke (ZZA 29) have studied atomic H by a microwave absorption technique which utilizes the mechanism of collision reduction of

Doppler width to provide 3 kc/sec wide resonance lines; they find

$$\Delta\nu_H = 1420 \cdot 40580 \pm 0 \cdot 00005 \text{ Mc/sec}.$$

More recently Kusch (ZZA 59) has repeated his earlier atomic hydrogen beam experiment and has found

$$\Delta\nu_H = 1420 \cdot 40573 \pm 0 \cdot 00005 \text{ Mc/sec},$$

in agreement with Dicke, while

$$\Delta\nu_D = 327 \cdot 384302 \pm 0 \cdot 000030 \text{ Mc/sec}.$$

Heberle, Reich, and Kusch (ZZA 33) have recently measured the $\Delta\nu$ of atomic hydrogen in the metastable $2\,{}^2S_{\frac{1}{2}}$ atomic state and find a value of $177 \cdot 5566 \pm 0 \cdot 003$ Mc/sec. This result is consistent with the first of Eqs. (IX. 26) and the assumption that $\Delta\nu(2S) = \Delta\nu(1S)/(8-5\alpha^2)$.

Page 280. Additional values of ${}^\alpha\Delta^\beta$ have been reported for $Cs^{133,134,135,137}$ (ZZA 55).

REFERENCES

AKH 48   A. Akheiser and J. Pomeranchuck, *J. Exper. and Theor. Phys. USSR*, **18**, 475 (1948).
ALV 40   L. W. Alvarez and F. Bloch, *Phys. Rev.* **57**, 111 (1940).
AMD 40   I. Amdur and H. Pearlman, *J. Chem. Phys.* **8**, 7 (1940).
AME 54   American Radio Relay League, *Radio Amateur's Handbook*, West Hartford, Connecticut.
AND 49   H. L. Anderson, *Phys. Rev.* **76**, 1460 (1949).
APP 54   Applied Physics vibrating reed electrometer, Applied Physics Corp., Pasadena, Calif., and Brown Instrument Co.
ARM 45   *Armco Magnetic Ingot Iron*, American Rolling Mill Co., Middletown, Ohio (1945).
ARN 47   W. R. Arnold and A. Roberts, *Phys. Rev.* **71**, 878 (1947).
ART 42   K. Artmann, *Z. Phys.* **118**, 624 and 659 (1942); **119**, 49 and 137 (1942).
AUS 51   N. Austern and R. Sachs, *Phys. Rev.* **78**, 292 (1950) and **81**, 705 and 710 (1951).
AVE 48   R. Avery and R. G. Sachs, *Phys. Rev.* **74**, 1320 (1948); 433 (1948).
BAR 52   R. G. Barnes, Ph.D. Thesis, Harvard University (1952).
BAR 53   M. Baranger, H. A. Bethe, and R. P. Feynman, *Phys. Rev.* **92**, 482 (1953).
BAR 54   R. G. Barnes, P. J. Bray, and N. F. Ramsey, *Phys. Rev.* **94**, 893 (1954).
BEC 48   G. E. Becker and P. Kusch, *Phys. Rev.* **73**, 584 (1948).
BEL 51   E. H. Bellamy, *Nature*, **168**, 556 (1951).
BEL 53   E. H. Bellamy and K. F. Smith, *Phil. Mag.* **44**, 33 (1953).
BEN 54   A. Bennett, *Phys. Rev.* **95**, 608 (1954).
BER 42   P. G. Bergmann, *Introduction to the Theory of Relativity*, Prentice-Hall, Inc., New York (1942).
BER 50   A. Berman, P. Kusch, and A. K. Mann, *Phys. Rev.* **77**, 140 (1950).
BER 52   A. Berman, *Phys. Rev.* **86**, 1005 (1952).
BES 42   W. H. Bessey and O. C. Simpson, *Chem. Rev.* **30**, 234 (1942).
BET 50   H. A. Bethe, L. M. Brown, and J. R. Stehn, *Phys. Rev.* **72**, 339 (1947); **77**, 370 (1950).
BIE 25   F. Bielz, *Zeits. f. Physik*, **32**, 81 (1925).
BIE 52   L. C. Biedenharn, J. M. Blatt, and M. E. Rose, *Rev. Mod. Phys.* **24**, 249 (1952).
BIT 49   F. Bitter, *Phys. Rev.* **75**, 1326 (1949).
BIT 49a   F. Bitter, *Phys. Rev.* **75**, 1326 (1949); **76**, 150 (1949).
BIT 49b   F. Bitter, A. Kastler, and J. Brossel, *Comptes Rendus*, **229**, 1213 (1949); *Phys. Rev.* **46**, 583 (1934); **77**, 136 (1949); **76**, 833 (1949); **79**, 196, 225 (1950); **85**, 1051 (1952); **86**, 308 (1952); **94**, 885 (1954); *J. Phys. et Radium*, **13**, 668 (1952); **11**, 255 (1950); *Rev. Mod. Phys.* **25**, 174 (1953).
BLA 52   J. M. Blatt and V. F. Weisskopf, *Theoretical Nuclear Physics*, John Wiley and Sons, New York (1952).
BLO 36   F. Bloch, *Phys. Rev.* **50**, 259 (1936).
BLO 37   F. Bloch, *Phys. Rev.* **51**, 994 (1937).
BLO 40   F. Bloch and A. Siegert, *Phys. Rev.* **57**, 522 (1940).
BLO 43   F. Bloch, M. Hammermesh, and H. Staub, *Phys. Rev.* **64**, 47 (1943).
BLO 45   F. Bloch and I. I. Rabi, *Rev. Mod. Phys.* **17**, 237 (1945).

REFERENCES

BLO 46   F. Bloch, W. Hansen, and M. E. Packard, *Phys. Rev.* **69**, 127 (1946); **70**, 474 (1946).
BLO 46a   F. Bloch, R. I. Condit, and H. H. Staub, *Phys. Rev.* **70**, 927 (1946).
BLO 48   F. Bloch, D. Nicodemus, and H. Staub, *Phys. Rev.* **74**, 1025 (1948).
BLO 50   F. Bloch and C. D. Jeffries, *Phys. Rev.* **80**, 305 (1950).
BLO 51   F. Bloch, *Phys. Rev.* **83**, 839 (1951).
BOH 48   A. Bohr, *Phys. Rev.* **73**, 1109 (1948).
BOH 50   A. Bohr and V. F. Weisskopf, *Phys. Rev.* **77**, 94 (1950).
BOH 51   A. Bohr, *Phys. Rev.* **81**, 134 and 331 (1951).
BOH 53   A. Bohr and B. Mottelson, *Dan. Mat. Fys. Medd.* **27**, no. 16 (1953) and **26**, no 14 (1952); *Phys. Rev.* **89**, 316 (1953) and **90**, 717 (1953); *Arkiv f. Fysik*, **4**, 455 (1952); and A. Bohr, *Rotational States of Atomic Nuclei*, Ejnar Munksgaards Forlag, Copenhagen, Denmark (1954).
BOL 52   D. I. Boleff and H. J. Zeiger, *Phys. Rev.* **85**, 799 (1952).
BOR 20   M. Born, *Physikal. Zeits.* **21**, 578 (1920).
BOZ 47   R. M. Bozorth, *Rev. Mod. Phys.* **19**, 29 (1947) and *Ferromagnetism*, D. Van Nostrand Co., New York (1951).
BRA 42   J. G. Brainerd, G. Koehler, H. J. Reich, and L. F. Woodruff, *Ultra-High-Frequency Techniques*, D. Van Nostrand Co., New York (1942).
BRA 48   J. K. Bragg, *Phys. Rev.* **74**, 533 (1948).
BRA 51   R. Braunstein and J. W. Trischka, *Phys. Rev.* **82**, 319 (1951).
BRA 52   P. J. Bray, R. S. Barnes, N. J. Harrick, and N. F. Ramsey, *Phys. Rev.* **87**, 220 (1952).
BRA 52a   P. J. Bray, Ph.D. Thesis, Harvard University (1952).
BRE 28   G. Breit, *Nature*, **122**, 649 (1928).
BRE 30   G. Breit, *Phys. Rev.* **35**, 1447 (1930).
BRE 31   G. Breit and I. I. Rabi, *Phys. Rev.* **38**, 2082 (1931).
BRE 32   G. Breit and J. Rosenthal, *Phys. Rev.* **42**, 348 (1932); **41**, 459 (1932).
BRE 33   G. Breit and L. Wills, *Phys. Rev.* **44**, 470 (1933).
BRE 47   G. Breit and F. Bloch, *Phys. Rev.* **72**, 135 (1947).
BRE 47a   G. Breit, *Phys. Rev.* **72**, 984 (1947).
BRE 47b   G. Breit and R. E. Meyerott, *Phys. Rev.* **72**, 1023 (1947); **75**, 1447 (1949).
BRE 48   G. Breit and G. E. Brown, *Phys. Rev.* **74**, 1278 (1948).
BRE 49   G. Breit, G. E. Brown, and G. B. Arfken, *Phys. Rev.* **76**, 1299 (1949).
BRE 50   G. Breit, *Phys. Rev.* **77**, 568 (1950); **78**, 300, 470 (1950); **79**, 891 (1950).
BRI 53   P. Brix, J. T. Eisinger, H. Lew, and G. Wessel, *Phys. Rev.* **92**, 647 (1953).
BRO 30   F. Brouwer, *Dissertation*, Amsterdam (1930).
BRO 33   L. F. Broadway, *Proc. Roy. Soc.* A **141**, 626 (1933).
BRO 35   L. F. Broadway, *Zeits. f. Physik*, **93**, 395 (1935).
BRO 41   H. Brooks, *Phys. Rev.* **59**, 925 (1941) and **60**, 168 (1941).
BRO 47   S. B. Brody, W. A. Nierenberg, and N. F. Ramsey, *Phys. Rev.* **72**, 258 (1947).
BRO 54   Brookhaven National Laboratory Molecular Beam Conference, August 1954. See also ZZA 45, ZZA 15, ZZA 16, NIE 54, HAM 53.
BUG 54   H. H. Buggie Company, Toledo, Ohio.
BYE 93   W. E. Byerly, *Fourier Series and Spherical Harmonics*, Ginn and Company, Boston, Mass. (1893).
CAL 39   P. Caldirola, *Nuovo Cimento*, **16**, 242 (1939).
CAR 52   R. O. Carlson, C. A. Lee, B. P. Fabricand, *Phys. Rev.* **85**, 784 (1952).
CAR 52a   H. Y. Carr and E. M. Purcell, *Phys. Rev.* **88**, 415 (1952).
CAS 32   H. B. G. Casimir, *Zeits. f. Physik*, **77**, 811 (1932).

CAS 35   H. B. G. Casimir, *Physica*, **2**, 719 (1935).
CAS 36   H. B. G. Casimir, *On the Interaction between Atomic Nuclei and Electrons*, Teylors, Tweede Genootschap, **11**, 36 (1936), Haarlem, Holland; also *Arch. du Musie Teylor*, iii. **8**, 201.
CAS 42   H. B. G. Casimir and G. Karreman, *Physica*, **9**, 494 (1942).
CLA 29   P. Clausing, *Physica*, **9**, 65 (1929), and *Zeits f. Physik*, **66**, 471 (1931).
CLA 30   P. Clausing, *Ann. Physik*, **7**, 489 (1930), and **8**, 289 (1928).
COB 44   A. Cobas and W. E. Lamb, *Phys. Rev.* **65**, 327 (1944).
COC 28   J. D. Cockcroft, *Proc. Roy. Soc.* **119**, 293 (1928), and **119**, 306 (1928).
COH 34   V. W. Cohen, *Phys. Rev.* **46**, 713 (1934).
COH 37   V. W. Cohen and A. Ellett, *Phys. Rev.* **51**, 65 (1937); **52**, 502 and 509 (1937).
COH 54   V. W. Cohen, D. A. Gilbert, S. Wexler, and L. S. Goodman, *Phys. Rev.* **95**, 569 and 570 (1954). See also ZZA 15 and ZZA 16.
CON 35   E. U. Condon and G. H. Shortley, *Theory of Atomic Spectra*, Cambridge University Press, Cambridge, England (1935).
CON 54   Consolidated Engineering Corporation helium leak detector has been found to be very satisfactory.
COP 35   M. J. Copley, T. E. Phipps, and J. Glasser, *Rev. Sci. Inst:* **6**, 371 (1935).
COP 37   M. J. Copley and V. Deitz, *Rev. Sci. Inst.* **8**, 314 (1937).
COR 53   N. Corngold, V. W. Cohen, and N. F. Ramsey, private communication (1953).
COR 54   N. Corngold, *The Neutron Magnetic Moment*, Ph.D. thesis, Harvard University (1954).
COT 53   R. E. Coté and P. Kusch, *Phys. Rev.* **90**, 103 (1953).
CRA 49   M. F. Crawford and A. L. Schawlow, *Phys. Rev.* **76**, 1310 (1949).
DAL 53   R. T. Daly, Jr., and J. R. Zacharias, *Phys. Rev.* **91**, 476A (1953).
DAL 54   R. T. Daly, Jr., and J. H. Holloway, private communication and *Phys. Rev.* (1954).
DAR 28   C. Darwin, *Proc. Roy. Soc.* **117**, 258 (1927).
DAV 48   L. Davis, B. T. Feld, C. W. Zabel, and J. R. Zacharias, *Phys. Rev.* **73**, 525 (1948).
DAV 48a L. Davis, *Phys. Rev.* **74**, 1193 (1948).
DAV 48b L. Davis, Report 88, Research Laboratory of Electronics, Massachusetts Institute of Technology (1948).
DAV 49   L. Davis, Jr., *Phys. Rev.* **76**, 435 (1949).
DAV 49a L. Davis, D. E. Nagle, and J. R. Zacharias, *Phys. Rev.* **76**, 1068 (1949).
DAV 49b L. Davis, B. T. Feld, C. W. Zabel, and J. R. Zacharias, *Phys. Rev.* **76**, 1076 (1949).
DAV 49c L. Davis, D. E. Nagle, and J. R. Zacharias, *Phys. Rev.* **76**, 1068 (1949).
DAV 53   N. Davy and N. H. Langton, *Quart. J. Mech. and Applied Math.* **6**, 115 (1953).
DEB 29   P. Debye, *Polar Molecules*, Dover Publications, New York (1929 and 1945).
DEH 50   H. G. Dehmelt and H. Krüger, *Naturwiss.* **37**, 111 (1950); *Zeits. f. Phys.* **129**, 401 (1951); *Naturwiss.* **37**, 398 (1950).
DEV 52   F. M. Devienne, *C. R. Acad. Sci.* (Paris), **234**, 80 (1952), **238**, 2397 (1954), and *J. Phys. Radium*, **13**, 53 (1952), and **14**, 257 (1953).
DIC 46   R. H. Dicke, *Rev. Sci. Inst.* **17**, 268 (1946).
DIC 50   W. C. Dickinson, *Phys. Rev.* **76**, 1414 (1949).

DIC 50a	W. C. Dickinson, *Phys. Rev.* **80**, 563 (1950).
DIC 51	W. C. Dickinson, *Phys. Rev.* **81**, 717 (1951).
DIC 54	R. H. Dicke, *Phys. Rev.* (1954) and private communication.
DIR 47	P. A. M. Dirac, *The Principles of Quantum Mechanics*, Oxford University Press, Oxford, England (1947).
DIR 48	P. A. M. Dirac, *Phys. Rev.* **74**, 817 (1948).
DIS 54	Distillation Products, Inc. and Consolidated Engineering Corporation, Rochester, New York.
DIT 29	R. W. Ditchburn and F. L. Arnot, *Proc. Roy. Soc.* **123**, 516 (1929).
DOB 28	Dobrezov and Terenin, *Naturwiss.* **33**, 656 (1928).
DOR 38	R. Dorrestein and J. A. Smit, *Proc. K. Akad. Amsterdam*, **41**, 725 (1938).
DOR 42	R. Dorrestein, *Physics*, **9**, 433 and 447 (1942).
DRE 52	S. Drell, private communication (1952).
DUB 33	L. A. DuBridge and H. Brown, *Rev. Sci. Inst.* **4**, 532 (1933); **4**, 665 (1933); **6**, 115 (1935).
DUM 50	J. W. M. DuMond and E. R. Cohen, *Phys. Rev.* **77**, 411 (1950).
DUM 53	J. W. M. DuMond and E. R. Cohen, *Rev. Mod. Phys.* **25**, 691 (1953).
DUN 11	L. Dunoyer, *Comptes Rendus*, **152**, 594 (1911); *Le Radium*, **8**, 142 (1911).
DUN 13	L. Dunoyer, *Comptes Rendus* **157**, 1068 (1913); *Le Radium* **10**, 400 (1913).
DUN 32	J. L. Dunham, *Phys. Rev.* **41**, 713, 721 (1932); **49**, 797 (1936); **34**, 446 (1929).
ECK 49	E. C. Eckstein, *Phys. Rev.* **76**, 1328 (1949); **78**, 731 (1950).
EDD 30	A. S. Eddington, *Mathematical Theory of Relativity*, Cambridge University Press, Cambridge, England (1930).
EDM 27	W. Edmondson and A. Egerton, *Proc. Roy. Soc.* A **113**, 520 (1927).
EIS 52	J. T. Eisinger, B. Bederson, and B. T. Feld, *Phys. Rev.* **86**, 73 (1952), and Mass. Inst. of Tech. Electronics Laboratory Technical Report number 212 (28 Jan. 1952).
ELD 27	J. A. Eldridge, *Phys. Rev.* **30**, 931 (1927).
ELL 29	A. Ellett, Olson, and H. A. Zahl, *Phys. Rev.* **34**, 493 (1929); **35**, 293 (1930); and **36**, 893 (1930).
ELL 31	A. Ellett and R. M. Zabel, *Phys. Rev.* **37**, 1112 (1931).
ELL 37	A. Ellett and V. W. Cohen, *Phys. Rev.* **52**, 509 (1937).
ELM 49	W. C. Elmore and M. Sands, *Electronics*, McGraw-Hill Book Company, New York City (1949).
EPP 53	Eppley Laboratory, Newport, Rhode Island, commercially manufactures a lock-in amplifier for the Golay infra-red detector that has been successfully used in molecular beams (WES 53).
EPS 24	P. S. Epstein, *Phys. Rev.* **23**, 710 (1924).
ESH 52	J. R. Eshbach and M. W. P. Strandberg, *Phys. Rev.* **85**, 24 (1952).
EST 23	L. Estermann and O. Stern, *Z. physikal. Chem.* **106**, 399 and 403 (1923).
EST 28	L. Estermann, *Z. physikal. Chem.* B **1**, 161 (1928), and **2**, 287 (1929).
EST 29	I. Estermann and M. Wohlwill, *Zeits. f. phys. Chem.* **2**, 287 (1929) and **20**, 195 (1933).
EST 30	L. Estermann and O. Stern, *Zeits. f. Physik*, **61**, 95 (1930) U.z.M.15.
EST 31	L. Estermann, O. Frisch, and O. Stern, *Zeits f. Physik*, **73**, 348 (1931).
EST 31a	I. Esterman, R. Frisch, and O. Stern, *Phys. Zeits.* **32**, 670 (1931).
EST 33	I. Estermann and R. G. J. Fraser, *J. Chem. Phys.* **1**, 390 (1933).
EST 33a	I. Estermann and O. Stern, *Zeits. f. Physik*, **85**, 17 (1933).
EST 33b	I. Estermann and O. Stern, *Zeits. f. Physik*, **85**, 135 (1933).
EST 33c	L. Estermann and O. Stern, *Zeits. f. Physik*, **86**, 132 (1933).

EST 37	I. Estermann, O. C. Simpson, and O. Stern, *Phys. Rev.* **52**, 535 (1937).
EST 46	I. Estermann, *Rev. Mod. Phys.* **18**, 300 (1946).
EST 47	I. Estermann, O. C. Simpson, and O. Stern, *Phys. Rev.* **71**, 238 (1947).
EST 47a	I. Estermann, S. Foner, and O. Stern, *Phys. Rev.* **71**, 250 (1947).
FAL 50	D. L. Falkoff and G. E. Uhlenbeck, *Phys. Rev.* **79**, 323 (1950).
FAN 48	U. Fano, *J. Research Nat. Bur. Stand.* **40**, RP 1866 (1948).
FEE 49	E. Feenberg and K. C. Hammock, *Phys. Rev.* **75**, 1877 (1949).
FEE 53	E. Feenberg and G. E. Pake, *Notes on the Quantum Theory of Angular Momentum*, Addison-Wesley Publishing Co., Inc., Cambridge, Mass. (1953).
FEL 45	B. T. Feld and W. E. Lamb, *Phys. Rev.* **67**, 15 (1945).
FEL 47	B. T. Feld, *Phys. Rev.* **72**, 1116 (1947).
FEL 52	B. T. Feld, *Ann. Rev. Nucl. Science* (1952).
FER 30	E. Fermi, *Zeits. f. Physik*, **60**, 320 (1930).
FER 33	E. Fermi and E. Segré, *Zeits. f. Physik*, **82**, 729 (1933).
FER 48	E. Fermi and E. Teller, Notes on Pocono Conference of Physics, sponsored by National Academy of Science, 1 April 1948.
FER 50	E. Fermi, J. Orear, A. H. Rosenfeld, R. H. Schuter, *Nuclear Physics*, University of Chicago Press (1950).
FES 51	H. Feshbach and J. Schwinger, *Phys. Rev.* **84**, 194 (1951).
FOL 47	H. M. Foley, *Phys. Rev.* **71**, 747 (1947).
FOL 47a	H. M. Foley, *Phys. Rev.* **72**, 504 (1947).
FOL 48	H. M. Foley and P. Kusch, *Phys. Rev.* **73**, 412 (1948).
FOL 50	H. M. Foley, *Phys. Rev.* **80**, 288 (1950).
FOL 54	H. M. Foley, R. M. Sternheimer, D. Tycko, *Phys. Rev.* **93**, 734 (1954).
FOL 54a	H. M. Foley, private communication to be published in *Phys. Rev.*
FOW 36	R. H. Fowler, *Statistical Mechanics*, Cambridge University Press, Cambridge, England (1936).
FOX 35	M. Fox and I. I. Rabi, *Phys. Rev.* **48**, 746 (1935).
FRA 26	R. G. J. Fraser, *Phil. Mag.* **1**, 885 (1926), and *Proc. Roy. Soc.* A**114**, 212 (1927).
FRA 31	R. G. J. Fraser, *Molecular Rays*, Cambridge University Press, Cambridge, England, (1931).
FRA 33	R. G. J. Fraser and L. F. Broadway, *Proc. Roy. Soc.* A**141**, 626 (1933).
FRA 34	R. G. J. Fraser, *Trans. Faraday Soc.* **30**, 182 (1934).
FRA 35	R. G. J. Fraser, H. S. W. Massey, and C. B. O. Mohr, *Zeits. f. Physik*, **97**, 740 (1935).
FRA 36	R. G. J. Fraser and J. U. Hughes, *J. Chem. Phys.* **4**, 730 (1936).
FRA 37	R. G. J. Fraser, *Molecular Beams*, Methuen and Co., Ltd., London (1937).
FRA 37a	R. G. J. Fraser and Jewett, *Proc. Roy. Soc.* (1937).
FRA 50	H. Frauenfelder, *Helv. Phys. Acta*, **23**, 347 (1950).
FRA 52	P. Franken and S. Koenig, *Phys. Rev.* **88**, 199 (1952).
FRI 33	R. O. Frisch and E. Segré, *Zeits. f. Physik*, **80**, 610 (1933).
FRI 33a	R. O. Frisch and O. Stern, *Zeits. f. Physik*, **84**, 430 (1933).
FRI 33b	R. O. Frisch, *Zeits. f. Physik*, **84**, 443 (1933).
FRI 33c	R. O. Frisch and O. Stern, *Zeits. f. Physik*, **85**, 4 (1933).
FRI 33d	R. O. Frisch, *Zeits. f. Physik*, **86**, 42 (1933), U.z.M.30.
FRI 50	H. Friedburg and W. Paul, *Naturwissenschaften* **37**, 20 (1950).
FRI 51	H. Friedburg and W. Paul, *Naturwissenschaften*, **38**, 159 (1951), and *Z. Phys.* **130**, 493 (1951).
FRO 52	R. A. Frosch and H. M. Foley, *Phys. Rev.* **88**, 1337 (1952).
FUN 30	U. Funk, *Ann. der Physik*, **4**, 149 (1930).

REFERENCES

GAE 13   W. Gaede, *Ann. Physik*, **41**, 331 (1913).
GAR 49   J. H. Gardner and E. M. Purcell, *Phys. Rev.* **76**, 1262 (1949).
GAR 54   Garlock Packing Company, Palmyra, New York.
GER 24   Gerlach and O. Stern, *Ann. Physik*, **74**, 673 (1924), and **76**, 163 (1925).
GES 51   S. Geschwind, R. Gunther-Mohr, and C. H. Townes, *Phys. Rev.* **81**, 288 (1951).
GOL 52   M. Goldhaber and R. D. Hill, *Rev. Mod. Phys.* **24**, 179 (1952).
GOL 53   J. N. Goldberg, J. W. Trischka, and P. G. Bergmann, *Phys. Rev.* **89**, 278 (1953).
GOR 36   C. J. Gorter, *Physica*, **3**, 503 and 995 (1936).
GOR 38   J. E. Gorham, *Phys. Rev.* **53**, 563 (1938).
GOR 49   W. Gordy, *Phys. Rev.* **76**, 139 (1949).
GOR 53   W. Gordy, W. V. Smith, R. F. Trambarulo, *Microwave Spectroscopy*, John Wiley and Sons, Inc., New York (1953).
GOU 29   S. Goudsmit and R. Bacher, *Phys. Rev.* **34**, 1499 (1929); **37**, 663 (1931).
GOU 33   S. Goudsmit, *Phys. Rev.* **43**, 636 (1933).
GOU 33a   S. Goudsmit and R. Bacher, *Phys. Rev.* **43**, 894 (1933).
GRA 50   L. Grabner and V. Hughes, *Phys. Rev.* **79**, 819 (1950).
GRA 51   L. Grabner and V. Hughes, *Phys. Rev.* **82**, 561 (1951).
GUN 51   R. Gunther-Mohr, S. Geschwind, and C. H. Townes, *Phys. Rev.* **81**, 289 (1951).
GUT 31   P. Guttinger, *Zeits. f. Physik*, **73**, 169 (1931).
GUT 49   A. Guthrie and R. K. Wakerling, *Vacuum Equipment and Techniques*, McGraw-Hill Book Co., New York City (1949).
GUT 51   H. Gutowsky and R. E. McClure, *Phys. Rev.* **81**, 276 (1951).
GUT 51a   H. Gutowsky, McCall, C. Slichter, and McNeil, *Phys. Rev.* **82**, 748 (1951), and **84**, 589 and 1246 (1951).
HAG 53   H. D. Hagstrum, *Phys. Rev.* **89**, 244 (1953).
HAH 51   E. L. Hahn and D. E. Maxwell, *Phys. Rev.* **84**, 1246 (1951), and **88**, 1070 (1952).
HAL 11   Hale, *Trans. Amer. Electro. Chem. Soc.* **20**, 243 (1911).
HAL 34   Halpern and Wasser, *Phys. Rev.* **46**, 176 (1934).
HAL 41   O. Halpern and L. Holstein, *Phys. Rev.* **59**, 960 (1941); O. Halpern and M. Johnson, *Phys. Rev.* **55**, 898 (1938); O. Halpern, H. Hammermesh, and M. Johnson, *Phys. Rev.* **59**, 981 (1941); M. Hammermesh, *Phys. Rev.* **61**, 17 (1942).
HAL 47   O. Halpern, *Phys. Rev.* **72**, 245 (1947).
HAL 49   O. Halpern, *Phys. Rev.* **76**, 1130 (1949).
HAM 39   D. R. Hamilton, *Phys. Rev.* **56**, 30 (1939).
HAM 41   D. R. Hamilton, *Am. Journ. of Physics*, **9**, 319 (1941).
HAM 47   M. Hammermesh, *Nuclear Physics Notes*, New York University Physics Department (1947).
HAM 53   D. R. Hamilton, private communication (1953), and Hamilton, Lemonick, Pipkin, and Reynolds, *Phys. Rev.* **95**, 1356 (1954).
HAN 32   R. R. Hancox, *Phys. Rev.* **42**, 864 (1932).
HAR 42   T. C. Hardy and S. Millman, *Phys. Rev.* **61**, 459 (1942).
HAR 52   N. J. Harrick and N. F. Ramsey, *Phys. Rev.* **88**, 228 (1952).
HAR 53   N. J. Harrick, R. G. Barnes, P. J. Bray, and N. F. Ramsey, *Phys. Rev.* **90**, 260 (1953).
HAV 33   G. G. Havens, *Phys. Rev.* **43**, 992 (1933).
HAX 48   O. Haxel, J. H. D. Jensen, and H. E. Suess, *Naturwiss.* **35**, 376 (1948); **36**, 153 (1949); *Phys. Rev.* **75**, 1766 (1949); *Zeits. f. Physik*, **128**, 295 (1950).

REFERENCES

HAY 41   R. H. Hay, *Phys. Rev.* **60**, 75 (1941).
HEI 53   W. Heitler, *Quantum Theory of Radiation*, Oxford University Press, Oxford, England (1953).
HEI 43   H. Heil, *Zeits. f. Physik*, **120**, 212 (1943).
HEN 37   J. O. Hendricks, T. E. Phipps, and M. J. Copley, *J. Chem. Phys.* **5**, 868 (1937).
HER 45   G. Herzberg, *Infra Red and Raman Spectra of Polyatomic Molecules*, D. Van Nostrand Company, New York (1950).
HER 50   G. Herzberg, *Can. J. Res.* A **28**, 144 (1950).
HER 50a   G. Herzberg, *Spectra of Diatomic Molecules*, D. Van Nostrand Company, New York (1950).
HEY 38   M. Heyden and R. Ritschl, *Zeits. f. Physik*, **108**, 739 (1938).
HIB 51   J. W. Hiby and M. Pahl, *Z. Phys.* **129**, 517 (1951), and **130**, 348 (1951).
HIP 49   J. A. Hipple, H. Sommer, and H. A. Thomas, *Phys. Rev.* **76**, 1877 (1949), and **80**, 487 (1950), and **82**, 697 (1951).
HOD 54   C. D. Hodgman, *Handbook of Chemistry and Physics*, Chemical Rubber Publishing Company, Cleveland, Ohio, U.S.A. (1954).
HUG 47   H. K. Hughes, *Phys. Rev.* **72**, 614 (1947).
HUG 48   D. Hughes, J. R. Wallace, and R. H. Holtzman, *Phys. Rev.* **73**, 1277 (1948).
HUG 49   H. K. Hughes, *Phys. Rev.* **76**, 1675 (1949).
HUG 50   V. Hughes and L. Grabner, *Phys. Rev.* **79**, 314 (1950).
HUG 50a   V. Hughes and L. Grabner, *Phys. Rev.* **79**, 829 (1950).
HUG 51   D. J. Hughes and M. T. Burgy, *Phys. Rev.* **81**, 498 (1951).
HUG 53   V. Hughes, G. Tucker, E. Rhoderick, and G. Weinreich, *Phys. Rev.* **91**, 828 (1953).
HUG 53a   D. J. Hughes, *Pile Neutron Research*, Addison-Wesley Publishing Co., Cambridge, Mass. (1953).
HUG 53b   V. Hughes and G. Weinreich, *Phys. Rev.* **91**, 196 (1953), and **95**, 1451 (1954).
HUN 36   R. D. Huntoon and A. Ellett, *Phys. Rev.* **49**, 381 (1936).
HUR 54   F. C. Hurlbut, *Phys. Rev.* **94**, 754 (A) (1954).
HYL 50   E. Hylleraas and P. R. Skavlem, *Phys. Rev.* **79**, 117 (1950).
IND 54   *Indiana Permanent Magnet Manual No. 4*, The Indiana Steel Products Co., Valparaiso, Indiana (1954).
ING 38   D. Inglis, *Phys. Rev.* **53**, 470 (1938); **53**, 880 (1938); **55**, 329 (1939); **60**, 837 (1941).
ION 48   N. I. Ionov, *Dokl. Akad. Nauk, SSSR*, **59**, 467 (1948).
ISH 54   E. Ishiguro and S. Koide, *Phys. Rev.* **94**, 350 (1954).
JAC 34   L. Jackson and H. Kuhn, *Nature*, **134**, 25 (1934), and *Proc. Roy. Soc.* A **148**, 335 (1935).
JAC 51   V. Jaccarino and J. G. King, *Phys. Rev.* **83**, 471 (1951).
JAC 52   V. Jaccarino, B. Bederson, and H. H. Stroke, *Phys. Rev.* **87**, 676 (1952).
JAC 54   V. Jaccarino and J. R. Zacharias, private communication; V. Jaccarino, J. G. King, R. A. Satten, and H. H. Stroke, *Phys. Rev.* **94**, 1798 (1954).
JAH 45   E. Jahnke and F. Emde, *Funktionentafeln*, Dover Publications, New York City (1945).
JAW 52   W. Jawtusch, R. Jaeckel, and G. Schuster, *Zeits. f. Physik*, **133**, 541 (1952) and **141**, 146 (1955); *Z. Naturforsch.* **9a**, 475 and 905 (1954).
JEA 25   J. Jeans, *The Dynamical Theory of Gases*, Cambridge University Press, Cambridge, England (1925).

JEF 51   C. D. Jeffries, *Phys. Rev.* **81**, 1040 (1951).
JEN 48   J. H. D. Jensen, H. E. Suess, and O. Haxel, *Naturwiss.* **33**, 249 (1946), **35**, 376 (1948); **36**, 153 (1949); *Phys. Rev.* **75**, 1766 (1949); *Zeits. f. Physik*, **128**, 295 (1950).
JEN 53   F. A. Jenkins, *J. Opt. Soc. Am.* **43**, 425 (1953).
JEW 34   Jewitt, *Phys. Rev.* **46**, 616 (1934).
JOH 28   T. H. Johnson, *J. Franklin Inst.* **206**, 308 (1928); **207**, 635 (1929); **210**, 145 (1930).
JOH 28a T. H. Johnson, *Phys. Rev.* **31**, 103 (1928).
JOH 29   T. H. Johnson, *J. Franklin Inst.* **207**, 629 (1929), and **210**, 141 (1930).
JOH 30   T. H. Johnson, *Phys. Rev.* **35**, 1299 (1930); *J. Franklin Inst.* **210**, 135 (1930).
JOH 31   T. H. Johnson, *J. Franklin Inst.* **212**, 507 (1931).
JOH 33   M. C. Johnson and T. V. Starkey, *Proc. Roy. Soc.* **140**, 126 (1933).
JOS 33   B. Josephy, *Zeits. f. Physik*, **80**, 755 (1933).
JUL 47   R. S. Julian, Ph.D. thesis, Mass. Inst. of Tech., Cambridge, Mass. (1947).
KAN 51   A. Kantrowitz and J. Grey, *Rev. Sci. Instr.* **22**, 328 (1951).
KAR 50   R. Karplus and N. Kroll, *Phys. Rev.* **77**, 536 (1950).
KAR 52   R. Karplus and A. Klein, *Phys. Rev.* **85**, 972 (1952).
KAR 52a R. Karplus, A. Klein, and J. Schwinger, *Phys. Rev.* **86**, 288 (1952).
KAR 52b R. Karplus and A. Klein, *Phys. Rev.* **87**, 848 (1952).
KEL 32   J. M. B. Kellogg, *Phys. Rev.* **41**, 635 (1932).
KEL 36   J. M. B. Kellogg, I. I. Rabi, and J. R. Zacharias, *Phys. Rev.* **50**, 472 (1936).
KEL 38   J. M. B. Kellogg and N. F. Ramsey, *Phys. Rev.* **53**, 331 (1938).
KEL 39   J. M. B. Kellogg, I. I. Rabi, N. F. Ramsey, and J. R. Zacharias, *Phys. Rev.* **55**, 318 (1939).
KEL 39a J. M. B. Kellogg, I. I. Rabi, N. F. Ramsey, and J. R. Zacharias, *Phys. Rev.* **56**, 728 (1939).
KEL 40   J. M. B. Kellogg, I. I. Rabi, N. F. Ramsey, and J. R. Zacharias, *Phys. Rev.* **57**, 677 (1940).
KEL 46   J. M. B. Kellogg and S. Millman, *Rev. Mod. Phys.* **18**, 323 (1946).
KEM 37   E. C. Kemble, *Fundamental Principles of Quantum Mechanics*, McGraw-Hill Publishing Company, New York (1937).
KEN 38   E. H. Kennard, *Kinetic Theory of Gases*, McGraw-Hill Book Company, Inc., New York (1938).
KIN 23   K. H. Kingdon, *Phys. Rev.* **21**, 408 (1923).
KIN 54   J. G. King and V. Jaccarino, *Phys. Rev.* **94**, 1610 (1954); **95**, 1706 (1954).
KIS 51   G. B. Kistiakowsky and W. P. Slichter, *Rev. Sci. Instr.* **22**, 333 (1951).
KLI 48   J. Kligman, *J. Exp. Theor. Phys. USSR*, **18**, 346 (1948).
KLI 52   P. F. A. Klinkenberg, *Rev. Mod. Phys.* **24**, 63 (1952).
KNA 26   F. Knauer and O. Stern, *Zeits. f. Physik*, **39**, 764 (1926). U.z.M.2.
KNA 26a F. Knauer and O. Stern, *Zeits. f. Physik*, **39**, 780 (1926). U.z.M.3.
KNA 29   F. Knauer and O. Stern, *Zeits. f. Physik*, **53**, 766 (1929). U.z.M.10.
KNA 29a F. Knauer and O. Stern, *Zeits. f. Physik*, **53**, 799 (1929). U.z.M.11.
KNA 30   F. Knauer and O. Stern, *Zeits. f. Physik*, **60**, 414 (1930).
KNA 33   F. Knauer, *Zeits. f. Physik*, **80**, 80 (1933); see HIB 51.
KNA 34   F. Knauer, *Zeits. f. Physik*, **90**, 559 (1934).
KNA 35   F. Knauer, *Zeits. f. Physik*, **93**, 397 (1935).
KNA 49   F. Knauer, *Zeits. f. Physik*, **126**, 310 (1949).
KNA 49a F. Knauer, *Zeits. f. Physik*, **126**, 319 (1949).

KNI 49	W. D. Knight, *Phys. Rev.* **76**, 1259 (1949).
KNI 49a	G. Knight and B. T. Feld, Report 123, Res. Lab. for Electronics, Mass. Inst. of Tech. (1949).
KNU 09	M. Knudsen, *Ann. Physik*, **28**, 76 and 999 (1909); **29**, 179 (1909); **48**, 1113 (1915).
KNU 10	M. Knudsen, *Ann. Physik*, **31**, 205 (1910) and **34**, 593 (1911).
KNU 15	M. Knudsen, *Ann. Physik*, **48**, 1113 (1915).
KOD 39	K. Kodera, *Chem. Soc. Japan, Bull.* **14**, 114 and 141 (1939).
KOE 51	S. Koenig, A. S. Prodell, and P. Kusch, *Phys. Rev.* **83**, 687 (1951); **88**, 191 (1952).
KOL 50	H. G. Kolsky, T. E. Phipps, N. F. Ramsey, and H. B. Silsbee, *Phys. Rev.* **79**, 883 (1950).
KOL 50a	H. G. Kolsky, T. E. Phipps, N. F. Ramsey, and H. B. Silsbee, *Phys. Rev.* **80**, 483 (1950).
KOL 50b	H. G. Kolsky, *The Radiofrequency Spectrum of H_2 in a Magnetic Field*, Ph.D. thesis, Harvard University (1950).
KOL 51	H. G. Kolsky, T. E. Phipps, N. F. Ramsey, and H. B. Silsbee, *Phys. Rev.* **81**, 1061 (1951).
KOL 52	H. G. Kolsky, T. E. Phipps, N. F. Ramsey, and H. B. Silsbee, *Phys. Rev.* **87**, 395 (1952).
KOP 40	H. Kopfermann, *Kernmomente*, Akademisch Verlagsgesellschaft, M.B.H., Leipzig, 1940, and Edwards Bros., Ann Arbor, Michigan, and new editions in German and English (1955).
KOP 49	H. Kopfermann, *Zeits. f. Physik*, **26**, 344 (1949).
KOP 49a	H. Kopfermann, H. Krüger, and H. Öhlmann, *Zeits. f. Phys.* **126**, 760 (1949).
KOR 51	M. I. Korsunskii and Ya. M. Fogel, *J. Exp. Theor. Phys.* (USSR), **21**, 25 and 38 (1951).
KOS 52	G. F. Koster, *Phys. Rev.* **86**, 148 (1952).
KRA 35	M. Kratzenstein, *Zeits. f. Physik*, **93**, 279 (1935).
KRO 49	N. M. Kroll and W. E. Lamb, *Phys. Rev.* **75**, 388 (1949).
KRO 51	N. M. Kroll and F. Polluck, *Phys. Rev.* **84**, 594 (1951).
KRU 51	U. E. Kruse and N. F. Ramsey, *J. Math. Phys.* **30**, 40 (1951).
KUR 29	U. Kurt and T. E. Phipps, *Phys. Rev.* **34**, 1357 (1929).
KUS 39	P. Kusch, S. Millman, and I. I. Rabi, *Phys. Rev.* **55**, 666 (1939).
KUS 39a	P. Kusch, S. Millman, and I. I. Rabi, *Phys. Rev.* **55**, 1176 (1939).
KUS 39b	P. Kusch and S. Millman, *Phys. Rev.* **56**, 527 (1939).
KUS 40	P. Kusch, S. Millman, and I. I. Rabi, *Phys. Rev.* **57**, 765 (1940).
KUS 48	P. Kusch and H. M. Foley, *Phys. Rev.* **74**, 250 (1948); **72**, 1256 (1947).
KUS 49	P. Kusch, *Phys. Rev.* **75**, 887 (1949).
KUS 49a	P. Kusch and H. Taub, *Phys. Rev.* **75**, 1477 (1949).
KUS 49b	P. Kusch, *Phys. Rev.* **76**, 138 (1949).
KUS 49c	P. Kusch and A. K. Mann, *Phys. Rev.* **76**, 707 (1949).
KUS 49d	P. Kusch, *Phys. Rev.* **76**, 161 (1949).
KUS 50	P. Kusch, *Lecture Notes on Molecular Beams*, Physics Department, Columbia University (1950).
KUS 50a	P. Kusch, *Phys. Rev.* **78**, 615 (1950).
KUS 51	P. Kusch, *Physica*, **17**, 339 (1951).
KUS 53	P. Kusch, *Phys. Rev.* **92**, 268 (1953).
KUS 53a	P. Kusch, *J. Chem. Phys.* **21**, 1424 (1953).
KUS 54	P. Kusch, and T. G. Eck, *Phys. Rev.* **94**, 1799 (1954).
KUS 54a	P. Kusch, *Phys. Rev.* **93**, 1022 (1954).

LAM 41   W. E. Lamb, *Phys. Rev.* **60**, 817 (1941).
LAM 50   W. E. Lamb and M. Skinner, *Phys. Rev.* **78**, 539 (1950).
LAM 50a W. E. Lamb, R. C. Retherford, S. Triebwasser, and E. S. Dayhoff, Part I, *Phys. Rev.* **79**, 549 (1950); Part II, *Phys. Rev.* **81**, 222 (1951); Part III, *Phys. Rev.* **85**, 259 (1952); Part IV, *Phys. Rev.* **86**, 1014 (1952); Part V, *Phys. Rev.* **89**, 98 (1953); Part VI, *Phys. Rev.* **89**, 106 (1953); and *Phys. Rev.* **72**, 241 (1947).
LAM 51   W. E. Lamb, *Reports on Progress in Physics*, **14**, 19 (1951).
LAM 29   Lammert, *Zeits. f. Physik*, **56**, 244 (1929). U.z.M.13. BF67.
LAN 25   I. Langmuir and K. H. Kingdon, *Proc. Roy. Soc.* A **107**, 61 (1925).
LEE 52   C. A. Lee, R. O. Carlson, B. P. Fabricand, and I. I. Rabi, *Phys. Rev.* **86**, 607 (1952); **91**, 1395 (1953); **91**, 1403 (1953).
LEN 29   W. Lenz, *Zeits. f. Physik*, **56**, 778 (1929).
LEN 34   W. Lenz and E. Brandt, *Zeits. f. Physik*, **92**, 631 and 640 (1934).
LEN 35   J. E. Lennard-Jones and C. Strachan, *Proc. Roy. Soc.* A **150**, 442 and 456 (1935).
LEN 36   J. E. Lennard-Jones and A. F. Devonshire, *Nature*, **137**, 1069 (1936), and *Proc. Roy. Soc.* A **156**, 6 (1935).
LEU 27   Leu, *Zeits. f. Physik*, **41**, 551 (1927). U.z.M.4.
LEU 28   Leu and R. Fraser, *Zeits. f. Physik*, **49**, 498 (1928). U.z.M.8.
LEW 48   H. Lew, *Phys. Rev.* **74**, 1550 (1948).
LEW 49   H. Lew, *Phys. Rev.* **76**, 1086 (1949).
LEW 53   H. Lew and G. Wessell, *Phys. Rev.* **90**, 1 (1953).
LEW 53a H. Lew, *Phys. Rev.* **91**, 619 (1953) and private communication.
LIN 50   G. Lindstrom, *Phys. Rev.* **78**, 817 (1950); *Physica*, **17**, 412 (1951); and *Ark. för Fysik*, Stockholm, **4**, 1 (1951).
LIV 51   R. Livingston, *Phys. Rev.* **82**, 289 (1951).
LOE 34   L. Loeb, *The Kinetic Theory of Gases*, McGraw-Hill Book Co., Inc., New York (1934).
LOG 51   R. A. Logan and P. Kusch, *Phys. Rev.* **81**, 280 (1951).
LOG 52   R. A. Logan, R. E. Coté, and P. Kusch, *Phys. Rev.* **86**, 280 (1952).
LOW 51   F. Low and E. E. Salpeter, *Phys. Rev.* **83**, 478 (1951); **77**, 361 (1950).
LUC 53   R. G. Luce and J. W. Trishka, *J. Chem. Phys.* **21**, 105 (1953).
LYO 52   H. Lyons, *Annals of New York Academy of Sciences*, **55**, 831 (1952).
MAC 47   J. E. Mack and N. Austern, *Phys. Rev.* **72**, 972 (1947); **77**, 745 (1950).
MAI 34   W. H. Mais, *Phys. Rev.* **45**, 773 (1934).
MAJ 32   E. Majorana, *Nuovo Cimento*, **9**, 43 (1932).
MAN 36   J. H. Manley. *Phys. Rev.* **49**, 921 (1936).
MAN 37   J. H. Manley and S. Millman, *Phys. Rev.* **51**, 19 (1937).
MAN 50   A. K. Mann and P. Kusch, *Phys. Rev.* **77**, 427 (1950).
MAN 50a A. K. Mann and P. Kusch, *Phys. Rev.* **77**, 435 (1950).
MAR 40   H. Margenau, *Phys. Rev.* **57**, 383 (1940).
MAR 40a H. Margenau and E. P. Wigner, *Phys. Rev.* **58**, 103 (1940).
MAR 43   H. Margenau and G. M. Murphy, *Mathematics of Physics and Chemistry*, D. van Nostrand Company, New York (1943).
MAR 50   A. M. Markus, I. M. Podgomyi, M. I. Korsunskii, and L. I. Pivovar, *J. Exp. Theor. Phys. USSR*, **20**, 860 (1950), and *J. Tech. Phys. USSR*, **21**, 155 (1951).
MAR 52   H. Martin and M. J. Meyer, *Naturwissenschaften*, **39**, 85 (1952).
MAR 52a R. Marshak, *Meson Physics*, McGraw-Hill Book Co., Inc. (1952).
MAS 33   H. S. W. Massey and C. B. O. Mohr, *Proc. Roy. Soc.* A **141**, 434 (1933). and A **144**, 188 (1934).

MAS 36	H. S. W. Massey and R. A. Buckingham, *Nature*, **138**, 77 (1936).
MAS 50	H. S. W. Massey, *Negative Ions*, Cambridge University Press, Cambridge, England (1950).
MAS 52	H. S. W. Massey and E. H. S. Burhop, *Electronic and Ionic Impact Phenomena*, Oxford University Press, Oxford, England (1952).
MAY 29	J. E. Mayer, *Zeits. f. Physik*, **58**, 373 (1929).
MAY 35	J. E. Mayer, *Phys. Zeits.* **36**, 845 (1935).
MAY 40	J. E. Mayer and M. G. Mayer, *Statistical Mechanics*, John Wiley and Sons, Inc., New York (1940).
MAY 48	M. Mayer, *Phys. Rev.* **74**, 235 (1948), and **78**, 16 (1950).
MCN 51	McNeil, C. Slichter, and H. Gutowsky, *Phys. Rev.* **84**, 589 (1951).
MEI 33	W. Meissner and H. Scheffers, *Phys. Zeits.* **34**, 48 (1933).
MEI 33a	W. Meissner and H. Scheffers, *Phys. Zeits.* **34**, 245 (1933).
MEI 37	K. W. Meissner and K. Luft, *Ann. Physik*, **28**, 667 (1937).
MEL 52	N. T. Melnikova, E. D. Shchukin, and M. M. Umonskii, *Zh. Eksper. Teor. Fiz.* **22**, 775 (1952).
MIL 35	S. Millman, *Phys. Rev.* **47**, 739 (1935).
MIL 36	S. Millman and M. Fox, *Phys. Rev.* **50**, 220 (1936).
MIL 37	S. Millman and J. R. Zacharias, *Phys. Rev.* **51**, 1049 (1937).
MIL 38	S. Millman, I. I. Rabi, and J. R. Zacharias, *Phys. Rev.* **53**, 384 (1938).
MIL 39	S. Millman, *Phys. Rev.* **55**, 628 (1939).
MIL 39a	S. Millman, P. Kusch, and I. I. Rabi, *Phys. Rev.* **56**, 165 (1939).
MIL 39b	S. Millman and P. Kusch, *Phys. Rev.* **56**, 303 (1939).
MIL 40	S. Millman and P. Kusch, *Phys. Rev.* **58**, 438 (1940).
MIL 41	S. Millman and P. Kusch, *Phys. Rev.* **60**, 91 (1941).
MIN 35	R. Minkowski and Brook, *Zeits. f. Physik*, **95**, 274 and 284 (1935).
MIY 51	H. Miyazawa, *Prog. Theor. Phys.* **6**, 263 (1951).
MIZ 31	S. Mizushima, *Phys. Zeits.* **32**, 798 (1931); see HIB 51.
MIZ 48	S. Mizushima, *Phys. Rev.* **74**, 705 (1948).
MOO 53	P. B. Moon, *Brit. Journ. Appl. Phys.* **4**, 97 (1953).
MOR 29	P. M. Morse, *Phys. Rev.* **34**, 57 (1929); J. L. Dunham, *Phys. Rev.* **34**, 446 (1929).
MOR 53	P. M. Morse and H. Feshbach, *Methods of Theoretical Physics*, McGraw-Hill Book Co., New York (1953).
MOS 54	S. A. Moszkowski and C. H. Townes, *Phys. Rev.* **93**, 306 (1954).
MOT 33	N. F. Mott and H. S. W. Massey, *Theory of Atomic Collisions*, Oxford University Press, Oxford, England (1933 and 1949).
MOT 36	L. Motz and M. Rose, *Phys. Rev.* **50**, 348 (1936).
MUR 49	K. Murakawa, S. Suwa, and T. Kamer, *Phys. Rev.* **76**, 1721 (1949).
NAF 48	J. E. Nafe, E. B. Nelson, and I. I. Rabi, *Phys. Rev.* **71**, 914 (1947); **73**, 718 (1948).
NAG 47	D. E. Nagle, R. S. Julian, and J. R. Zacharias, *Phys. Rev.* **72**, 971 (1947).
NAG 49	D. E. Nagle, *Phys. Rev.* **76**, 847 (1949).
NAT 54	National Research Corporation, Cambridge, Mass.
NEL 49	E. B. Nelson and J. E. Nafe, *Phys. Rev.* **75**, 1194 (1949).
NEL 49a	E. B. Nelson and J. E. Nafe, *Phys. Rev.* **76**, 1858 (1949).
NEW 50	G. F. Newell, *Phys. Rev.* **78**, 711 (1950).
NEW 50a	G. F. Newell, *Phys. Rev.* **80**, 476 (1950).
NIE 46	W. A. Nierenberg, N. F. Ramsey, and S. B. Brody, *Phys. Rev.* **70**, 773 (1946).
NIE 47	W. A. Nierenberg and N. F. Ramsey, *Phys. Rev.* **72**, 1075 (1947).
NIE 48	W. A. Nierenberg, I. I. Rabi, and M. Slotnick, *Phys. Rev.* **73**, 1430 (1948).

NIE 50	W. A. Nierenberg, *Phys. Rev.* **80**, 1102 (1950).
NIE 54	W. A. Nierenberg, J. C. Hubbs, J. Hobson, and H. Silsbee, *Phys. Rev.* **96**, 1450 (1954).
NOL 49	H. G. Nöller, G. W. Oetjen, and R. Jaeckel, *Z. Naturforsch.* **4a**, 101 (1949).
NOR 40	A. Nordsieck, *Phys. Rev.* **58**, 310 (1940).
NOR 49	L. Nordheim, *Phys. Rev.* **75**, 1894 (1949).
OBI 54	S. Obi, T. Ishidzu, H. Horie, S. Yanagawa, Y. Tanabe, and M. Sato, *Annals of Tokyo Astronomical Observatory*, Series 2, vol. iii, no. 3, Mitaka, Tokyo (1953).
OCH 50	S. A. Ochs, R. A. Logan, and P. Kusch, *Phys. Rev.* **78**, 184 (1950).
OCH 52	S. A. Ochs and P. Kusch, *Phys. Rev.* **85**, 145 (1951).
OCH 53	S. A. Ochs, R. E. Coté, and P. Kusch, *J. Chem. Phys.* **21**, 459 (1953).
PAG 31	See ZZA 51.
PAS 38	S. Pasternack, *Phys. Rev.* **54**, 1113 (1938).
PAU 24	W. Pauli, *Naturwiss.* **12**, 741 (1924).
PAU 32	W. Pauli, *Handbuch der Physik*, vol. 24, Part 1, Julius Springer, Berlin, and Edwards Bros., Ann Arbor, Michigan.
PAU 35	L. Pauling and E. B. Wilson, *Introduction to Quantum Mechanics*, McGraw-Hill Book Company, New York (1935).
PAU 40	W. Pauli, *Phys. Rev.* **58**, 716 (1940).
PAU 41	W. Paul, *Zeits. f. Physik*, **117**, 774 (1941).
PAU 41a	L. Pauling, *Nature of the Chemical Bond*, Cornell University Press, Ithaca, New York (1941).
PAU 48	W. Paul and G. Wessell, *Z. Phys.* **124**, 691 (1948).
PEA 51	R. Pease and H. Feshbach, *Phys. Rev.* **78**, 322 (1950); **81**, 142 (1951).
PEI 99	B. O. Peirce, *A Short Table of Integrals*, Ginn and Co., Boston (1899).
PER 53	W. Perl and V. Hughes, *Phys. Rev.* **91**, 842 (1953).
PER 53a	W. Perl, *Phys. Rev.* **91**, 852 (1953).
PET 52	A. S. Petschek and R. E. Marshak, *Phys. Rev.* **85**, 698 (1952).
PHI 27	T. E. Phipps and J. B. Taylor, *Phys. Rev.* **29**, 309 (1927).
PHI 31	T. E. Phipps and O. Stern, *Zeits. f. Physik*, **73**, 185 (1931). U.z.M.17.
PHI 41	M. Phillips, *Phys. Rev.* **60**, 100 (1941).
PHI 50	T. E. Phipps, Jr., Ph.D. thesis, Harvard University (1950).
PHI 52	M. Phillips, *Phys. Rev.* **88**, 202 (1952).
PHY 52	Physics Report of Symposium on Beta and Gamma Radioactivity, *Physica*, **18**, 989–1306 (1952).
PIR 06	M. v. Pirani, *Deutsch. Phys. Gesell. Verb.* **6**, 684 (1906).
POH 51	W. B. Pohlman, B. Bederson, and J. T. Eisinger, *Phys. Rev.* **83**, 475 (1951).
POU 48	R. V. Pound, *Phys. Rev.* **73**, 523 (1948), and **74**, 228 (1948).
POU 50	R. V. Pound, *Phys. Rev.* **79**, 685 (1950).
POW 38	P. Powers, *Phys. Rev.* **54**, 827 (1938).
POW 46	W. M. Powell, *D. C. Magnet Design*, pp. 49–55 of *Nuclear Physics for Engineers*, Report BP–18, Radiation Laboratory, University of California (1946).
PRI 47	H. Primakoff, *Phys. Rev.* **72**, 118 (1947).
PRO 50	A. G. Prodell and P. Kusch, *Phys. Rev.* **79**, 1009 (1950); **88**, 184 (1952).
PRO 50a	W. G. Proctor and F. C. Yu, *Phys. Rev.* **77**, 717 (1950).
PUR 46	E. M. Purcell, H. G. Torrey, and R. V. Pound, *Phys. Rev.* **69**, 37 (1946); **73**, 679 (1948).

REFERENCES

PUR 50	E. M. Purcell and N. F. Ramsey, *Phys. Rev.* **78**, 699 (1950).
PUR 52	E. M. Purcell and G. Benedek, private communication (1952).
RAB 29	I. I. Rabi, *Nature*, **123**, 163 (1929).
RAB 29a	I. I. Rabi, *Zeits. f. Physik*, **54**, 190 (1929). U.z.M.12.
RAB 33	I. I. Rabi and V. W. Cohen, *Phys. Rev.* **43**, 582 (1933).
RAB 34	I. I. Rabi, J. M. B. Kellogg, and J. R. Zacharias, *Phys. Rev.* **46**, 157 (1934).
RAB 34a	I. I. Rabi, J. M. B. Kellogg, and J. R. Zacharias, *Phys. Rev.* **46**, 163 (1934).
RAB 34b	I. I. Rabi and V. W. Cohen, *Phys. Rev.* **46**, 707 (1934).
RAB 36	I. I. Rabi, *Phys. Rev.* **49**, 324 (1936).
RAB 37	I. I. Rabi, *Phys. Rev.* **51**, 652 (1937).
RAB 38	I. I. Rabi, J. R. Zacharias, S. Millman, and P. Kusch, *Phys. Rev.* **53**, 318 (1938).
RAB 39	I. I. Rabi, S. Millman, P. Kusch, and J. R. Zacharias, *Phys. Rev.* **55**, 526 (1939).
RAB 52	I. I. Rabi, *Phys. Rev.* **87**, 379 (1952).
RAB 54	I. I. Rabi, N. F. Ramsey, and J. Schwinger, *Rev. Mod. Phys.* **26**, 167 (1954).
RAC 31	G. Racah, *Zeits. f. Physik*, **76**, 431 (1931).
RAC 42	G. Racah, *Phys. Rev.* **62**, 438 (1942).
RAD 47	Radiation Laboratory, *Microwave Series*, twenty-seven volumes, especially volumes 7, 8, 11, 15, and 16, McGraw-Hill Book Co., New York (1947).
RAI 50	J. Rainwater, *Phys. Rev.* **79**, 432 (1950).
RAM 40	N. F. Ramsey, *Phys. Rev.* **58**, 226 (1940).
RAM 48	N. F. Ramsey, *Phys. Rev.* **74**, 286 (1948).
RAM 49	N. F. Ramsey, *Phys. Rev.* **76**, 996 (1949).
RAM 50	N. F. Ramsey, *Atomic and Nuclear Moments*, Collier's Encyclopedia (1950).
RAM 50a	N. F. Ramsey, *Phys. Rev.* **77**, 567 (1950).
RAM 50b	N. F. Ramsey, *Phys. Rev.* **78**, 221 (1950).
RAM 50c	N. F. Ramsey, *Phys. Rev.* **78**, 695 (1950).
RAM 50d	N. F. Ramsey, *Phys. Rev.* **78**, 699 (1950).
RAM 50e	N. F. Ramsey, private communications and lectures (1950); see ZZA 60.
RAM 51	N. F. Ramsey, *Physica*, **17**, 303 (1951).
RAM 51a	N. F. Ramsey and R. V. Pound, *Phys. Rev.* **81**, 278 (1951).
RAM 51b	N. F. Ramsey, *Physica*, **17**, 388 (1951).
RAM 51c	N. F. Ramsey, *Phys. Rev.* **83**, 540 (1951).
RAM 51d	N. F. Ramsey, and H. B. Silsbee, *Phys. Rev.* **84**, 506 (1951).
RAM 52	N. F. Ramsey, *Nuclear Moments*, Annual Reviews of Nuclear Science, i. 97 (1952).
RAM 52a	N. F. Ramsey, *Phys. Rev.* **85**, 60 (1952).
RAM 52b	N. F. Ramsey and E. M. Purcell, *Phys. Rev.* **85**, 143 (1952).
RAM 52c	N. F. Ramsey, *Phys. Rev.* **85**, 688 (1952).
RAM 52d	N. F. Ramsey, *Phys. Rev.* **85**, 937 (1952).
RAM 52e	N. F. Ramsey, *Phys. Rev.* **86**, 243 (1952).
RAM 52f	N. F. Ramsey, *Phys. Rev.* **87**, 1075 (1952).
RAM 53	N. F. Ramsey, *Phys. Rev.* **89**, 527 (1953).
RAM 53a	N. F. Ramsey, *Phys. Rev.* **90**, 232 (1953).
RAM 53b	N. F. Ramsey, *Nuclear Moments and Statistics*, Part III of *Experimental Nuclear Physics*, John Wiley and Sons, Inc., New York (1953).

REFERENCES

RAM 53c   N. F. Ramsey, *Science*, **117**, 470 (1953).
RAM 53d   N. F. Ramsey, *Phys. Rev.* **91**, 303 (1953).
RAM 53e   N. F. Ramsey, *Nuclear Moments*, John Wiley and Sons, Inc., New York (1953).
RAM 53f   N. F. Ramsey, B. J. Malenka, and U. E. Kruse, *Phys. Rev.* **91**, 1162 (1953).
RAM 53g   N. F. Ramsey, *Nuclear Two Body Problems*, Part IV of *Experimental Nuclear Physics*, John Wiley and Sons, Inc., New York (1953).
RAM 54   N. F. Ramsey, private communication.
RAR 41   W. Rarita and J. Schwinger, *Phys. Rev.* **59**, 436 (1941).
REN 40   N. A. Renzetti, *Phys. Rev.* **57**, 753 (1940).
ROB 33   J. K. Roberts, *Heat and Thermodynamics*, Blackie and Son, Ltd., London (1933).
ROD 27   W. H. Rodebush and de Vries, *J. Amer. Chem. Soc.* **49**, 656 (1927).
ROD 32   W. H. Rodebush and U. Henry, *Phys. Rev.* **39**, 386 (1932).
ROD 36   W. H. Rodebush, V. Murray, and U. Bixler, *J. Chem. Phys.* **4**, 372 and 536 (1936).
ROG 49   E. H. Rogers and H. H. Staub, *Phys. Rev.* **76**, 480 (1949).
ROS 32   J. E. Rosenthal and G. Breit, *Phys. Rev.* **41**, 459 (1932).
ROS 35   S. Rosin and I. I. Rabi, *Phys. Rev.* **48**, 373 (1935).
ROS 39   P. Rosenberg, *Phys. Rev.* **55**, 1267 (1939).
ROS 48   L. Rosenfeld, *Nuclear Forces*, vols. i and ii, Interscience Publishers, Inc., New York (1948).
SAC 46   R. Sachs, *Phys. Rev.* **69** 611 (1946).
SAC 46a   R. Sachs and J. Schwinger, *Phys. Rev.* **70**, 41 (1946).
SAC 47   R. Sachs, *Phys. Rev.* **72**, 91 (1947).
SAC 53   R. Sachs, *Nuclear Theory*, Addison-Wesley Press, Cambridge, Mass. (1953).
SAG 54   P. L. Sagalyn, *Phys. Rev.* **94**, 885 (1954).
SAL 53   E. Salpeter, *Phys. Rev.* **89**, 92 (1953).
SAS 38   N. Saski and M. Fukuda, *Proc. Imp. Acad. Tokyo*, **14**, 166 (1938).
SCH 26   U. Schutz, *Zeits. f. Physik*, **38**, 854 (1926).
SCH 33   R. Schnurmann, *Zeits. f. Physik*, **85**, 212 (1933).
SCH 34   H. Scheffers, *Phys. Zeits.* **35**, 425 (1934).
SCH 34a   H. Scheffers and J. Stark, *Phys. Zeits.* **35**, 625 (1934).
SCH 36   H. Scheffers and J. Stark, *Phys. Zeits.* **37**, 217 and 220 (1936).
SCH 37   J. Schwinger, *Phys. Rev.* **51**, 545 (1937).
SCH 37a   J. Schwinger, *Phys. Rev.* **51**, 648 (1937).
SCH 37b   T. Schmidt, *Zeits. f. Physik*, **106**, 358 (1937).
SCH 39   H. Scheffers, *Phys. Zeits.* **40**, 1 (1939).
SCH 40   H. Scheffers, *Phys. Zeits.* **41**, 89 (1940).
SCH 40a   H. Scheffers, *Phys. Zeits.* **41**, 98 (1940).
SCH 46   J. Schwinger and J. Blatt, *Lecture Notes on Nuclear Physics*, Harvard University (1946)—reprinted by Boston University Physics Department (1952).
SCH 48   J. Schwinger, *Phys. Rev.* **73**, 416 (1948); **76**, 790 (1949).
SCH 49   L. Schiff, *Quantum Mechanics*, McGraw-Hill Book Co., Inc., New York (1949).
SCH 51   N. A. Schuster, *Rev. Sc. Inst.* **22**, 254 (1951).
SCH 51a   A. L. Schawlow and C. H. Townes, *Phys. Rev.* **82**, 268 (1951).
SCH 52   J. Schwinger, *On Angular Momentum*, Nuclear Development Associates (1952).

REFERENCES

SCH 53 R. Schwartz, Ph.D. thesis, Harvard University (1953).
SCH 54 C. Schwartz, private communication and *Phys. Rev.* **97**, 380 (1955).
SEG 53 E. Segré, J. Ashkin, H. A. Bethe, K. T. Bainbridge, N. F. Ramsey, H. H. Staub, B. T. Feld, and P. Morrison, *Experimental Nuclear Physics*, vols. i, ii, and iii, John Wiley and Sons, Inc., New York (1953).
SER 50 G. W. Series and H. Kuhn, *Proc. Roy. Soc.* A **202**, 127 (1950); A **208**, 277 (1951); and *Nature*, **166**, 961 (1950).
SHA 51 A. de Shalit, *Helv. Phys. Acta*, **24**, 296 (1951).
SHE 52 J. E. Sherwood, H. Lyons, R. H. McCracken, and P. Kusch, *Bull. Am. Phys. Soc.* **27** (1), 43 (1952).
SHE 54 J. E. Sherwood, T. E. Stephenson, C. P. Stanford, and S. Bernstein, *Phys. Rev.* **94**, 791 (1954) and **96**, 1546 (1954).
SHR 40 E. F. Shrader, S. Millman, and P. Kusch, *Phys. Rev.* **58**, 925 (1940).
SHU 51 C. G. Shull, E. O. Wollan, and J. Koehler, *Phys. Rev.* **81**, 626 (1951); **84**, 912 (1951).
SIL 50 H. B. Silsbee, Ph.D. thesis, Harvard University (1950).
SIM 54 A. Simon, *Numerical Table of Clebsch-Gordon Coefficients*, Report ORNL–1718, Oak Ridge National Laboratory, Tennessee.
SMA 51 B. Smaller, E. Yasaitis, and H. L. Anderson, *Phys. Rev.* **81**, 896 (1951), and **83**, 812 (1951).
SMA 52 B. Smaller, E. Yasaitis, E. C. Avery, and D. A. Hutchinson, *Phys. Rev.* **88**, 414 (1952).
SMI 51 K. F. Smith, *Nature*, **167**, 942 (1951).
SMI 51a J. Smith, E. M. Purcell, and N. F. Ramsey, private communication.
SMI 51b J. Smith, Thesis, Harvard University (1951).
SMI 54 K. F. Smith, private communication and report by C. S. Wu in *Glasgow Conference on Nuclear Physics*, Pergamon Press, London (1954).
SMO 10 M. v. Smoluchowski, *Ann. der Physik u. Chemie*, **33**, 1567 (1910).
SMY 22 H. D. Smyth, *Proc. Roy. Soc.* **102**, 283 (1922).
SPE 35 O. Specchia, *Nuovo Cimento*, **12**, 541 (1935).
STA 34 T. V. Starkey, *Phil. Mag.* **18**, 241 (1934).
STA 49 S. V. Starodubtsev, *J. Exp. Theor. Phys. USSR*, **19**, 215 (1949).
STA 54 R. H. Stahl, C. Swenson, and N. F. Ramsey, *Phys. Rev.* **96**, 1310 (1954) and *Rev. Sci. Inst.* **25**, 608 (1954).
STE 20 O. Stern, *Zeits. f. Physik*, **2**, 49 (1920), and **3**, 417 (1920).
STE 20a O. Stern, *Zeits. f. Physik*, **3**, 418 (1920).
STE 21 O. Stern, *Zeits. Physik*, **7**, 249 (1921).
STE 26 O. Stern, *Zeits. f. Physik*, **39**, 751 (1926) U.z.M.1.
STE 27 O. Stern, *Zeits. f. Physik*, **41**, 563 (1927), U.z.M.5.
STE 29 O. Stern, *Naturwiss.* **17**, 391 (1929).
STE 34 W. E. Stephens, *Phys. Rev.* **45**, 513 (1934).
STE 37 O. Stern, *Phys. Rev.* **51**, 563 (1937).
STE 40 A. F. Stevenson, *Phys. Rev.* **58**, 1061 (1940).
STE 50 R. Sternheimer, *Phys. Rev.* **80**, 102 (1950); **84**, 244 (1951); **86**, 316 (1952).
STE 53 R. M. Sternheimer and H. M. Foley, *Phys. Rev.* **92**, 1460 (1953).
STE 54 R. M. Sternheimer, *Phys. Rev.* **95**, 736 (1954), and private communication.
STR 38 J. Strong, *Procedures in Experimental Physics*, Prentice-Hall Inc., New York (1938).
STR 54 M. W. P. Strandberg and H. Dreicer, *Phys. Rev.* **94**, 1393 (1954) and **99**, 667 (A) (1955).

SUR 51	G. Suryan and S. Ramaseshan, *Curr. Sci.* **20**, 264 (1951).
SWA 52	J. C. Swartz and J. W. Trischka, *Phys. Rev.* **88**, 1085 (1952).
TAU 49	H. Taub and P. Kusch, *Phys. Rev.* **75**, 1481 (1949).
TAY 26	J. B. Taylor, *Phys. Rev.* **28**, 576 (1928).
TAY 29	J. B. Taylor, *Zeits. f. Physik*, **52**, 846 (1929). U.z.M.9.
TAY 29a	J. B. Taylor, *Zeits. f. Physik*, **57**, 242 (1929). U.z.M.14.
TAY 30	J. B. Taylor, *Phys. Rev.* **35**, 375 (1930).
TER 37	F. E. Terman, *Radio Engineering*, McGraw-Hill Book Co., New York (1937).
TEU 54	W. B. Teutsch and V. W. Hughes, *Phys. Rev.* **95**, 1461 (1954).
THO 49	H. A. Thomas, Driscoll, and J. A. Hipple, *Phys. Rev.* **75**, 902, 922 (1949), and **78**, 787 (1950).
THO 50	H. A. Thomas, *Phys. Rev.* **80**, 901 (1950).
TOL 39	S. Tolansky, *Proc. Roy. Soc.* A **170**, 205 (1939).
TOL 48	S. Tolansky, *Fine Structure in Line Spectra and Nuclear Spin*, Methuen and Co. Ltd., London (1948).
TOR 37	H. C. Torrey, *Phys. Rev.* **51**, 501 (1937).
TOR 41	H. C. Torrey, *Phys. Rev.* **59**, 293 (1941).
TOW 47	C. H. Townes, *Phys. Rev.* **71**, 909 (1947).
TOW 49	C. H. Townes, H. M. Foley, and F. Low, *Phys. Rev.* **76**, 1415 (1949).
TOW 49a	C. H. Townes, and B. P. Dailey, *J. Chem. Phys.* **17**, 782 (1949).
TOW 54	C. H. Townes, *Microwave Spectroscopy*, McGraw-Hill Book Co., New York (1955).
TRI 48	J. W. Trischka, *Phys. Rev.* **74**, 718 (1948).
TRI 49	J. W. Trischka, *Phys. Rev.* **76**, 1365 (1949).
TRI 52	J. W. Trischka, D. T. F. Marple, and A. White, *Phys. Rev.* **85**, 136 (1952).
TRI 54	J. W. Trischka and R. Braunstein, private communication and *Phys. Rev.* **96**, 968 (1954).
UND 54	E. M. Underhill, *Electronics Magazine*, McGraw-Hill Publ. Co., p. 126, Dec. 1943; p. 118, Jan. 1944; p. 126, Feb. 1944; Apr. 1944; and Jan. 1948. Also reprinted by Crucible Steel Co. of America under title, *Permanent Magnet Designs and Alloys*.
VAN 32	J. H. Van Vleck, *Electric and Magnetic Susceptibilities*, Oxford University Press, Oxford, England (1932).
VAN 51	J. H. Van Vleck, *Rev. Mod. Phys.* **23**, 213 (1951).
VAU 49	R. Vauthier, *C.R. Acad. Sci.* Paris, **228**, 1113 (1949).
VES 38	G. Veszi, *Zeits. f. physikal. Chem.* **38**, 424 (1938).
VIL 48	F. Villars, *Phys. Rev.* **72**, 256 (1948), and **86**, 476 (1952); F. Villars and A. Thellung, *Phys. Rev.* **73**, 924 (1948).
WAL 53	H. E. Walchli, Oak Ridge National Laboratory, Report ORNL-1469, *A Table of Nuclear Moment Data*.
WAN 52	Wang, Townes, Schawlow, and Holden, *Phys. Rev.* **86**, 809 (1952).
WEA 44	C. E. Weatherburn, *Vector Analysis* (Elementary and Advanced), G. Bell and Sons, Ltd., London (1944).
WEI 26	Weide and Bichowsky, *J. Amer. Chem. Soc.* **48**, 2529 (1926).
WEI 49	V. W. Weisskopf and T. Welton, *Rev. Mod. Phys.* **21**, 305 (1949); *Phys. Rev.* **75**, 1240 (1949); *Phys. Rev.* **74**, 1157 (1948).
WEI 50	V. Weisskopf, *Helv. Phys. Acta*, **23**, 187 (1950); *Science*, **113**, 101 (1951).
WEI 53	G. Weinreich, G. M. Grosof, and V. W. Hughes, private communication, and *Phys. Rev.* **91**, 195 (1953), and *Phys. Rev.* **95**, 1451 (1954).

REFERENCES

WEI 54	V. W. Weisskopf, *Quantum Mechanics Notes*, Mass. Inst. of Tech. (1954).
WES 53	G. Wessel and H. Lew, *Phys. Rev.* **92**, 641 (1953).
WEY 39	H. Weyl, *The Classical Groups*, Princeton University Press, Princeton (1939), pp. 149 ff.
WIC 33	G. C. Wick, *Zeits. f. Physik*, **85**, 25 (1933).
WIC 33a	G. C. Wick, *Nuovo Cimento*, **10**, 118 (1933).
WIC 48	G. C. Wick, *Phys. Rev.* **73**, 51 (1948).
WIG 31	E. Wigner, *Gruppentheorie und ihre Anwendung auf die Quantenmechanik der Atomspektren*, Friedr. Vieweg und Sohn Akt. Ges., Braunschweig, Germany (1931), and Edwards Bros., Ann Arbor, Michigan (1944).
WIG 41	E. Wigner, *Proc. Nat. Acad. of Science, U.S.*, **27**, 282 (1941).
WIL 31	A. P. Wills, *Vector and Tensor Analysis*, Prentice-Hall, Inc., New York (1931).
WIM 53	T. F. Wimett, private communication and *Phys. Rev.* **91**, 476A (1953).
WOH 33	M. Wohlwill, *Zeits. f. Physik*, **80**, 67 (1933).
WOO 15	R. W. Wood, *Phil. Mag.* **30**, 300 (1915), and **32**, 364 (1916).
WOO 20	R. W. Wood, *Proc. Roy. Soc.* **97**, 455 (1920); **102**, 1 (1922).
WRE 27	E. Wrede, *Zeits. f. Physik*, **41**, 569 (1927). U.z.M.6.
WRE 27a	E. Wrede, *Zeits. f. Physik*, **44**, 261 (1927). U.z.M.7.
ZAB 32	R. M. Zabel, *Phys. Rev.* **42**, 218 (1932).
ZAB 33	R. M. Zabel, *Phys. Rev.* **44**, 53 (1933); see HIB 51.
ZAB 34	R. M. Zabel, *Phys. Rev.* **46**, 411 (1934).
ZAC 40	J. R. Zacharias and J. M. B. Kellogg, *Phys. Rev.* **57**, 570 (1940).
ZAC 42	J. R. Zacharias, *Phys. Rev.* **61**, 270 (1942).
ZAC 53	J. R. Zacharias, private communication (1953) and *Phys. Rev.* **94**, 751T (1954).
ZAH 38	H. A. Zahl and A. Ellett, *Phys. Rev.* **38**, 977 (1931).
ZEI 50	H. J. Zeiger, D. I. Bolef, and I. I. Rabi, *Phys. Rev.* **78**, 340A (1950).
ZEI 52	H. J. Zeiger and D. I. Bolef, *Phys. Rev.* **85**, 788 (1952).
ZEN 32	C. Zener, *Phys. Rev.* **40**, 178 (1932).
ZXW 54	No complete or extensive references are here provided for the subject concerned. Further discussion and additional references are given in the author's *Nuclear Moments* (RAM 53e).
ZXX 54	No complete or extensive references are here provided for the subject concerned. Additional references are listed in the papers explicitly referred to.
ZXY 54	No complete or extensive references are here provided for the subject concerned. Further discussion and additional references are given elsewhere in this book.
ZXZ 54	Private communication.
ZZ	The best known moment value is here recorded, regardless of its method of measurement, but specific references are given only to the molecular-beam measurements. For references to the non-molecular-beam measurement see RAM 53, WAL 53, and *Science Abstracts*, A55–A57 (1952–4).
ZZA 1	N. Saski and K. Kodera, *Mem. Coll. Sci. Kyoto*, A **25**, 83 (1949).
ZZA 2	J. Brossel, B. Cagnac, and A. Kastler, *J. Phys. Radium*, **15**, 6 (1954).
ZZA 3	H. Bucka, *Zeits. f. Phys.* **141**, 49 (1955).
ZZA 4	W. B. Hawkins and R. H. Dicke, *Phys. Rev.* **91**, 1008 (1953) and **96**, 532 (1954).

REFERENCES

ZZA 5   W. T. Sharp, J. M. Kennedy, B. J. Sears, and M. G. Hoyle, *Table of Coefficients for Angular Distribution Analysis*, Atomic Energy of Canada, Ltd., Chalk River, Ontario, Report CRT-556.

ZZA 6   L. J. F. Broer, *Physica*, **12**, 49 (1946).

ZZA 7   J. Brossel, B. Cagnac, and A. Kastler, *Compt. Rend. Acad. Sci. Paris*, **237**, 984 (1953).

ZZA 8   C. Besset, J. Horowitz, A. Messiah, and J. Winter, *J. Phys. Radium*, **15**, 251 (1954).

ZZA 9   K. Althoff, *Zeits. f. Physik*, **141**, 33 (1955).

ZZA 10   U. Meyer-Berkhout, *Zeits. f. Physik*, 141, 185 (1955).

ZZA 11   G. Bennewitz and W. Paul, *Zeits. f. Physik*, **139**, 489 (1954).

ZZA 12   H. Woodgate and J. Hellworth, private communication.

ZZA 13   G. Wessel, *Phys. Rev.* **92**, 1581 (1953).

ZZA 14   P. Hobson, J. C. Hubbs, W. A. Nierenberg, and H. B. Silsbee, *Phys. Rev.* **96**, 1450 (1954).

ZZA 15   L. S. Goodman and S. Wexler, *Phys. Rev.* **95**, 570 (1954); **97**, 242 (1955); **99**, 192 (1955); and to be published.

ZZA 16   A. Gilbert and V. W. Cohen, *Phys. Rev.* **95**, 569 (1954) and **97**, 243 (1955).

ZZA 17   B. H. Flowers, *Phil. Mag.* **43**, 1330 (1952) and *Proc. Roy. Soc.* A **212**, 249 (1952); A **215**, 398 (1952); A **214**, 515 (1952); and A **215**, 120 (1952).

ZZA 18   M. G. Mayer and J. H. D. Jensen, *Elementary Theory of Nuclear Shell Structure*, John Wiley and Sons, Inc., New York (1955).

ZZA 19   A. B. Volkov, *Phys. Rev.* **94**, 1664 (1954).

ZZA 20   C. P. Stanford, T. E. Stephenson, and S. Bernstein, *Phys. Rev.* **96**, 983 (1954).

ZZA 21   G. Bemski, W. A. Nierenberg, and H. B. Silsbee, *Phys. Rev.* **98**, 470 (1955).

ZZA 22   R. Braunstein and J. W. Trischka, *Phys. Rev.* **98**, 1092 (1955). This paper on Li^7F^{19} appeared too late for the results to be included in the tables.

ZZA 23   L. Perl, I. I. Rabi, and B. Senitzky, *Phys. Rev.* **97**, 838 (1955) and **98**, 611 (1955).

ZZA 24   B. Senitzky, I. I. Rabi, and M. L. Perl, *Phys. Rev.* **98**, 1537 (A) (1955).

ZZA 25   N. F. Ramsey, *Phys. Rev.* **98**, 1853 (1955).

ZZA 26   R. L. White, *Rev. Mod. Phys.* **27**, 276 (1955).

ZZA 27   A. M. Sessler and H. M. Foley, *Phys. Rev.* **98**, 6 (1955).

ZZA 28   P. Kusch, private communication, to be published in *Phys. Rev.* (1955).

ZZA 29   J. P. Wittke and R. H. Dicke, *Phys. Rev.* **96**, 530 (1954).

ZZA 30   G. Wessel, *Phys. Rev.* **92**, 1581 (1953).

ZZA 31   R. Arnowitt, *Phys. Rev.* **92**, 1002 (1953).

ZZA 32   W. W. Clendenin, *Phys. Rev.* **94**, 1590 (1954).

ZZA 33   J. Heberle, H. Reich, and P. Kusch, *Phys. Rev.* **98**, 1194 (A) (1955).

ZZA 34   R. Beringer and M. A. Heald, *Phys. Rev.* **95**, 1474 (1954).

ZZA 35   L. D. White, *Phys. Rev.* **98**, 1194 (1955).

ZZA 36   J. P. Gordon, H. J. Zeiger, and C. H. Townes, *Phys. Rev.* **95**, 282 (1954) and **99**, 1264 (1955).

ZZA 37   J. G. King and V. Jaccarino, *Phys. Rev.* **84**, 852 (1951).

ZZA 38   J. R. Zacharias, J. G. Yates, and R. D. Haun, Mass. Inst. of Tech. Res. Lab. for Electronics *Quarterly Progress Report*, January 1955, p. 30 and *Proc. Inst. Radio Eng.* **43**, No. 3, 364 (1955).

REFERENCES

ZZA 39   L. Essen, J. V. L. Parry, and E. C. Bullard, *Nature*, **176**, 280 and 284 (1955).
ZZA 40   G. Bennewitz, W. Paul, and C. Schlier, *Zeits. f. Physik*, **141**, 6 (1955).
ZZA 41   J. Weneser, R. Bersohn, and N. M. Kroll, *Phys. Rev.* **91**, 1257 (1953).
ZZA 42   T. H. Maiman and W. E. Lamb, Jr., *Phys. Rev.* **96**, 968 (1954).
ZZA 42   E. W. Becker and K. Bier, *Z. Naturforsch.* **99**, 975 (1954).
ZZA 44   W. Nierenberg (private communication) finds that tantalum heater wires are much less apt to break than wolfram and the same heaters can be used repeatedly.
ZZA 45   R. C. Miller and P. Kusch, *Phys. Rev.* **99**, 1314 (1955).
ZZA 46   N. F. Ramsey, *Phys. Rev.* **100**, 1191 (1955).
ZZA 47   H. Salwen, *Phys. Rev.* **99**, 1274 (1955).
ZZA 48   W. Paul and H. Steinwedel, *Z. Naturforsch.* **8 a**, 448 (1953), and W. Paul and M. Raether, *Zeits. f. Physik*, **140**, 262 (1955).
ZZA 49   G. Fricke, *Zeits. f. Physik*, **141**, 171 (1955).
ZZA 50   G. Fricke, *Zeits. f. Physik*, **141**, 166 (1955).
ZZA 51   L. Page and N. A. Adams, *Principles of Electricity*, D. Van Nostrand Co., New York (1931) and W. R. Smythe, *Static and Dynamic Electricity*, McGraw-Hill Publication Co., New York City (1950).
ZZA 52   H. A. Jones, I. Langmuir, and G. M. Mackay, *Phys. Rev.* **30**, 201 (1927) and C. J. Smithells, *Tungsten*, D. Van Nostrand Co., New York City (1940).
ZZA 53   Rockwood ball valves with 'O' ring seals, Rockwood Sprinkler Co., Worcester, Mass.
ZZA 54   Hoke globe pattern toggle valves with 'O' ring seal helium leak tested and other Hoke valves, Hoke Incorporated, Englewood, New Jersey.
ZZA 55   H. H. Stroke, Ph.D. Thesis, Mass. Inst. of Tech., and private communication, J. R. Zacharias.
ZZA 56   J. B. Reynolds, R. L. Christensen, D. R. Hamilton, A. Lemonick, F. M. Pipkin, and H. H. Stroke, *Phys. Rev.* **99**, 613 (A) (1955).
ZZA 57   A. Lurio and A. G. Prodell, *Phys. Rev.* **99**, 613 (A) (1955).
ZZA 58   J. R. Zacharias, *Advances in Electronics* (1955).
ZZA 59   P. Kusch, *Phys. Rev.* **100**, 1188 (1955).
ZZA 60   N. F. Ramsey, *Phys. Rev.* **100**, 1191 (1955).
ZZA 61   C. Sherwin, *Phys. Rev.* **93**, 1429 (1954).
ZZA 62   J. P. Auffray, *Phys. Rev. Let.* **6**, 120 (1961).
ZZA 63   I. F. Zartman, *Phys. Rev.* **37**, 383 (1931); C. C. Ko, *J. Frank. Inst.* **217**, 173 (1934); R. C. Miller and P. Kusch, *Phys. Rev.* **99**, 1314 (1955).
ZZA 64   L. D. Landau and E. M. Lifshitz, *Quantum Mechanics*, Pergamon Press, London (1958), p. 416.
ZZA 65   H. A. Zahl and A. E. Ellett, *Phys. Rev.* **38**, 977 (1931).
ZZA 66   E. M. McMillan, *Phys. Rev.* **38**, 1568 (1931), and **42**, 905 (1932).
ZZB 1   I. Koba, *Prog. Theoret. Phys.* **4**, 319 (1949).
ZZB 2   P. Kusch and V. W. Hughes, *Encyclopedia of Physics* (S. Flugge, Editor) **37**, 1 (1959)
ZZB 3   W. A. Nierenberg, *A. Rev. Nucl. Sci.* **7**, 349 (1957).
ZZB 4   J. G. King and J. R. Zacharias, *Adv. Electronics Electron Phys.* **8**, 2 (1956).
ZZB 5   I. Estermann, *Rev. Mod. Phys.* **18**, 300 (1946).
ZZB 6   J. W. Trischka, *Meth. Exptl Phys.* (L. Marton and D. Williams, Editors) **3**, 589 (1961).

REFERENCES

ZZB 7   G. H. Fuller and V. W. Cohen, *Nuclear Moments*, Oak Ridge National Laboratory (1965).
ZZB 8   J. T. La Tourette, W. E. Quinn, and N. F. Ramsey, *Phys. Rev.* **107**, 1202 (1957).
ZZB 9   C. M. Sommerfield, *Phys. Rev.* **107**, 328 (1957).
ZZB 10   S. J. Brodsky and G. W. Erickson, *Phys. Rev.* **148**, 26 (1966).
ZZB 11   J. P. Auffray, *Phys. Rev. Lett.* **6**, 120 (1961) and P. Narum, *Bull. Am. Phys. Soc.* **9** (11), 10 (1964).
ZZB 12   R. Novick, *Phys. Rev.* **100**, 1153 (1955).
ZZB 13   E. R. Cohen and J. W. M. DuMond, *Rev. Mod. Phys.* **37**, 537 (1965).

AUTHOR INDEX

Adams, N. A., 431.
Akheiser, A., 193.
Althoff, K., 50, 286.
Alvarez, L. W., 8, 172, 195, 196.
Ampere, A. M., 162.
Anderson, H. L., 165, 172, 180, 279.
Arfkin, G. B., 75.
Arnold, W. R., 172, 195.
Arnot, F. L., 48.
Arnowitt, R., 268.
Artmann, K., 40.
Ashkin, J., 189.
Austern, N., 180.
Avery, E. C., 208.
Avery, R., 208, 270.

Bacher, R., 74, 75, 76.
Back, E., 78.
Bainbridge, K. T., 189.
Baranger, M., 340, 344, 345.
Bardeen, J., 65.
Barnes, R. S., 159, 164–8, 172, 177, 205, 209, 215, 216, 230, 233, 235, 236, 238, 248, 311, 319, 323, 347.
Becker, E. W., 363.
Becker, G. E., 86, 87, 174, 254, 255, 258, 270, 271, 272, 313.
Bedersen, B., 175, 250, 251, 253, 254, 255, 259, 261, 270, 280, 281, 282, 366, 384–6, 409.
Bellamy, E. H., 173–5, 250, 251, 254–6, 393.
Bemski, G., 209.
Benedek, G., 200.
Bennet, A., 24.
Bennewitz, G., 114, 135, 251, 255, 285, 298, 404.
Bergmann, P. G., 285, 307.
Beringer, R., 259, 260, 261, 266, 433 f.
Berman, A., 253–5, 280, 283, 367.
Bernstein, S., 192, 197, 433 f.
Bersohn, R., 344, 345.
Besset, C., 131.
Bessey, W. H., 2.

Bethe, H. A., 184, 189, 340, 341–5.
Bichowsky, J., 47.
Biedenharn, L. C., 67.
Bielz, F., 25, 32.
Bier, K., 363.
Bitter, F., 270, 271, 279, 280, 286.
Bixler, U., 298.
Blatt, J., 32, 51, 67, 178, 179, 184, 185, 189, 322, 323.
Bloch, F., 8, 116, 118, 122, 143, 147, 154, 158, 172, 180, 183, 190, 191, 195–7, 206, 248, 427.
Bohr, A., 52, 70, 73, 86, 166, 184, 188, 262, 267, 269, 281–3, 325–7, 328, 343.
Bolef, D. I., 9, 209–11, 222–5, 230, 231, 233, 306, 311, 312, 316.
Born, M., 25, 33.
Bose, S. N., 51, 178.
Bozorth, R. M., 401, 402.
Bragg, J. K., 65, 190, 192.
Brainerd, J. G., 407.
Brandt, E., 40.
Braunstein, R., 174, 209, 303, 307, 309, 311, 312, 378, 449 f.
Bray, P. J., 159, 164, 165, 167, 168, 172, 177, 205, 209, 215, 216, 230, 233, 235, 236, 238, 248, 311, 319, 323, 347.
Breit, G., 52, 75, 84, 104, 180, 247, 253, 254, 257, 260, 261, 263, 267–70, 283, 314.
Brix, P., 173, 261.
Broadway, L. F., 26, 34.
Brody, S. B., 159, 174, 219.
Broer, L. J. F., 89.
Brooks, H., 50, 169, 211.
Brossel, J., 50, 131, 286.
Brouwer, F., 26, 292, 293.
Brown, G. E., 75.
Brown, H., 381.
Brown, L. M., 340, 341, 343, 344.
Bucka, H., 50, 286.

Buckingham, R. A., 30.
Bullard, E. C., 255, 284.
Burhop, E. H. S., 2, 30, 31, 34, 35, 38, 39, 43–46.
Burgy, M. T., 193, 194.
Byerly, W. E., 399.

Cagnac, B., 50, 131, 286.
Caldirola, P., 118.
Carlson, R. O., 9, 173, 174, 209, 211, 238, 296, 299, 301, 303, 307, 308, 311, 312, 402, 403, 407.
Carr, H. Y., 208.
Casimir, H. B. G., 60, 65, 68, 71, 73, 75, 76, 272, 277, 280, 314.
Christensen, R. L., 255.
Claussing, P., 13–15, 46.
Clendenin, W. W., 271.
Cobas, A., 260, 378.
Cockcroft, J. D., 46.
Cohen, E. R., 172, 249, 418.
Cohen, V. W., 8, 24, 40, 106, 172, 175, 194, 197–201, 250, 251, 254, 255, 394, 402, 415.
Condit, R. I., 191.
Condon, E. U., 51, 54, 59, 60, 63, 64, 78, 81, 84, 88, 101, 179, 292, 294, 320, 322, 343, 424.
Copley, M. J., 378, 381.
Corngold, N., 172, 194, 197–201, 402.
Coté, R. E., 161, 162, 172–5, 209, 211, 224–7, 239, 240, 311, 312, 325, 371, 415, 433 f.
Crawford, M. F., 283.

Dailey, B. P., 315.
Daly, R. F., 71, 140, 142, 174, 272, 278, 313.
Darwin, C., 115.
Davis, L., 7, 173, 175, 250, 253–6, 272, 274, 280, 313, 314, 319, 368–70, 374, 375, 381–6, 408.
Davy, N., 398.
Dayhoff, E. S., 2, 9, 121, 327, 329–78.
Debye, P., 292, 294, 298.

AUTHOR INDEX

Dehmelt, H. G., 319.
Deitz, V., 378.
Devienne, F. M., 46.
Devonshire, A. F., 42–46.
De Vries, U., 47.
Dicke, R. H., 50, 200, 246, 255, 266, 284, 387, 433 f.
Dickinson, W. C., 162, 165, 175.
Dirac, P. A. M., 51, 59, 178, 179, 262, 328, 329, 332, 336, 340.
Ditchburn, R. W., 48.
Dobrezov, U., 50.
Dorrestein, R., 34, 48, 260, 378.
Dreiser, H., 50.
Drell, S., 206.
Drinkwater, U., 328.
Driscoll, R. L., 158, 172, 248.
Du Bridge, L. A., 381.
Du Mond, J. W. M., 172, 249, 418.
Dunham, J. L., 231, 291.
Dunoyer, L., 1, 48, 50, 376.

Eck, T. G., 71, 175, 274, 278, 313.
Eckstein, E. C., 193.
Eddington, A. S., 285.
Edmondson, W., 47.
Egerton, A., 47.
Einstein, A., 51, 178.
Eisinger, J. T., 173, 250, 253–5, 259, 261, 270, 280–2, 366, 384–6, 409.
Eldridge, J. A., 22, 23.
Ellett, A., 24, 40, 378.
Elmore, W. C., 361.
Elsasser, W. M., 185.
Emde, F., 98.
Epstein, P. S., 45.
Eshbach, J. R., 169.
Essen, L., 255,' 284.
Estermann, I., 2, 24, 25, 27, 28, 34–37, 40, 46, 102, 103, 297, 298, 346, 361, 364, 376, 377, 378, 381, 382.

Fabricand, B. P., 9, 173, 174, 209, 211, 233, 296, 299, 301, 303, 307, 308, 311, 312, 402, 403, 407.
Falkoff, D. L., 61.
Fano, U., 296.
Feenberg, E., 51, 184, 185.

Feld, B. T., 65, 173, 189, 215, 218–20, 250, 253–6, 270, 272, 274, 280–2, 313, 314, 319, 366, 370, 374, 375, 381, 384–6, 409, 423.
Fermi, E., 51, 69, 74–76, 162, 178, 192, 193, 254, 267, 268, 270, 281, 314, 316.
Feshbach, H., 53, 322, 323.
Feynman, R. P., 340, 341, 344, 345.
Flowers, B. H., 184.
Fock, V., 162.
Fogel, Y. M., 135, 404.
Foley, H., 76, 172, 174, 203, 211, 219, 225, 247, 248, 254, 255, 257–9, 261, 269–71, 282, 315–18, 324, 325.
Foner, S., 27, 28, 34, 35.
Fowler, R. H., 28.
Fox, M., 107, 172, 174.
Franken, P., 259–61.
Fraser, R., v, 2, 12, 15, 23, 24, 26, 30, 34, 39, 46–48, 99, 101, 102, 297, 298, 346, 363, 364, 371, 376–9, 394, 397, 415.
Fraunfelder, H., 46.
Fricke, G., 377, 387, 389.
Friedburg, H., 135, 404–6.
Frisch, R. O., 8, 40–43, 49, 102, 115.
Fröhlich, H., 340.
Frosch, R. A., 203, 254.
Fukuda, U., 24.
Funk, U., 48.

Gaede, W., 45.
Gardner, J. H., 158, 172, 247, 248, 259–61.
Gerlach, W., 8, 100, 376, 394, 395.
Geschwind, S., 52, 319.
Gilbert, A., 175, 250, 251, 254, 255, 394, 415.
Glasser, J., 378.
Goldberg, J. N., 307.
Goldhaber, M., 188.
Goodman, L. S., 174, 175, 250, 255, 394, 415.
Gordon, J. P., 285, 298, 309.
Gordy, W., 65, 289, 290, 296, 298, 324.
Gorham, J. E., 114.
Gorter, C. J., 116.

Goudsmit, S., 74, 75, 76, 328.
Grabner, L., 174, 209, 210, 233, 295, 296, 303, 305–7, 311, 312.
Grey, J., 363.
Grosof, G. M., 253–5, 263, 409.
Gunther-Mohr, R., 52, 319.
Guthrie, A., 350, 361, 410, 411.
Gutlinger, P., 115.
Gutowsky, H., 165, 206.

Hagstrum, H. D., 260, 378.
Hahn, E. L., 206.
Hale, U., 389.
Halpern, O., 24, 191, 193, 269.
Hamilton, D. R., 2, 13, 107, 110, 111, 135–7, 173, 174, 250, 255, 272, 305, 313, 320, 364, 394, 399, 406, 407.
Hammermesh, M., 154, 191, 193.
Hammock, K. C., 185.
Hancox, R. R., 38, 46.
Hansen, W., 116, 195, 197, 206.
Hardy, T. C., 175, 254–6, 270, 271, 313.
Harrick, N. J., 8, 9, 159, 164, 165, 167, 172, 177, 205, 209, 215, 216, 230, 233, 235, 236, 238, 248, 311, 319, 323.
Hartree, D. R., 162, 184.
Haun, R. D., 284, 370.
Havens, G. G., 170.
Hawkins, W. B., 50.
Haxel, O., 185–7.
Hay, R. H., 159, 172, 175, 380.
Heald, M. A., 259, 260, 261, 266, 433 f.
Heberle, J., 255, 266, 433 f.
Heil, H., 387.
Heitler, W., 340, 341.
Hellworth, U., 142.
Hendricks, J. O., 381.
Henry, U., 379.
Herzberg, G., 203, 205, 231, 289–91, 296.
Heyden, M., 50.
Hiby, J. W., 31, 34.
Hill, R. D., 188.
Hipple, J. A., 158, 172, 248,

AUTHOR INDEX

Hirschberg, J. G., 329.
Hobson, J., 174, 251, 255, 394.
Hodgman, C. D., 371, 372, 379, 380.
Holden, A. N., 319.
Holloway, J. H., 71, 272, 278, 313.
Holstein, L., 191, 193.
Holtzmann, R. H., 191.
Horie, H., 67.
Horowitz, J., 131.
Houston, W. V., 328.
Hoyle, M. G., 67.
Hubbs, J. C., Jr., 174, 251, 255, 394.
Hughes, D. J., 189–91, 193, 194.
Hughes, H. K., 9, 292–4, 299–303, 402.
Hughes, V. W., 49, 174, 198, 209, 211, 233, 253–5, 260, 261, 263, 295, 296, 298, 376, 378, 303, 305–7, 311, 312, 408, 409.
Huntoon, R. D., 378.
Hurlbut, F. C., 46.
Hutchinson, D. A., 208.
Hylleraas, E., 165.

Inglis, D., 183, 184.
Ionov, N. I., 387.
Ishidzu, T., 67.
Ishiguro, E., 165.

Jaccarino, V., 52, 71, 142, 173, 175, 251, 254, 255, 274, 276, 277, 280, 283, 313, 314, 319, 363, 370, 384.
Jackson, L., 50.
Jaeckel, R., 98, 363.
Jahnke, E., 98.
Jawtusch, W., 28.
Jeans, J., 28.
Jeffries, C. D., 158, 172, 248.
Jenkins, F. A., 290.
Jensen, J., 178, 184–7.
Jewitt, U., 48, 378.
Johnson, M. C., 46, 191, 193.
Johnson, T. H., 36, 40, 377, 378.
Jones, H. A., 379.
Josephy, B., 37, 38.
Julian, R. S., 246, 254, 255, 257, 264, 374, 391, 392.

Kahn, B., 340.
Kamer, T., 329.
Kantrowitz, A., 363.
Karplus, R., 247, 263, 268, 340, 341, 344, 345.
Karreman, G., 68, 71, 277.
Kastler, A., 50, 131, 286.
Kellogg, J. M. B., 1, 6, 8, 9, 40, 65, 66, 80, 104, 106, 107, 111, 112, 114–16, 158–60, 172, 173, 204, 208, 209, 213, 225, 239, 311, 319, 346, 363, 373, 377, 391, 392, 397, 398, 400, 407, 423.
Kemble, E. C., 151, 340.
Kennard, E. H., 11, 19.
Kennedy, J. M., 67.
King, J. G., 52, 71, 142, 173, 175, 255, 274, 276, 277, 280, 283, 313, 314, 319, 384.
Kingdon, K. H., 378, 379.
Kistiakowsky, G. B., 363.
Klein, A., 268, 340, 341, 344, 345.
Kligman, J., 324.
Klinkenberg, P. F. A., 188.
Knauer, F., 1, 12, 24, 26, 33, 34, 36, 38, 102, 346, 377–9.
Knight, G., 65.
Knight, W. D., 165.
Knudsen, M., 13, 15, 45–7, 391.
Kodera, K., 379.
Koehler, J., 192, 407.
Koenig, S., 9, 247, 257, 259–61, 266.
Koide, S., 165.
Kolsky, H. G., v, 9, 131, 172, 205, 206, 209, 215, 230, 235, 236, 238, 311, 316, 319, 320, 323, 354, 356, 360, 363, 391, 392, 393, 401, 407, 415, 416.
Kopfermann, H., 2, 50, 52, 74, 75, 280, 314, 329.
Korsunskii, M. I., 135, 377, 404.
Koster, G. F., 75, 76, 173–5, 255, 314, 315.
Kramers, H. A., 340.
Kratzenstein, M., 12.
Kroll, N. M., 247, 263, 268, 336, 341, 344, 345.
Krüger, H., 319, 329.

Kruse, U. E., 52, 60, 118, 123.
Kuhn, H., 50, 329.
Kurt, U., 101.
Kusch, P., v, 1, 2, 8, 9, 24, 71, 86, 87, 115, 116, 122, 131, 141, 158, 159, 161, 162, 172–5, 209, 211, 215, 216, 224–7, 239–48, 251, 252, 254, 255, 257–61, 266–74, 278–80, 284, 311–13, 325, 346, 364, 365, 371, 372, 407, 408, 412, 416, 433 f.

Lamb, W. E., v, 2, 9, 121, 162, 163, 215, 219, 260, 262, 292, 293, 327, 328, 329, 331–41, 344, 345, 372, 378, 423.
Lammert, U., 24.
Langmuir, I., 379.
Langton, N. H., 398.
Larmor, J., 161, 162, 164.
Lee, C. A., 9, 173, 174, 209, 211, 233, 296, 299, 301, 303, 307, 308, 311, 312, 402, 403, 407.
Lemonick, A., 13, 135, 136, 251, 255.
Lennard-Jones, J. E., 42–46.
Lenz, W., 24, 40.
Leu, U., 48, 101.
Lew, H., v, 2, 48, 173–6, 250, 253–5, 261, 262, 272, 274–6, 313, 364, 367, 368, 371, 380, 382, 384–9, 410–14.
Lindstrom, G., 158.
Livingston, R., 173, 319.
Loeb, L., 28.
Logan, R. A., 172–5, 225, 226, 254, 255, 270, 271, 280, 311, 312, 325, 415.
Low, F., 75, 267–9, 281, 324, 325.
Luce, R. G., 296, 303, 307.
Luft, K., 50.
Lurio, A., 255.
Lyons, H., 255, 284, 285.

Mack, J. E., 329.
Mackay, G. M., 379.
Maiman, T. H., 340.
Mais, W. H., 34.
Majorana, E., 115, 121, 153, 154, 194.

AUTHOR INDEX

Malenka, B. J., 52, 60.
Manley, J. H., 107, 172, 352.
Mann, A. K., 159, 172, 175, 247, 254, 255, 259, 261, 272–4, 280, 313.
Margenau, H., 53, 75, 183, 184, 258, 260.
Markus, A. M., 377.
Marple, D. T. F., 381.
Marshak, R., 442 f.
Martin, H., 48, 179.
Massey, H. S. W., 2, 30–32, 34, 35, 38, 39, 43, 44–46.
Maxwell, D. E., 206, 405.
Mayer, J. E., 12, 13, 377.
Mayer, M., 53, 178, 179, 184, 185, 187.
McCall, D. W., 206.
McClure, R. E., 165.
McCracken, R. H., 255, 284.
McNeil, C. P., 206.
Meissner, W., 50, 101, 104.
Melnikova, N. T., 46.
Messiah, A., 131.
Meyer-Berkhout, U., 50.
Meyerott, R. E., 269.
Michelson, A. A., 327.
Miller, R. C., 24, 240, 365.
Millman, S., 1, 8, 9, 84, 107–11, 114–16, 139, 155–9, 161, 172–5, 215, 216, 241–6, 252, 254–6, 267, 270, 271, 280, 311, 313, 346, 364, 371, 399, 400, 407, 408.
Minkowski, R., 50.
Miyazawa, H., 183.
Mizushima, S., 31, 298.
Mohr, C. B. O., 30, 31, 32.
Moon, P. B., 24.
Morley, E. W., 327.
Morrison, P., 189.
Morse, P. M., 53, 231, 290
Moszkowski, S. A., 325.
Mott, N. F., 30.
Mottleson, B., 70, 184, 188, 269, 281, 325, 326.
Motz, L., 116.
Murakawa, K., 329.
Murphy, G. M., 53.
Murray, V., 298.

Nafe, J. E., 246, 254, 255, 257, 259, 261, 264–6, 269, 279, 280, 333.
Nagle, D. E., 173, 175, 246, 250, 253–5, 264, 374, 382, 384.

Nelson, E. B., 246, 254, 255, 257, 259, 261, 264–6, 269, 279, 280, 333.
Newell, G. F., 164, 172, 230, 231, 315, 316.
Nicodemus, D., 172, 195–7.
Nierenberg, W. A., 9, 60, 65, 159, 174, 209, 210, 219–22, 239, 240, 251, 255, 296, 307, 311, 312, 366, 394.
Nöller, H. G., 363.
Nordheim, L., 185, 187.
Nordsieck, A., 65, 172, 230, 315, 316.

Obi, S., 67.
Ochs, S. A., 211, 239, 240, 254, 255, 270, 280, 371, 433 f.
Oetjen, G., 363.
Öhlmann, H., 329.
Olsen, U., 40.
Orear, J., 193.

Packard, M. E., 116, 195, 197, 206.
Page, L., 431.
Pahl, M., 31, 34.
Pake, G. E., 51.
Parry, J. V. L., 255, 284.
Paschen, F., 78.
Pasternack, S., 328, 340.
Paul, W., 25, 49, 50, 114, 135, 298, 315, 378, 384, 404–6.
Pauli, W., 51, 52, 71, 153, 178, 179, 183, 187, 320, 427.
Pauling, L., 103, 321, 343.
Pease, R., 322, 323.
Peirce, B. O., 64.
Perl, M. L., 173, 174, 255, 283, 313.
Perl, W., 260, 261, 263.
Petschek, A. G., 184.
Phillips, M., 258, 259.
Phipps, T. E., 101, 115, 378, 381.
Phipps, T. E., Jr., 9, 131, 141, 172, 205, 206, 209, 215, 230, 235, 236, 238, 311, 316, 319, 320, 323, 354, 356, 363, 392, 401, 407.
Pipkin, F. M., 13, 135, 136, 251, 255.
Pirani, M. v., 374, 378, 379, 389–93.

Pivovar, L. I., 377.
Podgomyi, I. M., 377.
Pohlman, W. B., 259, 261.
Polluck, F., 268.
Pomeranchuck, J., 193.
Pound, R. V., 55, 116, 124, 195, 206, 270, 271.
Powell, W. M., 401.
Powers, P., 191.
Present, R., 340.
Primakoff, H., 180.
Proctor, W. G., 165.
Prodell, A. G., 9, 246, 247, 254, 255, 257, 259–61, 266–9, 279, 280.
Purcell, E. M., 58, 83, 116, 158, 172, 195, 200–2, 206, 208, 233, 247, 248, 259–61.

Rabi, I. I., v, 1, 2, 6, 8, 9, 26, 27, 29, 34, 49, 65, 66, 79, 80, 84, 104–12, 114–16, 118, 147, 148, 154, 158, 159–61, 172–5, 195, 204, 206–11, 213, 215, 216, 225, 233, 239, 241–6, 253–5, 257, 264–6, 270, 279, 280, 285, 286, 296, 297, 299, 301, 303, 307, 308, 311, 312, 313, 316, 319, 349, 363, 364, 373, 377, 391, 392, 395–9, 402, 403, 407, 423, 427, 429.
Racah, G., 51, 54, 55, 60, 66, 67, 75, 298, 314.
Raether, M., 384.
Rainwater, J., 325.
Ramaseshan, S., 50.
Ramsey, N. F., 1, 2, 6, 8, 9, 51, 52, 57–62, 65, 66, 69, 74, 76, 80, 83, 88, 103, 104, 115, 116, 118, 122–4, 126, 130–4, 140–3, 147, 148, 150, 154, 158–78, 182, 185, 189, 193–5, 197–202, 204–13, 215, 216, 219–22, 225, 227–40, 248, 284, 298, 307, 311, 312, 316, 319, 320, 323, 349, 354, 356, 362, 363, 371, 391, 392, 402, 407, 408, 416, 423, 429, 433 f.
Rarita, W., 179, 320, 322, 323.
Reich, H. J., 255, 266, 407, 433 f.
Renzetti, N. A., 87, 107, 111, 174, 255, 313, 364, 397.

AUTHOR INDEX

Retherford, S., 2, 9, 262, 327, 329-36, 372, 378.
Reynolds, J. B., 13, 135, 136, 251, 255.
Rhoderick, E., 49, 198, 253, 260, 261, 263, 376, 378, 408.
Richardson, U., 328.
Ritschel, G., 50.
Roberts, A., 172, 195.
Roberts, J. K., 11, 19.
Rodebush, W. H., 47, 298, 379.
Rogers, E. H., 155, 172, 197.
Rose, M., 67, 116.
Rosenberg, P., 27, 34.
Rosenfeld, A. H., 193.
Rosenfeld, L., 68, 178, 183, 184.
Rosenthal, J. E., 52, 283.
Rosin, S., 26, 29, 34.
Rydberg, J. R., 74.

Sachs, R., 178, 180, 184, 185, 270.
Sagalyn, P. L., 286.
Salpeter, E., 75, 267-9, 281, 341, 343-5.
Salwen, H., 122, 131, 142.
Sands, M., 361.
Saski, N., 24, 379.
Sato, M., 67.
Satten, R. A., 52, 70, 142, 175, 274, 277, 313, 384.
Schawlow, A. L., 183, 283, 319.
Scheffers, H., 35, 101, 104, 292, 298.
Schiff, L., 71, 164, 186, 187, 341.
Schlier, C., 298.
Schmidt, T., 181-3, 187, 188.
Schurmann, R., 101.
Schuster, G., 28.
Schuster, N. A., 387.
Schuter, R. H., 193.
Schutz, U., 101.
Schwartz, C., 70, 71, 73, 75, 76, 174, 175, 253, 254, 271, 272, 277-9, 281, 313.
Schwartz, R., 169.
Schwinger, J., 51, 68, 69, 116, 118, 147, 148, 154, 179, 180, 184, 191, 247, 262, 263, 268, 320, 322, 323, 340, 341, 344, 345, 427.

Sears, B. J., 67.
Segré, E., 75, 76, 115, 189, 254, 281, 314.
Senitzky, B., 173, 174, 255, 286, 313.
Series, G. W., 329.
Sessler, A. M., 270, 282.
Shalit, A. de, 183.
Sharp, W. T., 67.
Shchukin, E. D., 46.
Sherwood, J. E., 192, 255, 284.
Shortley, G. H., 51, 54, 59, 60, 63, 64, 78, 81, 84, 88, 101, 179, 292, 294, 320, 321, 343, 424.
Shrader, E. F., 173.
Shull, C. G., 192.
Siegert, A., 118, 122, 143.
Silsbee, H. B., 9, 14, 15, 118, 124, 131-4, 141, 172, 174, 205, 206, 209, 215, 230, 235, 236, 238, 251, 255, 311, 316, 319, 320, 323, 354, 356, 363, 391, 392, 407.
Simon, A., 67.
Simpson, O. C., 2, 24, 25, 102, 103.
Skavlem, P. R., 165.
Skinner, M., 327, 336-40.
Slichter, C., 206, 363.
Slotnik, M., 296.
Smaller, B., 164, 165, 172, 208, 268, 279, 280.
Smit, J. A., 34, 48.
Smith, J., 58, 201, 202.
Smith, K. F., 173-6, 184, 250, 251, 253-6, 322, 393.
Smith, W. V., 65, 289, 290, 296, 298.
Smoluchowski, M. v., 14.
Smyth, H. D., 48.
Smythe, W. R., 431.
Sommer, H., 158, 172, 248.
Sommerfeld, A., 327, 328.
Specchia, O., 101.
Stahl, R. H., 363.
Stanford, C. P., 192, 197, 433 f.
Stark, J., 35, 298, 331, 334.
Starkey, T. V., 46.
Starodubtsev, S. V., 371, 379.
Staub, H. H., 155, 172, 189, 191, 195-7.
Stehn, J. R., 340, 341, 343, 344.

Steinwedel, H., 384.
Stephens, W. E., 384.
Stephenson, T. E., 192, 197, 433 f.
Stern, O., 1, 8, 12, 21, 22, 24-28, 33-38, 40-42, 46, 90, 93, 100-3, 115, 346, 361, 371, 376-8, 389, 394, 395.
Sternheimer, R. M., 172, 315-18, 325.
Stevenson, A. F., 118, 122.
Strachman, U., 45.
Strandberg, M. W. P., 50, 169.
Stroke, H. H., 52, 70, 142, 175, 251, 255, 274, 277, 280, 313, 384, 433 f.
Strong, J., 349, 350, 351, 361, 381, 410, 411.
Suess, H. E., 185-7.
Suryan, G., 50.
Suwa, S., 329.
Swartz, J. C., 209, 211, 294-6, 303, 307, 403, 404.
Swenson, C., 363.

Tanabe, Y., 67.
Taub, H., 158, 172, 247, 248, 251, 254, 255, 258, 259, 261, 270, 271, 364, 408.
Taylor, J. B., 19, 37, 40, 45, 46, 101, 379.
Teller, E., 270.
Terenin, U., 50.
Terman, F. E., 361, 407.
Teutsch, W. B., 255, 263.
Thomas, H. A. 158, 162, 165, 172, 177, 248, 316, 328.
Thomas, L. H., 212, 213.
Tolansky, S., 50, 71, 277.
Torrey, H. C., 114, 116, 118, 120, 121, 123, 195, 206.
Townes, C. H., 52, 65, 183, 203, 285, 289-93, 296, 298, 309, 315, 316, 319, 324, 325.
Trambarulo, R. F., 65, 289, 290, 296, 298.
Triebwasser, S., 2, 9, 121, 327, 329-31, 333-6, 378.
Trischka, J. W., 9, 174, 209, 211, 294-6, 301, 303-5, 307, 309, 311, 312, 378, 381, 402, 403, 404, 409, 410, 449 f.

AUTHOR INDEX

Tucker, G., 49, 198, 253, 260, 261, 263, 376, 378, 408.
Tycko, D., 315–18.

Uehling, E. A., 340.
Uhlenbeck, G. E., 61, 328.
Umanskiĭ, M. M., 46.
Underhill, E. M., 401, 402.

Van Vleck, J. H., 163, 169, 170, 203, 225, 229, 240, 288, 292, 294, 296.
Vauthier, R., 135, 404.
Veszi, G., 40.
Villars, F., 180.
Volkov, A. B., 183.

Wakerling, R. E., 350, 361, 410, 411.
Walchli, H. E., 171, 172 f., 202, 248.
Wallace, J. R., 191.
Wang, T. C., 319.
Wasser, U., 24.
Weatherburn, C. E., 419, 431.
Weide, U., 47.
Weinreich, G., 49, 198, 253–5, 260, 261, 263, 376, 378, 408, 409.

Weisskopf, V. F., 32, 70, 71, 73, 178, 179, 184, 185, 189, 269, 281–3, 340, 341, 344.
Welton, T., 340, 341, 344.
Weneser, J., 344, 345.
Wessel, G., 2, 25, 48, 172–4, 176, 250, 253, 254, 255, 261, 262, 313, 378, 387–9, 410.
Wexler, 255.
Weyl, H., 60, 61.
White, A., 381.
White, L. D., 259, 260, 261, 266, 433 f.
White, R. L., 211, 433 f.
Wick, G. C., 168, 170, 209, 211.
Wigner, E., 51, 60, 66, 183, 184, 320.
Williams, R. C., 328.
Wills, A. P., 75, 314, 419.
Wilson, E. B., 103, 321, 343.
Wimett, T. F., 208.
Winter, J., 131.
Wittke, J. B., 246, 255, 266, 433 f.
Wohlwill, M., 297, 378.
Wollan, E. O., 192.
Wood, R. W., 45, 329, 372.

Woodgate, U., 142.
Woodrull, L. F., 407.
Wrede, E., 101, 297.
Wu, C. S., 322.

Yates, J. G. 284.

Zabel, R. M., 34, 40, 173, 254–6, 270, 274, 280, 313, 314, 319, 370, 374, 375, 378, 381.
Zacharias, J. R., v, 1, 2, 6, 8, 9, 52, 65, 66, 71, 80, 84, 106–12, 114–16, 134, 138, 140, 142, 158–61, 172–5, 204, 208, 209, 213, 239, 246, 250, 253–7, 264, 272, 274, 277, 280, 284, 285, 311, 313, 314, 319, 363, 364, 370, 373–5, 377, 381, 382, 384, 391, 392, 397–9, 407, 423.
Zahl, H. A., 40.
Zeeman, P., 76, 331, 332, 333, 336, 338.
Zeiger, H. J., 9, 209–11, 222–5, 230, 231, 233, 285, 298, 306, 309, 311, 312, 316.
Zener, C., 46.

SUBJECT INDEX

A, 288 f.
a, 73, 119, 290 f., 398.
a_0, 291.
A-magnet, 3.
Absolute values of moments, 158, 245 f.
Acceleration interaction, 212 f.
Accommodation coefficient, 44.
Across diagram transitions, 245.
Adiabatic transitions, 149.
Adsorption, 44 f.; selective, 42 f.
Alignment: beam, 415 f.; collision, 128, 138 f., 168, 433.
α, 20, 74, 336, 429.
α', 44.
α_e, 291.
Alpha, 20.
Amplifier, molecular, 298, 309.
Angle, critical, 193.
Angular momentum, 51 f.
Anomalous electron moment, 247 f., 256 f., 262 f., 268.
Anomalous hyperfine structure, 269 f., 279 f., 319.
Anomaly quadrupole, 319.
Antishielding, quadrupole, 317 f.
Association, 47 f.
Atomic beams, 1 f.
Atomic clocks, 283 f.
Atomic core, 316 f.
Atomic fields, 51 f.
Atomic hydrogen, 263 f.
Atomic hyperfine structure, 241 f., 251 f., 254 f., 327 f.
Atomic magnetic moments, 100 f., 241 f., 256 f.
Atomic states, excited, 285 f.
Atoms, 241 f.
Atoms, dissociated, sources of, 372 f.
Attenuation of beam, 16, 25 f.
Averaged functions, 426.

b, 118, 272.
B_e, 291.
B-magnet, 3.
B_v, 291.
Batteries, storage, 358.
Beam alignment, 416 f.
Beam intensity, 16 f., 94 f., 347, 349; shape, 17 f., 92 f., 348 f.
Beams, atomic and molecular, 1 f.
β_e, 291.
Bohr frequency, 139.
Bohr magneton, 248.

Bohr–Mottelson model, 188, 325 f.
Bohr–Weisskopf theory, 281 f.
Bolometer detectors, 378.
Bombardment ionizer, 387 f.
Bonds, chemical, 315 f.
Bose–Einstein statistics, 51.
Bragg reflection, 192.
Breit–Rabi formula, 80, 84 f., 104 f.

C, 73.
c, 277.
c_1, 208.
$C_q^{(k)}$, 54.
C-magnet, 3.
Caesium clock, 284 f.
Centrifugal stretching, 205, 212, 222, 224, 230, 291 f., 316.
Characteristic velocities, 20 f.
Characteristics, apparatus, 356.
Charge renormalization, 340.
Chemical bonds, 315 f.
Chemical detectors, 377; equilibria, 47; shift, 162 f., 165 f.
Clocks, atomic, 283 f.
Collective nuclear model, 188, 325 f.
Collimator, 4, 417; position, 351 f.
Collision alignment, 128, 138 f., 168, 433.
Collisions, 25 f.; inelastic, 44.
Complex nuclei, 273 f.
Condensation coefficient, 46.
— detectors, 376.
Conjugate nuclei, 184.
Constants, fundamental, 245 f., 418.
Core, atomic, 316 f.
Coupling, 184 f.; intermediate, 234.
— rules, 187.
Covalent bonds, 315 f.
Critical angle, 193.
Cross sections, 28 f., 34.
Current density, 71.
Cyclotron frequency, 248.

d, 17.
D, 290 f.
D_e, 291.
D_v, 291.
δ, 207.
δ_L, 257.
δ_S, 257.
ΔE, 336.
$^\alpha \Delta^\beta$, 269, 280.
$(\nabla E^e)_{ij}$, 57.
$\Delta \nu$, 76.

ΔW, 76.
Debye unit, 288 f.
Deflecting field, electrostatic, 402 f.
Deflecting magnetic fields, 394 f., 431 f.
Deflexion, 348; electric, 288 f., 296 f.
— measurements, 114.
— relations, 89 f.
Defocusing states, 110.
Design: illustration, 354; mechanical, 410 f.; principles of, 346 f.
Detectors, 3; chemical, 377; condensation, 376; electron bombardment ionizer, 387 f.; electron ejection, 378; ionization gauge, 377; miscellaneous, 374 f.; Pirani, 3, 389 f.; radioactivity, 393 f.; semiconductor, 377; surface ionization, 379 f.
Deuterium, 160, 167 f., 230 f., 235, 238 f., 259, 264 f., 269 f., 279 f., 331 f.
Deuteron, 160, 179, 319 f.
Diamagnetic correction, 162 f.; interactions, 228 f.; moments, 104; susceptibility, 169 f., 233.
Diamagnetism, 259.
Diatomic molecules, 289 f., 293 f.
Diffraction, 38 f.
Diffusion pumps, 410 f.
Dimers, 240.
Dipole moment: electric, 52, 202, 288 f., 298 f., 303; magnetic, 52, 195 f., 202, 233, 262 f., 270 f.
Dirac theory, 262, 328.
Direct nuclear moment measurements, 270 f.
Dissociated atoms, sources of, 372 f.
Dissociation, 47.
Distortions of resonances, 139 f., 334 f.
Distribution: electron, 229 f.; velocity, 240, 348.
Doppler broadening, 50, 140.
Double resonance experiments, 50.
Doublets, 110, 244, 270 f.
Dunham series, 291.

ϵ, 281.
Effective field, 148.
Effective magnetic moment, 89.
Effusion, molecular, 11 f., 47, 347.
Electric deflexion, 288 f., 296 f.
Electric dipole moment, 202, 288 f., 298 f.
Electric field gradients, 62 f., 313.
Electric polarizability, 298 f.
Electric quadrupole interaction, 213 f., 271 f., 290 f., 293 f., 305 f., 310 f., 319 f., 421 f.
Electric quadrupole moment, 52, 56 f., 59 f., 111, 172 f., 310, 313, 319 f.; sign of, 225.
Electric resonance, 288 f., 298 f.

Electrodynamics, quantum, 262 f.
Electrometers, 381 f.
Electron bombardment ionization, 48 f., 387 f.
Electron coupled interactions, 206 f.
Electron distribution, 170, 229 f.
Electron ejection detectors, 378.
Electron magnetic moment, 247 f., 262 f., 433 f.
— — — anomalous, 247 f., 256 f., 268.
Electron multipliers, 381 f., 385 f.
Electronegativity, 315.
Electrostatic fields, 402 f.
Electrostatic interaction, 52 f.
Elementary constants, 418.
Elements, matrix, 234.
Emission, stimulated, 298, 309.
Excessive oscillatory fields, 143.
Exchange coupling, 207.
Excitation, 48.
Excited atomic states, 285 f.
Experimental techniques, 361 f.

f, 228.
F, 66.
$F_q^{(k)}$, 55.
Fermi–Dirac statistics, 51.
Fermi–Segré formula, 254.
Ferromagnetic mirrors, 192 f.
Field gradients, 313 f.
Field variations, 140, 143.
Fields: atomic and molecular, 51 f.; electrostatic, 402 f.; focusing, 298, 404 f.; magnetic, 394 f., 431 f.; oscillatory, 407.
Filters, 190.
Fine structure, atomic, 327 f.
Fine structure constant, 268, 336.
— — separation, 75.
Flop-in experiments, 7, 249, 417.
Flop-out experiments, 7.
Flop-out on flop-in, 276 f.
Focusing, 114, 134 f.; fields, 298, 404 f.; space, 114; states, 110.
Force, tensor, 320.
Formula, Fermi–Segré, 254.
Formulae, molecular beam, 347 f.
Fountain experiment, 138, 285.
Free electron, 260 f.
Free-free collision, 44.
Free molecules, 203 f.
Free radicals, 48.
Frequency, cyclotron, 248.
— standards, 283 f.
Full half-width, 124, 129.
Fundamental constants, 245 f., 418.

g, 52, 101.
γ_e, 291.

SUBJECT INDEX

γ_I, 52.
$\gamma(r)$, 317.
Gaskets, rubber, 411.
Gas kinetics, 118.
Gauges, vacuum, 415.
Gradient of electric field, 62 f., 313 f.
Group theory, 60 f.
Gyromagnetic ratio, 52, 72 f., 145 f., 244, 260.

H_1, 146.
H_0, 72, 145.
$\mathfrak{H}_{Ek}$, 52 f.
$\mathfrak{H}_{Mk}$, 68 f.
$\mathfrak{H}_S$, 80 f.
Hamburg deflecting fields, 394 f.
Hamiltonian, 233, 272 f.; molecular, 289 f., 293 f.
Hartree model, 184.
Heaters, oven, 366.
Heights, beam, 352 f.
Helium, 260, 336 f.
Heteronuclear molecules, 215.
High frequency terms, 163.
Homonuclear molecules, 82 f., 215.
Hybridization of bonds, 316.
Hydrogen, 158, 160, 167 f., 230 f., 235, 238 f., 246, 259, 263 f., 327 f., 331 f., 433 f.
Hydrogen deuteride, 160, 167 f., 230 f., 235, 238 f.
Hyperfine structure, 73 f., 241 f., 246, 251 f., 254 f., 293 f., 305 f.
— —, anomalous, 73, 269 f., 279, 433 f.
— — of hydrogen, 263 f.
— — separation, 112 f., 433 f.

I, 16, 51.
I_0, 18.
$I(s)$, 96.
$I(x)$, 123.
Inelastic collisions, 44.
Inhomogeneities of field, 140.
Intensity, beam, 16 f., 94 f., 347, 349.
Interaction: acceleration, 212 f.; diamagnetic, 228 f.; electric quadrupole, 213 f., 271 f., 290 f., 293 f., 305 f., 310 f., 319 f., 421 f.; molecular, 289 f., 293 f.; nuclear and molecular, 203 f., 289 f., 293 f.; nuclear multipole, 313; nucleus and atom, 51 f.; pseudo-quadrupole, 227; spin-rotational, 208 f., 222, 290, 293 f., 305 f.; spin-spin, 204 f.
Interchangeable source, 362, 368.
Intermediate coupling, 78 f., 83 f., 104 f., 234.
Ionic bonds, 315 f.
Ionization, 48 f.

Ionization gauge detectors, 378.
Ionizer, electron bombardment, 387 f.
Iron magnets, 399 f., 431 f.
Isotope shift, 52.
Isotopes, radioactive, 250.

k, 30.
$K(x)$, 123.
κ, 14.
Kappa factor, 14.
Kinetics, gas, 11 f.
Knudsen's law, 45 f.

l_i, 3.
λ, 30.
$\bar{\lambda}$, 128.
λ_{Ms}, 12.
λ_p, 25.
Lamb–Retherford experiment, 327 f.
Lamb shift, 336 f.
Landé g-factor, 101.
Large spins, 153 f.
Larmor frequency, 117.
— theorem, 147 f., 164.
Leak detector, He, 411.
— hunting, 411.
Legendre polynomials, 53 f.
Lengths, beam, 352 f.
Line shapes, 120, 127 f., 139 f., 215 f., 334 f.
Lines, Schmidt, 279.
Liquid drop model, 184.

m^0_{zk}, 163.
μ', 290.
μ_ϵ, 287 f.
μ_{eff}, 89.
μ_I, 52, 72.
μ_J, 167.
μ_N, 52.
μ_0, 52.
Magic numbers, 185, 324.
Magnet, permanent, 199.
Magnetic fields, 394 f., 431 f.
Magnetic fields, uniform, 401.
Magnetic moment, 51 f., 68 f., 145 f.
— — atomic, 256 f., 267 f.; dipole, 52; electron, 262 f.; multipole, 68; neutron, 195 f., 202; nuclear, 102 f., 203 f., 254 f., 270 f.; octupole, 52, 172 f., 274 f., 277 f.; rotational, 102 f., 166 f., 233.
— — measurements, 159.
— — table of, 172 f.
Magnetic resonance methods, 115 f.
Magnetic scattering, 191, 193.
— shielding, 162 f., 177, 233.
Magneton: Bohr, 52, 248; nuclear, 52, 246.
Magnets, 4.

Magnets, iron, 399 f., 431 f.
Majorana formula, 121, 154, 428 f.
Majorana transitions, 115 f., 401.
Maser, 285, 298, 309.
Mass number, 51.
Mass renormalization, 341.
Mass spectrometers, 381 f.
Matrix elements, 234 f., 421 f.
Mean free path, 12, 25 f., 347.
Mechanical design, 410 f.
Metastable states, 285 f.
Microwave spectroscopy, 50.
Migration, 46.
Millman effect, 139, 155 f.
Mirror nuclei, 184.
Models, nuclear, 184 f.
Moderator, 190.
Molecular amplifier, 298, 309; beam resonance methods, 1 f.; beams, 1 f.; clocks, 283 f.; data, 303; energy, 289 f., 293 f.; fields, 51 f.; hydrogens, 238 f.; interactions, 80 f., 203 f., 289 f., 293 f.; oscillator, 285; polymerization, 239 f.; rotational moments, 166 f.; scattering, 25; vibration, 205, 212, 222, 224, 230 f., 291 f., 316.
Molecules, 203 f.
— polar, 287 f.
Moment: absolute scale of, 245 f.
— anomalous, 256 f., 268.
— atomic, 256 f.
— electric dipole, 202; electric quadrupole, 52 f., 56 f., 59 f., 213 f., 271 f., 310 f., 313, 319 f.
— magnetic dipole, 52 f., 72 f., 145 f., 254, 313; magnetic octupole, 52 f., 274 f., 277 f., 313.
— multipole, 52 f., 66 f., 272.
— neutron, 195 f.
— nuclear magnetic, 51 f., 102 f., 110, 145 f., 170 f., 203 f., 244.
— nuclear, table of, 172 f.; theory of, 178 f.
— nucleon, 179, 183 f.
— rotational, 203.
— signs of, 256.
— table of, 172 f.
Morse potential, 290 f.
Multiple oscillatory fields, 122, 143.
Multiple quanta transitions, 122, 130, 141, 242, 306 f.
Multiplier, electron, 381 f., 385 f.
Multipole moments, 52 f., 66 f., 272.

n, 11, 193.
v_0, 6, 117.
Neutron: beam, 189 f.; electric dipole moment, 202; magnetic moment, 195 f., 202; mirror, 193; moderator, 190;
polarization, 190 f., 194 f.; reflection, 192 f.; resonance, 201; spin, 433.
Non-condensable gas, 361 f.
Non-focusing states, 110.
Non-linearity, 183 f.
Nuclear interactions, see Interaction.
— magneton, 52, 248.
— models, 184 f.
— moments, see Moment.
— polarizability, 319.
— shell models, 181 f., 185 f.
— spins, 114, 249; table of, 172 f.
Nuclear spin-spin interactions, 204 f.
Nucleon moments, see Moment.

ω, 118.
ω_e, 291.
ω_0, 119.
Ω, 172, 277.
Octupole moment, 52, 172 f., 274 f., 277 f.
Optical alignment, 415 f.
— spectroscopy, 50.
Optimal designs, 351.
Orientation dependent diamagnetism, 228 f.
Oscillator, 407 f.; molecular, 285, 298, 309.
Oscillatory fields, 407.
Oven, 47, 364 f.
— materials, 364.
— temperature, 372.

p, 17.
p_r, 55.
P_k, 53.
$P_{p,q}$, 119.
Paramagnetic atoms, 241 f.
Paramagnetism, second order, 163.
Parity, 321.
Perturbations, 222, 253, 258 f., 271 f.
Perturbation theory, 87, 234.
Phase shifts, 131 f., 142.
Photo-ionization, 48.
Physical constants, 245 f., 418.
Pirani detector, 3, 389 f.
Polarizability, electric: molecular, 287, 298; nuclear, 52, 319.
Polarization: collision, 138 f., 168, 433; neutron, 189 f., 194 f.; vacuum, 340.
Polar molecules, 287 f.
Polyatomic molecules, 289 f., 298.
Polymerization, 48, 239 f.
Position, collimator, 351 f.
Precession, Thomas, 212 f.
Pressure gauges, 415.
Pressures for molecular beams, 3, 356 f.
Principles, design, 346 f.
Probable velocity, 20.
Proton, 158; magnetic moment of, 248.

Pseudo-quadrupole interaction, 225.
Pump speed, 350.
Pumps, vacuum, 358, 410 f.

q, 62 f., 64, 313 f.
Q, 12, 61.
Q_e, 229.
Q_{ij}, 57.
$Q_q^{(k)}$, 54.
q_J, 62, 313 f.
$\bar{q}$, 224.
$q^{(J)}$, 224.
$q^{(v)}$, 224.
Quadrupole anomaly, 319.
Quadrupole interaction, 213 f., 271 f., 290 f., 305 f., 319 f., 421.
Quadrupole moment, 52, 56 f., 59 f., 310 f., 313, 319 f.; of electron distribution, 229 f.; signs of, 225; table of, 172 f.
Quantum electrodynamics, 262 f.
Quenching, 331.
Quenching asymmetry, 334.

R_e, 290 f.
R_S, 318.
R_Y, 74.
Rabi deflecting field, 395 f.
— method, 118 f., 349.
Radioactive nuclei, 184 f., 250, 393 f.
Radioactivity detector, 46, 393 f.
Radiometer detectors, 378.
Ramsey method, 124, 350, 354.
Recoil, 49.
Red shift, relativistic, 285.
Reduced mass correction, 268.
Reflection, 35 f.
— neutron, 192 f.
— polarization, 192.
Refocusing, 5.
— electric, 299.
— method, 111 f.
Refraction, neutron, 193.
Relativistic effects, 258 f.
Relativistic red shift, 285.
Renormalization, 340 f.
Representation: F, m, 423 f.; m_I, m_J, 423 f.
Resistance, pumping, 351.
Resonance at twice Larmor frequency, 227.
— electric, 288 f., 298 f.
— for excited states, 285 f.
— frequency, 5.
— line shapes, 215 f., 334 f.
— methods, 1 f., 115 f., 349 f.
— neutron, 201.
Restitution, 45.
Restrictions on multipoles, 58 f.

Rheostats, 401.
Rotating coordinates, 146 f.
Rotation, 291 f., 316.
Rotational magnetic moments, 102 f., 166 f., 203 f., 233.
Rotational-spin interaction, 208 f., 222.
Rules, selection, 118, 242, 301.
Russell-Saunders coupling, 258.

s_α, 92.
$S(z)$, 217.
S_{12}, 320.
S_H, 336.
σ, 28, 162.
σ^{HF}, 163.
Scattering, 25 f.
— attenuation, 347.
— experiments, 33 f.
— magnetic, 193.
Schmidt lines, 181 f., 279.
Second moments of electron distributions, 230.
Second-order paramagnetism, 163.
Secular equation, 84, 234.
Selection rules, 118, 242, 301.
Selective adsorption, 42 f.
Self-conjugate nuclei, 184.
Semi-conductor detectors, 377.
Separated oscillatory field method, 124.
Shape, beam, 17 f., 92 f., 348, 349 f.
Shapes, resonance lines, 215 f.
Shell models, 181 f., 185 f., 324.
Shielding, magnetic, 177, 162 f., 233; electric quadrupole, 316 f.
Signs: of nuclear moments, 113 f., 155 f., 256; of quadrupole moments, 225.
Single oscillatory field method, 118 f.
Six pole fields, 135, 406 f.
Slow molecules, 138, 285.
Source, 11 f., 47, 50, 361 f., 369 f.
— of dissociated atoms, 372 f.
— slit, 347.
Space focusing, 114, 134 f.
Spectroscopy, 50.
Specular reflection, 35 f.
Speed, pump, 350.
Spin, 51 f., 114.
— measurement of, 114.
— neutron, 197, 433.
— nuclear, 249.
Spin, table of, 172 f.
Spin-orbital coupling, 185.
Spin-rotational interaction, 164, 208 f., 222, 290, 293 f., 305 f., 433.
Spin-spin interaction, 80, 204 f.
Standards, frequency, 283 f.
Stark effect, 292 f.
Statistics, 51, 178.

Stern–Gerlach experiment, 100 f.; field, 100, 394 f.
Stern–Pirani detector, 3, 389 f.
Sticking coefficient, 46.
Stimulated emission, 298, 309.
Stop wire, 298.
Stretching, centrifugal, 205, 212, 222, 224, 230 f., 291 f., 316.
Structure, hyperfine, 241 f.
Surface ionization detector, 3, 379.
Susceptibility, diamagnetic, 169 f.
Sylphon, 411.
Systematics, 181.

Table of atomic hyperfine structure, 255; atomic magnetic moments, 261; collision cross-section, 34; deuterium molecule, 238; effusion through canals, 15; electric resonance, 303; fundamental constants, 418; hydrogen-deuteride, 239; hydrogen molecule, 235; hyperfine anomaly, 280; $\mathscr{F}(z)$, 29; magnetic shielding, 177; matrix elements, 235, 423 f.; molecular data, 303; nuclear electric quadrupole, 172; nuclear magnetic dipole, 172, 303; nuclear magnetic octupole, 172; nuclear moments, 172; nuclear multipole interactions, 313; nuclear quadrupole interactions, 311, 312; nuclear spins, 172; oven temperatures, 372; physical constants, 418; spin-rotation interaction, 209; typical apparatus, 356; vector and tensor relations, 419; velocity averages, 425.
Techniques, 361 f.
Temperatures, oven, 372.
Tensor force, 320.
Tensor relations, 419.
Theory of diamagnetism, 229; hyperfine structure of hydrogen, 267 f.; Lamb shift, 340 f.; line shapes, 215 f.; nuclear interactions in molecules, 51; nuclear moments, 178 f., 319 f.; quadrupole moments, 319 f.
Thermopile detectors, 378.
Θ, 119.
Thomas precession, 212 f.

Transition: multiple quanta, 306 f.; probabilities, 118 f., 127 f., 150, 153 f., 428 f.; π and σ, 242.
Transmission polarization, 192.
Twice Larmor resonance, 161, 227.
Two-quanta transition, 306 f.
Two-wire deflecting field, 397 f., 431 f.
Typical molecular beam, 3, 354.

Uniform magnetic fields, 401.
Universal detector, 387 f.

v, 391.
Vacuum fluctuations, 268; gauges, 415; polarization, 340; pumps, 358, 410 f.; system, 410 f.
Valves, 413.
Vapour pressure, 47.
Variations in field, 140.
Vector relations, 419.
Velocity-averaged functions, 426; distribution, 19 f., 93 f., 123 f., 128 f., 240, 348; measurements, 21 f.; selection, 415.
Vibrating rotator, 290 f., 316.
Vibration, molecular, 164, 205, 212, 222, 224, 230 f., 291 f., 316.

W, 66.
w'_i, 95, 349.
$W_{J,v}$, 291.
van der Waals force, 33.
Weak fields, 219.
Wide beams, 138.
Width, beam, 352 f.; resonance line, 350.

x, 86.
x_e, 291.
ξ, 169, 291.
$\xi_{\sigma,\pi}$, 228.

y_e, 291.
Y_{lj}, 291 f.
$Y_q^{(k)}$, 54.

Zeeman splitting, 332.
Zero-moment method, 107.
Zero-point vibration, 230 f.